W9-BPL-699

Nature's Building Blocks

Praise for *Nature's Building Blocks*

'A readable and entertaining guide...Doubles as both an accessible reference source and an enjoyable and fascinating 'dip into' read.'
Materials World

'A succinct history of everything...this marvelous reference work...is the kind of book people consult in the pursuit of a single fact, but this fact will lead to another and another, drawing the reader in an enjoyable chase from naturally occurring nuclear reactors to human zinc deficiency and on to the number of elements named for one small town in Sweden (four).'
New York Times Book Review

'Emsley's design, layout and presentation is logical, clear and beautifully written. The introduction itself is both informative and full of unexpected, yet valuable information.'
Chemistry in Britain

'An engaging gadgeteer of the elements.'
George Johnson, *New York Times*

'A delightful, idiosyncratic survey of the known elements, this guide also includes many nuggets of surprising information—for example, the use of fluorine (found in our bones and teeth) in the development of the nonstick frying pan.'
Natural History

'This book is like a bar of Cadbury's chocolate: You can't eat just one square...I think this is a wonderful book for scientists of all persuasions.'
Andrew R. Barron, *Chemical and Engineering News*

Nature's Building Blocks

Everything you need to know about the elements

NEW EDITION

JOHN EMSLEY

OXFORD
UNIVERSITY PRESS

ABINGTON FREE LIBRARY
1030 OLD YORK ROAD
ABINGTON, PA 19001

OXFORD

UNIVERSITY PRESS

Great Clarendon Street, Oxford OX2 6DP

Oxford University Press is a department of the University of Oxford.
It furthers the University's objective of excellence in research, scholarship,
and education by publishing worldwide in

Oxford New York

Auckland Cape Town Dar es Salaam Hong Kong Karachi
Kuala Lumpur Madrid Melbourne Mexico City Nairobi
New Delhi Shanghai Taipei Toronto

With offices in

Argentina Austria Brazil Chile Czech Republic France Greece
Guatemala Hungary Italy Japan Poland Portugal Singapore
South Korea Switzerland Thailand Turkey Ukraine Vietnam

Oxford is a registered trade mark of Oxford University Press
in the UK and in certain other countries

Published in the United States
by Oxford University Press Inc., New York

© John Emsley, 2011

The moral rights of the author have been asserted
Database right Oxford University Press (maker)

First published 2001
First published in paperback 2003
First published as new edition 2011

All rights reserved. No part of this publication may be reproduced,
stored in a retrieval system, or transmitted, in any form or by any means,
without the prior permission in writing of Oxford University Press,
or as expressly permitted by law, or under terms agreed with the appropriate
reprographics rights organization. Enquiries concerning reproduction
outside the scope of the above should be sent to the Rights Department,
Oxford University Press, at the address above

You must not circulate this book in any other binding or cover
and you must impose the same condition on any acquirer

British Library Cataloguing in Publication Data
Data available

Library of Congress Cataloging in Publication Data
Data available

Typeset by SPI Publisher Services, Pondicherry, India
Printed in Great Britain
on acid-free paper by
Clays Ltd, St Ives plc

ISBN 978-0-19-960563-7

10 9 8 7 6 5 4 3 2 1

NOV 16 2011

ABINGTON FREE LIBRARY
1030 OLD YORK ROAD
ABINGTON, PA 19001

Contents

Preface

Feedback from readers who bought the first edition of *Nature's Building Blocks* was most helpful and resulted in improved subsequent printings of the book. Even so, it became necessary to write a second edition because of the way things have changed during the past 10 years. New uses have been found for lots of the elements while others are now universally condemned, such as lead and mercury. Nor are the radioactive elements seen as quite so threatening, and indeed they are an essential part of medical treatments, like the fluorine-18 and arsenic-74 of positron emission tomography, while thorium could be a source of boundless energy.

It comes as a surprise to many people to realise that the Earth is a naturally radioactive planet. Every cell of our body contains millions of radioactive atoms and these are decaying at the rate of hundreds per day. They come from natural sources and are present in the food we eat and the air we breathe. They are carbon-14, potassium-40, lead-210, polonium-210, radium-226, rubidium-87, thorium-232, and uranium-238.[1]

In the first edition there was one metal for which I could find nothing to say in the *Element of surprise* section which comes at the end of each element and that was thulium, element 69. That is no longer the case—even thulium has become important in several applications.

Each element's entry begins with an *Essential element* headline to highlight the reason why that element is important and tempt you to read more.

Of course *Nature's Building Blocks* can only tell part of the story, but I hope there is enough to supply all that the average reader requires. We know so much about some elements that whole books have been written about them, even multivolume works in the case of the most important ones. For some elements, of course, there is so little known about them that it could all be written on a Post-it note.

I should also like to acknowledge the invaluable help of Walter Saxon, of Jackson Heights New York, who has provided most of the information about the radioactive elements which come at the end of the periodic table, and especially the transfermium ones. His input has resulted in many elements that received only a brief mention in the first edition now having sections to themselves, and becoming part of the alphabetical sequencing of this book.

---

[1] There are also minute traces of caesium-137, plutonium-239, plutonium-240, and strontium-90, which are there as a result of environmental contamination due to above ground nuclear weapons testing.

Introduction

There are many ways of arranging the elements. In *Nature's Building Blocks* they are in alphabetical order, or at least up to copernicium, the element with atomic number 112. Those of higher atomic numbers have yet to be named and these are grouped together under 'Un-named elements' for the simple reason that their names all begin with 'un' for reasons you will discover.

Why put the elements in alphabetical order? The answer is simple: it is for the convenience of the general reader. There is a price to pay, however, because this leads to their being arranged in random order, so that any similarities between elements are lost, and that is what interests chemists like me.

In textbooks of the elements they are more likely to be arranged in the most scientifically logical order, grouping elements of similar chemistry together. It would also have been possible to arrange *Nature's Building Blocks* in order of atomic numbers, starting with hydrogen (number 1), then helium (number 2), lithium (number 3), beryllium (number 4), and some authors have chosen to do this, such as Theodore Gray for his book *The Elements: A visual exploration of every known atom in the universe*, which is the most stunning book about the elements ever produced. Nevertheless, for the general reader for whom this book was devised, the alphabetical sequence is probably more helpful.

There are other ways of listing the elements, but none so entertaining as the poem that Tom Lehrer (1928–) wrote in 1959. He amused a whole generation of chemists when he sang his poem to the tune of the song 'I Am the Very Model of a Modern Major General' from Gilbert and Sullivan's operetta *The Pirates of Penzance*. It goes as follows:

> There's antimony, arsenic, aluminum, selenium,
> And hydrogen and oxygen and nitrogen and rhenium,
> And nickel, neodymium, neptunium, germanium,
> And iron, americium, ruthenium, uranium,
> Europium, zirconium, lutetium, vanadium,
> And lanthanum and osmium and astatine and radium,
> And gold, protactinium and indium and gallium,
> And iodine and thorium and thulium and thallium.
> There's yttrium, ytterbium, actinium, rubidium,
> And boron, gadolinium, niobium, iridium,
> And strontium and silicon and silver and samarium,
> And bismuth, bromine, lithium, beryllium and barium.
> There's holmium and helium and hafnium and erbium,
> And phosphorus and francium and fluorine and terbium,
> And manganese and mercury, molybdenum, magnesium,
> Dysprosium and scandium and cerium and cesium,
> And lead, praseodymium, platinum, plutonium,
> Palladium, promethium, potassium, polonium,
> Tantalum, technetium, titanium, tellurium,
> And cadmium and calcium and chromium and curium.

> There's sulfur, californium and fermium, berkelium,
> And also mendelevium, einsteinium and nobelium,
> And argon, krypton, neon, radon, xenon, zinc and rhodium,
> And chlorine, cobalt, carbon, copper, tungsten, tin and sodium.
> These are the only ones of which the news has come to Harvard,
> And there may be many others but they haven't been 'discarvard'.

Lehrer wrote his poem at a time when new elements were being synthesised thanks to advances in atomic and nuclear physics and it includes every element up to nobelium (atomic number 102). I am sure news has reached Harvard of the 20 new elements that have been 'discarvard' since then.

There are 78 essentially stable elements and these have atomic numbers 1–42, 44–60, 62–73, and 75–81. The others are radioactive and atom-by-atom, they are slowly decaying away. In recent years elements like tungsten, lead, and bismuth, which were long considered totally stable, have been found to be radioactive although they have half-lives so incredibly long that for all intents and purposes they still can be regarded as stable. Their half-lives are in excess of 10^{18} years, which is a billion billion years, and millions of times longer than the age of the universe. The number of elements considered to be completely stable is likely to change. Theoretical calculations indicate that all elements above atomic number 66 (dysprosium) should be radioactive. If this really is the case, then there are only 64 stable elements and these have atomic numbers 1–42, 44–60, and 62–66. The highest stable isotope would in fact be dysprosium-164.

The number of elements that might once have occurred naturally on planet Earth is not known, but some physicists say that it might be as high as element 126. Research has demonstrated that some isotope isomers can have half-lives above 10^8 years (that is, 100 million), a possibility which means that traces of superheavy elements may still remain. Several experiments have claimed their detection in minerals, and that these have atomic numbers 108, 111, and 122, but more research is needed to confirm them.

There are several sections under each element and some of these may need a little explanation. The sections are as follows, although not all are given with every element:

Name: how it is pronounced and where it came from.

Cosmic element: its production and occurrence in the stars.

Element of life: the amount in the average person and what it does.

Food element: this covers those that are essential nutrients, and a few others that we take in with our food, yet which serve no useful purpose.

Dangerous element: explains the toxicity of certain elements.

Medical element: how the element has been used by doctors through the ages.

Element of history: who discovered it, how, when, and where.

Element at war: the role of elements in weaponry, explosives, munitions, and armour.

Economic element: annual production, where it comes from, and what it is used for.

Environmental element: the amounts in the Earth's crust, soils, seas, air and the threat, if any, that it poses.

Chemical element: main properties, position in the periodic table, chemical reactivity, preferred oxidation states, and isotopes.

Element of surprises: curious uses or unexpected features of the element.

Finally the book ends with a chapter on the periodic table.

Names

In *Nature's Building Blocks* I have tried to avoid words and phrases that only a chemist would understand, and I have often hyphenated complex chemical names as an aid to understanding.

Although we think of chemistry as a perfectly logical science there are a few illogicalities in its use of language and symbols. One aspect which seems puzzling is the choice of chemical symbols for the elements. Why is gold Au, and silver Ag? The answer lies in the past: Au comes from the Latin *aurum*, and Ag from *argentum*. In fact, there are only 10 common elements with symbols that cannot be guessed from their names, and these are:

Antimony, Sb

Copper, Cu

Gold, Au

Lead, Pb

Mercury, Hg

Potassium, K

Silver, Ag

Sodium, Na

Tin, Sn

Tungsten, W—although its alternative name is wolfram.

The symbol for other elements uses the initial letter as a capital and the second letter as lower case; if this symbol is already being used then you take the next letter and so on. So we have Ca for calcium, Cd for cadmium, Ce for cerium, and Cs for caesium. This rule is not always adhered to and chlorine is Cl not Ch, magnesium is Mg not Ma, and radon is Rn not Ra, Rd, or Ro. The first of these possibilities had already been used for radium. Such inconsistencies are not likely to confuse you. The common elements of organic chemistry have single letters: C for carbon, H for hydrogen, O for oxygen, N for nitrogen, P for phosphorus, S for sulfur and F for fluorine, as do a few non-metals, namely B for boron, and I for iodine. Even some metals have single letter symbols, namely uranium (U), tungsten (W), vanadium (V), and yttrium (Y). However, all new symbols must now have two letters. It is preferred that the second letter be a consonant; usually the first letter of the second syllable in multisyllable names. If this conflicts with a known symbol, another consonant is chosen. No symbol can have the same letter repeated, and a vowel is only used if there are no available consonants.

Three elements have different spellings of their English names: aluminium is the preferred spelling, but in the USA it is spelt aluminum; caesium is the preferred spelling, but in the USA it is spelt cesium; sulfur is the preferred spelling, but in the British Commonwealth of Nations it is often spelt sulphur.

And who says which name is preferred? That is the remit of the International Union of Pure and Applied Chemistry (IUPAC) which was founded in 1919. This is a worldwide association of chemists from all countries, and it has a committee that meets to decide such weighty issues—and they were weighty issues causing not a little embarrassment as in the naming of element 102. This was first claimed and named by the Nobel Institute yet the research on which their claim was based later proved to be incorrect. Other man-made elements also caused controversies between American and Russian physicists and were a feature of the Cold War. Rival names were given to the new elements that they created at their nuclear research facilities.

In 1974, IUPAC and IUPAP (International Union of Pure and Applied Physics) set up a committee to study these claims. It consisted of three members from the USA, three from the USSR, and three from 'neutral countries'. In fact the group never met although several of its members did engage in correspondence. A new committee was set up in 1985 called the Transfermium Working Group (TWG), the name coming from the fact that the controversial elements came after fermium (element 100) in the periodic table, and eventually an agreed system of names was established. A fuller account of the naming controversies is given under Transfermium elements.

The committee which today decides on names is called the Joint Working Party (JWP) and that was set up in 1998 specifically to study claims for elements 110–112 but its remit includes later elements up to 118. The JWP bases it decisions on very cautious criteria and it has awarded discovery priority on the basis of complete and convincing data, even when that element was probably discovered earlier by another laboratory whose data was not as compete. In some cases elements had been known for years, with many laboratories regularly producing and studying their isotopes, and yet no discoverer had been recognised and no official name been given.

The pronunciation of the name in English is based on the following:

term	as in	term	as in	term	as in	term	as in
a	cat	er	term	n	not	sh	shop
ah	part	f	fat	ng	sing	t	top
ai	hair	g	go	o	dog	th	thin
aw	saw	h	hat	oh	go	u	rum
ay	day	i	pit	oo	soon	uh	your
b	big	iy	lie	ow	cow	v	vat
ch	chin	j	joy	oy	joy	w	win
d	dig	k	kit	p	pin	y	yet
e	set	l	lid	r	red	z	zip
ee	see	m	mat	s	set	zh	treasure

Cosmic element

The section *Cosmic element* is included for several elements, and could in theory have been included for all of them, but I have limited this entry to those which seem particularly interesting.

Stars like the sun are converting hydrogen and helium into carbon, oxygen, nitrogen, neon, magnesium, silicon, and iron, but nothing heavier than iron (atomic number 26). Red giants make most of the other elements, while it falls to supernovae to create the heaviest atomic nuclei of all such as gold, iodine, and uranium. Because the human body needs elements heavier than iron, such as copper, zinc, and selenium, we know these must have been created by a red giant at some time in the past, and because we also need iodine, then part of us must once have been scattered into the cosmos by a supernova. Truly, we are made of star dust.

The Universe came into existence 13.7 billion years ago when, in the incredible temperature of the big bang, protons and neutrons were able to combine to form the nuclei of elements. A proton by itself is simply hydrogen, and adding a neutron to it gives so-called 'heavy' hydrogen isotope, also known as deuterium or hydrogen-2. The '2' is the mass number of one proton plus one neutron. When two of these nuclei combine they form helium-4, one of the most stable of all nuclei and one whose formation releases a vast amount of energy. It is this energy that continues to fuel stars like the Sun.

Hydrogen, helium, and traces of lithium were the only elements produced during the first five minutes of the big bang. Traces of beryllium (element 4) and boron (element 5) were produced a few minutes later by various nuclear reactions. All elements above atomic number 5 have been produced only in stars. This did not start until about 200 million years after the big bang, when these first formed. Then elements began to be produced by nuclear-fusion reactions (aka nuclear burning) and by reactions brought about by the slow capture of neutrons (aka the s-process) which produced elements of up to atomic number 83 (bismuth). Elements of higher atomic number are produced only in Type II supernovae by reactions and decay brought about by the rapid capture of neutrons (aka the r-process) and in merging/accreting neutron stars by the rapid capture of protons (aka the rp-process).

Elemental matter comprises only about 5 percent of the composition of the Universe; 22 percent is dark matter, whose composition has yet to be determined; and 73 percent is dark energy, which is also yet to be determined. Of the elemental matter, about 75 percent is hydrogen, and 24 percent is helium, with all the other elements together comprising no more than 1 percent.

As the Universe expanded, galaxies and stars formed as matter condensed under the pull of gravity. Given enough matter, a star can have sufficient pressure at its centre to create temperatures of several million degrees. When this happens it allows the hydrogen to helium reaction to begin again. Indeed, inside such a star the temperature is hot enough to cause other nuclear reactions to occur, such as the combination of helium-4 nuclei to form other elements such as carbon-12, oxygen-16, and neon-20. These in their turn can combine to form silicon, sulfur, magnesium, and iron, with all of these nuclear fusions releasing yet more energy. However, with iron the energy-releasing processes come to an end. Iron-56 is a particularly stable nucleus and turning this into heavier elements results in energy being absorbed instead of released and consequently normal stars do not form heavier elements. These are only formed in a Type II supernova when a massive star rapidly collapses and then rebounds to expel and scatter most of its contents throughout interstellar space. Later this debris may condense under the pull of gravity to form another solar-type system.

We know this is what happens because the light from stars can be analysed to show the elements they contain. So-called astrochemistry began 200 years ago when, in 1802, William Wollaston examined the spectrum of sunlight and noticed that it contained many dark bands, but he could make nothing of them. In 1814 Joseph Fraunhofer carried out a more careful analysis of the Sun's spectrum and counted some 600 dark lines which became known as the Fraunhofer lines. Then, in 1864, Sir William Huggins matched some of these with lines in the atomic spectra of elements and rightly deduced that these elements were present in the Sun. So began the study of other stars and these became classified according to their spectra which indicates the temperature at the surface of the star, the elements that are present, and its age.

Thus O-type stars are blue and are the hottest, having surface temperatures above 25,000 degrees, and consist only of hydrogen and helium. Cooler are B-type stars, which are blue-white and have temperatures of between 11,000 and 25,000 degrees. Cooler still are A-type stars, which shine white and are 7,500–11,000 degrees. Sirius and Vega are A-type stars. Next come F-type, which are yellow-white and have surface temperature of around 6,000–7,500 degrees, and now lines due to metallic elements are noticeable in their spectrum.

The Sun is cooler than these, and is a yellow G-type star, which means it has a surface temperature of 5,000–6,000 degrees, and with metallic lines aplenty in its spectrum. After that come K-type orange stars at 3,500–5,000 degrees, where metallic lines dominate, and M-type red stars with temperatures below 3,500 degrees and noticeable bands of titanium oxide in the spectrum. Betelgeuse is one such star. Some of these latter types of stars have outer atmospheres enriched with heavier elements. Some K- and M-type stars even show spectral features of carbon compounds, C_2, CN, and CH. L-dwarfs and T-dwarfs have now been added to the list. These were discovered and named in the 1990s. L-type stars have temperatures below 2,000 degrees with certain metal hydrides detected. T-type stars have temperatures below 960 degrees and contain methane molecules.[2]

The composition of the Sun is roughly 94 percent hydrogen, 6 percent helium, and less than 0.2 percent other elements. Its 10 most abundant elements are given in the table below. These are ranked on a comparative basis, showing the number of atoms compared to a million atoms of hydrogen.

---

[2] If this letter code for star types seems somewhat arbitrary it is for historical reasons. However, the list of O, B, A, F, G, K, M, L, T can be remembered as follows: Oh Be A Fine Girl, Kiss My Lips Tonight! (or Guy can be substituted for Girl).

The top 10 elements in the Sun and their relative abundance*

Ranking	Element	Proportions
1	Hydrogen	1,000,000
2	Helium	63,000
3	Oxygen	690
4	Carbon	410
5	Nitrogen	87
6	Silicon	45
7	Magnesium	40
8	Neon	37
9	Iron	32
10	Sulfur	16

* Taken from my book *The Elements*.

Element of life

At the head of this section, which is included for every element that occurs naturally on Earth, there is a table giving the amount of an element in the main components of the human body: blood, bone, and tissue (tissue generally meaning muscle tissue). Sometimes a single figure is quoted which means that there is little variation, with the body maintaining an equilibrium, balancing absorption against loss. Sometimes a range of values is given when research has shown that the particular element varies considerably from person to person. There is also a figure for the total amount in the average person, and this is taken to be an adult weighing 70 kg.[3] Often the amounts in this table are tiny and given in units of p.p.m. (parts per million) or p.p.b. (parts per billion) and if you find such small numbers difficult to visualise then you might want to consult the section 'Quantities and numbers, large and small' on page 17.

The human body requires 24 elements if it is to function properly; these are called the *essential* elements and are listed in the table below as fractions of the total body mass. In the tables for individual elements, the actual weights are given. Those tables also include the non-essential elements as well, because we contain traces of all of the elements that exist naturally on Earth.

The fact that the human body is more than 60 percent water (H_2O) explains why oxygen and hydrogen feature so prominently. Oxygen also features in bone which is formed from calcium and phosphate (PO_4^{3-}), and because the skeleton of the average person amounts to 9 kg this explains why calcium and phosphorus come high up in the list.

The element that is most important is carbon, which is the basis of all the myriad of different molecules that make up the living cell. The amino acids, which account

---

[3] This is the same as 154 pounds (USA) or 11 stones (UK).

for most body tissue, are made up of carbon, hydrogen, oxygen, nitrogen and sulfur and again this explains why these elements come high in the list.

Essential human elements	Fraction of total body mass
oxygen	61%
carbon	23%
hydrogen	10%
nitrogen	2.6%
calcium	1.4%
phosphorus	1.1%
sulfur	0.2%
potassium	0.2%
sodium	0.14%
chlorine	0.12%
magnesium	0.027% = 270 p.p.m.
silicon	0.026% = 260 p.p.m.
iron	60 p.p.m.
fluorine	37 p.p.m.
zinc	33 p.p.m.
copper	1 p.p.m.
manganese	0.2 p.p.m.
tin	0.2 p.p.m.
iodine	0.2 p.p.m.
nickel	0.1 p.p.m.
molybdenum	0.1 p.p.m.
vanadium	0.1 p.p.m.
chromium	0.03 p.p.m. = 30 p.p.b.
cobalt	0.02 p.p.m. = 20 p.p.b.

Many elements serve no known purpose, but they come with the food we eat, the water we drink, and the air we breathe. Our body absorbs them, perhaps mistaking them for more useful elements. As a result we find that the average adult contains significant amounts of aluminium, barium, cadmium, lead, and strontium, and we each contain trace amounts of many others, including silver, gold, and uranium.

Because strontium so closely resembles calcium, we absorb a lot of this element and the average person has about 320 mg in their body, far more than many of the essential elements. On the other hand the weight of gold in the average person is only 7 milligrams, and the weight of uranium is a mere 0.07 milligrams (which is 70 micrograms) but in terms of the number of atoms this is around a billion trillion.

Our body tends to retain some of these unwanted intruders and deposits them either in our skeleton or our liver. In the case of uranium, this has a special propensity to bind to phosphate and as bone contains a large amount of this, this is where most uranium ends up. The liver, on the other hand, contains proteins that can trap metals like cadmium and gold.

Food element

Some elements need to be taken in on a daily basis because the body needs a regular supply to replace that which is being continually lost. For these elements this section explains why the body needs them, how much it needs each day, and the foods that are richest in them. It also explains how some elements which we don't need also get into the food chain and thence into us.

Dangerous element

Some elements are intrinsically toxic such as arsenic, lead, and mercury while others are toxic when they are part of certain molecules which are poisonous. Even apparently harmless elements like argon can kill given the right conditions. This section will be found for many elements but only insofar as these have been a threat at some time, or maybe they present a risk to health when exposure is excessive.

Medical element

Down the centuries doctors have prescribed all kinds of medicines and many of these were compounds of elements that we now regard as dangerous. This section discusses those that were given to treat diseases, and explains why some elements are still used in the fight against disease. These are now used mainly on a diagnostic basis. Some metals, such as titanium, are inserted into the body as implants.

Historical element

This section deals with the discovery of the elements, by whom this was made, where, when, and how. Where possible, I have included the dates of the discoverers, some of whom are still with us.

For some elements this data is lost in the mists of time, because they were simply there to be found, such as gold and sulfur. Even when we know that someone must have been the first to extract a new metal from its ores, who that person is cannot now be discovered. In the 1600s, elements began to be discovered by certain individuals although often they were not aware that they were discovering a chemical element as such. The concept of a *chemical* element was then still not understood because of the continuing theory put forward by the Greek philosophers 2,000 years earlier, which said that there were just four elements, namely earth, air, fire, and water. Even the forerunners of chemistry, the alchemists, believed that all metals were composed of the elements salt, sulfur, or mercury.

James Marshall of the University of North Texas and his wife Virginia Marshall have done some remarkable research into the discoverers of the elements and in some cases this has led to a reappraising of the person to whom a particular element's discovery should be attributed. An example of this is radon, where almost all published work (including the first edition of this book) assigns the discovery to Friedrich Ernst Dorn whereas it was really the work of Ernest Rutherford which uncovered this element.

In his book *Traité Élémentaire de Chimie,* published in 1789, the great French chemist Antoine Lavoisier (1743–94) introduced the concept of *chemical* elements and listed those he knew. From that time on it was possible to give full credit to those who discovered them, because they now realised what they had found. The 1790s and the early 1800s saw many such discoveries, which petered out only in 1925 when rhenium, the last stable element to be discovered, was found. However, this was not the end of the story. The remainder of the 1900s saw the synthesis of new elements, rather than their discovery in nature. Today new elements can be made using particle accelerators even if it's only a few atoms lasting only a fraction of a second.

During the course of history, the same element may have been discovered by different chemists at different times, while some prolific chemists discovered several elements during the course of their research. The Appendix lists the order in which elements were probably discovered, and the person to whom discovery can be attributed, and where this occurred.

Element of war

This may appear a strange category to include in a book on the elements but I felt it merited a separate section rather than being categorised under the Economic element section, which was where I first put them if they had military applications. Clearly weapons of war can be thought of as 'uses' but they are intrinsically detrimental to human life so I felt they detracted from the positive information that the Economic element section sought to convey.

Twenty-two elements have a War section and these are: antimony (a component of Greek fire); arsenic (mustard gas); beryllium (hydrogen bombs), carbon (gunpowder); chlorine (war gas); cobalt (magnetic mines), fluorine (nerve gases), gallium (secret radar scanners), hydrogen (nuclear weapons); iodine (Napoleonic wars); iron (swords); lithium (hydrogen bombs), magnesium (incendiary bombs); manganese (armour); molybdenum (tank armour), nitrogen (explosives); phosphorus (bombs, nerve gases); plutonium (Nagasaki bomb); sulfur (gunpowder); tungsten (armour); uranium (atomic bombs); vanadium (armour) and zirconium (cluster bombs).

Economic element

This section begins with the common minerals of an element, where they are to be found, the extent to which they are extracted, and the reserves that have been estimated to be economically extractable. In many cases the chemical formula of the mineral is given, but some minerals contain such a complex mixture that there is little point in providing that data. I have tried to reduce the formula to the simplest form possible— and without including molecules of water that are often associated with them.

The tonnage of the various elements that are mined every year is not easy to obtain because not all countries file reliable returns, and for this reason the figures given in this section should be taken as a rough guide only. The same goes for estimate of reserves. The production figures of the 10 most commonly mined elements are given in the table below.

World mining and/or production of the most important elements*

Ranking	Elements	Million tonnes**
1	Iron (steel)	1,344
2	Aluminium metal	38
3	Manganese ore	34
5	Chromium ores	24
5	Copper metal	15
6	Zinc metal	11
7	Lead metal	8
8	Barium (barytes, $BaSO_4$)	8
9	Titanium (TiO_2)	7
10	Fluorine (fluorspar, CaF_2)	6

* World Mineral Production 2003–2007, British Geological Survey, 2009.
** Aka metric tons; a tonne is 1,000 kg.

Environmental element

The table at the head of this section gives the proportion of an element in the Earth's crust, the order in which it comes in terms of ranking, its concentration in soil, in seawater and in the atmosphere. Again the quantities range from the very large to the very small and if the units are unfamiliar then please consult the section 'Quantities and numbers, large and small' on page 17 for guidance. The following table gives the top 10 most abundant elements in the Earth's crust:

The top 10 elements in the Earth's crust and their abundance*

Ranking	Elements	Proportions**
1	Oxygen	466,000
2	Silicon	277,000
3	Aluminium	82,000
4	Iron	50,000
5	Calcium	41,000
6	Sodium	28,000
7	Potassium	26,000
8	Magnesium	21,000
9	Titanium	4,400
10	Hydrogen	1,400

* Taken from *Macmillan's Chemical and Physical Data*.
** Tonnes per million tonnes of the Earth's crust.

Introduction

The elements dissolved in the oceans can be equally vast, and even when the concentration is low the total can still be quite surprising. For example, although an element is present at only 5 p.p.m. in seawater, its total mass amounts to almost 7 trillion tonnes.[4] The lure of such vast amounts has tantalised people down the ages, and the amount of gold in the sea particularly so. At a concentration of only 10 parts per trillion (p.p.t.), even this amounts to more than 13 million tonnes. But gold is not the rarest metal in the sea. For example, the concentration of the metal holmium is only 0.4 p.p.t., amounting to only 550,000 tonnes in all. Even so, this is five times as much holmium as is estimated to be in land-based reserves in ores.

The 10 most abundant elements in the oceans are given in the following table:

The top 10 elements dissolved in the oceans and their abundance*

Ranking	Elements	Proportions (p.p.m.)
1	Chlorine	19,400
2	Sodium	10,800
3	Magnesium	1,300
4	Sulfur	904
5	Calcium	411
6	Potassium	392
7	Bromine	67
8	Carbon	28
9	Strontium	8
10	Boron	5

* Taken from *Macmillan's Chemical and Physical Data*.

The least weighty part of the environment is the atmosphere, although it has a disproportionate role to play in human affairs. The next table lists the top 10 gaseous elements in the atmosphere. There are other gases present in more abundance than some of these, such as carbon dioxide (c.390 p.p.m.), methane (1.8 p.p.m.), nitrous oxide (0.5 p.p.m.), and ozone (0.4 p.p.m.) although this last one is also a form of elemental oxygen. In addition, there are trace amounts of many more gases. There is also air-borne dust and sea-spray in the atmosphere and these carry traces of most other elements.

---

[4] The volume of the oceans is 1.37 *billion* billion cubic metres (1.37×10^{18}) hence 5 p.p.m., which is 5×10^{-6}, results in a total mass of 6.85 a 10^{12} tonnes (a cubic metre of water weighs a tonne).

The top 10 gaseous elements in the atmosphere and their abundance*

Ranking	Elements	Proportions (p.p.m.)**
1	Nitrogen	780,900
2	Oxygen	209,500
3	Argon	9,300
4	Neon	18
5	Helium	5.2
6	Krypton	1.14
7	Hydrogen	0.5
8	Xenon	0.086
9	Radon	traces
10	Chlorine	virtually nil, but measurable in the stratosphere

* Data taken from *Kaye and Laby's Table of Physical and Chemical Constants*.
** The values are for dry air. Normally air also contains water vapour and this can vary considerably from place to place, and from day to day accounting for up to 4%.

Finally, the environmental element section will also concern itself with actual and potential pollution by a particular element.

Chemical element

At the head of this section is a data file giving the most important properties of an element, including its chemical symbol. This is followed by its atomic number, which starts at 1, which is hydrogen, and goes up to 118 (and may be even more by the time you are reading this). The atomic number is the number of positive particles, or protons, which the element has in its nucleus.

The atomic weight of an element is a physical measurement and this is generally accurately known, which for historical reasons are all *relative* weights compared to the weight of an atom of carbon which is taken as exactly 12.0000.[5]

Melting and boiling points are reported on the Celsius (C) scale, in which the freezing point of water is 0°C and the boiling point is 100°C. If you want to convert these to absolute temperatures, which is the Kelvin (K) scale then add 273. If you prefer the Fahrenheit (F) scale in which the freezing point of water is 32°F and its boiling point 212°F then multiply the temperature in C by 1.8 and add 32.

Density is reported in grams per cubic centimetre (g/cm^3) which is the same as kilograms per litre, which are the standard units of weight and volume.

The chemical formula of an element's oxides reflects its valences or oxidation numbers. The former term refers to the number of bonds a compound forms, whereas the latter term is a way of expressing its electronic state, and is more general when

---

[5] The actual weight of a carbon atom is 12 g divided by the number of atoms it contains, which is 6.022×10^{23}, and that comes to 1.91×10^{-23} g. In other words we are talking of weights in the range of a *trillion* trillionth of a gram.

talking about metals. Two elements even have zero valency, which means they are reluctant to form any compounds, and these are helium and neon.

The *Chemical element* section also describes the physical characteristics of the element, the groups into which it falls in the periodic table, its stability in air and water, and its reaction with acids and alkalies and its various oxidation states and their relative stabilities. The terms used by chemists to describe the various groups of elements are as follows:

The alkali metals, which constitute group 1 of the periodic table.

The alkaline earth metals, which constitute group 2.

The transition metals, which are the d-block element of the periodic table.

The platinum group metals are a sub-set of the d-block and are the heavier elements of groups 8, 9, and 10.

The coinage metals (copper, silver, gold) are those of group 11.

The typical metals are the metal of the p-block.

The metalloids are those elements that lie on the border between metals and non-metals in the p-block.

The non-metals are those of the 1s and p blocks.

The halogens are the elements of group 17.

The noble gases are the elements of group 18, plus helium.

The lanthanoids are the first row of the f-block elements. These, along with the elements of group 3, used to be known as the rare earths; this term is still in use.

The actinides are the lower row of the f-block.

The transuranium elements and those elements with atomic numbers greater than 92 (uranium).

The transfermium elements are those with atomic numbers greater than 100 (fermium).

The *Chemical elements* section also lists those isotopes of an element that make up the bulk of the element as it is generally encountered on Earth, noting when these isotopes are radioactive. I should point out that many radioactive isotopes, with atomic numbers 28–64, occur in the Earth's crust and come from the fission of uranium atoms, but these are in such tiny amounts that there are no estimates of how much is present, nor do they have any effect of the overall isotopic composition of the elements as given in this section.

So what do radioactive isotopes of the elements emit? There are about 20 modes of radioactive decay, with a variety of particles, nuclei, and radiation being ejected. Some are exotic and were only discovered in the 1980s, occurring in synthesised, extremely neutron-rich and proton-rich isotopes. Normally, radioactive decay involves three main types of radiation: alpha rays, beta rays (minus and plus) and gamma rays. Alpha rays are really alpha particles, being the nuclei of helium atoms. Beta rays are electrons (beta minus decay) or positrons (beta plus decay). Gamma rays are high-energy electromagnetic radiation which often accompany alpha and beta emissions. Gamma rays are like powerful X-rays and are highly penetrating.

When an atom undergoes alpha decay it is transformed into an atom that is two atomic units less and four mass units less. For example, when uranium-238 (atomic number 92, mass 238) emits an alpha particle it becomes thorium-234 (atomic number 90, mass 234).

In beta minus decay, an electron is emitted from the nucleus thereby changing a neutron to a proton and increasing the atomic number by one while the atomic mass remains unchanged. Thus thorium-234 (atomic number 90) becomes protoactinium-234 (atomic number 91).[6]

In beta plus decay, a positron is emitted from the nucleus, changing a proton into a neutron and thereby decreasing the atomic number by one while leaving the atomic mass unchanged. For example potassium-40 becomes argon-40. Often accompanying alpha- and beta-emissions are gamma-emissions, these are high energy rays of electromagnetic radiation which are like x-rays and highly penetrating.

Radioactivity can vary from very weak to very dangerous. The common stable elements, carbon and potassium, exhibit mild radioactivity because of the presence of radioactive isotopes carbon-14 and potassium-40. Uranium is much more radioactive, and the elements that are made in nuclear reactors, such as plutonium, are highly radioactive.

Element of surprises

These tailpieces contain information that could easily have come under one of the other sections, but which I have separated out because they reveal something quite unexpected about an element, such as its role in history, its use in modern life, or just a curious feature worthy of note.

How many elements can there be?

Although more new elements have been announced since the first edition of *Nature's Building Blocks* was published in 2001, these are all highly unstable and many have yet to be confirmed before they can be officially named. Three of the elements which had not been named in the first edition, elements 110, 111, and 112, are now called darmstadtium (Ds), roentgenium (Rg), and copernicium (Cn) respectively, and as such they now justify their own slots in the alphabetical ordering. The elements which previously were all grouped together as the transfermium elements, namely 101 mendelevium, 102 nobelium, 103 lawrencium, 104 rutherfordium, 105 dubnium, 106 seaborgium, 107 bohrium, 108 hassium, and 109 meitnerium are still grouped together under this heading which explains the significance of these elements.

Those elements which have not been officially named are given names based on a numbering system with uses ancient *Greek* or *Latin* terms so that element 113 is un-un-tri-um, element 126 is un-bi-hexium. When an element has one of these inelegant

---

[6] The newly formed protactinium now needs an extra electron for its outer orbit if it is to be a neutral atom, and this it easily acquires from its surroundings.

titles then they are to be found in the 'un-named' section of the book under U and immediately preceding uranium.

The most important elements?

I became aware, while writing the first edition of *Nature's Building Blocks*, that some elements covered several pages of text while others struggled to fill one page—but was this really a reflection of their relative importance? And has it changed? In fact, I believe it is, and that it has. However, before looking at the ranking of elements in order of word numbers, let us look at the importance of elements in other ways. Of course there are two aspects to be considered: abundance and importance. An element can be abundant without its being important, but whereas abundance is a scientific property that we can measure, importance is a subjective thing and what may be important to one person or their culture may be of little significance to others on this planet. Gold and silver spring to mind.

Clearly certain elements are far more important than others. For example, it goes without saying that the five elements that make up DNA (carbon, hydrogen, nitrogen, oxygen, and phosphorus) are all equally important. The same is true for the other essential elements that the human body requires, such as calcium, iron, magnesium, potassium, zinc, etc.

Then, of course, there are metals which may not be important to our metabolism, but nevertheless are essential to a modern lifestyle, such as aluminium. The same is true of uranium, without which a significant portion of the world's electricity needs could not be generated. There is also a lot to write about when it comes to dangerous elements, such as lead and mercury, and about elements with a long history such as copper and tin. Judged then by the number of words written about them in *Nature's Building Blocks*, the most important elements would appear to be the following:

The main elements in *Nature's Building Blocks*

Ranking	Element	Total words
1=	Hydrogen	5,700
1=	Mercury	5,700
3	Lead	5,000
4	Phosphorus	4,500
5	Silicon	4,400
6	Sulfur	4,300
7	Nitrogen	4,200
8	Oxygen	4,100
9	Arsenic	4,000
10	Iron	3,600
11=	Carbon	3,300
11=	Zinc	3,300

The list comprises eight non-metals and four metals, of which the highest two are lead and mercury, both metals with shady pasts because of their damaging health and environmental implications—and the same is true for the non-metal arsenic which is also high in the list. The other two metals are iron and zinc both of which have important dietary requirements as well as being industrially important.

Other elements with large sections are: aluminium, antimony, arsenic, cadmium, calcium, carbon, cerium, chlorine, copper, fluorine, gold, iodine, lithium, magnesium, nickel, platinum, plutonium, potassium, selenium, silver, sodium, thorium, tin, titanium, tungsten, and uranium.

Quantities and numbers, large and small

The most common quantities when talking about elements are weights, volumes, and proportions. On the everyday scale Europeans speak of weights in grams and kilograms, volumes in ccs (or mls) and litres, and percentages (or fractions of a percent). In the USA, the common units of weight are ounces and pounds, and volumes are pints and gallons.[7] However, *Nature's Building Blocks* is basically a science book and the quantities used are those of the *Système International des Unites*, or SI system, in which the basic unit of weight is the kilogram, and of volume is the cubic metre. This latter quantity is large (a cubic metre of water weighs a tonne) and more useful units are the litre, which is a thousandth part of a cubic metre, and the cubic centimetre (cc), which is a thousandth part of a litre.

Nature's Building Blocks will deal with both the very small, such as the concentration of a rare element in the sea, and the very large, such as the total amount of an element on Earth. To talk about tiny proportions we will need to speak of parts per million (p.p.m.), parts per billion (p.p.b.) and even parts per trillion (p.p.t.)[8] and at the other extreme millions, billions, and trillions of tonnes will be used when talking of global quantities. Each of these is a thousand times larger than the previous one, so that a trillion tonnes (10^{12}) is a thousand times larger than a billion tonnes (10^9), which is a thousand times larger than a million tonnes (10^6). While such amounts are hard to visualise, small quantities are easier to imagine: 1 p.p.m. is equivalent to 1 milligram (mg) which is about a single crystal of sugar or salt in a litre of water; 1 p.p.b. is 1 microgram (µg) which is about a speck of dust in the same volume; while 1 p.p.t. is 1 nanogram (ng) in this volume and this tiny amount would be almost invisible to the naked eye.

A trillionth of anything is difficult to imagine but one second in time is a trillionth of 30,000 years and the span of a hand (15 cm) is a trillionth part of the distance from the Earth to the Sun. When talking about some of the radioactive elements that have been made in only the tiniest of quantities then the units may be as small as a trillionth of a gram, a quantity known as a nanogram (10^{-9}), or even one a thousand time smaller, namely a picogram (10^{-12}). Analytical chemists can even detect amounts a thousand times smaller than this.

---

[7] A pint is just over half a litre, 0.568 to be exact, and an ounce is roughly 30 grams, 28.35 to be exact. A US gallon is 3.79 litres.

[8] 1 p.p.t. is a thousand times smaller than 1 p.p.b., which is a thousand times smaller than 1 p.p.m.

A note of caution needs to be sounded here because two scales of measurement overlap: p.p.m. and percentages. When talking about small quantities I could say 0.01 percent or 100 p.p.m., or for larger concentrations talk of 0.1 percent rather than 1,000 p.p.m. Thus there comes a point where it may be more informative to the reader to use one or the other. The context will determine which it is to be although in some cases I will use both for the sake of clarity and say for example that the concentration of an element in the Earth's crust is 10,000 p.p.m. (1 percent).

Finally a word of caution. My earlier book, *The Elements* is a compilation of numerical data that is scientifically derived and, one hopes, is unlikely ever to change in a significant way. The same can be said for some of the data in *Nature's Building Blocks* such as that in the various tables at the head of each section. Other data in this book is less secure and liable to change with time, such as figures for mining and amounts used for various processes. For this reason I have often rounded figures up or down.

The elements (A–Z)

Actinium

Pronounced ak-tin-ee-um, the name comes from the Greek word *aktinos* meaning a ray.
French *actinium*; German *Actinium*; Italian *attinio*; Spanish *actinio*; Portuguese *actínio*.
Actinium is the first member of a series of 14 radioactive elements collectively known as the actinides, with atomic numbers ranging from 89 to 102 (nobelium).

Essential element

Actinium has possible medical applications against cancer.

Element of life

Actinium is never encountered outside nuclear facilities or research laboratories, and it has no role to play in living things.

Dangerous element

Actinium-227 is powerfully radioactive and those working with it have to ensure they are well protected.

Medical element

The isotope actinium-225 (half-life 10.0 days) is used to produce radioactive bismuth-213 (half-life 45 minutes) which has potential medical applications. It can be incorporated into a monoclonal antibody, and this can be targeted on cancer cells and will destroy them by the powerful alpha particles it emits.

Actinium-225 was first produced in 2000 using a cyclotron. However, there is an alternative source in unwanted stores of fuel pellet thorium which was intended to be used in a nuclear reactor in the 1960s. These pellets contained 3 percent uranium-233 and this has been decaying via thorium-229 to actinium-225 to the extent that they now contain extractable amounts of this isotope.

Element of history

Actinium was discovered in 1899 by André Debierne in Paris, France, who extracted it from the uranium ore pitchblende in which it occurs naturally in trace amounts. Three years later, the German chemist Friedrich Otto Giesel independently extracted it, and unaware it had already been discovered, he named it 'emanium' derived from the Latin *emanare* meaning to flow out or emanate.

Economic element

Actinium extracted from uranium ores is the isotope actinium-227 which is a beta-emitter with a half-life of 21.7 years and it decays to form a sequence of isotopes, the first of which is thorium-227. The actinium used for research purposes is made by bombarding radium-226 with neutrons which produces radium-227 and this decays with a half-life of 42 minutes to give actinium-227. The tiny amounts of actinium-227 so produced have little economic value beyond research purposes, where it is used as a source of neutrons, being about 150 times more active than radium.

Environmental element

Actinium in the environment	
EARTH'S CRUST	present but only as a trace element.
	Actinium is among the 10 least abundant elements.
SEAWATER	minute trace.
ATMOSPHERE	nil.

Actinium occurs naturally in the Earth's crust, produced as one of the sequence of isotopes that starts with the decay of uranium-235. The half-life of actinium-227 means that a tonne of the uranium ore pitchblende contains only around 150 mg.

Chemical element

Data file	
CHEMICAL SYMBOL	Ac
ATOMIC NUMBER	89
ATOMIC WEIGHT	227.0 (isotope Ac-227)
MELTING POINT	1047°C
BOILING POINT	av. 3200°C (est.)
DENSITY	10.06 g/cm³
OXIDE	Ac_2O_3

Actinium is a soft, silvery-white, radioactive metal of group 3 (row 6d) of the periodic table. Pure samples of the metal have been obtained only with difficulty. Actinium reacts with water to release hydrogen gas. Its preferred oxidation state is +3 as in compounds like AcF_3 and $AcCl_3$.

There are 31 known isotopes of actinium ranging from actinium-206 to actinium-236, and 7 known isomers. The three naturally occurring isotopes are actinium-227 which has the longest half-life of 21.7 years, actinium-225 (10.0 days), and actinium-228 (6.15 days). These are produced by the radioactive decay of other elements. Although 99 percent of actinium-227 decays to thorium-227, which it does by emitting beta particles (that is, electrons from the nucleus), 1 percent decays to francium-221 by emitting alpha particles (which are equivalent to helium nuclei). Actinium also emits intense gamma-radiation and it is this which makes it dangerous to work with.

Element of surprises

All actinium samples glow blue in the dark because its intense radioactivity excites the air around it.

Aluminium
or Aluminum (USA)

Pronounced al-yoo-min-ee-um, or ah-loo-min-um (USA), and the name is derived from *alumen* meaning 'bitter salt' which is the Latin name for alum (potassium aluminium sulfate). Aluminum was the original name given to the element by Humphry Davy (1778–1829), having first thought to name it alumium. Others in Europe combined the two words and it became aluminium, but those in the USA preferred Davy's name and when the American Chemical Society debated on the issue, in 1925, it decided to stick with aluminum. It was only in 1990 that IUPAC approved the spelling aluminium and in 1993 it also approved aluminum as an acceptable alternative.

French *aluminium*; German *Aluminium*; Italian *alluminio*; Spanish *aluminio*; Portuguese *alumínio*.

Essential element

Aluminium is the second most used metal after iron, and aluminium sulfate is the key to water purification.

Element of life

Aluminium in the human body	
BLOOD	0.4 mg per litre
BONE	between 4 and 27 p.p.m.
TISSUE	between 1 and 28 p.p.m.
TOTAL AMOUNT IN BODY	about 60 mg.

No living species has been discovered to need aluminium as an essential element. Because it is so abundant in soil, all plants absorb it, and grasses can accumulate it, to the extent of its being more than 1 percent of their dry weight. Some plants, such as tea bushes, absorb a lot of aluminium and alum is used as a fertilizer in tea plantations. Acid soils allow aluminium to become more available to plants and this can reduce growth. In neutral soils it is not a problem. Wheat has adapted to tolerate higher levels of aluminium by releasing molecules which bind to it and render it harmless. The gene responsible for wheat's aluminium tolerance has been identified.

Food element

Food crops also take up aluminium and levels are highest in spinach (104 p.p.m., dry weight), oats (82 p.p.m.), lettuce (73 p.p.m.), onions (63 p.p.m.) and potatoes (45 p.p.m.). The average daily intake of aluminium is about 5 milligrams, but can be double this depending on the diet. Only a small fraction is absorbed, probably less than 10 micrograms per day.

Most aluminium in food passes through the gut without being absorbed, and absorption is prevented by the presence of silicon, although it appears to be encouraged by citric acid since this forms a soluble compound with aluminium. Once inside the blood stream, the body finds it very difficult to remove aluminium.

Foods with most aluminium are processed cheese, sponge cake mix, lentils, chickpeas, and basmati rice. Cakes and biscuits are often made with the leavening agent sodium aluminium phosphate. One of the richest sources in food is processed cheese which can have almost 700 parts per million (p.p.m.). Cooking in aluminium pans does not greatly increase the amount of aluminium in food except when cooking acid foods such as rhubarb.

Beverages account for 50 percent of the average daily intake of aluminium, especially among tea drinkers, while bread and cereals together make up 25 percent.

Dangerous element

Aluminium was once regarded as a serious threat to health—wrongly as it turned out. It all began in the 1970s when patients on kidney machines began to suffer progressive brain damage resulting in premature deaths. The cause of this so-called dialysis dementia was eventually traced to high levels of aluminium leaching from the equipment that was used. Once in the blood, the aluminium attached itself to molecules called transferrin, and in this way it gained entry to the brain.

Consequently it was thought that aluminium was the primary cause of Alzheimer's disease because it was reported that the plaques which form in the brain of people with this disease had high levels of aluminium. However, later analysis by other chemists in the 1990s, showed that there was much less aluminium than had been supposed, and in the earlier analyses samples has been inadvertently contaminated by aluminium in the dyes used to stain the brain tissue to identify the plaques. Even so, later analysis has since then shown some brain plaques do contain traces of aluminium, although this is no longer regarded as the cause of Alzheimer's disease.

Medical element

As long ago as the first century AD, the Roman army doctor Dioscorides wrote a medical text, *De materia medica,* in which he recommended alum to stop bleeding—which it undoubtedly does—and for various skin conditions, such as eczema, ulcers and dandruff—for which it would be of little benefit. Today it is still available as an over-the-counter treatment in the form of styptic pencils for stopping bleeding from shaving cuts.

Aluminium hydroxide is used as a long-acting antacid.

Aluminium oxyhydroxide, AlO(OH), and aluminium hydroxyphosphate $Al(OH)_x(PO_4)_y$ (x and y can vary) have been added to vaccines for many decades because the aluminium acts to boost the immune system's response to the antigen in the vaccine. It is thought to trigger a receptor that prompts the white blood cells to produce more antibodies.

Element of history

The analysis of a curious metal ornament removed from the tomb of Chou-Chu, a military leader in third-century China, turned out to be 85 percent aluminium. How it was produced remains a mystery.

The ancient world used aluminium in the form of alum, which is potassium aluminium sulfate, a mixed salt which occurs naturally in Greece, Turkey and Italy where it was used for centuries as a fixing agent for dyeing. It was exported mainly from Turkey via Constantinople, but in the Middle Ages it was discovered that it could also be made from clay and sulfuric acid. However, the discovery of alunite, a potassium aluminium sulfate rock, at Tolfa in territories controlled by the Pope, led to a Papal monopoly of its manufacture in Europe from the 1460s onwards. The industry employed 8,000 men and produced 1,500 tonnes of alum a year. This state of affairs continued until English alum production started in the 1600s in North Yorkshire where a large deposit of alum shale was discovered. Production there broke the Papal monopoly and the price of alum fell dramatically. Production continued in Yorkshire for more than 250 years.

Alum was used by paper-makers as a preservative, by doctors to stop bleeding and to preserve anatomical preparations, and as a mordant by dyers whereby it fixed natural dyes to fabrics. Despite alum-making being the first true chemical industry, there was little understanding of the nature of the product or the process of its manufacture.

By the end of the 1700s, aluminium oxide was known to contain a metal, but it defied decomposition by conventional means, such as heating with carbon. Humphry Davy (1778–1829), who had used electric current to decompose other oxides to get the metals sodium and potassium, found his new process did not release aluminium.

The first person to produce aluminium, in 1825, was probably Hans Christian Oersted (1777–1851) in Copenhagen, Denmark, and he did this by heating aluminium chloride with potassium metal, but the sample he produced was impure. It fell to the German chemist Friedrich Wöhler (1800–82) to perfect the method. Using sodium instead of potassium, he obtained the first pure samples of the new metal. Naturally it was expensive to produce by this method since sodium metal was costly, and consequently aluminium was for many years treated as a novelty of the rich and powerful. In the 1860s the Emperor Napoleon III of France impressed visiting heads of state with special cutlery made of the metal. For another 20 years or so, aluminium continued to be much admired but its cost of production meant there were few uses made of it.

Then on 23 February 1886, a 21-year-old US student, Charles M. Hall (1863–1914), of Oberlin College, Ohio, finally developed a method for extracting aluminium from molten aluminium salts using an electric current. The process had been shown to be chemically feasible 30 years earlier but, at the time, it was not commercially viable. Hall perfected his process in a makeshift laboratory in the family woodshed using a homemade battery, and on 9 July 1886 he was granted a patent for the method. However, his patent came under attack from two directions.

Hall had tried to get a local company interested in his work and for a short time they employed him while he carried out his research. This allowed them to claim his invention, as happens with all employees whose work leads to a patent. Thankfully, the courts did not uphold their claim. The second attack was far more serious: a young Frenchman had applied for a patent on 23 April 1886 for the same process, several weeks before Hall. The inventor was 23-year-old Paul-Louis-Toussaint Héroult (1863–1914), another young student who had also worked alone on the problem. The US Patent Office accepted Hall's claim to be the first, on the basis of two letters he had written to his brother the day after he had first made aluminium describing what he had done, and which were still in their postmarked envelopes. Nevertheless litigation continued until finally an agreement was reached which gave Hall the American rights and Héroult the European rights.

Both men had discovered that in order to reduce aluminium oxide to aluminium electrolytically it had to be dissolved in molten cryolite, in which it was quite soluble, and that carbon electrodes had to be used. Cryolite is sodium aluminium fluoride, Na_3AlF_6. It occurs naturally and was discovered at the end of the 1700s in Greenland from where it was exported as raw material for making sodium carbonate, aluminium sulfate, and aluminium oxide. Cryolite's most remarkable property is that it melts easily on heating to give a liquid which will dissolve aluminium oxide.

As a result of the Hall-Héroult process, the price of aluminium fell until it cost one thousand of the price when Napoleon III had purchased his precious spoons.

Economic element

Today aluminium is made by the Hall-Héroult process but the cryolite is synthesised instead of being mined. The electrodes are made of graphite, the cathode being the one at which aluminium forms and where it collects as molten metal, while the anode removes the oxygen by converting it to carbon dioxide, a process that also produces a lot of heat which helps keep the cell at its operating temperature of 960°C. Anodes last about three weeks before needing to be replaced. A typical aluminium plant producing about 120,000 tonnes a year requires more than 200 MW of electricity, enough to meet the needs of a small town. Around two thirds of this is used to heat the cell and a third to convert the oxide to aluminium metal. Aluminium production consumes 5 percent of the electricity generated in the USA.

Aluminium is the most used metal after iron and global production is in excess of 38 million tonnes a year,[9] and a similar amount comes from recycling aluminium products. Aluminium has one third the density and stiffness of steel and it can be machined, cast, and extruded. Aluminium is difficult to solder or weld because it does not glow red before melting.

---

[9] Iron production exceeds 1 billion tonnes per year.

Known reserves of aluminium oxide ores are 30 billion tonnes, mainly in the form of bauxite, AlO(OH), and gibbsite, Al(OH)$_3$. Originally these were mined in tropical countries where they were created geologically as a result of the exhaustive weathering of clays which removed all other elements but insoluble aluminium oxide, and this explains the large deposits in Australia, Brazil, Guinea, and Jamaica. The production of alumina (aluminium oxide, Al$_2$O$_3$) totals around 78 million tonnes a year with the main producers being China (20 million), Australia (19 million), Brazil (7 million), Jamaica (4 million), and the USA (4 million).

The main producers of new aluminium metal are China (12.5 million tonnes), Russia (4 million tonnes), Canada (3 million tonnes), USA (2.5 million tonnes), and Australia (2 million tonnes), Smelting is also carried out in countries where supplies of cheaper electricity are available because the cost of this can account for up to 40 percent of the cost of refining, so there is production in Norway (1.3 million tonnes), the Gulf states (1.5 million tonnes), and even in New Zealand (300,000 tonnes).

Aluminium is easy to recycle, and with aluminium drink cans this is done simply by putting them into a molten salt flux of sodium and potassium chlorides which absorbs the impurities in the scrap, leaving a pool of molten aluminium which sinks to the bottom of the furnace and can be drawn off. Recycled aluminium requires only 5 percent of the energy needed to extract aluminium from aluminium oxide. Other sources of aluminium for recycling are vehicles and building materials.

Aluminium metal is used in hundreds of ways in the home, industry, and transport, and in products as diverse as street lamps, power cables, boat masts, car bodies, motor engines, beer kegs, cooking foil, door handles, and drink cans. It is widely used in all aircraft except high performance airforce planes.

Aluminium is lightweight, strong, and protected from reacting with air and water by a micron-thick oxide film which forms rapidly on the surface. This can be deliberately made thicker by anodising, a process in which the aluminium acts as the anode of an electrolytic cell with sulfuric acid as the electrolyte.

Some alloys of aluminium, such as duralumin, are much stronger than aluminium itself. This alloy is mainly aluminium with 4 percent copper, plus small amounts of magnesium, manganese, iron, and silicon. Duralumin can be made even stronger by quenching in water after heating the alloy to 600°C. Curiously it continues to gain strength for about a week thereafter, a process known as 'ageing'. Although pure aluminium is a weak metal, adding a few percent of magnesium and silicon renders it as strong as steel and especially after it has been baked for 30 minutes at domestic oven temperatures. The silicon and magnesium form nano-scale structures within the aluminium and it is these which give it strength.

Aluminium's ability to conduct electricity makes it ideal for cables, and its equally good reflectance properties find use in insulation, heat-reflecting blankets and solar mirrors. Aluminium has a high reflectance and will retain this even when ground to powder and used to make 'silver' paint.

Useful aluminium compounds are aluminium ammonium sulfate Al(NH$_4$)(SO$_4$)$_2$ (aka alum), aluminium oxide Al$_2$O$_3$ (aka alumina, corundum), and aluminium sulfate, Al$_2$(SO$_4$)$_3$. Aluminium oxide occurs naturally as rubies and sapphires or is man-made, such as that used in lasers. Aluminium sulfate is used as a mordant in dyeing and in water purification. For over a century, water engineers have been using aluminium sulfate as a flocculating agent. When water is cloudy with silt and bacteria it can be made sparkling clean by adding a small amount of lime (calcium hydroxide) and aluminium sulfate. This combination precipitates aluminium hydroxide which carries the impurities down with it. Aluminium hydroxide is so insoluble that it leaves behind only 0.05 p.p.m. of dissolved aluminium in the water, well below the 0.2 p.p.m. limit set for drinking water. Surprisingly, flocculation can even remove excess aluminium which may be present naturally.

The most recent development in aluminium technology is the production of aluminium foam, produced by adding to the molten metal a compound (a metal hydride) which releases hydrogen gas. The molten aluminium has to be thickened before this is done so that the bubbles of hydrogen gas don't escape, and this is achieved by adding aluminium oxide or silicon carbide fibres. The result is a solid foam which is used to line traffic tunnels because it acts to muffle sounds and is very durable. Aluminium foam was also used in the space shuttle.

Warships which once had aluminium superstructures have them no longer, preferring steel instead. Their vulnerability was shown during the Falklands War of 1982 when the aluminium parts of ships of the Royal Navy caught fire when struck with Exocet missiles.

Environmental element

Aluminium in the environment	
EARTH'S CRUST	82,000 p.p.m. (8.2%)
	Aluminium is the 3rd most abundant element.
SOIL	main constituent, varying between 0.5 and 10%.
SEAWATER	*av*.0.5 p.p.b. The Atlantic has levels 50 times those of the Pacific.
ATMOSPHERE	traces as dust particles.

Aluminium is the most abundant metal in the Earth's crust, where it exists as the ion Al^{3+}. Aluminium contributes greatly to soil properties where it is present mainly as insoluble aluminium hydroxide, with the result that soil water generally has only 0.4 p.p.m. of aluminum. However, if the soil becomes acidic and the pH drops below 4.5, then solubility increases sharply to more than 5 p.p.m. The result greatly affects plants and crops by reducing root growth and phosphate uptake.

Chemical element

Data file	
CHEMICAL FORMULA	Al
ATOMIC NUMBER	13
ATOMIC WEIGHT	26.981538
MELTING POINT	661°C

Aluminium

BOILING POINT	2467°C
DENSITY	2.7 g/cm³
OXIDE	Al_2O_3

Pure aluminium is a soft and malleable metal, and a member of group 13 (row 3p) of the periodic table. Aluminium is soluble in concentrated hydrochloric acid and sodium hydroxide solution. Its preferred oxidation state is +3, such as Al_2Cl_6 and the ion Al^{3+}. Lithium aluminium hydride, $Li[AlH_4]$, is a particularly powerful reducing reagent used in chemical reactions.

There are 23 known isotopes of aluminium with mass numbers 21 to 43, and four known isomers. There is only one main naturally occurring isotope, aluminium-27, which is stable. The longest-lived radioactive isotope is aluminium-26 which has a half-life of 717,000 years. This is produced by cosmic rays striking argon in the upper atmosphere and it has been used to date marine sediments, and to study meteorites and moon rocks which have relatively high levels of this isotope because of their exposure to cosmic rays. The heavy isotope, aluminium-42, turned up unexpectedly when physicists at Michigan State University were trying to make magnesium-40, an isotope with 12 protons and 18 neutrons. Aluminium-42 has 13 protons and 17 neutrons and existed for a millisecond.

Element of surprises

1. Several gem stones are made of the clear crystal form of aluminium oxide known as corundum; the presence of traces of other metals creates various colours: iron produces yellow topaz; cobalt creates blue sapphires, and chromium makes red rubies. All these are now easy and cheap to manufacture artificially.

2. In 2006 a superatom cluster Al_7^- was reported and this behaves chemically as if it were just a single atom and is more like germanium than aluminium.

3. Powdered aluminium is highly reactive and mixed with ammonium nitrate it makes a powerful explosive. Because it is dangerous there are limits on its sale to the public. A paste made of nano-scale aluminium particles and ice (aka ALICE) was shown to be capable of acting as a fuel in 2009. When ignited, the water reacts with the aluminium to form hydrogen gas and as this burns it ignites more aluminium to generate temperatures of 3,000°C. ALICE has been used to launch a rocket but is unlikely to be used as such. Aluminium powder itself is used in fireworks and in solid-fuel rockets such as those of the space-shuttle boosters.

4. Aluminium also occurs as a native metal in the form of tiny grains in recently produced volcanic muds and rocks with low oxygen contents. These are from geologically special regions in Siberia and China. Such grains have also been found in meteorites.

5. Aluminium was the cause of the Hindenberg disaster in 1937, not hydrogen gas as many assumed. Powdered aluminium was used in the paint for the airship's hull and this was ignited by static electricity. Within seconds it burned intensively across the 803-foot length of the hull. For more of the Hindenberg story see hydrogen.

Americium

Pronounced amer-iss-ee-um, it is named after America where it was first made.
French *américium*; German *Americium*; Italian *americio*; Spanish *americio*; Portuguese *amerício*.

Essential element

This lethal radioactive element is a key part of safety devices found in most homes.

Element of life

Although americium is a radioactive element, it poses little threat and is unlikely to get into the human food chain. It can have no role to play in living things.

Dangerous element

Where americium is handled, in nuclear facilities and research laboratories, stringent precautions have to be taken because it is an alpha-emitter, making it particularly dangerous if it gets into the body, where it tends to concentrate in the skeleton. Inside a living cell an alpha particle can cause irreparable damage and even if the cell survives it may become cancerous. When americium compounds are handled in gram amounts its gamma radiation is also a problem. See also *Element of surprises*.

Element of history

Americium was first made late in 1944 at the University of Chicago, Illinois, USA, by a team which included Glenn T. Seaborg (1912–99), Ralph A. James, Leon O. Morgan, and Albert Ghiorso (1915–). Seaborg was clearly not affected by americium or any of the radioactive elements he worked with because he lived to be 86. The americium was produced as a result of the bombardment of plutonium with neutrons in a nuclear reactor.[10] This element was in fact discovered after curium, the element which follows it in the periodic table.

---

[10] The overall reaction is: $^{239}Pu + 2n \rightarrow {}^{241}Am + e$.

Economic element

Americium-241 is extracted from nuclear reactors and has a half life of 432 years, decaying to neptunium-237 which has a half-life of 2.1 million years. Most americium is used for research purposes, but a little of the 241 isotope is employed as a source of radiation for gamma-radiography and in smoke detectors.

A smoke detector contains 150 micrograms (μg) of americium oxide, and it relies on americium's alpha radiation to ionise the air in a gap between two electrodes, thereby causing a tiny current to flow between them. This flow is monitored by an electronic circuit which sounds an alarm if the current drops below a certain level. This happens when smoke gets between the electrodes because particles of soot absorb the ions causing the current to fall. When the smoke is wafted away the current rises again, and the alarm stops.

Although 33,000 americium atoms per *second* undergo radioactive decay in a smoke detector, none of the α-particles escapes because these cannot penetrate solid matter—a sheet of paper, and skin, will stop them. Even in air they travel only a few inches before colliding with an oxygen or nitrogen molecule, and as they do, the alpha particle grabs electrons and becomes an atom of helium gas.

The US Atomic Energy Commission first offered americium oxide for sale in March 1962, at a price of $1,500 per gram. Today it costs $160 per milligram, roughly a hundred times as much. A gram will supply enough americium for more than 5,000 detectors. Smoke detectors must conform to standards set out by the National Radiological Protection Board but their disposal is not supervised. However, the radiation hazard they pose is insignificant compared to the benefit which they confer by saving lives.

Americium can also be made to produce neutrons, and these can be used in analytical probes. When an alpha particle hits a beryllium atom it transmutes this element into carbon, and as it does so it emits a neutron. The stream of neutrons from such americium-beryllium sources is used to test flasks designed to hold radioactive materials, to ensure they are completely radiation proof.

The gamma rays that americium gives off have shorter wavelengths than X-rays, and so are more penetrating. They were once used in radiography to determine the mineral content of bones and the fat content of soft tissue, but are now only used to determine the thickness of plate glass and metal sheeting. The transmission of the rays shows how thick the material is.

Environmental element

Americium was produced on planet Earth for hundreds of thousands of years in natural nuclear reactors. Starting about 1.8 billion years ago, geological processes concentrated enough uranium to start sustained nuclear reactors at 16 locations in what is now Oklo, Gabon, in Africa. These functioned continuously, and by neutron capture and beta decay processes they produced the transuranium elements from atomic numbers 83 (neptunium) to 100 (fermium).

Americium still occurs naturally on Earth, but only in trace amounts in uranium minerals.

Chemical element

Data file	
CHEMICAL SYMBOL	Am
ATOMIC NUMBER	95
ATOMIC WEIGHT	243.1 (isotope Am-243)
MELTING POINT	994°C
BOILING POINT	2607°C
DENSITY	13.7 g/cm³
OXIDES	Am_2O_3 and AmO_2

Americium is a silvery, shiny, radioactive metal and one of the actinide group (row 5f) of the periodic table. The metal itself is made from americium fluoride by reaction with molten barium at 1,100°C. Americium is attacked by O_2, steam, and acids, but not alkalis. It is denser than lead. Its preferred oxidation state is +3, but it can display a range of states, namely +2, +4, +5, +6, and +7. Several compounds of americium have been prepared and most of them are coloured such as the more stable americium(IV) oxide AmO_2 which is black, and the less stable americium(III) oxide Am_2O_3 which is red-brown. The halides are also coloured: AmF_3 and $AmCl_3$ are pink, while AmF_4 is light brown, and AmI_4 is yellow. In solution the preferred state is Am^{3+}.

There are 16 known isotopes of americium with mass numbers 232 to 247, and 8 known isomers. The longest-lived isotopes are americium-243, which has a half-life of 7,370 years, and americium-241, which has a half-life of 432 years. Of the rest, none has a half-life longer than 51 hours (americium-240).

There are 5 isotopes with mass numbers 241 to 245 which are present in nature and they are produced by a series of neutron capture and beta-decay processes of uranium and plutonium isotopes.[11]

Element of surprises

1. The discoverers of americium chose an unusual way to announce the new element. They did it on a children's radio show in the USA called Quiz Kids which was broadcast on 11 November 1945. The guest scientist on the panel that week was a 33-year-old chemist, Glenn T. Seaborg, who had worked on the top-secret atomic weapons programme that had produced two new elements, curium and americium. Americium came to light as part of the Allied project to develop nuclear weapons, and so its discovery was kept secret until the end of World War II. Its

---

[11] Uranium-238 captures a neutron to become uranium-239 which undergoes two beta decays, first to nepturnium-239 and thence to plutonium-239. Through subsequent neutron captures this becomes plutonium-240 to 243. Plutonium-241 and 243 undergo beta decay to americium-241 and 243. Successive neutron captures, starting with americium-241 produce americium-242 to 244. Naturally occurring primordial plutonium-244 undergoes beta decay to americium-244. Americium-245 can be occasionally produced by beta decay of the rare infrequently occurring isotope berkelium-249, which can occur in extremely minute trace amounts.

existence was officially announced a few days after the broadcast, and the follow-ing year Seaborg proposed naming it americium, after the continent on which it was first produced.

2. People wishing to commit suicide have swallowed the capsule containing the americium in smoke-detectors, only to discover that nothing happened to them. The gold foil, which prevents the leakage of alpha particles from these detectors, protects the capsule from being attacked by the acid of the stomach so it passes through the gut and emerge with the americium intact.

3. Americium-241 is fissile and could be used for nuclear weapons. Although several countries have considered this possibility, no test explosion has ever taken place.

	51
	Sb

Antimony

Pronounced anti-mony, the name comes from the Greek *anti-monos* meaning not-alone, although there were suggestions that it derived from anti-monk on account of its supposed use in punishing monks for drunkenness—see below. The chemical symbol, Sb, comes from the Latin word *stibium* which was the name of the mineral by which antimony sulfide was known in ancient times. Constantine of Africa, who died in 1078 AD, first used the word antimony in his work *De Gradibus Simplicibus* and he is believed to have coined the name, although he was not referring to the element itself.

French *antimoine*; German *Antimon*; Italian *antimonio*; Spanish *antimonio*; Portuguese *antimónio*.

Essential element

Antimony compounds find use as flame-retardants and semiconductors.

Element of life

Antimony in the human body	
BLOOD	3.3 p.p.b.
BONE	0.01–0.6 p.p.m.
TISSUE	0.01–0.2 p.p.m.
TOTAL AMOUNT IN BODY	about 2 mg.
DANGEROUS INTAKE	more than 100 mg.

Antimony has no biological role. Only tiny amounts of this element are present in soils although when it is in soluble form it is taken up by plants. The level in food crops is of the order of parts per trillion and so poses no threat to human health. Levels in humans are very low although the liver and kidneys have around 0.2 p.p.m. and the hair around 0.7 p.p.m.

When the level of antimony in the blood of healthy babies was measured in the UK in the 1990s it was found to be higher than expected at 13 p.p.m. This came from the dust of their domestic environment; the average infant consumes about 100 mg of house dust a day. In some older houses the levels of antimony in house dust exceeded 1,800 p.p.m. The most likely source was from lead, which often contains traces of antimony, and which is still present in older homes as layers of lead-based paints.

Dangerous element

Antimony can kill. People who have taken a large dose of antimony accidentally have suffered the expected vomiting, appeared to recover, and then died a few days later. Mass poisonings occurred when fruit drinks were prepared in enamel vessels, from which antimony has leached out, although rarely with fatal consequences.

Some Victorian doctors even used antimony to dispose of their unwanted wives and relatives, the symptoms being disguised as general gastric disorders. In this way Drs Palmer (1855), Smethurst (1859) and Pritchard (1865) carried out their murders— and were all eventually hanged. Other murderers were the society hostess Florence Bravo who poisoned her husband in 1876 but was not brought to trial, and innkeeper George Chapman, who disposed of a few tiresome girlfriends and was executed in 1902.

Antimony attaches itself to particular enzymes because of its attraction to sulfur atoms at the enzyme's active site. Once attached, it cannot easily be dislodged. Far more deadly is the gas stibine, which is antimony hydride (SbH_3).

Although many of its former uses have now been discontinued because of its toxicity, antimony oxide is still used a flame retardant at levels of 2 to 8 percent. It was this use which led to its being accused in the 1990s of causing cot deaths in babies. The theory was that antimony was converted to volatile methyl stibine (CH_3SbH_2) by fungi in the mattress, and that babies were breathing in enough of the vapour to kill them. The fungus, *scopulariopsis brevicaulis,* was fingered as the likely culprit, and it is indeed capable of reducing antimony to stibine. Support seemed to come from the analysis of tissue from cot death victims which had higher than expected levels of antimony in their bodies. A panel set up to enquire into the matter reported in 1998, saying that there was no scientific evidence to support the theory. The fungus was only found in a tiny number of cot mattresses, and even when it was present there was no evidence of the toxic gas being emitted from them.

Some accidental deaths have been attributed to antimony, and the French playwright Molière was convinced that his only son had been killed by an antimony-containing medicine. Another death from excess antimony may have been Mozart's—see box.

The inexplicable early death of Mozart

In October 1791, the 35-year-old composer Wolfgang Amadeus Mozart was living in Vienna and fell ill. Slowly he got worse and died on 5 December of that year. He believed he was being poisoned and his rival, Antonio Salieri, confessed to his murder many years later, although when he did so he was suffering from senile dementia and the claim is not taken seriously.

More likely Mozart had been poisoned accidentally by antimony which was prescribed by his doctors, and he was fond of dosing himself with it in the form of tartar emetic (see below). This was a popular hangover cure, and Mozart was known to be a heavy drinker. On 20 November 1791 he developed a fever, his hands, feet and stomach became swollen and he had sudden attacks of vomiting. Miliary fever, an illness we no longer recognise, was diagnosed and he died of this two weeks later. Its symptoms are the same as those of acute antimony poisoning.

The antidote for antimony poisoning is a chelating agent like Dimercaprol which will scavenge atoms of the element from the blood and organs of the body so that it can be excreted.

Medical element

Small doses of antimony of around 5 mg produce profuse sweating while doses of 10 times this amount will cause vomiting, both symptoms which were regarded as useful in treating various complaints in earlier times. In 1657 King Louis XIV of France, who was only 19 years old, fell ill with typhoid fever and was treated with an antimony medication known as Earl of Warwick's Powder—its active ingredient was antimony oxide—and was thereby apparently cured. His dramatic recovery ensured the popularity of this type of medication for the next 200 years and all kinds of medicines were devised which contained antimony compounds. A popular one was Dr James's Powder which consisted of calcium phosphate and antimony oxide and was prescribed to cure 'fevers'. This patented cure was introduced in 1747 by Robert James MD (1705–76). Antimony Wine was also sold in the mid-1700s and consisted of antimony oxide sulfide dissolved in Madeira wine. It was recommended as a tonic, a slimming aid, and a sweating agent.

The Roman physician Dioscorides, who lived in the second half of the first century AD, was familiar with antimony sulfide (Sb_2S_3) or *stibi* as it was then known, and in his book *De materia medica* he recommended it for skin complaints and burns. Abulcasis (Kalaf ibn-Abbas al-Zahrawi) who died in 1013 was a distinguished doctor of Arabic Spain and he was probably the first to advocate the use of antimony sulfide as a medicament. Antimony became more popular in the 1500s when the famous physician Paracelsus (1493–1541) advocated its use for all manner of ailments.

A popular medication was tartar emetic, which was the antimony potassium salt of tartaric acid. As its name indicates, it produces rapid vomiting and was seen as a way of expelling bad 'humours' from the body. One way of generating this compound was to leave some wine standing overnight in a cup made of antimony. By morning it had usually dissolved enough of the metal to produce the desired effect. Such cups were popular in Germany and were reputed to be what monks were forced to drink to cure them of overindulgence. Others could experience the same benefit by dissolving a little tartar emetic in wine.

Oswald Croll (1580–1609) wrote *Basilica Chymica* which contained 23 recipes that included antimony, and especially 'butter of antimony' (antimony chloride, $SbCl_3$) made by reacting the sulfide with mercury chloride ($HgCl_2$). The 1600s and 1700s were the heyday for antimony treatments, although many doctors opposed the use of antimony, knowing how dangerous it could be.

There was a revival of interest in antimony salts by doctors and especially vets early last century after it was discovered, in 1915, to be effective against parasitic infections such as bilharziasis, trichinosis, and schistosomiasis. It was prescribed in doses that would kill these organisms yet not poison the patient. The medicine given was as a complex molecule known as stibophen, and injected daily for several days; the amount of antimony delivered would be around 500 mg. People treated this way then had higher levels of antimony in their blood and urine for many months. (Unlike its sister element, arsenic, antimony is not readily excreted from the body.)

The parasitic infection leishmaniasis is caused by a protozoa which invades the skin and is a widespread childhood disease in hot countries where it is spread by sand flies. The best treatment for this is antimony sodium gluconate (aka pentostam) which has antimony in its higher oxidation state.

Element of history

Antimony and its compounds were known to the ancient world and there is a 5,000-year-old vase in the Louvre which is almost the pure metal. Antimony sulfide (Sb_2S_3) is mentioned in an Egyptian papyrus of the sixteenth century BC, and from that time onwards the black form of this pigment, which occurs naturally as the mineral stibnite, has been used as a type of mascara known as *khol*. The most famous user was the temptress Jezebel whose exploits are recorded in *The Bible*.[12]

Another pigment known to the Chaldean civilisation, which flourished in the sixth and seventh centuries BC, was yellow lead antimonate. This is found in the glaze of the ornamental bricks which Nebuchadnezzar (604–561 BC) used to adorn the walls of his capital, Babylon. This pigment was still being made in the last century (from lead oxide or carbonate and antimony oxide) and was known as Naples yellow.

Antimony intrigued the alchemists because it resembled gold in one respect, in that it was not dissolved by the most powerful acid then known: aqua regia. Antimony was regarded by them as a possible candidate for transmutation into gold. In 1604, the influential book *The Triumphal Chariot of Antimony* appeared which accurately described many antimony compounds. This book was reputed to have been written by a monk called Basil Valentine in 1400 but was actually written by a German named Johann Thölde.

Element of war

Greek fire may have contained antimony sulfide in the form of the natural mineral stibnite (Sb_2S_3). This incendiary liquid was fired from the warships of the Byzantine navy and brought terror to those exposed to it because it was impossible to extinguish and it even burned on the surface of water. How it was made has remained a secret to this day and it was a capital offence to reveal it. It was last used in the final and unsuccessful defence of the capital, Constantinople, in 1453. The most likely composition of Greek fire was crude oil, stibnite and saltpetre (KNO_3), a combination that would be

---

[12] When antimony sulfide is made by precipitation it is orange-red in colour and this form was used to make the heads of household matches in the 1800s.

highly flammable and almost impossible to extinguish with water. Once it is ignited, antimony sulfide also generates a lot of heat.

Antimony trisulfide has a different role to play in modern warfare. It is used in camouflage paints because it reflects infrared radiation in the same manner as green vegetation.

Economic element

Antimony is still an important metal. Annual production is around 185,000 tonnes per year, with 85 percent coming from China. Other producers are Bolivia, Russia, and South Africa. World reserves total around 4 million tonnes. The chief ores are stibnite and tetrahedrite, the latter being mainly a copper ore, but it yields antimony as a by-product. In the USA about half the antimony that is used is recovered from lead batteries and although there are large deposits of antimony in North America these are not currently being mined.

Antimony metal has a curious property in that it expands when it solidifies. This property was appreciated in the ancient world where it was used to produce finely cast objects. It really came into its own with the development of printing when an alloy consisting of 60 percent lead, 10 percent tin, and 30 percent antimony was cast to give a sharp typeface.

Antimony is not quite a true metal—it is often referred to as a metalloid—and although in its metallic form it resembles lead, it is a poor conductor of electricity. Nevertheless it can be alloyed with other metals and this accounts for its major use. Antimony-lead alloys, apart from being used for storage batteries, are also used in bearings and cable sheathing, although these uses are declining as lead gives way to other materials. When lead-based pewter fell out of favour because of its toxic properties, it was replaced by an alloy of composition 89 percent tin, 7 percent antimony, with 2 percent each of copper and bismuth.

Babbitt metal, patented by Isaac Babbitt in 1839, has remarkable anti-friction properties and was originally a lead-antimony alloy but is now available as a tin-antimony alloy. It is cast onto metal surfaces which come into contact while moving, as in gas turbines and crankshafts, and the alloy adopts its shape to ensure minimum friction between the two surfaces.

Antimony compounds have several uses. An early one was of antimony trichloride in hydrochloric acid, and known as bronzing fluid, which was used to give a bronze-like finish to cast iron objects. Antimony oxide, Sb_2O_3, is the form in which antimony is added as a flame retardant to plastics used in cars, toys, seat covers, cables, roof sheeting, and cot mattresses. This use accounts for about two thirds of all consumption although for some uses molybdenum oxide is now the preferred fire retardant. These oxides quench flames by reacting chemically with the hot plastic to form a viscous layer that smothers the flame. Antimony oxychloride (SbOCl) is a fire-retardant used in textiles. Some antimony chemicals such as SbF_5 are used in the manufacture of fluorine compounds and as a catalyst.

Research into antimony still continues, with chemists using it to make semi-conductors such as gallium arsenide antimonide with the result that antimony is now finding use in electronic devices.

Environmental element

Antimony in the environment	
EARTH'S CRUST	av. 0.2 p.p.m.
	Antimony is the 63rd most abundant element.
SOIL	av. 1 p.p.m.
SEAWATER	av. 0.3 p.p.b.
ATMOSPHERE	tiny but increasing.

A little uncombined antimony occurs naturally as granular masses or nodules, generally in silver-bearing lodes, and these have been found in Borneo, Sweden, Germany, Portugal, Italy, Canada, and the USA. Antimony is not a mobile element—it generally occurs as antimony sulfide which is very insoluble—and samples taken from sediments and bogs corresponding to the pre-historical environment showed antimony levels that were almost too small to be measured. Bill Shotyk of the University of Heidelberg is the world expert on environmental antimony. His analyses of snow and ice in the Canadian arctic shows background levels of antimony that are around 2 p.p.t. Analysis of remote peat bogs in Scotland shows that antimony levels have increased over the centuries in parallel with lead levels, and they have been increasing more rapidly in the last century as more uses were found for this element. Levels in 2004 were around a hundred times higher than in the mid 1800s.

Chemical element

Data file	
CHEMICAL SYMBOL	Sb
ATOMIC NUMBER	51
ATOMIC WEIGHT	121.760
MELTING POINT	631°C
BOILING POINT	1635°C
DENSITY	6.67 g/cm³
OXIDE	Sb_2O_3

Antimony is a metalloid element and a member of group 15 (row 5p) of the periodic table. It can exist in two forms: the metallic form is bright, silvery, hard, and brittle; the non-metallic form is a grey powder. Antimony is stable in dry air, and is not attacked by dilute acids or alkalis. Its compounds can exist in two oxidation states: +3 (as in the butter-like solid $SbCl_3$) and +5 (as in liquid $SbCl_5$), the former being the more stable.

There are 37 known isotopes of antimony with masses 103 to 139, and 29 known isomers. There are two stable naturally occurring isotopes: antimony-121 which accounts for 57% and antimony-123 (43%).[13] Antimony-125 is the longest-lived radio-active isotope with a half-life of 2.76 years.

Element of surprises

1. In the Middle Ages antimony pills were sold as reusable laxatives. Those suffering constipation would swallow one, about the size of a pea, and then wait for its toxic action to aggravate the intestines, causing them to become more active in order to expel the irritant pill. This was eventually retrieved from the expelled excrement and stored for future use. There are even reports of such pills being passed from generation to generation.

2. Higher than expected levels of antimony were found in bottled drinking waters in 2006. The antimony had leached from the plastic containers which were made of PET (polyethylene terephthalate) and for which antimony is used as a catalyst in its manufacture. The water could have up to 375 p.p.t. of antimony, and double this amount after standing for several months, compared to water in polypropyl-ene bottle which had only 8 p.p.t. Antimony is not involved in its manufacture. However, the levels in the water were far below the maximum level set by the World Health Organization which is 20 micrograms per litre (20,000 p.p.t.) or even the low US level of 6 micrograms per litre.

[13] There are a number of radioactive isotopes produced as fission products in uranium-bearing deposits.

Argon

Pronounced ar-gon, the name is derived from the Greek word *argos* meaning idle.
French *argon*; German *Argon*; Italian *argo*; Spanish *argón*; Portuguese *argônio*.

Essential element

This gas provides inert atmospheres that protect sensitive materials and hot metals.

Element of life

Argon has no biological role, although bacteria in the nodules of certain plants like beans, which have the ability to absorb nitrogen from the air and convert it to ammonia, can also absorb argon, although they are unable to process it further.

Food element

As yet argon has no role to play in food preservation, although one day liquid argon may be used as a preservative, and argon gas as a non-oxidising atmosphere inside containers of oxidisable products.

Dangerous element

Argon can kill if it excludes air from the lungs and is used for the mass slaughter of poultry either when an indoor flock is being affected by disease, or simply as a means to kill them humanely. In fact as it replaces oxygen in the body of the bird it gives the carcass a longer shelf life.

Element of history

Although it is fairly abundant in the Earth's atmosphere (nearly 1 percent), argon evaded discovery until 1894. Its discovery came about as a result of the collaboration of two eminent scientists: the physicist John Strutt, who was the third Lord Rayleigh (1842–1919) and professor of experimental physics at Cambridge University, and the chemist Sir William Ramsay (1852–1916), who was professor of chemistry at University College London. They reported its existence to the annual meeting of the British Association for the Advancement of Science at Oxford. Rather strangely, they then fell silent about their remarkable discovery, declining to publish their results. The reason for this was so that they could enter a competition for 'some new…discovery about atmospheric air' which was organised by the Smithsonian Institution in Washington DC. A condition of the competition was that the discovery should not have been previously disclosed before the closing date at the end of 1894. Rayleigh and Ramsay eventually won the $10,000 prize (equivalent to $300,000 today) in competition with 218 other entrants.

Then on 31 January 1895, before an audience of more than 800 people in the main lecture theatre of University College London, Professor Ramsay read a paper which he and Rayleigh had submitted to the Royal Society reporting the discovery. In 1904 Rayleigh won the Nobel Prize for physics and Ramsay won the Nobel Prize for chemistry.

In fact they were not the first to discover argon although they were the first to identify the gas. Argon was first isolated in 1785 in Clapham, London, by the wealthy eccentric, Henry Cavendish (1731–1810) who had a private laboratory. There he set about investigating the chemistry of the atmosphere. He passed electric sparks through a sample of air and hydrogen and trapped out the gases which formed, but he was puzzled that there remained about 1 percent of its volume that would not combine chemically. He did not realise that he had stumbled on a new gaseous element, not that he would have been able to identify it using the primitive techniques of his day. For over a century Cavendish's observations were not understood—but they were not forgotten.

The second encounter with argon occurred around 1882 when Hugh Frank Newall (1857–1944) and Walter Noel Hartley (1846–1913) independently recorded some new lines in the spectrum of air photographed at low pressures. They were unable to identify the element that was giving rise to these lines, although they realised there was something, as yet unidentified, that was present in the atmosphere. However, it was not this observation which motivated Rayleigh and Ramsay. They discovered argon as a result of trying to solve a puzzling feature of nitrogen gas: why did the density of this depend on how it was obtained?

The nitrogen which was extracted from air had a density of 1.257 grams per litre, whereas that obtained by decomposing ammonia had a density of 1.251. Rayleigh had reported this in the prestigous journal *Nature* in 1892 but no one could offer a convincing reason why this should be so. Either atmospheric nitrogen contained a heavier gas, or chemically-derived nitrogen was diluted with a lighter gas.

Ramsay wrote to Rayleigh suggesting that each of them should investigate one of the two kinds of nitrogen: he would look for a heavier gas in that from air, while Rayleigh would look for a lighter gas in the chemically-derived nitrogen. Consequently Ramsay devised an experiment to remove all the nitrogen from his sample of the gas obtained from air. He repeatedly passed it over heated magnesium, with which nitrogen reacts to form a solid, magnesium nitride. Like Cavendish more than a hundred years earlier, he was left with about 1 percent of the volume which would not react, but unlike his predecessor he measured the density of the residual gas, and found it was denser than nitrogen at 1.784 grams per litre.

More importantly, Ramsay could examine its spectrum and there he saw new groups of red and green lines, confirming that he had isolated a hitherto unidentified gas. The great spectroscopist Sir William Crookes (1832–1919) carried out a more detailed analysis of the spectrum and recorded nearly 200 lines. When Newall and Hartley re-examined the photographic plates they had taken 12 years earlier, the same lines were present.

Medical use

Blue argon lasers are used in surgery to correct defects of the eye, to weld arteries, and to destroy tumours. Cryosurgery, which uses very low temperatures to destroy cancer cells, can be done using liquid argon.

Economic element

Argon is an important industrial gas extracted from liquid air and world production is around 700,000 tonnes per year. The supply is virtually inexhaustible, with an estimated 66×10^{12} tonnes (= 66 trillion tonnes) circulating the planet. Hundreds of industrial plants around the world produce it. A typical plant processes 375 tonnes of air a day, separating it into oxygen, nitrogen and argon, which are shipped out as liquids in tankers holding 20 tonnes per load.

Argon is used as an inert atmosphere in most kinds of electric lighting, and was the gas in the old-fashioned incandescent light bulbs. In fluorescent tubes, the argon makes the lamps easier to start. This use relies on argon's chemical inertness—nothing will induce it to react with any other material, no matter how high the temperature to which it is heated, nor how strong the electrical discharge passed through it. Even when it is ionised by radiation, as it is in some types of Geiger counter, the gas still refuses to react.

Argon is particularly important for the metals industry. Most of it goes to making steel where it is blown through the molten metal, along with oxygen. The argon acts to stir the steel while the oxygen reacts with carbon to form carbon dioxide, thereby reducing the level of carbon. Argon is also used as a blanket gas when air must be excluded to prevent oxidation of hot metals, as in welding aluminium and the production of titanium metal. Welding aluminium is done with an electric arc welder which uses direct current to create a spark that melts the welding rod which is surrounded by argon flowing at a rate of 10–20 litres per minute. Atomic energy fuel elements are protected with argon during refining and reprocessing. Argon is also used in graphite electrical furnaces to prevent the graphite (which is carbon) from burning.

The alloys used for making high grade tools start as metal powders, and these are produced by directing a jet of liquid argon, temperature –190°C, at a jet of the molten metal. The result is an ultrafine powder with a clean surface.

Some smelters prevent toxic metal dusts escaping to the environment by venting them through an argon plasma torch. In this, argon atoms are electrically charged to reach temperatures of 10,000°C and the toxic dust particles passing through it are turned into a blob of molten scrap.

Some consumer products contain argon, such as sealed double glazing. In the 14 mm gap between the panes of glass it provides better insulation because it is a poorer conductor of heat than ordinary air. The thermal conductivity of argon is $18\,mWm^{-1}K^{-1}$ at normal temperatures compared to air which is $26\,mWm^{-1}K^{-1}$. For this reason argon is also used to inflate diving suits. Illuminated signs glow blue if they contain argon and bright blue if a little mercury vapour is also present. The most exotic use of argon is in the tyres of luxury cars. Not only does it protect the rubber from attack by oxygen, but it ensures less tyre noise when the car is moving at speed.

Blue argon lasers are used by chemists to probe molecular states which exist for only a trillionth of a second. Argon is also the gas used in various methods for analysing minute samples of material by gas chromatography and mass spectrometry.

Replacing air with argon in bottles and ampoules of pharmaceutical products, high purity chemicals, and in barrels of wine, protects the contents against oxidation. Rare old documents and other artefacts that are susceptible to oxidation can be protected by storing them in an atmosphere of argon rather than helium, which used to be used, because there is less seepage of the gas from containers.

Environmental element

Argon in the environment	
EARTH'S CRUST	1.2 p.p.m.
	Argon is the 56th most abundant element.
SEAWATER	0.45 p.p.m. and its average residence time is 28,000 years.
ATMOSPHERE	0.93% by volume.

The argon in the Earth's atmosphere has slowly built up over billions of years, coming from the decay of potassium-40 which has a half-life of 12.7 billion years. Of a million potassium atoms, 117 are potassium-40 and of these 12 are able to seize one of the electrons surrounding the nucleus—a decay mode known as electron capture—to give argon-40. The rest either decays by beta ray emission and converts to calcium-40, or by beta-plus emission also to give argon-40. By measuring the ratio of potassium to argon in a mineral, it is possible to date the mineral sample. This provides a valuable dating technique for geologists, known as the potassium-argon dating method.

Argon-40 accounts for more than 99 percent of the argon in the atmosphere, the remainder being traces of argon-36 and argon-38 which were present when the Earth formed.

The atmosphere of Mars also contains argon, to the extent of 1.6 percent, and this too is almost all argon-40. Mercury's thin atmosphere is mainly argon. Titan, the largest moon of Saturn, has a dense atmosphere consisting mainly of nitrogen with some argon.

Chemical element

Data file	
CHEMICAL SYMBOL	Ar*
ATOMIC NUMBER	18
ATOMIC WEIGHT	39.948
MELTING POINT	−189°C
BOILING POINT	−186°C
DENSITY OF THE GAS	1.8 g per litre
OXIDES	none

* Originally the symbol for argon was simply 'A' but was changed to Ar by IUPAC in 1957.

Argon is one of a group of elements called the noble gases which constitute group 18 of the periodic table; the others are helium, neon, krypton, xenon, radon and the yet unnamed element 118. Like its fellow gases, it is colourless and odourless, but unlike them it is not rare. Argon is as soluble in water as oxygen gas, and more than twice as soluble in water as nitrogen gas.

Argon was thought to be inert towards all other elements and chemicals, and until recently it resisted all attempts to make it bond to other elements and preferred to exist as single atoms. It turned out to be not quite so chemically inert as had always been assumed. In 2000 chemists at the University of Helsinki, Finland, announced in the magazine *Nature* (24 August 2000) that they had made the first ever compound, argon fluoro-hydride (HArF). This they obtained by freezing a mixture of argon and hydrogen fluoride on to caesium iodide at −265°C and exposing the mixture to UV radiation, which caused the gases to react to form HArF. They identified the new compound by looking at its infrared spectrum. At temperatures below −246°C the compound clearly exists, albeit with extremely weak bonds, but on warming it reverts back to argon and hydrogen fluoride.

In 2009 a molecular ion, $ArSiF_2^{2+}$, which has an argon-to-silicon chemical bond, was shown to form when streams of SiF_3^{2+}, made by ionising SiF_4 gas, and argon collided, but it existed only in the gas phase.

There are some chemical compounds which contain argon, the so-called argon clathrates, but in these it is present merely as atoms trapped in holes in the lattice of a larger molecule and is chemically unbound. As soon as the clathrates dissolve in water the argon escapes.

There are 24 known isotopes of argon with masses 30 to 53, and 1 known isomer. Naturally occurring argon is not radioactive. The argon in the atmosphere consists mainly of argon-40 (99.60 percent), argon-36 (0.34 percent), and argon-38 (0.06 percent). The longest-lived, radioactive isotope is argon-39 with a half-life of 269 years.

Element of surprises

1. Argon originally puzzled chemists because it turned out to be *too* heavy. Its relative atomic weight ought to be smaller than that of potassium, element 19, which follows it in the periodic table, but argon's relative atomic weight is 39.95, which is greater than that of potassium which is 39.10. The reason lies in the isotopic distribution of the two: argon is mainly composed of the isotope argon-40, with 18 protons and 22 neutrons, while potassium is predominantly potassium-39, with 19 protons but only 20 neutrons.

2. In 2005 the Scripps Institute of Oceanography at La Jolla, California showed that argon tends to be slightly more abundant in the air which is nearer the Earth's surface than in the air at the top of mountains. This was done by analysing air in the Borrego Sink, a depression in the California desert, and comparing it with air at higher elevations.

3. Compressed argon cools as it expands and is used to cool the heat-seeking detector in the heads of air-to-air missiles such as the US Sidewinder.

Arsenic

Pronounced ars-nik, the name probably comes from *arsenikon*, the Greek name for the arsenic mineral yellow orpiment, although at one time it was thought to have been derived from the word *arsenikos* meaning male. The Arabic for orpiment was *al-zarnik*.

The first recorded use of the English word arsenic was in 1310, and it was mentioned by Chaucer in his *Canterbury Tales* written in 1386.

French *arsenic*; German *Arsen*; Italian *arsenico*; Spanish *arsénico*; Portuguese *arsénic*.

Essential element

This deadly element is now treating people with leukaemia as well as providing new kinds of microchips.

Element of life

Arsenic in the human body	
BLOOD	2–9 µg per litre of blood
BONE	0.1–1.6 p.p.m.
TISSUE	0.1–1.6 p.p.m.
HAIR	1 p.p.m.
TOTAL AMOUNT IN BODY	*av.* 7 mg but varies between 0.5–15 mg.

Arsenic is an essential trace element for some animals, and maybe even for humans; if so the necessary intake may be as low as 0.01 mg per day. Chickens and rats that were fed an arsenic-free diet had stunted growths. The need for arsenic is linked to the metabolism of the essential amino acid arginine. In 2009, it was revealed that a sponge from New Caledonia in the southwest Pacific Ocean takes in arsenic to form a molecule known as arsenicin-A which incorporates four arsenic atoms and has the formula $As_4O_3(CH_2)_3$.

Arsenic permeates all body tissue including bone, hair, and nails. In the hair it becomes chemically bonded to the sulfur atoms of the keratin of which hair is made, and this explains why hair analysis reveals the extent to which a person has been exposed to arsenic.

Food element

The daily intake of arsenic in a normal diet can vary considerably, even being as high as 1 mg if certain foods are consumed. Levels of arsenic in rice are higher than in most other foods because this plant takes up the element from both water and soil, and in the inorganic form. Sea creatures also extract arsenic from their environ-ment so that plaice and oysters have 4 p.p.m., mussels 120 p.p.m., and prawns as much as 175 p.p.m., but the arsenic clearly does them no harm, nor does it affect those who eat them. The arsenic is in an organic form known as arsenobetaine,[14] and although this is readily absorbed it is also rapidly excreted in the urine. Organic arsenic can also be in the form of sugar molecules and polyunsaturated fatty acids, depending on the species. The acids are of the omega-3 type such as docosahexae-noic acid (DHA).

It has long been known that those who were prescribed medication containing arsenic tended to put on weight. Consequently arsenic has been used to fatten poultry and pigs and the compound used for that purpose is roxarsone, which is a deriva-tive of arsenic acid.[15] Roxarsone has been added to chicken feed to the extent of 900 tonnes a year in the USA for more than 30 years and in addition to promoting growth it kills the parasites which cause diarrhoea. It is not added to the animals' feed for several days before they are slaughtered to ensure that all of the element is excreted from their flesh. Although some producers no longer use it, more than 4 billion broiler chickens raised annually in the USA are fed it. When the manure from these birds is used to fertilize farmland it breaks down into inorganic arsenic and so raises the arsenic level in crops that are grown thereafter.

Medical element

Arsenic can have an apparent tonic effect which it does by activating enzymes. The arsenite ion (AsO_3^{2-}) allows oxidation to proceed at an increased rate. In small doses inorganic arsenic like this stimulates metabolism and boosts the formation of red blood cells, but prolonged exposure causes chronic dermatitis and is likely eventually to lead to skin cancer. Arsenic is present in several spring waters, some of which were reputed to have tonic effects. The famous Vichy water may contain 2 p.p.m. of arsenic, and other spa waters have been found to contain even more. Zam Zam holy water comes from a particular well not far from the Kaaba in Mecca, and where a spring miraculously appeared to Abraham's wife around 2,000 BC. Tests on bottled Zam Zam water in the UK have shown arsenic levels of three times the legal limit.

Hippocrates (460–377 BC) said that the red ore realgar was a remedy for ulcers on the body. Orpiment and realgar were used to treat abscesses and scrofula in ancient China and were mentioned in medical texts of 200 BC. In Europe, arsenic trioxide was used in the 1110s as an anti-malarial ('tertian fever') medicine. Other physicians down the centuries favoured the more common yellow orpiment, which was taken to treat all sorts of conditions, such as arthritis, asthma, malaria, TB, diabetes, and venereal diseases, for which it could have had only marginal benefits.

Arsenic became popular with the introduction of Dr Fowler's Solution. He con-cocted this in 1780 and found it beneficial when treating patients with fevers. It is essentially a solution of potassium arsenite plus lavender water to make it easily

---

[14] This has the formula $(CH_3)_3As^+CH_2CO_2^-$.   [15] Roxarsone is 4-hydroxy-3-nitrobenzene-arsonic acid.

recognisable and so prevent accidents.[16] A few drops of the medicine were added to a glass of water or taken with wine. The maximum single dose recommended in the British Pharmacopoeia was 0.5 ml of Fowler's solution which would deliver 5 mg of arsenic. It was regarded as a popular cure-all, a general tonic, and an aphrodisiac. Charles Darwin's mysterious illness may well have been due to arsenic poisoning. He was known to take Fowler's solution regularly to treat a tremor in his hands. Fowler's solution was often prescribed by doctors to aid convalescence.

In 1909 a real breakthrough in arsenic-based medication came when Paul Ehrlich reported a chemical that was capable of curing syphilis. He had undertaken a systematic study of arsenic compounds guided by the belief that he might discover one that was toxic to the syphilis spirochete yet not to the patient. On his 606th attempt he finally discovered the one he was looking for (arsphenamine) and it quickly became the recognised treatment for this disease, and was named Salvarsan. Eventually it was superseded by penicillin, although it continued to be used in the treatment of sleeping sickness and dysentery.

Inorganic arsenic may cause cancer—see below—but it can also help in the fight against it. Although arsenic disappeared from medicine for many years, it has reappeared in a quite dramatic way, and via Chinese medicine which has traditionally found a use for arsenic sulfide minerals. Arsenic trioxide (As_2O_3), in the form of the drug Trisenox, is used to treat leukaemia, and it acts by stimulating the production of normal blood cells which have become crowded out by cancerous white blood cells. A combination of Trisenox and retinoic acid is particularly effective and is thought to re-programme cancerous cells into behaving like normal cells again. It was known many years ago that arsenic stimulated the formation of blood and it was once a recognised treatment for anaemia.

Radioactive arsenic-74, half-life 18 days, has been used to label antibodies that bind to cancer cells thereby allowing these to be pinpointed by positron emission tomography (PET) scans. Arsenic-74 is one of only a few radioactive isotopes which emit positrons (positive electrons) and which has a half-life suitable for the investigative technique. The arsenic is combined with an antibody which binds strongly to unique lipids (fat molecules) which are found on the surface of tumours.

Dangerous element

Arsenic exhibits the property known as hormesis. In other words, in tiny amounts it may actually be beneficial whereas in larger doses it becomes poisonous. A lethal dose of arsenic for an adult is about 200 mg. Its effects are vomiting, colic, diarrhoea, dehydration, coma—and death within a day or so. Not surprisingly, arsenic was the agent chosen by many poisoners in the 1800s and early 1900s, when it was readily available as weed-killer or could easily be extracted from flypapers.

Earlier ages had also made use of it, especially in Italy and France where there was a flourishing underground trade in so-called 'succession powders' that removed dukes, kings and even popes. Those who perpetrated such crimes usually escaped

[16] The recipe—in modern units—consists of 10 g of arsenic trioxide, 7.6 g of potassium hydrogen carbonate, and 30 ml of alcohol, in 1 litre of water.

prosecution since there was then no reliable way to detect arsenic in the victim's body. This state of affairs changed radically when James Marsh devised his famous test in 1836 which could measure the element in minute amounts.

Sometimes mass poisonings occurred, such as that which afflicted 6,000 beer-drinkers in Manchester, England, in 1900, of whom 70 died. The beer contained 15 p.p.m. of arsenic, and had been contaminated by the invert sugar used to brew the beer. This had been made using sulfuric acid manufactured from the sulfur of a particular type of iron pyrites (FeS_2) which contained a lot of arsenic.

Accidental arsenic poisoning appeared to be a threat in the 1800s when wallpapers were often printed with arsenical dyes which were copper arsenites. When walls became damp they could give off trimethylarsine gas, $As(CH_3)_3$, produced by moulds growing on them. Bacteria, fungi, and yeasts have the ability to methylate arsenic, in other words to replace the oxygen atoms normally attached to arsenic with up to three methyl groups (CH_3). Breathing the air in such rooms as in a bedroom, was capable of producing chronic arsenic poisoning—or so it has long been supposed. The theory was finally laid to rest in 2005 by William Cullen and Ronald Bentley in their paper 'The toxicity of trimethylarsine: an urban myth'.

The vapour supposedly given off by damp wallpaper had been identified correctly as $As(CH_3)_3$ in 1932 by Professor Challenger who claimed that this was what had afflicted so many people who lived in damp dwellings. However, tests with rats failed to show that it was fatal to them. Indeed the accepted tolerable level of this gas in a working environment, based on animal studies, could well be around 20,000 p.p.m. (2 percent).

Other arsenic gases are more dangerous and in particular arsine, AsH_3, for which the maximum acceptable level in air is 5 p.p.m. (0.0005 percent). Indeed the statutory limit in the USA for arsine is 0.05 p.p.m. The German chemist A. F. Gehlen (1775–1815) discovered how dangerous arsine could be when he made some in 1815. Within an hour of sniffing the gas he became ill, displaying symptoms of vomiting and shivering. He took to his bed but became more and more ill and died nine days later.

It has been speculated Napoleon was poisoned by his wallpaper when held captive on St Helena. This now seems unlikely and neutron activation analysis of his hair shows that while excess arsenic was present compared to hair of today, the amount was typical of the hair from other famous people at that time such as the King of Rome, the Empress Josephine, and even King George III.[17] In 2005 a manuscript written by one of the doctors who attended Napoleon's post-mortem turned up in Scotland and shows quite clearly that he died of cancer of the stomach, as had his father before him.

The evidence now points to *inorganic* arsenic as the forms to avoid; in other words arsenite (AsO_3^{2-}), which is arsenic in oxidation state(III), and arsenate (AsO_4^{3-}), which is in oxidation state(V). Arsenic in the lower oxidation state is the more toxic. There is also epidemiological evidence which links inorganic arsenic to bladder, lung, and liver cancer. Organic arsenic is relatively safe.

---

[17] King George's hair had 17 p.p.m. of arsenic but this might well have come from the tartar emetic he was prescribed—see antimony. Often he was given three grains (180 mg) of this per day, and given it forcibly. At that time tartar emetic could contain as much as 5% arsenic which was present in the antimony from which tartar emetic was made.

Element of history

Contact with arsenic goes back more than 5,000 years and we know this because the hair of the Iceman, who was preserved in a glacier in the mountains of the Italian Alps, contained high levels of the element. His exposure to arsenic is thought to indicate that he was a coppersmith by trade since the smelting of this metal is often from ores that are rich in arsenic. Arsenic was certainly known to the ancient Egyptian civilisation—it is mentioned in the Leyden papyrus as a way of gilding—although not as the element itself. Theophrastus, Aristotle's pupil and successor, recognised two arsenic sulfide minerals: orpiment (As_2S_3) and realgar (As_4S_4). The ancient Chinese also knew of arsenic and the encyclopedic work of Pen Ts'ao Kan-Mu mentions it, noting its toxicity and use as a pesticide in rice fields.

A more dangerous form of arsenic, called white arsenic (As_2O_3), was also known: this was the trioxide. Mixed with vegetable oil and heated it yielded arsenic metal itself, which when applied to copper gave it a silver colour. However, the discovery of the element arsenic is attributed to Albertus Magnus (1193–1280) who may have made it this way, or perhaps by using charcoal instead of oil.

Element of war

An arsenic derivative, named Lewisite,[18] was developed in World War I as a chemical weapon although it was rarely used. It acted by forming terrible blisters on exposed skin and damaging the lungs. Its chemical name is 2-chlorovinyldichlorarsine and its common name came from the American chemist, Lewis, who developed it. Other arsenical chemical warfare agents were Sneeze Gas and Adamsite.[19] Sneeze Gas was used in World War I and produced unbearable irritation of the respiratory tract.

The antidote for Lewisite is BAL (British Anti-Lewisite) which has to be injected and which formed a compound with the arsenic and thereby removed it from the body. BAL is 2,3-dimercaptopropan-1-ol, and it is still used to treat people who have been poisoned by arsenic, mercury and other heavy metals. The two sulfur atoms in this molecule attach themselves to the arsenic and detach it from the proteins and enzymes it is blocking.

The dictator Sadam Hussein used Lewisite in the Iraq-Iran war of the 1980s, as did the Sudan Government against the Sudan People's Liberation Army. In 1999 planes dropped 16 exploding canisters of it on the towns of Lainya and Kaya. A group of UN World Food Programme workers who visited Lainya a few days later were also affected.

Economic element

A little uncombined arsenic occurs naturally as microcrystalline masses, but the vast majority is found in conjunction with sulfur as red realgar (As_4S_4), yellow orpiment (As_2S_3), silvery arsenopyrite (FeAsS), and iron grey enargite (Cu_3AsS_4). None is

---

[18] Chemical formula ClCH=CHAsCl$_2$.

[19] Their chemical names are phenyldichlorarsine, and diphenylaminechlorarsine respectively.

Abington Free Library
1030 Old York Road
Abington, PA 19001-4530

mined as such because enough arsenic is produced as a by-product of refining the ores of other metals, such as copper and lead. World production of arsenic, in the form of its oxide, is around 50,000 tonnes per year, far in excess of that required by industry, and China is the chief exporting country. Other arsenic producing countries are Chile and Mexico. World resources of arsenic in copper and lead ores exceed 10 million tonnes.

One of the few productive uses to which arsenic was put in its early years was as the pigment orpiment. Its bright yellow colour (called 'royal yellow') was much favoured by Dutch painters of the 1600s. In 1775 it was joined by a beautiful green pigment called Scheele's green, named after Karl Scheele (1742–86). This was copper arsenite ($CuHAsO_3$) which continued to be popular for the next 100 years; Manet used it in the 1860s.

Industrial uses of arsenic were also found, and it is still required for certain kinds of glass, such as that used for the screens of liquid crystal displays. A little arsenic added to glass neutralises the green tint which traces of iron produce. Other uses have been in gunshot, embalming, taxidermy, weed control, rat killer, pyrotechnics, and cosmetics. Most of these no longer involve arsenic, but some still do.

The addition of a little arsenic (0.4 percent) to molten lead ensured the formation of perfectly spherical pellets of shot when the melt was poured down a shot tower. Arsenic was widely used as a wood preserver, especially in the USA. Wood was pressure treated with chromium copper arsenate (CCA) to prevent it rotting and being eaten by termites and 70 percent of treated timber products contained CCA. As of 2004 the use of this preservative started being phased out in favour of a copper boron compound.

Arsenic has become a key component of microchips. It is added to silicon and germanium semiconductors where it acts to supply electrons to the crystal lattice. Arsenic is the V component of the III-V series of semiconductors,[20] generally in the form of gallium arsenide and indium arsenide. Integrated circuits made with gallium arsenide perform much faster than those made with semiconductor silicon, although they are more expensive.

Environmental element

Amounts of arsenic in the environment	
EARTH'S CRUST	1.5 p.p.m.
	Arsenic is the 53rd most abundant element.
SOILS	1–10 p.p.m.
SEAWATER	1.6 p.p.b.
ATMOSPHERE	trace amounts.

[20] The III and V are based on the older notation for groups 13 and 15 of the periodic table.

Arsenic in the atmosphere comes from various sources: volcanoes release about 3,000 tonnes per year; micro-organisms release volatile methylarsines to the extent of 20,000 tonnes per year; but much more than these natural sources comes from human activity, which releases 80,000 tonnes per year from the burning of fossil fuels. Natural gas can contain traces of arsenic mainly in the form of trimethylarsine. Although these human additions to the environment seem threatening, they add only marginally to the planetary burden and in any case environmental arsenic is rendered safe once it meets up with a sulfur atom and forms an insoluble, immobile, arsenic sulfide.

A much more widespread threat from arsenic afflicts many in developing countries, and this has inadvertently come about as a result of improving water supplies. In countries like India, Bangladesh, Thailand, Cambodia, and Sumatra, villagers were encouraged to dig wells and tap into the subterranean water table which was free of disease pathogens. Tens of thousands of such wells were dug but unbeknown to those who relied on this supply of water, arsenic levels were often above 50 p.p.b. and a significant fraction had levels in excess of 300 p.p.b. More than 20 million people were drinking water of the former kind and as many as 5 million were drinking water of the latter kind.[21] (The World Health Organization's recommended maximum level of arsenic in drinking water is 0.01 mg per litre, that is, 0.01 p.p.m.) The result has been to expose millions of people to chronic arsenic poisoning over many years, exacerbated in hot countries where people drink several litres of water per day when working in the heat. This has caused widespread arsenicosis, in which the skin erupts in disfiguring, leprosy-like, lesions. After many years of such exposure cancerous growths begin to appear.

Removing arsenic from water is not difficult. It can be passed over alumina (aluminium oxide), which absorbs it and forms an insoluble aluminium arsenic salt, or over rusty iron filings, which form insoluble iron arsenite. A simple filter consisting of layers of sand, charcoal, and iron filings, costs very little, and will last five years. More than 20,000 such units are now being used in Bangladesh. It is even possible to decontaminate arsenic-containing water using the dried roots of the prolific aquatic weed called water hyacinth.

Contaminated drinking water is not the only source of arsenic pollution. In parts of India, Nepal, and Bangaladesh, dried cow pats are often used as fuel for cooking and if these are from cows which have been fed contaminated rice straw, grown in water with high arsenic levels, then burning releases the arsenic as arsenic trioxide putting at risk those who are cooking in unventilated kitchens. Cheap coal briquettes, which are used by 3 million people in China's Guizhou province, may also generate arsenic trioxide in the kitchens as the coal burns. The Chinese Government provides such homes with stoves that vent the smoke to the outside.

---

[21] The problem is that these aquifers are geologically very young and natural contaminants in the form of decaying organic matter are still present. This creates a reducing environment which solubilises any arsenic which is present.

Chemical element

Data file	
CHEMICAL SYMBOL	As
ATOMIC NUMBER	33
ATOMIC WEIGHT	74.92160
MELTING POINT	arsenic does not melt—it sublimes at 616°C
DENSITY	see below
OXIDE	As_2O_3 and As_2O_5

Arsenic is a metalloid element and a member of group 15 (row 4p) of the peri-odic table. It has two forms: grey, or α-arsenic, which is metallic with a density of $5.8\,g/cm^3$; and yellow, or β-arsenic, which is a non-metal with density $2.0\,g/cm^3$. The metallic form is brittle, tarnishes, and burns in oxygen, while the non-metal form is less reactive but will dissolve when heated with strong oxidising acids and alkalis. Arsenic exhibits the oxidation states –3, as in AsH_3, +3, as in $AsCl_3$ and As_2O_3, and +5, as in $AsCl_5$ and As_2O_5. When arsenic oxides are dissolved in alkaline solutions, the lower oxide forms arsenite (AsO_3^{3-}) ions and the higher oxide forms arsenate (AsO_4^{3-}) ions.

Arsenic has 33 known isotopes with mass numbers 60 to 92, and 10 known iso-mers. Naturally occurring arsenic consists of a single isotope, arsenic-75, which is not radioactive.[22] The longest-lived radioactive isotope is arsenic-73 with a half-life of 80 days.

Element of surprises

1. The arsenic-eaters of Austria were not an urban myth—they really did consume deadly arsenic oxide on a daily basis. In 1875 at a meeting of the Association of German Scientists and Physicians in Graz, one such peasant consumed 400 mg of arsenic trioxide before a large audience and analysis of his urine subsequently confirmed that he was able to tolerate such a dose which is twice the fatal dose for a normal adult.

   Regular ingestion of arsenic can allow the body to tolerate quite large doses, well in excess of that which would kill a normal person. The arsenic-eaters of the Styrian Alps in the 1700s and 1800s were reputed to consume doses of 250 mg of arsenic trioxide twice a week, and this they obtained from the flues of small metal refineries where it had sublimed. They claimed it produced a fresh complexion in women and enabled men, and horses, to work better at high altitudes. Some peas-ants were reputed to have eaten arsenic regularly for most of their adult life but many suffered from skin cancers.

2. The inhospitable salt-laden Searle's Lake in the Mojave Desert of south-eastern Cali-fornia has a pH of 10, high levels of arsenic, and very little oxygen. Nevertheless, in

---

[22] There are a number of radioactive isotopes produced as fission products in uranium-bearing deposits.

2005 a new bacterium (named SLAS-1) was discovered living happily in its deadly waters and using arsenate rather than oxygen to generate energy.

In 2010 a team from the NASA Astrobiology Institute in the USA took another bacterium (named GFAJ-1) from the briny Mono Lake, located in the Eastern Sierra region of California, and showed that it could use arsenate in place of phosphate, maybe even to the extent of replacing the phosphate of its DNA with arsenate.

3. Some plants thrive on arsenic. The Chinese Ladder fern *Pteris Vittata*, also known as the brake fern, grows rapidly and absorbs arsenic. This plant has the remarkable ability to absorb arsenic to the extent of its making up 2 percent of its weight (5 percent of its dry weight).

4. The water which the villagers of San Pedro de Atacama, drink—they live in an isolated region of modern day Chile—contains 500 p.p.m. of arsenic. However, they show no signs of arsenic-related diseases. Newer residents to the area are much more likely to suffer from skin disease and cancers. Those whose ancestors have lived a long time in that region appear not to be affected by it, whereas among the newcomers it appears that 10 percent of all deaths of persons over 30 are due to bladder and lung cancer and that is linked to their arsenic intake.

5. The stimulatory effect of arsenic used to be a way for dishonest racehorse trainers to improve an animal's chances of winning.

85
At

Astatine

Pronounced as-tah-teen, the name comes from *astatos* the Greek word meaning unstable. French *astate*; German *Astat*; Italian *astato*; Spanish *astato*; Portuguese *astato*.

Essential element

Even though its longest-lived isotope last only a few hours, chemists have shown that it resembles iodine by making similar compounds and thereby confirming its position in the periodic table.

Element of life

Astatine has no biological role and is never encountered outside nuclear facilities or research laboratories. It is highly dangerous because of its intense radioactivity.

Historical element

In 1931, American physicist Fred Allison used a magneto-optical machine to analyse minerals, searching for element 85, the unknown element below iodine in the periodic table. He claimed to have discovered it and proposed the name alabamine (symbol Ab) after the state of Alabama where he worked. Other researchers showed that the technique he was using could not detect elements, and we now know he must have been wrong because all isotopes of astatine are particularly short-lived.

In 1937, chemist Jajendralal De, working in Dacca in British India, claimed discovery of element 85 after analysing minerals. He proposed the name dakine, based on one of the variations of spelling Dacca. He too was mistaken.

Slightly more convincing evidence was that produced by others in 1939. Romanian physicist Horia Hulubei (1896–1972) and French chemist Yvette Cauchois (1908–99) analysed mineral samples using a high-resolution X-ray apparatus and detected new spectroscopy lines in the emission spectrum of radon, claiming this to be element 85. They re-ran the experiment, but because of the disruption caused by World War II, they did not propose a name until 1944 and they chose dor (symbol Do) based on a Romanian word implying 'better' and said to be looking forward to the end of the war and better times.

Also in 1939, Swiss chemist Walter Minder (1905–92) observed weak beta decay in radium. Chemical tests confirmed the analogies of the substance with iodine. This was the new element 85. He announced it the following year and proposed the name helvetium (symbol Hv) based on Helvetia, the official name for Switzerland.

Element 85 was convincingly produced for the first time at the University of California in 1940 by Dale R. Corson (1914–), K.R. Mackenzie (1912–2002), and Emilio

Segré (1905–89). (Segré was an Italian physicist who had fled from Mussolini's Italy. He had previously synthesised technetium—see page 527). The astatine was made by bombarding bismuth with alpha particles (helium nuclei). Although they reported their discovery, they were unable to carry on with their research due to World War II and the demands of the Manhattan project which diverted all such work into the making of atomic weapons. Only in 1947 were they able to name it astatine.

Medical element

Astatine-211 is the second longest-lived isotope and has a half-life of only 7.2 hours. It has been tested as a possible use in radiation therapy; so far only cancers in mice have been treated successfully.

Environmental element

Although astatine is a radioactive element, it presents no threat to the planet or life because the total amount present in the Earth's crust is estimated to be less than 30 grams at any one time.

Chemical element

Data file

CHEMICAL SYMBOL	At
ATOMIC NUMBER	85
ATOMIC WEIGHT	210.0 (isotope At-210)
MELTING POINT	*av.* 300°C (*est.*)
BOILING POINT	*av.* 340°C (*est.*)
DENSITY	not reported

Astatine is a reactive, radioactive non-metal element which resembles iodine, and is likewise one of the halogen group (group 17) of the periodic table and in row 6p. Its chemical behaviour was predicted from that of iodine and observed by studying its chemistry in extremely dilute solutions. It has been investigated chemically despite all its isotopes having short half-lives. Research carried out at the Brookhaven National Laboratory, at Upton, New York on Long Island, has produced some compounds, fleetingly synthesised, confirming that astatine is like iodine in forming hydrogen astatide (HAt), ionic compounds like sodium astatide (NaAt), and covalent compounds like carbon tetraastatide (CAt_4).

There are 31 known isotopes of astatine with mass numbers ranging from 193 to 223, and 23 known isomers. All are radioactive, of fleeting existence, and even the longest-lived, astatine-210, has a half-life of only 8.3 hours. (The shortest-lived, astatine-213, has a half-life of less than a microsecond.) Isotopes 209, 210, and 211 can be produced by bombarding bismuth with high-energy alpha-particles and they are easily separated because astatine is volatile.

There are four astatine isotopes that occur naturally, astatine-215, and astatine-217 to 219, and these are continuously produced by the various radioactive decay series of uranium, neptunium, and plutonium. The last of these, astatine-219, has a half-life of only 56 seconds.

Element of surprises

1. Total man-made production of astatine to date is estimated to be less than a millionth of a gram, and virtually all of this has now decayed away. A piece of this element has never been seen with the naked eye, nor is enough ever likely to be made to make this possible. Even if it were, the heat generated by its intense radioactivity would immediately result in its complete evaporation.

2. Astatine was long thought to be the least abundant naturally occurring element on planet Earth. However, it has lost this claim to fame to berkelium, a few atoms of which can be produced in very highly-concentrated uranium-bearing deposits.

Barium

Pronounced bair-ee-um, the name comes from the Greek *barys* meaning heavy, and derives from the heaviness associated with barium minerals.

French *baryum*; German *Barium*; Italian *bario*; Spanish *bario*; Portuguese *bário*.

Essential element

Barium is an abundant toxic metal which has some surprising uses, including preserving works of art.

Element of life

Barium in the human body	
BLOOD	70 p.p.b.
BONE	3–70 p.p.m.
TISSUE	0.1 p.p.m.
TOTAL AMOUNT IN BODY	approx. 22 mg.

Barium has no biological role, or so it was generally believed, although for some species it now appears to be essential—see *Element of surprises*. As long ago as the 1770s, it was detected in the ash from plants and vegetables, and in 1865 it was found to be present in seaweed and other sea creatures.

Food element

Barium is quite an abundant element, so perhaps it is not surprising to find it in food to the extent that we probably take in about 1 mg per day. Because barium is so like calcium, we absorb some of it into our system, but it serves no useful purpose.

The amount of barium in soils ranges from as little as 20 p.p.m. to as high as 2,300 p.p.m. (0.23 percent). Plants take up some barium, the highest concentrations being found in certain trees and shrubs which can have as much as 1 percent (dry weight).

The food plants with most barium are carrots (13 p.p.m. dry weight), onions (12 p.p.m.), lettuce (9 p.p.m.), beans (8 p.p.m.) and cereal grains (6 p.p.m.), while the lowest are fruits such as apples (1 p.p.m.) and oranges (3 p.p.m.). Brazil nuts have been reported with as much as 10,000 p.p.m. (1 percent) of barium.

Medical element

The barium of a diagnostic 'barium meal' is barium sulfate, $BaSO_4$ which is amongst the most insoluble of all salts, so it is safe to ingest and does not react with stomach acid. This form of barium was given to patients suffering gastric and intestinal disorders and as it moved through the gut it could be monitored. The insoluble barium sulfate was highlighted because it absorbed X-rays whereas the surrounding tissue allowed them to pass through. Any abnormal constrictions due to cancerous growths were thereby located.

Dangerous element

Barium can stimulate metabolism to the extent that it will cause the heart to beat erratically (known as ventricular fibrillation) and its soluble salts are dangerously toxic if taken. The symptoms of barium poisoning are vomiting, colic, diarrhoea, tremors and paralysis. Although barium carbonate is insoluble, it is toxic because it dissolves in the acid of the stomach, and it has been used as a rat poison. Barium has occasionally killed patients who were given the wrong barium compound instead of barium sulfate prior to being X-rayed.

Sometimes death occurs within 10 minutes of taking a large dose although most who were poisoned this way survived for 24 hours or more. A fatal dose of barium carbonate is around a gram but a typical barium meal would be several grams of barium sulfate and if the equivalent amount of barium carbonate, $BaCO_3$, was consumed then there would be little hope of saving the patient's life unless immediate action were taken. (The antidote is sodium sulfate which forms insoluble barium sulfate, but this has to be administered very quickly.)

There have been very few cases in which murders have been committed with barium salts. In 1994 in Mansfield, Texas, a 16-year-old schoolgirl named Marie Robards murdered her father by feeding him barium acetate which she had stolen from her chemistry laboratory. His death was diagnosed as due to heart failure and Marie would have got away with her crime had she not told a class mate what she had done. She is reputed to have confessed to a friend and was apparently driven to doing this as a result of appearing in the school's production of Shakespeare's *Hamlet,* which is all about the murder of a father and the evil that results. In 1996 Marie was tried, convicted, and sentenced to 28 years in jail.

Element of history

Barium sulfate occurs naturally as the mineral barite and it was first investigated by a shoemaker who dabbled in alchemy, Vincenzo Casciarolo, who found some shiny pebbles near his home in Bologna, Italy, in the early 1600s. Heating these stones to redness bestowed upon them a rather curious property: they shone in the dark. Indeed, provided it was exposed to the sun in the daytime then Bologna stone, as it became known, would shine for quite some time the following night.

Another Italian, Ulisse Aldrovandi (1522–1605), published an account of the phenomenon, and brought it to the notice of a wider public. In the 1700s, Bologna stone was investigated by eminent chemist Carl Wilhelm Scheele (1742–86) who analysed it and realised it was the sulfate of an unknown element. Meanwhile a mineralogist, Dr William Withering (1741–99), had found another curiously heavy mineral in a lead mine in Cumberland, England, which clearly was not a lead ore and he called it witherite; it was later shown to be barium carbonate.

Separating the metal itself was beyond the method which chemists used in the 1700s, that relied on heating the oxide with carbon, but early in 1808 Humphry Davy (1778–1829) at the Royal Institution in London, was able to produce it by the electrolysis of molten barium hydroxide.

Economic element

The chief mined ores are barite (aka baryte, barium sulfate, $BaSO_4$), which is also the most common, and witherite (barium carbonate, $BaCO_3$). More than 8 million tonnes are mined each year, with half being produced in China, followed by India, Morocco, and the USA, each of which produces in excess of half a million tonnes. It is also mined in smaller amounts in 35 other countries. Reserves of barite are reported to exceed 400 million tonnes. There is also a rare barium mineral called benitoite (barium titanium silicate) which is a translucent blue colour and used as a gemstone.

Barium metal is made by the electrolysis of molten barium chloride, or by heating barium oxide with aluminium powder at 1,200°C and condensing the barium vapour into solid metal. Barium has few uses as a metal beyond being inserted as a 'getter' which scavenges the last traces of gas from a vacuum tube so as to produce a perfect vacuum. Some is used to deoxidise copper, and there are alloys such as that with nickel. This readily emits electrons when heated and is used in electron tubes and for spark plugs. Frary metal is a calcium-barium-lead alloy used for bearings.

Some barium compounds are useful. Barium sulfate is employed on a large scale as a finely ground lubricating slurry in drilling for oil and gas exploration. The sulfate also finds use as coating for photographic and art paper, and a filler and delustrant for textiles. The brilliant white pigment known as lithopore is a mixture of barium sulfate and zinc sulfide, and it has the advantage of not darkening in the presence of sulfides. Barium hydroxide, $Ba(OH)_2$, is used in several ways: in the manufacture of oil and grease additives; in the refining of beet sugar; in the manufacture of special types of glass; and as a depilatory agent. Barium nitrate and chlorate give fireworks a green colour.

Old frescoes can be preserved by spraying them with a solution of barium nitrate followed by a solution of ammonia. Together these form barium hydroxide which fills the cracks and defects in the plaster. Over time the barium hydroxide is slowly converted to permanently insoluble barium carbonate.

Environmental element

Barium in the environment	
EARTH'S CRUST	500 p.p.m.
	Barium is the 14th most abundant element.
SOILS	*approx.* 500 p.p.m.
SEAWATER	10 p.p.b. in the Atlantic, 20 p.p.b. in the Pacific.
ATMOSPHERE	traces.

Barium is surprisingly abundant in the Earth's crust, despite the fact that it is a heavy element. It is more abundant than carbon, sulfur, and zinc. Because it forms insoluble salts with other common components of the environment, such as carbonate and sulfate, it is not mobile and poses little risk. In ocean depths barite crystallises out and is part of the sediment.

Chemical element

Data file	
CHEMICAL SYMBOL	Ba
ATOMIC NUMBER	56
ATOMIC WEIGHT	137.327
MELTING POINT	729°C
BOILING POINT	1637°C
DENSITY	3.6 g/cm³
OXIDES	BaO, BaO$_2$ (barium peroxide)

Although its ores are heavy, metallic barium is unexpectedly light as is shown by its density, which is half that of iron. Barium is a soft, silvery metal of the group known as the alkaline earths, which comprise group 2 (row 6s) of the periodic table. It rapidly tarnishes in air and reacts with water. Its only oxidation state is +2 as in compounds like BaCl$_2$, and in solution it is present as the barium ion Ba^{2+}.

There are 40 known isotopes of barium with mass numbers 114 to 153, and 10 known isomers. There are seven main naturally occurring isotopes of barium[23]: barium-138, which comprises 71.7 percent, barium-137 (11.2 percent), barium-136 (7.9 percent), barium-135 (6.6 percent), barium-134 (2.4 percent), barium-130 (0.1 percent), and barium-132 (0.1 percent). Two of these are radioactive although with incredibly long half-lives: barium-130 whose half-life is 4×10^{21} years, and barium-132 with a half-life of 3×10^{20} years. The next longest-lived radioactive isotope

---

[23] There are also minute traces of other radioactive isotopes of barium in uranium ores where they have been produced by nuclear fission.

is barium-133 which has a half-life of 10.5 years and is used to calibrate gamma ray detectors.

Element of surprises

Large, single-cell algae known as desmids manage to thrive in inhospitable places such as pools on cold, upland peat bogs which are supplied mainly by rain water and where there are few nutrients. These curious plants, of which *Closterium* is the most common, can be up to a millimetre in length, and have prominent fluid-filled cavities containing tiny crystals of barium sulfate. The barium has to be scavenged from water with only parts per billion of the element. At the other extreme, desmids will happily grow in water which contains as much as 35 p.p.m. of this toxic metal. Nor is it there in place of calcium sulfate, the element which barium mimics. Desmids will produce barium sulfate crystals from water that may be 10,000 times richer in calcium than barium. What purpose the crystals serve is unclear, but one suggestion is that they use them as gravity sensors to orientate themselves. Desmids will not grow in the absence of barium which shows how important this metal is to them.

Berkelium

Pronounced berk-eel-ee-um, it was named after the town of Berkeley, California, where it was first made. The reason why this name was chosen was that the element above it in the periodic table, terbium, was named after the town, Ytterby, Sweden, where it was discovered. Berkeley itself was named after the English bishop and philosopher George Berkeley (1685–1735).

French *berkélium*; German *Berkelium*; Italian *berkelio*; Spanish *berkelio*; Portuguese *berquélio*.

Essential element

As yet, there is nothing worthy of note except its making.

Element of life

Berkelium has no role to play in living things, and it is never encountered outside nuclear facilities or research laboratories. It is highly dangerous because it is a powerful source of radiation; the maximum permitted body burden is 0.4 nanograms (0.4×10^{-9} grams) but so little is ever made that the chances of encountering even one atom of it are remote.

Element of history

Berkelium was first produced on Monday, 19 December 1949, at the University of California at Berkeley, and was made by Stanley G. Thompson (1894–1953), Albert Ghiorso (1915–), and Glenn T. Seaborg (1912–99). They bombarded 7 mg of pure americium-241 with helium ions for several hours in the 60-inch cyclotron. Although americium had been discovered five years earlier, it took some time to collect and purify enough americium to form a target large enough for the high energy helium ions to hit and to stand a chance of combining to form a new element of higher atomic number. The americium had been made by exposing plutonium to the neutrons inside a high flux nuclear reactor.

Later that day the Berkeley team dissolved the target and put the solution through a technique known as ion exchange to separate any new element that had been created. By late that Monday evening their hopes had been realised: a new element was definitely present. They had made the isotope berkelium-243 which has a half-life of about five hours. They were even able to deduce some of its chemistry, showing it had two oxidation states just like the element above it in the periodic table.

It took a further nine years before enough berkelium had been made to see with the naked eye and even this was only a few micrograms. The first chemical compound, berkelium dioxide, BkO_2, was made in 1962.

Economic element

No practical use of berkelium has so far emerged. It can be produced in nuclear reactors and in 100 mg quantities, as the isotope berkelium-249, by bombarding plutonium-239 with neutrons. This isotope has a half-life of 330 days and is the one that has been most investigated. It has been used as the target material in making other, heavier, elements.

Environmental element

Berkelium does occur naturally on Earth, but only in insignificant amounts—it is the least abundant of all naturally occurring elements. There are two berkelium isotopes: berkelium-249 and berkelium-250. They are produced only in very highly concentrated uranium-bearing deposits and this happens by a series of neutron-capture reactions followed by beta decay, and goes via americium and curium isotopes. Curium-249 undergoes beta decay to yield berkelium-249, which can capture a neutron to become berkelium-250. These isotopes have fairly short half-lives and quickly decay.

Berkelium was also produced for hundreds of thousands of years in natural nuclear reactors which started up about 1.8 billion years ago, following natural geological processes which concentrated uranium. This led to 16 natural nuclear reactors functioning in what is now Oklo, Gabon, in Africa. These operated continuously and by neutron capture and beta decay processes they produced the transuranium elements from atomic numbers 83 (neptunium) to 100 (fermium) including berkelium.

Chemical element

Data file	
CHEMICAL SYMBOL	Bk
ATOMIC NUMBER	97
ATOMIC WEIGHT	247.1 (isotope Bk-247)
MELTING POINT	1047°C
BOILING POINT	not known
DENSITY	13.25 g/cm³
OXIDES	BkO_2 and BkO_3

Berkelium is a radioactive metallic element that is a member of the actinide group (row 5f) of the periodic table. A visible sample of the metal itself, which is silver in colour, was made from berkelium-249 isotope by reacting the oxide with molten barium at 1,100°C. Its chemistry has been investigated to a limited extent and several compounds have been made. Berkelium metal is attacked by oxygen, steam, and acids, but not alkalis. It exhibits oxidation states +3 and +4 in its compounds and in solution. A weighable amount of $BkCl_3$ was made in 1962.

There are 20 known isotopes of berkelium with mass numbers ranging from 235 to 254, and 2 known isomers. The longest-lived isotope is berkelium-247 with a half-life of 8,179 years.

Element of surprises

The discovery of berkelium caught the attention of *The New Yorker* magazine which saw it as part of a growing number of new elements and jokingly suggested that the University of California at Berkeley had missed a golden opportunity to immortalise itself forever on the periodic table. The article said that they should have named the sequence of new transuranium elements 'universitium', 'ofium', 'californium' and 'berkelium'.

Glenn Seaborg replied to the article, saying that they had thought about this but it would open the door to someone in New York discovering the next two elements and naming them 'newium' and 'yorkium.' *The New Yorker* retorted that they were already hard at work on the new elements, but so far all they had was their names.

Beryllium

Pronounced ber-il-ee-um, the name comes from the Greek, *beryllo*, the word for the mineral beryl. When it was first discovered it was given the name glaucinium from the Greek word *glykys*, meaning sweet, because its compounds tasted sweet.

French *béryllium*; German *Beryllium*; Italian *berillio*; Spanish *berilio*; Portuguese *berílio*.

Essential element

Beryllium is needed for alloys with remarkable properties such as tools that must be used in oil refineries and scientific instruments located in outer space.

Cosmic element

Only hydrogen (atomic number 1), helium (2), and lithium (3) were formed by the Big Bang, but traces of beryllium and boron were produced shortly thereafter by radioactive decay and nuclear reactions among these first three elements. The next element, beryllium (4), is relatively rare in the universe because it is not even formed in the nuclear furnaces of stars. It is generated only in supernova events, when heavier nuclei decay into sub-atomic particles. Curiously, the analysis of the spectrum of the star G64-12 shows it contains far more beryllium that theory predicts, suggesting there is yet another way in which this element can be formed.

Radioactive beryllium-10 is produced by cosmic rays in the Earth's upper atmosphere interacting with oxygen and nitrogen, and this has been detected in Greenland ice cores and marine sediments. The amount in ice cores deposited over the past 200 years has increased and decreased in line with the Sun's activity, as shown by the frequency of sun-spots. The amount of this isotope in marine sediments laid down in the last ice age was 25 percent higher than that in post-glacial deposits, indicating that the Earth's magnetic field was much weaker than it is currently. A weaker magnetic field deflects cosmic rays less well and so leads to more beryllium-10 being formed.

In 1950 the popular science writer Isaac Asimov wrote a prophetic short story called 'Sucker Bait', which was one of a collection entitled *The Martian Way*. It is about a space expedition which went to colonise a seemingly fertile planet but after a few years the colonists began to die of a mysterious disease which made breathing progressively difficult. Investigations revealed that the planet had abundant plant life and the atmosphere was quite breathable. So what was wrong? The explanation was that there were high levels of beryllium in the planet's environment making it totally unsuitable for human habitation. Could such planets really exist? Very likely. In 1992, a group of European astronomers discovered six old stars in our galaxy whose spectral investigation showed they had large quantities of beryllium.

Element of life

Beryllium in the human body	
BLOOD	10 p.p.t.
BONE	3 p.p.b.
TISSUE	0.1 p.p.b.
TOTAL AMOUNT IN BODY	*approx.* 35 mg.

Beryllium has no known biological role, and indeed it is an insidious metal to take into the human body. However, the tiny amount in the average person does not affect their health.

Beryllium in soil can pass into the plants grown on it, provided it is in a soluble form, and typical levels in plants vary between 1 and 400 p.p.b.— too low to affect the animals that live off them—even when they eat plants that have an ability to concentrate the metal, such as legumes. Too much beryllium in the soil is toxic to plants and those affected have stunted foliage.

Dangerous element

If beryllium dust or fumes from the metal are breathed in they can cause chronic inflammation of the lungs and shortage of breath, and this is persistent even though most of the inhaled beryllium is carried by the blood to other sites in the body, generally to the bone where it concentrates. Brief exposure to a lot of beryllium, or long exposure to a little, will bring on this lung condition which is known as berylliosis. The disease may take up to five years to manifest itself and about a third of those who contract this condition die because of it, while the rest are permanently disabled. Sometimes lung cancer may result, and the metal has indeed been shown to be carcinogenic for laboratory animals.

Research at Los Alamos has shown that the beryllium disrupts the hydrogen bonds of small but important molecules especially in transferrin which moves iron around the body. Attached to transferrin, berylliumt then gains access to cells.

Taken by mouth, beryllium compounds may be less toxic but they are also dangerous and have been known to kill. Beryllium is related to magnesium, which is an essential element of human nutrition, and it can mimic this and displace it from certain key enzymes which then malfunction. The lungs are particularly sensitive. Beryllium may take decades to cause death. In the UK there were around 30 recorded deaths from berylliosis in the last century, including one man who died 29 years after he was first exposed to the metal while working as a machinist.

Workers in industries using beryllium alloys are most at risk, as were those making early types of fluorescent lamps which were coated inside with phosphors which contained up to 12 percent beryllium oxide. A survey in 1949 uncovered 400 cases of beryllium-related lung disease among workers in the industry in the USA, about 5 percent of those working in the industry, and in 1950 manufacture of this type of lamp was ended.

Other than industrial workers, few people have been exposed to high levels of beryllium, but this changed in 1990 when there was an explosion at a Russian military plant near the border with China. It was producing and processing beryllium for nuclear warheads. The blast threw a cloud containing four tonnes of beryllium oxide dust over the nearby town of Ust-Kamenogorsk and 120,000 people were exposed to the chemical. Subsequent monitoring of the inhabitants showed that only about 9 percent had been affected and even they had only slightly raised levels of beryllium in their bodies.[24]

The only murder committed with beryllium is a fictional one in Camille Minichino's detective novel *The Beryllium Murder*, published in 2000.

Element of history

Beryl and emerald, which are both forms of the aluminium silicate, $Be_3Al_2(SiO_3)_6$, were known as gemstones by the ancient Egyptians, Jews, and Romans. The Roman writer Pliny recognised them as forms of the same mineral. He said beryl was mined near the Red Sea and that emerald was imported from India, but see *Element of surprises* below. Beryl was also mined in the desert of Nubia at the time of Cleopatra. The emperor Nero used a large emerald, the better to view gladiatorial fights in the arena.

That these minerals might harbour a new element was suspected by the French mineralogist Abbé René-Just Haüy (1743–1822) who asked his compatriot Nicholas Louis Vauquelin (1763–1829) to analyse them. This he did and on 15 February 1798 he presented his results to the French Academy, announcing that they contained a new element although he was unable to separate this from its oxide. The metal itself was isolated in 1828 by Friedrich Wöhler (1800–82) in Berlin and independently by Antoine-Alexandere-Brutus Bussy (1794–1882) in Paris, both of whom extracted it from beryllium chloride ($BeCl_2$) by reacting this with potassium.

Beryllium was to play a historic role in advancing our knowledge of atomic theory since it helped uncover the fundamental particle, the neutron. This was discovered in 1932 by James Chadwick (1891–1974) who bombarded a sample of beryllium with alpha-rays (from radium) and observed that it emitted a new kind of subatomic particle that had mass but no charge. The combination of radium and beryllium can be used to generate neutrons for research purposes, although a million alpha-particles only manage to produce 30 neutrons.

Element of war

Beryllium has a rather unusual property: it does not absorb neutrons, it even reflects them, and for this reason it is used in nuclear weapons and in the nuclear energy industry. In a nuclear warhead, which relies on neutron bombardment releasing energy from uranium, a casing of beryllium ensures a higher neutron flux within in the bomb.

---

[24] The role of beryllium in the environment is dealt with in detail by M. Brisson and A. Ekechukwu (eds.) in *Beryllium: environmental analysis and monitoring*, RCS Publishing, 2009.

Economic element

Emerald, which is beryllium aluminium silicate, gets its beautiful green colour from around 2 percent of chromium. Aquamarine is also beryllium aluminium silicate but with less chromium.

The chief ores of beryllium are beryl and bertrandite, which is a beryllium silicate of formula $Be_4Si_2O_7(OH)_2$. Sometimes truly enormous crystals of the mineral turn up, one specimen found in Maine in the USA was over 5 metres in length and weighed almost 20 tonnes. The main areas where beryl and bertrandite are found are Brazil, USA, Argentina, Madagascar, Mozambique, and Zambia. Of the 3,000 tonnes of beryl mined in 2007, 2,500 were produced in the USA and 500 in China. Known reserves are in excess of 400,000 tonnes, most of which is in the USA.

Commercial production of beryllium metal began in 1957 but it never found the market that was once envisaged. Its lightness, high melting point, thermal conductivity, rigidity, and other desirable properties are countered by the dangers associated with the metal. However, when the products are of military importance, or are to reside in space, then beryllium has been used and its alloys played an important part in the space shuttle, and are part of space telescopes, the mirrors of which operate at temperatures of −240°C. Despite these applications, less than 500 tonnes of metal are refined each year and this is produced by heating beryllium fluoride with magnesium.

When copper and nickel are alloyed with beryllium they become not only much better at conducting electricity and heat, but they display remarkable elasticity. For this reason their alloys make good springs which were once used in wristwatches. The copper alloy is also used to make spark-proof tools, which are the only ones allowed in sensitive areas such as oil refineries and plants where flammable gases are stored. The copper beryllium alloy has been used for the heads of golf clubs.

Because beryllium metal and its alloy with copper have no magnetic properties, tools made from them are needed for special situations like defusing magnetic mines of the kind used in naval warfare, and when working with MRI scanners whose magnetic power is awesome and will send any iron or steel tool crashing into it.

Beryllium absorbs X-rays which is why it is used as windows for X-ray tubes, and some beryllium oxide goes into specialist ceramics.

Environmental element

Beryllium in the environment	
EARTH'S CRUST	2.6 p.p.m.
	Beryllium is the 47th most abundant element.
SOILS	6 p.p.m.
SEAWATER	0.2 p.p.t.
ATMOSPHERE	minute traces—see *Cosmic element* above.

There is some environmental contamination by beryllium from coal-burning industries.

Chemical element

Data file	
CHEMICAL SYMBOL	Be
ATOMIC NUMBER	4
ATOMIC WEIGHT	9.012182
MELTING POINT	1278°C
BOILING POINT	2970°C
DENSITY	1.8 g/cm³
OXIDE	BeO

Beryllium is a silvery-white, lustrous, relatively soft metal of group 2 (row 2s) of the periodic table, a group also known as the alkaline earth metals. It is unaffected by air or water, even at red heat. Its preferred oxidation state is +2 as in $BeCl_2$, and in solution it exists as Be^{2+} with four water molecules attached. Because it resembled aluminium, the chemists of the 1800s assumed it must have an oxidation state of +3 but careful analysis of its salts in the 1870s proved it was a divalent metal.

There are 12 known isotopes of beryllium with mass numbers 5 to 16, and no known isomers. Mined beryllium has but a single isotope, beryllium-9, which is not radioactive but beryllium-10 is radioactive with a half-life of 1.5 million years, and is the longest-lived radioactive isotope. This isotope is produced by cosmic rays in the upper atmosphere causing spallation of oxygen and nitrogen nuclei.[25]

Element of surprises

Emeralds are rarer than diamonds and most come from Columbia, Brazil, and Zambia. Emeralds have been traded since the time of the Ancient Egyptians and are to be found in early jewellery across Europe, including that of the Celts. Analysis of the oxygen in the gems enables their source to be identified because the isotope ratio of oxygen-18/oxygen-16 varies from deposit to deposit. The Romans and Celts got their emeralds mainly from Austria, although some came from as far away as Pakistan.

More surprising is the discovery that the Moguls rulers of India got some of theirs from South America. Analysis of gems from the treasure of the Nizam of Hyderabad which were acquired in the 1700s showed they were of Colombian origin, and this has been interpreted to mean that there was trade across the Pacific Ocean by Spanish vessels which carried emeralds to their colonies in the Philippines, from where they eventually reached India.

---

[25] Spallation is the technical term for the process in which a bombarded nucleus breaks up into many fragments.

Bismuth

Pronounced biz-muth, the name comes from the German *Bisemutum* which was derived from *Wismuth* and that, in its turn, was a corruption of *Weisse Masse* meaning 'white mass' which is a reference to its common ore bismuthinite.

French *bismuth*; German *Bismut*; Italian *bismuto*; Spanish *bismuto*; Portuguese *bismuto*.

Essential element

Bismuth can be protective over our heads in public buildings and soothing down in our stomach when we're feeling dyspeptic.

Element of life

Bismuth in the human body	
BLOOD	*approx.*16 p.p.b.
BONE	*approx.* 200 p.p.b.
TISSUE	30 p.p.b.
TOTAL AMOUNT IN BODY	less than 0.5 mg.

Bismuth has no biological role to play and it appears to be relatively benign for a heavy element. The amount we ingest each day is probably less than 20 micrograms since the element is not readily taken up by plants, and those levels that have been measured were a tiny 20 p.p.b.

Dangerous element

Only one person has ever been reported to have died from an overdose of bismuth, and compounds of this element are still used to treat stomach upsets. Even so, it is necessary to warn against taking it in excess because it can damage the liver.

Medical element

So-called 'bismuth mixture' appeared in the 1780s when doctors began to prescribe bismuth subnitrate, approximate composition $Bi(NO_3)(OH)_2$, as a treatment for gastric disorders, and especially peptic ulcers. This compound was superseded by bismuth subcarbonate, approximate composition $Bi_2O_2(CO_3)$, in the 1860s. Today, the preferred form is colloidal bismuth subcitrate (also known as CBS or De-Nol). Its mode of action is still unclear but it appears to act on the mucus which lines the stomach and protects it against the acidic gastric juices which the body produces to digest food. Bismuth also deactivates the pepsin enzyme which is particularly aggressive.

Nowadays, peptic and duodenal ulcers are thought to be due to the bacterium *Helicobacter pylori* which can survive the stomach's extremely acid conditions and digestive enzymes. Bismuth as CBS can be used in combination with powerful anti-biotics to eliminate these microbes.

Element of history

Bismuth was discovered by an unknown alchemist around 1400 AD. By 1450 it was being alloyed with lead to make cast type for printing, which began around 1440, and by 1480 objects such as decorated caskets were being crafted in the metal. In Germany, artisans working with bismuth formed their own guild. For three centuries or more, bismuth was confused with lead, although Georgius Agricola (1494–1555) speculated that it was distinctly different, as did the early chemist Caspar Neuman (1683–1737), but proof that it was so finally came in 1753 thanks to the work of Claude-François Geoffrey (1729–53).

Bismuth was used as an alloying metal in bronze by the Inca of South America around 1500 AD. A knife of the period has a llama-head handle whose bronze contains 18 percent bismuth (plus a small amount of tin), while the bronze blade of the knife contains no bismuth only tin, indicating that the light colour of the handle was purposefully made. The bismuth had probably been found as the native metal. The Inca appear to have been better metallurgists that previously credited.

Economic element

Bismuth occurs naturally as the metal itself and is found as crystals in the sulfide ores of nickel, cobalt, silver, and tin. Most bismuth occurs as the mineral bismuthinite which is bismuth sulfide (Bi_2S_3). Bismuth is mainly produced as a by-product from lead and copper smelters, especially in the USA, and this accounts for the larger part of the annual 7,000 tonnes or so that is refined. Mined bismuth accounts for around 5,500 tonnes with China, Mexico, and Peru being the main sources. China produces 1,900 tonnes (30 percent) of this with the other two each accounting for 25 percent. It is also mined in Japan, Bolivia, Peru, and Australia. There is no reliable estimate of how much bismuth is available to be mined, but it seems unlikely that there will be a shortage of this metal even in the distant future, and a lot could be recycled. However, recycled bismuth is not yet important and could never be recoverable from some of its uses.

Bismuth has a relatively low melting point for a metal (271°C) and when alloyed with tin and lead it melts even lower. Because of this, it was once commonly used for electric fuses, and Wood's metal, which is an alloy of bismuth with 12 percent cadmium that melts at 70°C, is still used in automatic fire-sprinkler systems. Bismuth is an important component of solders, this being one of its uses which is likely to increase as lead solders are phased out. Another alloy is Bismanol which consists of bismuth, manganese, and iron, and this was discovered by the US Naval Ordnance Laboratory in the 1950s. It makes powerful permanent magnets.

Bismuth

The chemical industry uses bismuth compounds as catalysts in manufacturing acrylonitrile, the starting material for synthetic fibres and rubbers.

Environmental element

Bismuth in the environment	
EARTH'S CRUST	48 p.p.b.
	Bismuth is the 70th most abundant element.
SOILS	*approx.* 0.25 p.p.m.
SEAWATER	400 p.p.t.
ATMOSPHERE	virtually none.

Bismuth appears to pose no environmental threat to any living species.

Chemical element

Data file	
CHEMICAL SYMBOL	Bi
ATOMIC NUMBER	83
ATOMIC WEIGHT	208.98038
MELTING POINT	272°C
BOILING POINT	1,560°C
DENSITY	9.7 g/cm³
OXIDE	Bi_2O_3

Bismuth is a heavy, silvery metal with a faint pink tinge and is a member of group 15 (row 6p) of the periodic table. Strictly speaking it is a radioactive element, but see below. As a pure metal it is too brittle to be used, and indeed is the least metal-like of all the metals, with one of the lowest thermal conductivities[26] and the highest electrical resistivity. Bismuth can even behave as a semiconductor in certain circumstances and bismuth telluride is just such a material used in infrared detectors. Like antimony, the metal above it in the periodic table, Bismuth also expands on solidifying, by about 3 percent.

Bismuth is stable to oxygen and water, but dissolves in concentrated HNO_3. Its preferred oxidation state is +3 although there are some +5 compounds known such as BiF_5. All bismuth salts form insoluble compounds when put into water. It also forms compounds in which there are clusters of bismuth atoms such as the cation Bi_9^{5+}.

There are 35 known isotopes of bismuth of mass numbers 184 to 218, and 50 known isomers. There is only one important naturally occurring isotope and that is bismuth-209, which was thought not to be radioactive until in 2002 when the

---

[26] The thermal conductivity of mercury is even lower.

Institut d'Astrophysique Spatiale of France proved that it was so and that it has a half-life of 1.9×10^{19} years (which is 19 *billion* billion years).[27] This is the longest-known half-life for an alpha-emitter. There are traces of six other bismuth isotopes: bismuth-210 to bismuth-215, which occur in thorium and uranium minerals where they are part of the decay chain of those elements. Bismuth-210 has a half-life of five days; bismuth-211 of two minutes, bismuth-212 of 60 minutes, bismuth-213 of 45.5 minutes, bismuth-214 of 20 minutes, and bimuth-215 of 7.6 minutes. Apart from bismuth-209, the longest-lived radioactive isotopes are bismuth-208 which has a half-life of 368,000 years and bismuth-210m whose half-life is 3 million years.

Bismuth has been used to synthesise transfermium elements by the 'cold fusion' method. It acts as the target which is bombarded with ions of lighter elements in the hope that they will fuse together and form the nucleus of a new element. Bohrium, meitnerium, and roentgenium were discovered this way. The cold fusion method was developed in 1973–4 by Yuri Ogannesian and Alexander Demin at the Joint Institute for Nuclear Research in Dubna, Russia, although the first person to use it was Amnon Marinov in 1971 (see copernicium).

Element of surprises

1. The pearl effect in cosmetics such as nail varnish and lipstick is produced with bismuth oxychloride, $BiOCl$, which is a lustrous crystalline powder. This is made from bismuth chloride, $BiCl_3$, which reacts with water to form a hydroxide and this, on heating, gives the oxychloride.

2. Bismuth could be used as a fuel to power spacecraft and satellites which use ion engines that rely on the heaviest of elements to be ejected as fuel—see xenon, page 611.

3. Bismuth strontium calcium copper oxide (aka 'bisco') has superconducting properties and retains these up to −164°C, which is much higher temperature than most other materials of this kind.

[27] Nuclear theory had long indicated that bismuth should be a radioactive element with a calculated half-life of 4.6×10^{19} years, in impressive agreement with the measured value.

Bohrium

Pronounced bor-ee-um, the element is named after Neils Bohr (1885–1962) the Danish atomic physicist and winner of the 1922 Nobel Prize for physics. He was the first person to combine atomic and quantum theory to give the correct explanation of atomic structure, which he did in 1913.

French, *bohrium*; German, *Bohrium*; Italian, *bohrio*; Spanish, *bohrio*; Portuguese, *bóhrio*.

Element of creation

In 1975 a team led by Yuri Oganessian at the Russian Joint Institute for Nuclear Research (JINR) in Dubna, bombarded bismuth-209 with chromium-54 and produced element 107 isotope 261. They also bombarded bismuth-205 with iron-58 but to no avail. They published the results of their successful run in 1976 and submitted a discovery claim.

In 1981, a team led by Peter Armbruster and Gottfried Münzenberg at the German nuclear research institute the Gesellschaft für Schwerionenforschung (GSI) bombarded bismuth-209 with chromium-54 in an attempt to get isotope 262, and they succeeded in making a single atom.

Naming element 107 became part of a wider controversy—see Transfermium elements. IUPAC, in a strangely worded decision, said that the GSI should be awarded the discovery because they had the more credible submission, but that the JINR had probably discovered it first. As a result they recommended that the two laboratories share in the discovery and should confer with each other in choosing a name. The two groups agreed to do this. The GSI let the JINR choose the name in recognition of their contribution of developing the 'cold fusion' method of element discovery, by which element 107 had been made. They proposed the name neilsbohrium in honour of Neils Bohr, the name that the JINR had originally proposed for element 105. IUPAC shortened this to bohrium on the basis that elements named after individuals were traditionally based on the surname alone. The name was formally approved in 1997, along with names for elements 104 to 109.

Chemical element

Data file	
CHEMICAL SYMBOL	Bh
ATOMIC NUMBER	107
ATOMIC WEIGHT	260 (longest-lived isotope currently known)
MELTING POINT	not known
BOILING POINT	*ca* 180°C (see below)

DENSITY	not known
OXIDE	likely to be Bh_2O_7

Bohrium is a man-made, radioactive element and a member of group 7 (row 6d) of the periodic table, coming below rhenium. So far not enough of this element has been made at one time to be visible to the naked eye, but predictions are that it would be a silvery metal susceptible to attack by air, steam, and acids. Its preferred oxidation state should be +7. Very little chemistry has been carried out on bohrium, but in 2000 its chemical behaviour was investigated using atoms generated in the Philips cyclotron at the Paul Scherrer Institute, in Villigen, Switzerland. Although only six atoms were made, three of these were swept into an atmosphere of oxygen and hydrogen chloride, there to react to form the oxychloride, BhO_3Cl, whose 'volatility' was assessed to be 180°C. In this respect it confirmed that bohrium really was a member of group 7 of the periodic table, coming below rhenium, whose oxychloride, ReO_3Cl, has a boiling point of 131°C.

Bohrium has 10 known isotopes of mass numbers 260 to 262, 264 to 267, 270, 272, and 274, and 1 known isomer. The longest-lived isotope is bohrium-260 with a calculated half-life of 61 seconds. (Bohrium-273 and 274 are theoretically calculated to have half–lives of 90 minutes.)

Boron

Pronounced bohr-on, the name was devised by Humphry Davy (1778–1829) who combined the words BORax and carbON on the basis that the element came from the mineral of that name yet appeared to be like carbon. Borax derived its name from the Arabic word *buraq*.

French *bore*; German *Bor*; Italian *boro*; Spanish *boro*; Portuguese *boro*.

Essential element

How it formed in the cosmos remains a mystery, but nevertheless it has many down-to-earth applications such as fertilising olive groves, making heat-proof glass, in semiconductors, and being added to cosmetics.

Cosmic element

Boron's very existence is surprising because there appears to be no way in which it can be created within stars. (The same is true of beryllium.) At the time of the Big Bang traces of boron and beryllium were produced shortly after the creation of the first three elements: hydrogen, helium, and lithium. It was radioactive decay and nuclear reactions among these which produced them.

However, if the Big Bang had been inhomogeneous then it could have created boron in neutron-rich regions. The existence of boron is taken as indicating that this is what must have happened. The only other way that boron could be formed would be within a Type II supernova or by interstellar cosmic rays shattering the atoms of other nuclei, some of which splinter to give boron. This process is known as spallation.

Element of life

Boron in the human body	
BLOOD	0.13 p.p.m.
BONE	*approx.* 2 p.p.m.
TISSUE	2 p.p.m.
TOTAL AMOUNT IN BODY	18 mg.

Boron is an essential element for the cell walls of plants, so not surprisingly we have a regular daily intake of about 2 mg and a body burden of 18 mg. Foods relatively rich in boron are fruits, in particular apples, grapes, peaches, cherries, and pears. Vegetables such as broccoli, cabbage, and onions, have more than 1 mg of boron per kg.

Dangerous element

Boron can upset the body's metabolism. It takes 5 g of boric acid to make a person ill, and 20 g or more to put their life in danger. Despite this moderate toxicity, boron compounds, such as borax, $Na_2B_2O_5(OH)_4.8H_2O$, and boric acid, H_3BO_3, were once used in medicines.

Boric acid acts as an insecticide especially for ants and cockroaches which are particularly susceptible to it and unable to detect its presence in baits such as sugar solution or cereal pellets.

Medical element

Boron compounds are being investigated as a treatment for brain tumours, using a process known as BNCT (boron neutron capture therapy). Patients are given a substance known to be attracted to cancer cells but to which a cluster of boron atoms has been attached. When these have been taken into the tumour cells they are then bombarded with neutrons which are absorbed by the boron atoms. The boron-10 isotope, which constitutes 20 percent of boron, is an excellent absorber of neutrons, and in so doing it undergoes immediate nuclear fission to form lithium and emits deadly alpha-rays which kill the cancer cells. (Alpha rays cannot penetrate far from the cell in which they are generated, so they don't threaten healthy cells of the body.)

Boric acid and sodium borate are two of the active ingredients in over-the-counter brands of eye wash.

Historical element

There is some evidence that the Etruscans of Ancient Italy, who pre-dated the Romans, used borates from local hot springs to produce the glazes for their black relief vases for which they were famed. Even though there are references to *borax* in the writings from ancient Babylon, Egypt, and Rome, we cannot know if this really was sodium borate or another salt. The only source of borax in the ancient world was to be found in Tibet. There it crystallised as deposits from Lake Yamdok Cho, south of Lhasa, and is found just below the soil surface. It is not certain when trade in borax from Tibet began, but by 1100 AD goldsmiths were using borax as a flux to make working with the molten metal easier.

Borax was mentioned in Geoffrey Chaucer's *Canterbury Tales,* written in the 1380s. In the prologue he introduces his 29 pilgrims who are going on a pilgrimage. Of the summoner, a lawyer for the church courts who was clearly suffering from acne, he says:

> *'No quicksilver, lead ointments, tartar creams,*
> *Boracic, no, nor brimstone, so it seems,*
> *Could make a salve that had the power to bite,*
> *Clean up or cure his whelks[28] of knobby white.'*

[28] A whelk was another name for a pimple.

Boracic was the old name for borax and it was clearly being used as a skin ointment (along with other elements mentioned, namely mercury, lead, and sulfur). Borax was also part of Queen Elizabeth I's collection of cosmetics 200 years later. For many years the Venetians had the monopoly for its importation into Europe which they profitably exploited.

The characteristic green flame which came to symbolise boron was first reported in 1732 by Geoffroy the Younger (1685–1752). It can be produced by treating borax, or any borate, with sulfuric acid to convert the boron to boric acid, and then adding alcohol to dissolve this and setting it alight. This simple test of analytical chemistry was to be used by mineral prospectors and eventually to result in the discovery of the massive deposit in Death Valley in the Californian desert.

In 1779 a large deposit of borate was discovered in Tuscany, Italy, but it was not mined until the early 1800s and then Italy became the main exporting country for more than 75 years. Slowly the use of boron became more widespread, and in 1873 the US deposits were discovered and soon became more profitable to work. As a result the Italian industry slowly declined and eventually ended.

In 1808, Louis-Josef Gay-Lussac (1778–1850) and Louis-Jacques Thénard (1777–1857) working in Paris, and Sir Humphry Davy (1778–1829) in London, independently extracted boron from borate by heating it with potassium metal. The two Frenchmen reported their results on 21 June that year and Davy on 30 June. Neither party had produced the pure element which is almost impossible to obtain. A purer type of boron was isolated in 1892 by Henri Moissan (1852–1907) but it was not until E. Weintraub, of the General Electric Company of America, produced it by sparking a mixture of boron chloride, BCl_3, vapour and hydrogen that it was finally obtained. Pure boron was found to have very different properties from those previously reported.

Economic element

Boron is only found as borates (boron-oxygen compounds) and the main ores are kernite and tincal (borax), which are both sodium borates, and ulexite and colemanite, which are calcium borates. Borates are mined in Turkey (mainly as colemanite), the USA (borax, kernite, and ulexite), Argentina (various ores) and Chile (ulexite) with world production being about 4.5 million tonnes a year. Known reserves are of the order 500 million tonnes, and of the workable deposits 70 percent are in Turkey.

Boron is an important industrial material although as the pure element it is little used except to make metal borides, or to be added to metals to improve their conductivity (aluminium), or to make them easier to refine (nickel), or to make them flow better (iron). Some borides conduct electricity many times better than the metal itself and they have very high melting points. Metal borides are used for turbine blades, rocket nozzles, high temperature reaction vessels, and inert electrodes. Ultrapure boron (99.9999 percent) is produced and in the form of boron trifluoride, BF_3, it is used as the dopant for making p-type semiconductors. Boron fibres, like carbon fibres, are strong and used in some products like fishing rods.

Boron as part of neodymium iron boride, $Nd_2Fe_{14}B$, forms one of the strongest magnets known.

The main boron compounds are borax, boric oxide, and boric acid. They are important in making glass, detergents, and for agriculture. Pyrex glass is tough and heat resistant because of the 12–15 percent boric oxide used in its manufacture. Glass fibre is also boro-silicate glass and this finds use in reinforcing plastics and as insulation in buildings. Ceramic glazes for tiles and kitchen equipment now account for a lot of the demand for boric acid.

An early use of borax was to make perborate, the bleaching agent once widely used in household detergents. Boron compounds also came into the average home in the guise of food preservatives, especially for margarine, fish and other foodstuffs, but it was eventually banned. Some is still used to preserve caviar.

Boron compounds such as sodium octaborate, $Na_2B_8O_{13}$, are used to fire-proof fabrics, wood, and the flakes of shredded paper that are used to simulate snow on film sets.

The isotope boron-10 is an excellent absorber of neutrons and is separated from the more abundant boron-11 for use in the nuclear industry. Incorporated into a polymer matrix it is used to regulate nuclear reactors and to provide a safety system for quenching a reactor should it threaten to go critical.

Environmental element

Boron in the environment	
EARTH'S CRUST	10 p.p.m.
	Boron is the 38th most abundant element.
SOILS	varies between 1–450 p.p.m., but generally is 10–30 p.p.m.
SEAWATER	4 p.p.m.
ATMOSPHERE	minute traces detectable in rain water.

Thousands of tonnes of borate are added to fertilizers each year because this element is vital to plants. If soil lacks boron then plants are stunted, don't pollinate or produce fruit, and are susceptible to diseases. Boron revitalised the age-old olive groves of Greece and Spain that were giving poor yields of fruit. Only 60 g of soluble sodium octaborate are needed per tree per year and not only does the crop increase but it eliminates pests such as the olive fly and the olive moth.

Plants take up boron in order to mobilise sugars, and it has been shown that the amount of sugar that sugar beet produces goes up in line with the amount of boron in the soil in which it is grown, although if this is too high then a toxic reaction sets in and yields begin to fall. Boron deficiency in several plant species is common and its lack can stunt the growth of crops such as sugar beet, celery, sunflowers, legumes, and apples. Trees appear to need more boron than most plants. Too much boron can have an adverse effect on some crops such as kidney beans and lemons, while at the other extreme there are plants that seem to be able to tolerate any amount, such as turnips, cotton, and clover.

Chemical element

Data file	
CHEMICAL SYMBOL	B
ATOMIC NUMBER	5
ATOMIC WEIGHT	10.811
MELTING POINT	2,300°C
BOILING POINT	3,658°C
DENSITY	2.3 g/cm³
OXIDE	B_2O_3

Boron is a non-metal element and the only non-metal of group 13 (row 2p) of the periodic table. It has several forms, the most common being amorphous boron, which is a dark powder, unreactive to oxygen, water, acids, and alkalis. It reacts with metals to form borides. Boron can display oxidation states such as –3 (as in the borane known as diborane, B_2H_6) and +3 which is the preferred state, as in BF_3 and BCl_3 and borates. There are many boranes and borates and these are noteworthy for having complex structures.

There are 14 known isotopes of boron with mass numbers 6 to 19, and no known isomers. Naturally occurring boron consists of two isotopes, boron-10 (20 percent) and boron-11 (80 percent), neither of which is radioactive. The longest-lived radioactive isotope is boron-8 which has a half-life of only 770 milliseconds.

Element of surprises

1. Boron nitride (BN) is a remarkable material, and like carbon it can exist in two very different forms, either as cubic-BN or hexagonal-BN. The former sparkles like diamond and is almost as strong, and it makes an excellent abrasive for industry. The latter is soft and smooth like graphite, but unlike graphite it is colourless, and it can be fabricated into items with high heat capacity, high electrical resistance, and can cope with extremely high temperatures and even molten metals. It enhances the performance of many other materials, including face powder.

2. Boron carbide is another remarkable material. It is both chemically inert and heat resistant and has a hardness approaching that of diamond. It is made by heating boron or boron oxide with carbon at 2500°C, and is manufactured in tonne quantities. It is used in bullet-proof vests and tank armour, and has even been turned into nanowires and fabrics. Its chemical composition is between $B_{13}C_2$ and B_4C, but its exact chemical structure has yet to be determined.

3. Magnesium diboride, MgB_2, is a superconductor below 40K (–233°C) and it can be fabricated into wires offering a cheap alternative to other materials that are needed for superconducting magnets.

4. The fragrance given off by hyacinths contains a boron compound which has not yet been identified.

Bromine

Pronounced broh-meen, the name comes from the Greek, *bromos* meaning stench.
French *brome*; German *Brom*; Italian *bromo*; Spanish *bromo*; Portuguese *bromo*.
Bromine and bromide: bromine is the name of the element which exists as the molecule Br_2, whereas bromide, as in potassium bromide, is a bromine atom that has acquired a negative electron and is written Br^-. Br_2 is reactive and dangerous, while Br^- is stable and relatively safe.

Essential element

Whatever bromine is used for—in sedatives, soil conditioners, and petrol additives—it ultimately appears to be disadvantageous, and yet production of the element continues to increase.

Element of life

Bromine in the human body	
BLOOD	5 p.p.m.
BONE	*av.* 7 p.p.m.
TISSUE	*av.* 7 p.p.m.
TOTAL AMOUNT IN BODY	260 mg.

Although a small amount of bromine is present as bromide in all living things, and some species even produce organo-bromine compounds, no biological role has been identified for this element in humans. It has been suggested that some bromide is essential to health, but this has not been proved.

Our average dietary intake varies widely depending on the foods we eat and may be as low as 1 mg or as high as 20 mg. A few species contain organo-bromine compounds, some of which are to be found in the eggs of sea birds which they probably accumulate from their marine diet. Marine plants can concentrate bromide to levels a thousand times higher that in the surrounding water. Land plants can do the same from soil and in the case of lettuces that were grown in a greenhouse fumigated with methyl bromide the level exceeded 0.1 percent of the weight.

Dangerous element

Bromine itself is corrosive and its vapour attacks the eyes and lungs. It is very toxic with as little as 100 mg being a fatal dose, whereas it requires more than 3,000 mg of bromide to provoke a toxic response and more than 10 times this amount to pose a threat to life.

Medical element

In the 1800s and early 1900s, bromide salts were prescribed by doctors for 'nervous' complaints and even epilepsy, and they were the Victorian equivalent of tranquilizers. Indeed, the word 'bromide' came to mean something that eased tension or was soothing, but too much leads to depression and loss of weight. The use of bromide salts as sedatives and to reduce sexual activity has now been discontinued because the element is slightly toxic. Some spa waters owed their popularity to their high bromide content.

Element of history

Antoine-Jérôme Balard (1802–76) discovered bromine while investigating residues from brine in Montpellier, France. By passing chlorine gas into them, he liberated an orange-red liquid which he deduced was a new element. He sent an account of his findings to the French Academy's journal in 1826 and proposed the name *muride* for it. The Academy's vetting committee did not like his choice of name, which he had derived from murex, the name of the deep purple natural dye, so they decided to change it to bromine to be consistent with the related elements chlorine and iodine, which this new element clearly resembled. Balard was at first upset by this but he eventually agreed.

A year earlier a student at Heidelberg, Carl Löwig (1803–90) had brought his professor a sample of bromine that he had produced from the waters of a natural spring near his home at Keruznach. He was asked to produce more of it, and while he was doing so Balard published his results and so became known as its discoverer.

Once bromine had been discovered, its chemistry was quickly investigated since it could be predicted on the basis of its relationship to chlorine and iodine. Equally quickly other chemists discovered it in many mineral water springs around Europe and showed it to be present in seawater, plants, and animals.

Economic element

Bromine used to be produced from naturally occurring brines by adding sulfuric acid and bleaching powder. Then Herbert Dow devised a method using electrolysis and began to extract it from bromine-rich brine deposits which underlay the US Midwest town of Midland. These contained up to 0.2 percent bromine. Its extraction from sea water began in 1934 at Kure Beach, North Carolina, by what had then become the Dow Chemical Company. The largest bromide-rich source in the USA is located in Arkansas.

World production of bromine is around 450,000 tonnes per year, and the three main producing countries are the USA, which produces 50 percent, and Israel, which produces 35 percent. China account for most of the remaining 15 percent, producing it from evaporation ponds in the Shandong province where there are 90 square kilometres of low grade bromide deposits. The third largest producer used to be the UK where it was extracted from seawater at a plant on the coast of Anglesey, Wales.

However, this plant closed in 2004. Israel's source is the Dead Sea which is particularly rich in bromide with levels as high as 0.5 percent.

Known reserves of bromine are virtually limitless because the oceans contain more than a trillion (10^{12}) tonnes of bromide. Bromine can be extracted from seawater, which contains 65 parts per million of bromide, by bubbling chlorine gas through the water which converts the dissolved bromide salts to bromine. This is then removed by blowing air through the water and the expelled bromine gas is collected.

Bromine, its salts, and its organic derivatives—known as organo-bromine compounds—find a wide range of uses such as for dyes, water disinfectants, pharmaceuticals, DNA sequencing, and as agrochemicals. Some of these are being phased out but the increasing production of bromine shows that other uses are more than compensating for this loss of market. The first commercial use of bromine was in the development of daguerreotype photographs in the 1840s.

Bromine can be used as a disinfectant for hot tubs. It can either be added as sodium bromide along with an oxidising activator, or it can be added as BCDMH (short for 3-bromo-1-chloro-5,5-dimethylhydantoin) which is self-activating. Either way it ends up as the hypobromite ion, BrO^-, which is the active agent. Bromine is preferred for hot tubs because it is stable under these conditions. It cannot be used in outdoor pools because it is deactivated by sunlight.

Chemists in pharmaceuticals companies use bromine intermediates during the manufacture of healing drugs, even though the element does not appear in the final product. Bromine is used in industry to make organo-bromo compounds although the demand for these has varied considerably over the years. A major outlet was dibromomethane, formerly as fuel additives for leaded petrol. This compound reacted with the lead during the combustion phase to form lead bromide which then passed out in the exhaust gases instead for forming a deposit of lead metal in the engine itself.

Other organo-bromines are used in the so-called halon fire extinguishers. Halon-1211 ($CBrClF_2$) and halon-1301 ($CBrF_3$) are the fire extinguishers of choice for tackling a blaze in a confined space or where water might do as much damage as the fire, for example in archives, museums, art galleries, aircraft, engine rooms of ships, military vehicles such as tanks, and large computer facilities. Although these vapours are highly effective at dousing flames, and are non-toxic, the halons are now frowned upon because the bromine they contain adds to the destruction of the ozone layer in the upper atmosphere.

One of the most widely used flame retardants was decaBDE (short for decabromodiphenyl ether) which was added to furniture foams, plastic casings, and some textiles. Although it was never shown to pose a threat, it is a persistent chemical in the environment and has been suspected by some of having hormone-like properties. The EPA in the USA has said it must be phased out by 2012. Bromine-containing flame-retardants generate hydrogen bromide when they are heated and this suppresses the chemical reactions on which fire depends to keep it going.

At one time methyl bromide (aka bromomethane, CH_3Br) was widely used as a soil fumigant to the extent of 50,000 tonnes per year, but this is being phased out for environmental reasons—see below.

Sodium and calcium bromides are very soluble in water and produce a very dense solution that is use in drilling.

Environmental element

Bromine in the environment	
EARTH'S CRUST	0.4 p.p.m.
	Bromine is the 62nd most abundant element.
SOILS	varies widely with a range of 5–40 p.p.m. with some volcanic soils being as high as 500 p.p.m.
SEAWATER	65 p.p.m.
ATMOSPHERE	a few p.p.t.

More than 1,500 naturally occurring organo-bromine compounds have been identified, the most abundant of which is methyl bromide (aka bromomethane). Marine algae produce this to the extent of around 55,000 tonnes per year. This molecule has also been produced on a similar scale by the chemical industry in the past. It was found to be an effective soil fumigant in the 1950s and has been widely used since to kill nematodes,[29] insects, bacteria, mites and fungi. Methyl bromide is particularly effective because the gas penetrates to areas that other pesticides can't reach, and it kills pests at all stages in their life cycles. An equally important observation was that during 40 years of use there been no indication of pest resistance evolving.

The crops which benefited most were tomatoes, celery, lettuce, strawberries, grapes, tobacco, and flowers such as carnations and chrysanthemums. Warehouses where crops were stored were easily protected against insects and rodents by its potent vapour, nor did it contaminate the food. Despite all these advantages, in 1992 it was added to the list of compounds that had to be phased out under the Montreal Protocol because of the threat it poses to the ozone layer. It ceased to be manufactured by 2010 and was replaced by methyl iodide (aka iodomethane, CH_3I)—see page 249.

Removing methyl bromide compounds from the atmosphere may not be as easily achieved as it sounds. Nature also produces volatile organo-bromine compounds in vast quantities from forest fires and marine plankton, and the salt pans of the bromide-rich Dead Sea increase the amount in the atmosphere. Yet Nature may also play a part in removing methyl bromide since the oceans themselves can dissolve it, bacteria in soil can convert it to bromide, the oxygen in the atmosphere can react with it, and the leaves of many plants will absorb it. In fact methyl bromide is not quite so threatening as it first appears. Environmental research uncovered the unexpected result that half the methyl bromide sprayed on soil never evaporates into the air because it is consumed by bacteria.

Marine plankton and algae release around half a million tonnes of various bromomethanes a year and in particularly tribromomethane (aka bromoform, $CHBr_3$). Even more surprising has been the discovery that something in the oceans is making pentabromodiphenyl ether. This has been used as a fire-retardant, and when, in 2005, it was found to be present in whale blubber it was at first thought to be the man-made variety. However, the carbon atoms it contained had detectable amounts

---

[29] These are small parasitic worms.

of the radioactive isotope carbon-14 meaning that they were of recent origin, whereas the fire retardant is made entirely from fossil resources and contains no carbon-14.

Even when it appears benign as bromide ions in water, this element can still pose a threat to health. Ozonising drinking water in order to sterilise it converts any bromide to bromate (BrO_3^-) which is a suspected carcinogen and so must not exceed 10 p.p.b. Some reservoirs in California where this has been exceeded have had to be drained because of it.

Chemical element

Data file	
CHEMICAL SYMBOL	Br
ATOMIC NUMBER	35
ATOMIC WEIGHT	79.904
MELTING POINT	−7°C
BOILING POINT	59°C
DENSITY OF LIQUID BROMINE	3.14 g/cm³
OXIDES	none stable

Bromine is a deep red, dense, sharp-smelling liquid (Br_2) that is highly reactive. It is a member of the halogens, the elements of group 17 (row 4p) of the periodic table. Bromine is soluble in organic solvents and in water, whence it forms 'bromine water' which is both a powerful oxidising agent and a convenient source of the element that was formerly used to test for unsaturated organic molecules, that is, those which had carbon-carbon double-bonds in their structure. These react rapidly and the bromine water goes colourless. An alternative to using bromine itself in chemical reactions is the reagent N-bromosuccinimide which is easier to handle.

There are 32 known isotopes of bromine with mass numbers 67 to 98, and 14 known isomers. Naturally occurring bromine consists of two isotopes, bromine-79 which comprises 51 percent, and bromine-81 which accounts for 41 percent.[30] Neither is radioactive. The longest-lived radioactive isotope is bromine-77 whose half–life is 2.4 days.

Element of surprises

The famous purple of the togas worn by the Roman Emperors was produced from a naturally-occurring dye that contained bromine atoms. It was called Tyrian purple and was extracted from the Mediterranean mollusc *Murex brandaris*. However, it was known in the Middle East well before the time of the Romans and it is mentioned in the *Bible* in the book of the prophet Ezekiel, verse 27, lines 7 and 16, the first of which describes a mythical luxury yacht with an awning that was 'purple from the

---

[30] In addition, there are traces of radioactive isotopes produced as fission products in uranium-bearing deposits.

isles of Elisha' and the second reference is to merchants from Syria who traded in 'emeralds, purple, fine linen, coral and rubies.' Both references show how highly valued the dye was.

The rich colour of Tyrian purple was not to be rivalled until an 18-year-old chemistry student, William Perkin (1838–1907), discovered the beautiful violet dye mauve, in 1865, while experimenting at his home in East London. He studied under August Wilhelm von Hofmann at the Royal College of Chemistry, now Imperial College London.

Cadmium

Pronounced cad-mee-um, the word is derived from the Latin *cadmia* the name for the mineral calamine, which is in fact *zinc* carbonate—see *Element of history* below.

French *cadmium*; German *Cadmium*; Italian *cadmio*; Spanish *cadmio*; Portuguese *cádmio*.

Essential element

This element is an accumulative poison and its many uses are now being phased out although ptarmigan birds will always suffer from it. As far as humans are concerned its usefulness is now confined to electrical and electronic devices.

Element of life

Cadmium in the human body	
BLOOD	5 p.p.t.
BONE	*approx.* 2 p.p.m.
TISSUE	*approx.* 2 p.p.m. It tends to accumulate in the kidneys and liver in which there can be 10 times this level.
TOTAL AMOUNT IN BODY	increases with age to around 20 mg at age 50.

Because it is such a toxic element, like mercury which comes below it in the periodic table, it had been assumed that no species would require this metal as an essential element. However, marine organisms readily absorb it, which is why there is so little cadmium in the surface layers of the oceans, although that is not proof that they require it—the same is true of mercury which is absorbed by fish like tuna. The strongest evidence of cadmium's being essential for one species was the discovery, in 2000, that the marine diatom *Thalassiosira weissflogii* produces a cadmium-specific enzyme which catalyses the interconversion of carbon dioxide and carbonic acid. Even so, the required amount is minute but it is essential and appears to have been part of the evolution of these creatures which live in a zinc-poor environment and so had to use cadmium instead for these particular enzymes. It has been suggested that other marine species also use this enzyme and this would explain the cadmium profile in the oceans because it is not evenly distributed and it increases as you descend to the ocean floor.

Cadmium can never be totally excluded from the diet; it is present in foods like offal, shell-fish, and rice. Some plants have an ability to absorb cadmium such as the mushroom *Amanita muscaria* which can have levels of 30 p.p.m. even growing on soil with as little as 0.3 p.p.m. cadmium. Other food crops which seem to absorb the metal are lettuce, spinach, cabbage, and turnip, while potatoes, maize, beans, and peas are among the least absorbing. Smoking adds to the body burden because tobacco plants absorb it. Kidney and liver from pigs, sheep, and cattle contain more than 0.2 mg per kg.

Plants growing on contaminated land, either near ancient mines, metal-processing industries, or from land fertilised with human sewage (once known as sewage farms) can all have high levels of the metal; for example, lettuce can reach levels of 70 p.p.m. Even when such land is no longer used to grow food and is turned over to sheep grazing it has been found that these animals can accumulate cadmium in their kidneys and livers.

The human daily intake of cadmium may be as low as 10 microgrammes (μg), or as high as 1000 μg, but the average is probably less than 25 μg. The World Health Organization has set a recommended safe daily intake of 70 μg.

Dangerous element

Cadmium is an accumulative poison which is very insidious and is on the UN Environmental Programme's list of top 10 hazardous pollutants. Inhaled cadmium oxide fumes are particularly dangerous and this has sometimes had fatal consequences for workers in industry, as in the case of a group of men who were working on the Severn Road Bridge in England in 1966. They had to remove part of a construction tower and they decided to use an oxyacetylene torch to melt away the steel bolts that held it together. Unfortunately there was no ventilation inside the tower and the job took all day. The next day when the men reported for work they all complained of feeling ill, were coughing a lot and had difficulty breathing. Within a few hours one man had to be taken to hospital, where he died a week later, and one by one the others were also admitted for treatment although they survived. And the cause of their illness? The bolts they were removing were coated with cadmium and this had been volatilised by the heat with the result that they all succumbed to cadmium poisoning.

Cadmium mimics the essential element zinc and so gains access to the body. Luckily the human gut, which needs to allow zinc to be absorbed, is able to keep out most of the cadmium, but as much as 10 percent of it may slip through, with the result that a few micrograms enters the system each day. There it triggers a defence mechanism by stimulating the production of the enzyme metallothionen which is a protein with lots of cysteine amino acid units. The sulfur atoms in these attract the cadmium to the extent that as many as seven metal atoms can attach themselves to every enzyme. These then transport the cadmium to the kidneys with the object of getting get rid of it, but cadmium binds so strongly to the enzyme (300 times stronger than zinc binds) that it tends to accumulate there. The result is that a cadmium atom remains in the human body on average for around 30 years, which is why there is so much concern about the effect of this metal on human health. Because zinc is also part of DNA polymerase, which is crucial to the production of sperm, its replacement by cadmium is particularly damaging to the testicles.

The body can go on storing cadmium but eventually there comes a point when the kidneys can cope no longer. If the level exceeds 200 p.p.m. it prevents reabsorption of proteins, glucose, and amino acids, and damages the filtering system and this has occasionally led to cases of kidney failure. Treatments with chelating agents that could remove cadmium from the system are risky because they would also remove zinc just as easily. In fact, low levels of cadmium in the kidneys do not appear to cause progressive damage and what damage there is may even be reversible.

Continued exposure to cadmium leads to a disease known in Japan as Itai-Itai,[31] which weakens the bones and joints making movement painful. It was at Fuchu, a town of 45,000 inhabitants which is 200 miles north-west of Tokyo, where the condition was first recognised and the cause was traced to a combination of high cadmium intake and low levels of vitamin D in the victims. The cadmium came from rice grown on contaminated land and watered from a source that was polluted by waste from the spoil heaps of a zinc mine. The rice had 10 times as much cadmium as normal rice, and this resulted in a daily intake of around 0.6 mg per day (220 mg per year) and 10 times the amount which the World Health Organization says is safe.

Cadmium caused cancer in special laboratory-bred rats (but not in mice or hamsters), and some of its compounds are definitely carcinogenic. There has been concern that it may cause cancer in humans, but this is debatable. An analysis of cancer cases, in 1988, among those who worked with it in Britain seemed to show this was so, and higher rates were also reported in the USA. However, a follow-up study of 7,000 people whose work involved cadmium could find no link between cancer and cadmium.

Element of history

The apothecaries of Hanover, Germany, in the early 1800s used to make zinc oxide by heating calamine, also known as cadmia, a naturally occurring form of zinc carbonate. Sometimes the product they ended up with was an unpleasant yellow colour instead of pure white. When the inspector of pharmacies, Professor Friedrich Stromeyer (1776–1835) of Göttingen University, looked into the problem he traced the discoloration to a component of the mineral that he could not identify and which he deduced must be a previously unknown element. This he separated as its brown oxide and by heating it with lampblack (carbon) he produced a sample of a blue-grey metal which he named cadmium after the name for the mineral. That was in 1817. Meanwhile two other Germans, Karl Meissner in Halle, and Karl Karsten (1782–1853) in Berlin, were working on the same problem and announced their discoveries of the element the following year.

Interest in the new element generated a spate of research and it was soon found to be present in all zinc ores, but as its presence seemed to have no effect on the performance of the zinc compounds that were derived from it there was little incentive to remove it. It was partly being removed when zinc metal was smelted because cadmium is more volatile and was lost to the atmosphere. (The boiling point of cadmium is 756°C which is appreciably lower than that of zinc's 907°C.)

---

[31] 'Ouch-ouch!' would be the equivalent in English.

Economic element

Specimens of native metallic cadmium have been found in Siberia, but they are very rare. The first true cadmium mineral to be discovered came to light when the Bishopton railway tunnel was being bored near Port Glasgow in Scotland in 1841. This mineral, cadmium sulfide (CdS), was named greenockite after Lord Greenock who was in charge of the project. In fact greenockite had been used for more than 2,000 years as a yellow pigment and then it had been mined in Greece and Bohemia.

Other cadmium ores are known, such as cadmoselite (CdSe) and otavite (CdCO$_3$) but no ore is mined specially for this metal. More than enough is produced as a by-product of the smelting of zinc from its ore sphalerite, ZnS, in which CdS is a significant impurity, making up as much as 3 percent. Consequently the main mining areas are those associated with zinc. World production is around 18,000 tonnes per year, and falling. The main producing country is China, which produces around 4,000 tonnes per year, with South Korea, Kazakhstan, Mexico, and Canada producing around 1,000 tonnes per year. The total amount of cadmium used now is less than 10 percent of that which was used 50 years ago, but production of cadmium is still significant because it has to be removed from zinc ores. Cadmium reserves total 1.5 million tonnes.

When used a pigment, cadmium sulfide was known as cadmium yellow, although it can be anything from yellow, through orange and red, to brown depending on the proportions of sulfur and selenium in the pigment. Bright yellow cadmium sulfide was added to paints, artists' colours (Monet used it), rubber, plastics, printing inks, and vitreous enamels. Almost all these uses have now been phased out.

Cadmium has been particularly used to electroplate steel where a film of cadmium only 0.05 mm thick will provide complete protection against that most corrosive of environments, the sea. This reacts to form an impervious layer of insoluble cadmium chloride. (Galvanising with zinc is less protective because any zinc chloride which forms dissolves away from the surface.)

While most uses of cadmium were being phased out as dangerous to the environment and human health, others were taking their place such as nickel-cadmium batteries, which were seen as environmentally friendly because they could be recharged a thousand times or more and still work well. In these lightweight batteries the cathode is nickel hydroxide, the anode is cadmium oxide, and the electrolyte is potassium hydroxide. These batteries account for 80 percent of the cadmium currently produced but they too are being phased out in favour of nickel metal hydride batteries which also deliver more current.

The Nissan car company designed an all-electric car with a range of 250 kilometres (150 miles) and it relied on a new style nickel-cadmium battery that can be recharged in only 15 minutes. Such batteries are more efficient for electric vehicles because they are only a third the weight of conventional lead-acid batteries. Cadmium recycling from batteries now recovers several thousand tonnes of the metal a year. Nevertheless, in motor vehicles all other roles for cadmium have been phased out, even that which formerly coated nuts and bolts that were exposed to the corrosive spray that comes from salted roads in winter.

Since the 1960s, when alarms were first sounded about cadmium, industry has found alternatives for most products that reach the consumer. Loss of cadmium to the environment is now negligible, and its continued use is only justified on the grounds that cadmium makes better batteries and more stable plastics. In the aerospace, mining, and off-shore oil industries it is needed to protect steel. Cadmium-plated steel is still used for critical components of aircraft such as the landing gear. Cadmium has the ability strongly to absorb neutrons and so is used in some nuclear reactors and for the control rods that regulate the self-sustaining chain reaction that provide the energy.

Environmental element

Cadmium in the environment	
EARTH'S CRUST	0.1 p.p.m.
	Cadmium is the 65th most abundant element.
SOILS	varies widely but on average is 1 p.p.m. with some polluted soils being as high as 1,500 p.p.m.
SEAWATER	0.1 p.p.b. with only 1 p.p.t. in the surface layers.
ATMOSPHERE	tiny with between 1 and 50 ng per cubic metre of air, depending on location, with industrial cities having the most.

Cadmium contamination comes from three sources: (i) old lead and zinc mines; (ii) zinc smelters which produce cadmium as a by-product; and (iii) natural outcrops, such as the Carboniferous marine black shales.

About 8,000 tonnes of cadmium enters the atmosphere each year of which 90 percent is from human activity. Some of this is washed out by rain on to soil used to grow crops and feed animals. Consequently we can trace the 30 µg of cadmium in a hamburger to the grass the cattle ate, and ultimately to the soil on which it grew. Some of this is natural, some as a result of air-borne contamination, and some from fertilizers added to the soil.

Phosphate rock from Morocco, which is no longer used in the EU, has more than 50 g per tonne of cadmium whereas Florida phosphate has much less than a tenth of this amount. Sewage sludge used as a fertilizer can raises cadmium levels in soils especially if it comes from industrial areas; in some countries millions of tonnes of sewage is added to the land each year. All types of fertilizers appear to increase the level of cadmium in farmed soils. Experiments at Rothamsted Experimental Station, England, showed that both phosphate fertilizers and farmyard manure increased cadmium. More surprising was the finding that the level in unfertilized plots went up from 0.5 to 0.8 p.p.m. over a period of 140 years. This was attributed to atmospheric deposition.

Geochemical atlases have been drawn up for some countries and these show that the amount of cadmium in most soils rarely exceeds 1 p.p.m. but there are local hot spots which exceed the 3 p.p.m. maximum recommended by the EU to protect soils that are fertilized with sewage sludge. Some hot spots have levels exceeding 40 p.p.m. Britain's worst polluted area is around Shipham in Somerset where the spoil from old zinc mines that operated up to about 1850 have led to soil levels of 500 p.p.m. This came to light in 1979 when children at a local school were trying to grow vegetables and found that their leaves turned yellow.

Chemical element

Data file	
CHEMICAL SYMBOL	Cd
ATOMIC NUMBER	48
ATOMIC WEIGHT	112.411
MELTING POINT	321°C
BOILING POINT	756°C
DENSITY	8.7 g/cm³
OXIDE	CdO

Cadmium is a silvery metal and a member of group 12 (row 4d) of the periodic table. Its surface has a bluish tinge and the metal is soft enough to cut with a knife, but it tarnishes in air. It is soluble in acids but not alkalis. Its preferred oxidation state it +2 as in $CdCl_2$ and Cd^{2+} in solution.

Cadmium has 39 known isotopes of mass numbers 95 to 133, and 12 known isomers. There are eight main naturally occurring cadmium isotopes: cadmium-114 is the most abundant and accounts for 29 percent of atoms, cadmium-112 (24 percent), cadmium-111 (13 percent), cadmium-110 (12.5 percent), cadmium-113 (12 percent), cadmium-116 (7.5 percent), cadmium-106 (1 percent), and cadmium-108 (1 percent).[32]

Cadmium was not considered to be radioactive, but in 1970 it was shown that cadmium-113 was a beta-emitter with an incredibly long half-life of 9 million billion years (9×10^{15}). This isotope entered the *Guinness Book of Records* as the longest half-life then known and remained there for several years. The consequence of this radioactivity is negligible; in a 10-gram piece of the metal only one atom of cadmium is lost per minute. Other isotopes of cadmium have since been found to be radioactive, namely cadmium-106 which has a half-life of 4×10^{21} years, cadmium-114 with a half-life of 6.4×10^{18} years, and cadmium-116 with a half-life of 2.9×10^{19} years. Clearly very few atoms of these isotopes will have decayed since they were created.

The next longest-lived radioactive isotope is cadmium-109 with a half-life of 462 days.

Element of surprises

1. Cadmium telluride might well be the way that this metal can find an approved role in modern society because it is an efficient material for solar panels. A square metre of such a panel can generate a power output of nearly 100 watts.

---

[32] There are also trace amounts of other short-lived radioactive isotopes produced as fission fragments in uranium-bearing deposits.

2. The ptarmigan birds of Colorado are a species of grouse whose plumage in winter is white. They have fragile bones and this is due to high levels of cadmium in their diet, according to research carried out at Oregon State University. The cadmium is coming from natural sources and is not due to environmental pollution. It comes from the buds and shoots of the willow trees on which the birds feed in winter and early spring. This tree absorbs cadmium from the soil and concentrates it in the buds and shoots. It then collects in the birds' kidneys and liver where it upsets their calcium metabolism, and causes this element to be leached from the bones.[33]

[33] Information supplied by John Dodds.

Caesium
or CESIUM (USA)

Pronounced seez-ee-um, the name comes from the Latin, *caesius* meaning sky blue, in reference to the colour its compounds impart to a flame.

French *césium*; German *Caesium*; Italian *cesio*; Spanish *cesio*; Portuguese *césio*.

Essential element

This element tells us the exact time in the form of caesium clocks which around the world are an essential part of the internet and mobile phone networks.

Element of life

Caesium in the human body	
BLOOD	4 p.p.b.
BONE	10–50 p.p.b.
TISSUE	*approx.* 1 p.p.m.
TOTAL AMOUNT IN BODY	*approx.* 6 mg.

Caesium has no known biological role, but it may partly replace the essential element potassium, which it resembles chemically. Although a lone atom of caesium is much larger than one of potassium, when they are dissolved in water the solvent molecules which cluster around them make them appear to be almost the same size.

The average daily intake of this metal is only 0.03 mg and it comes with foods that are richest in potassium. The caesium in some plants has been measured and can be up to 3 p.p.b. in vegetables and fruits.

Dangerous element

Rats fed a diet of caesium in place of potassium died within two weeks, so it can be regarded as a toxic element for that species at least, although caesium chloride is probably no more toxic than sodium chloride. Certainly it would have serious effects if it were to be taken in excess, and the tests on rats showed them to experience intense irritability and spasms.

Caesium-137 has a half-life of 30.17 years and this was once a worrying pollutant of the natural environment in the years when above-ground nuclear weapons tests were carried out. When dogs were given single doses of around 4 micrograms of this isotope its danger became apparent and most died within a few days. Nevertheless, caesium-137 has a place in medicine because its gamma-rays can be used to destroy certain gynaecological cancers.

Element of history

Caesium was almost discovered by Carl Plattner (1800–58) in 1846 when he investigated the mineral pollucite. He found he could only account for 93 percent of the elements is contained, but then he had no more mineral left to continue his analysis. In fact it was later discovered that he mistook most of the caesium for a mixture of sodium and potassium.

Caesium was eventually discovered in 1860 at Heidelberg, Germany, when it was shown to be present in mineral water from Durkheim. Atomic spectroscopy was the technique used to show that there was a new element present. The work was done jointly by Gustav Kirchhoff (1824–87) and Robert Bunsen (1811–99), who is best remembered for developing the bunsen burner. They took about 30,000 litres of the mineral water which they boiled down, and from which they removed the lithium, sodium, potassium, magnesium, calcium, and strontium salts. This left them with a liquor which they sprayed into a bunsen flame and analysed the light by means of their spectroscope. This showed two blue lines very close together, something that they had never previously seen and they immediately realised that they had stumbled on a hitherto unknown element. Eventually they produced around 7 grams of caesium chloride from the residue, but were unable to produce a sample of the new metal itself. That was done by the Swedish chemist Carl Theodor Setterberg (1853–after 1890) at the University of Bonn. He obtained it by the electrolysis of molten caesium cyanide, $CsCN$.

Economic element

Only a few caesium minerals are known. Pollucite ($Cs_4H_4Al_4Si_9O_{27}$) is the main one and even this is relatively rare. As the silicate magmas of the Earth cooled to form granite these left behind in the still molten rock higher levels of elements like lithium, rubidium and caesium (among others). In some cases the last fractions to solidify were rich in caesium and this mainly crystallised to form pollucite. Caesium is also present in lepidolite (see lithium). World production of caesium compounds is a mere 20 tonnes per year and the element comes mainly from Bernic Lake (Manitoba, Canada), with a little from Zimbabwe, and South-West Africa. Reserves of pollucite at Bernic Lake are estimated to be around 60,000 tonnes. The main company producing caesium compounds is Chemetall at Langelsheim in Germany.

Caesium is released from its ore by heating the mineral with hydrochloric acid, which leaves behind the silica, and then precipitating the caesium as a mixed chloride with another metal such as lead or tin. Caesium metal can obtained by the electrolysis of a concentrated salt solution, but a more general method is to heat caesium chloride with calcium. These react to form caesium and calcium chloride, and the caesium is then distilled off under vacuum.

Most caesium that is used commercially is in the form of drilling fluids and these contain caesium formate (HCO_2Cs). The important feature of these fluids is their relatively high density of $2.3 \, g/cm^3$ and low environmental impact. They are ideal as a lubricant for the drill bits and they perform well under very high pressures. A special plant at Bernic Lake converts most of the caesium mined there into this material.

High density liquids are also needed in molecular biology to isolate tiny organisms like viruses by means of ultracentrifuges. Solution of salts like caesium chloride are ideal for this purpose and are used to aid the separation and purification of DNA for biochemical research.

There are other outlets for caesium. Caesium nitrate is used to make optical glass. Other types of glass can be strengthened by dipping into molten caesium salts; the caesium ions exchange with sodium ions on the surface and thereby make it more resistant to etching or breakage. Caesium carbonate has been added to glass that is used as optical fibres because it reduces its electrical conductivity while making it more stable. Caesium makes a good catalyst promoter, boosting the performance of other metal compounds.

Caesium iodide and caesium fluoride absorb X-rays, gamma rays and other atomic particles; and as they do so, they give off light. This so-called scintillation effect is used in medical diagnostics and radiation monitoring, and in the prospecting for minerals like uranium.

Caesium metal will scavenge the last traces of gas in the production of vacuum tubes after they have been sealed. The caesium is made inside the tube by a chemical reaction which converts caesium chromate into caesium vapour which then reacts with any oxygen, nitrogen or other unwanted gases.

Caesium clocks are an important part of the internet and mobile (cell) phone networks. The caesium clock is the standard measure of time, and what this refers to is a resonant frequency between two states of the caesium atom in the gas phase. The frequency is exactly 9,192,631,770 cycles per second and the standard second is defined with respect to this number. There are hundreds of these clocks in operation around the world and some are accurate to one second in 15 million years.

Uranium fuel rods in nuclear power stations produce caesium-137, which gives off gamma-rays as it decays, and so it can be used for various gauges which measure thickness.

Environmental element

Caesium in the environment	
EARTH'S CRUST	3 p.p.m.
	Caesium is the 46th most abundant element.
SOILS	0.1–5 p.p.b.
SEAWATER	0.3 p.p.b.
ATMOSPHERE	virtually nil.

Although caesium is much less abundant than the other alkali metals of group 1, it is still more common than elements like arsenic, iodine, and tin. It has achieved little fame for its beneficial uses, but lots of ill-fame for its artificially produced radioactive isotopes, caesium-134 and caesium-137, which have been released during nuclear accidents. These isotopes were also released from the testing of nuclear weapons in the atmosphere as late as 1980, after which most countries agreed to comply with the 1968 ban on above-ground tests.

The half-life of caesium-137 is about 30 years, which means that it takes over 200 years to reduce it to 1 percent of its former level. For this reason an accident at a nuclear power plant can contaminate the environment around for generations, which is why the Chernobyl accident in the Ukraine in 1986 was such an environmental disaster. It released a large amount of radioactive caesium-137 which drifted all over Western Europe, affecting sheep farms as far west as Scotland, Ireland and Wales, over 1,500 miles from the accident. There it was washed to earth by heavy rain, taken up by the roots of plants and so became part of the vegetation that the sheep ate.

Because caesium is rapidly excreted from the body, an animal reared on contaminated grass can be decontaminated by being grazed on unaffected pasture for a few days before it is slaughtered. This does not solve the problem of affected land and research has been directed to finding ways to treat farms with chemicals that will lock up the caesium and prevent it being absorbed by plants. Some clay soils do this naturally and minerals like vermiculite are used specifically for this purpose.

Another method of decontamination of meat uses the traditional dye Prussian blue. This chemical contains potassium ions which can exchange for caesium and thereby trap it and render it inactive. In Germany, Austria, and Norway the Prussian blues are encapsulated into slow-release pills which are given to animals. There they remain in the animal's stomach for several weeks releasing the agent which binds the caesium so that it passes out of the animal without being absorbed.

Humans affected by caesium-137 have also been successfully treated with Prussian blue type agents. For most people who are exposed to low levels of radioactive caesium the simple expediency of taking in more potassium in their diet is sufficient to reduce absorption of the caesium.

Chemical element

Data file	
CHEMICAL SYMBOL	Cs
ATOMIC NUMBER	55
ATOMIC WEIGHT	132.90545
MELTING POINT	28°C
BOILING POINT	679°C
DENSITY	1.9 g/cm³
OXIDE	Cs_2O is the one which forms when caesium burns but there is also a superoxide, CsO_2, which is explosive, and an ozonide, CsO_3, plus several suboxides such as Cs_3O, etc.

Caesium is a soft, shiny, gold-coloured metal and a member of the alkali metals of group I (row 6s) of the periodic table. When it is dropped into water caesium reacts explosively to give off hydrogen gas which ignites spontaneously. It will even react rapidly with ice. Caesium metal quickly oxidises when exposed to the air and can form the dangerous superoxide on its surface. It is generally stored under oil in sealed cans. Its preferred oxidation state is +1 as in the ion Cs^+ which is how it occurs in salts and in solution. A lower oxidation state in the form of -1 is known and this occurs as the Cs^- ion in certain compounds. Caesium hydroxide is highly caustic, and has the ability to dissolve glass.

There are 40 known isotopes of caesium with mass numbers 112 to 151, and 22 known isomers. Naturally occurring caesium ores consist entirely of the caesium-133 isotope which is not radioactive. (In addition there are also minute traces of radioactive caesium isotopes produced as fission products in uranium-bearing deposits. There are seven isotopes produced, mainly as the stable caesium-133 and the radioactive isotopes caesium-134, caesium-135, and caesium-137.) The longest-lived radioactive isotope is caesium-135 with a half-life of 2.3 million years.

Element of surprises

1. Caesium 'thrusters' (aka ion engines) have been used to steer satellites. Caesium is ionised in a vacuum chamber and then the ions are accelerated through an electric field and ejected through a nozzle thereby giving a counter thrust to the satellite. The heavier the ions, the larger the impulse it delivers, and this is why caesium is used because it has a high atomic mass. A kilogram of caesium used this way will propel a space vehicle 140 times further than by using the same weight of any known fuel. See also xenon for more details.

2. When caesium metal is alloyed with sodium and potassium metals, in the atom ratio of 41:47:12, it produces a metal alloy which remains liquid down to $-78°C$, the lowest liquid temperature of any alloy.

3. In Japan, so-called 'caesium gel' is sold as a skin rejuvenating cream, supposedly working by boosting enzymes.

Calcium

Pronounced kal-sium, the name comes from the Latin *calx* meaning lime.
French *calcium*; German *Calcium*; Italian calcio; Spanish *calcio*; Portuguese *cálcio*

Essential element

Calcium salts are generally insoluble, which is why calcium phosphate is the basis of bone and calcium carbonate safely locks up most of the carbon on this planet.

Cosmic element

The visible spectra of the sun and of many stars show that calcium is present and indeed lines in its spectrum can indicate both magnetic activity and reveal the rate at which a star is rotating.

Element of life

Calcium in the human body	
BLOOD	61 p.p.m., with the plasma having 80 p.p.m. and red blood cells only 4 p.p.b.
BONE	3–17%
TISSUE	*av.* 120 p.p.m.
TOTAL AMOUNT IN BODY	1.2 kg.

Calcium is essential to practically all living things, except a few insects and bacteria, and it performs a variety of functions. As calcium carbonate, $CaCO_3$, it provides the skeleton for most marine creatures and the lens in eyes. As calcium phosphate, $Ca_5(PO_4)_3(OH)$, it is found as the bones and teeth of land-based animals. As calcium oxalate CaC_2O_4 it is abundant in plants where it acts as a calcium store and possibly as a deterrent partly because it is toxic and partly because the crystals it forms can be sharp-edged and quite abrasive, so that plant-eating creatures like caterpillars are deterred from consuming leaves which contain it.

Calcium is the most abundant metal in the human body, and bone accounts for 1,190 grams of the 1,200 grams of calcium in an average adult. The other 10 grams have five important metabolic functions: regulating the activity at membranes such as cementing cells together; controlling muscular contraction and nerve impulses; keeping the pH of the blood stable which aids clotting; controlling cell division; and triggering hormone release.

Food element

Pregnant women and children are encouraged to eat foods rich in calcium to promote the growth of teeth and bones. Traditionally they have been advised to eat cheese, milk, and white bread, but more recently the trend has been to recommend leafy green vegetables as well. People of all ages would be well advised to eat calcium-rich foods, although calcium deficiency is rarely a problem with a normal diet.[34] When it does occur, its most obvious effect is bone disease, but it is not lack of calcium that is generally the primary cause but some other factor. For example rickets in young children is due to lack of vitamin D.

The ideal calcium intake is probably about 1,000 mg per day but, during childhood, pregnancy, and old age when extra calcium is needed, it is best to boost this to 1,500 mg as insurance. In theory, an adult needs around 500–700 mg per day and teenage children around 800–1,000 mg. Sardines, egg yolk, almonds, cheddar cheese and milk chocolate are particularly good sources of calcium, providing more than 200 mg per 100 g, while cereal and dairy products will generally contain more than 100 mg per 100 g. Vegetables, and especially cabbage, broccoli, onions and red kidney beans, can also be a useful source. To utilise this calcium, the body needs plenty of vitamin D which is provided by foods like liver oils, many types of fish, butter, margarine, and eggs—and by sunshine.

Medical element

Calcium compounds have featured in medicine down the centuries and they continue to do so: calcium carbonate is used as an over-the-counter antacid for indigestion; calcium lactate can be prescribed to treat calcium deficiency; calcium chloride in small doses acts as a diuretic and it may be included in a saline drip; and calcium sulfate is used to make plaster casts.

The weakening of the bones in old age (known as osteoporosis) is not cured just by taking in extra calcium—see box.

Bone idle?

We tend to think of bone as somehow different from the other cells that make up our body. It is more like a mineral than living flesh; but it is just as 'alive' in that it is endlessly changing, being constructed and broken down at millions of sites throughout the skeleton by cells that are called osteoblasts and osetoclasts. In this way bone carries out an important function of keeping the level of calcium in the blood steady, thereby ensuring all its other needs for calcium are met.

When our diet does not provide enough calcium, the deficiency is made good from bone—ideally to be replaced later when there is an excess of calcium in our blood. As

---

[34] People on a high protein diet should be aware that this increases calcium loss via the urine.

we get older this replacement process does not completely compensate for the loss, so to check this depletion we need to make sure we get a daily dose of calcium and vitamin D. Even so, calcium loss cannot be entirely prevented.

Attention has focused on calcium, as ageing populations suffer weakened skeletons and bones that break easily. Our peak bone mass, as it is called, is reached when we are about 35 years old. It stays moderately stable for several years thereafter, but then begins to decline, with women losing bone faster than men, and particularly after the menopause. Then it might be as much as 1 percent per year. This can be reduced by hormone replacement therapy (HRT), as well as the drugs bisphosphonate and calcitonin. This latter drug supplements the natural calcitonin secreted by the thyroid gland, specifically to prevent bone loss. Fluoride therapy is used in some countries and, given at a rate of 20 milligrams a day, it strengthens bone by forming a tougher kind of calcium phosphate known as fluoroapatite, $Ca_5(PO_4)_3F$.

Calcium supplementation to the diet may even benefit children, as shown by Conrad Johnston of Indiana University, USA. He took 60 pairs of identical twins, aged 6–14, and gave one of the pair a tablet of calcium every day for three years. These children showed accelerated bone growth.

There are drugs which control the movement of calcium through cell walls. One such is verapamil which blocks this thereby promoting relaxation of the muscles lining blood vessels and is a way of treating high blood pressure, and angina.

Dangerous calcium

For most people calcium does not pose a threat no matter how much of it we take into our body, but for some people this element can kill. This was discovered in Australia where a student with a broken leg told the medical team who were about to operate on him that he was very worried because 10 of his relatives who had been anesthetised had died on the operating table. Consequently he was carefully monitored and it became obvious within minutes of his being given an anaesthetic that something was wrong. His heart began to race and his temperature soared. His life was saved. Investigation showed that he suffered a rare genetic condition now known as malignant hyperthermia. In this the body responded to chemicals like halothane anaesthetic by opening all the calcium channels. Calcium floods out of cells and triggers all kinds of bodily processes which run out of control.

Element of history

Lime (aka quicklime, calcium oxide, CaO) was used by all the ancient civilisations to make mortar for building and it is mentioned in several books of the Roman period. It was obtained by heating limestone (calcium carbonate, $CaCO_3$). Mortar was made by mixing lime with sand and water and for a while it would remain soft and workable, but then it would set hard, and its hardness increases over time as it absorbed carbon dioxide from the air to form calcium carbonate. Adding water to lime converted it to so-called slaked lime (calcium hydroxide, $Ca(OH)_2$) which was also a useful material because of its acid-neutralising ability and soil improving quality.

In 1755 Joseph Black (1728–99) of Edinburgh proved that when limestone was heated to form lime, it gave off carbon dioxide ('fixed air' was his name for this gas) and the great chemist Antoine Lavoisier (1743–94) listed lime as an element because it seemed impossible to reduce it to simpler components, but he also suspected that it was the oxide of an unknown element. This was proved in 1808 when Humphry Davy (1778–1829) of the Royal Institution in London, obtained calcium metal by electrolysis. He had produced sodium and potassium this way the previous year and confidently thought that he could do the same by mixing lime with potash, heating the mixture till it melted, and then passing an electric current through it, but all this produced was potassium. So he tried a mixture of molten lime and mercury oxide and this was a little more successful since it produced an amalgam of calcium and mercury, but not enough to do anything with.

In May 1808 he got a letter from Jöns Jacob Berzelius (1779–1848) who told of similar experiments he had conducted with Dr. M. Pontin, physician to the King of Sweden. They also produced the amalgam, but more of it. This prompted Davy to try again, this time using a higher ratio of lime to mercury oxide, and this produced enough of the amalgam for him to be able to distil off the mercury leaving the calcium behind, albeit still contaminated with a little mercury. However, it was possible for him to record some of its properties. Perhaps not surprisingly, it was not until the early 1900s that a commercial method of making calcium metal was perfected.

Economic element

Many calcium minerals are known: anhydrite, which is calcium sulfate ($CaSO_4$); calcite, limestone and aragonite, which are all forms of calcium carbonate ($CaCO_3$);[35] dolomite, which is a mixed calcium magnesium carbonate ($CaMg(CO_3)_2$); and gypsum, which is hydrated calcium sulfate ($CaSO_4.2H_2O$).

The chief mined ores are calcite, dolomite, gypsum (used in cement and plaster) and anhydrite (used to make sulfuric acid). World production of calcium metal is only 2,000 tonnes per year, but of lime it is around 120 million tonnes. Reserves of limestone are virtually unlimited. Calcium metal is produced by heating lime with aluminium metal in a vacuum.

When the bulk metal is exposed to the air it reacts also with nitrogen and a white coating of calcium nitride and oxide forms on the surface. For this reason it also serves as a 'getter' to remove residual amounts of air from vacuum tubes. The oxide/nitride coating serves to protect calcium and keep it stable, and stable enough even to be machined. The metal is used in the manufacture of zirconium, thorium and the rare earth metals. It is used as an alloying agent for aluminium, beryllium, copper, lead, and magnesium alloys.

Lime has many uses of which the key ones are in metallurgy, in water treatment, in the chemical industry, and for making cement. Most is used in steel making as a flux, where it reacts with the impurities in the molten iron ore to form a viscous slag

---

[35] Onyx is a type of calcite.

that can be tapped off. Another major outlet is in sewage and in pollution control to reduce the acidity of water.

Gypsum can also be used as a building material in the form of plaster, but first it has to be heated to remove its water molecules. In addition to covering walls and making plasterboard, it is used to protect broken limbs, when it is known as Plaster of Paris.[36] Alabaster is a crystalline form of gypsum that is soft and easy to carve, and is used to make sculptures, which, when polished, become semi-translucent which greatly adds to their beauty.

Cement is made by heating a mixture of limestone and clay in a kiln at 1,500°C, which drives off carbon dioxide leaving a mixture of calcium and aluminium silicates. These are then pulverized and gypsum added. When it is mixed with water, cement undergoes a complex series of reactions which have only recently been understood.

In 2008, Michel Barsoum announced that some of the 2.3 million blocks which make up the Great Pyramid of Giza were composed of a form of cement. He had analysed the limestone in some of the outermost blocks and this revealed components like calcium phosphate which were not present in the limestone from which most blocks had been cut. Whether the ancient Egyptians really had discovered cement seems unlikely but the ancient Romans certainly knew of it.

Calcium carbide (CaC_2) reacts with water to form acetylene (aka ethyne, C_2H_2) which was once widely used as an illuminating gas and in acetylene torches for welding.

Calcium hypochlorite is a stable solid which will generate hypochlorite ions (bleach) when added to water and this can provide an emergency supply of safe drinking water.

Environmental element

Calcium in the environment	
EARTH'S CRUST	41,000 p.p.m. (4.1%)
	Calcium is the 5th most abundant element.
SOILS	can be up to 5%, but on average is between 1–2% p.p.m.
SEAWATER	400 p.p.m.
ATMOSPHERE	minute traces.

Calcium is one of the most abundant elements in the Earth's crust, and the third most abundant metal. It is found mainly as sedimentary rocks formed by precipitation or by the accumulation of the shells of marine creatures over millions of years. Vast areas of calcium carbonate, in the crystalline form known as aragonite, are present as coral, and it has been estimated that coal reefs cover about 2 million square kilometres of the planet.

---

[36] Heating gypsum drives off most of its water and this is then ground to a fine powder. On adding water it forms a paste and slowly reverts to gypsum, eventually setting hard, which produces an ideal protective cast for broken limbs.

There is a so-called 'calcium cycle' which moves vast amounts of the element through the environment, from land to sea, to sediments, and then back to land as continents uplift. It is estimated that around 500 million tonnes of calcium move through the cycle each year. Analysing the ratio of the isotopes calcium-44 to calcium-40 in deposits that have precipitated from seawater has revealed that the calcium level in the oceans has varied over geological time.

Calcium carbonate and phosphate are not very soluble but some calcium dissolves in rain water that percolates through soil and rock and this leads to what is referred to as hard water. (If the water drips into an underground cave it slowly forms icicle-like stalactites.) Such water is thought to be beneficial for drinking since it acts as a source of this essential element. When used for cleaning purposes the hardness interferes with the cleaning agents, such a soap and detergents, in which case water-softening chemicals are needed to tie-up the calcium ions and prevent them interfering with the washing process.

Calcium is the fifth most abundant ion in seawater—sodium, chloride, magnesium, and sulfate are more abundant.

Chemical element

Data file	
CHEMICAL SYMBOL	Ca
ATOMIC NUMBER	20
ATOMIC WEIGHT	40.078
MELTING POINT	839°C
BOILING POINT	1484°C
DENSITY	1.6 g/cm³
OXIDE	CaO

Calcium is a silvery, relatively soft, metal and a member of group 2 (row 4s), aka the alkaline earth metal group, of the periodic table. Calcium metal is attacked by oxygen and water. Its preferred oxidation state is +2 as in compounds such as $CaCl_2$, and in solution it exists as Ca^{2+} ions. Its compounds are generally colourless.

Calcium has 24 known isotopes with mass numbers 34 to 57, and no known isomers. Two of the six calcium isotopes which occur in ores are radioactive. Calcium consists of various isotopes: calcium-40, which accounts for 97 percent, calcium-44 (2 percent); calcium-42 (0.65 percent), calcium-46 (0.2 percent), calcium-43 (0.15 percent), and calcium-48 (a mere 0.004 percent). This last isotope is also radioactive but with a half-life in excess of 5×10^{19} years (which is 50 *billion* billion years) and calcium-46 is radioactive with a half-life of 2.8×10^{15} years (which is 2.8 million billion years). The next longest-lived radioactive isotope is calcium-41 with a half-life of 103,000 years, and this finds use as a portable source of gamma rays.

Element of surprises

1. Limelight was developed by the Scottish engineer Thomas Drummond (1797–1846) in 1823 in order to help surveyors who wanted a source of light that could be seen over long distances. He found that when a jet of hydrogen, burning in oxygen, played upon a compact block of calcium oxide (lime) then it would glow brilliantly white. The light it emitted could be focussed by lenses and was visible for more than 60 miles. (The hydrogen was generated from sulfuric acid and zinc, and the oxygen produced by heating a mixture of manganese dioxide and potassium chlorate. The two gases were stored in bellows-shaped bags.)

   Drummond's invention was quickly seized upon by theatre managers and lighthouse keepers, but it was the former who brought it to the attention of the public when it was used to throw a sharp circle of light on to performers on the stage—and in this way calcium chemistry entered the English language in the phrase 'to be in the limelight', meaning to be the centre of attention.

2. Spraying calcium chloride onto peaches and nectarines increases their shelf-life by strengthening cell walls and this allows them to be stored for up to two weeks before being eaten.

3. Calcium has two 'doubly magic' isotopes: calcium-40 and calcium-48. Both have 'magic' numbers of protons, in this case 20, and magic numbers of neutrons, in this case 20 and 28 respectively. Tin is the only other element with two 'doubly magic' isotopes.

Californium

Pronounced kali-for-nee-um, and named after the university and state of California where it was first made.

French *californium*; German *Californium*; Italian *californio*; Spanish *californio*; Portuguese *califórnio*.

Essential element

Although it is highly radioactive, this element has been used to treat cancer, inspect the contents of sealed containers, and in prospecting for gold.

Cosmic element

Californium has been observed in outer space. It was produced by a Type II supernova, and was detected by researchers analysing the spectra lines of the material ejected during that event.

Element of life

Californium has no role to play in living things, nor could it ever have had such a role because of its rarity and intense radioactivity.

Element of history

Californium was first made in February 1950 at Berkeley, California, by a team consisting of Stanley G. Thompson (1894–1953), Kenneth Street Jr., Albert Ghiorso (1915–) and Glenn T. Seaborg (1912–99). They made it by firing helium at curium-242, which itself had only been made for the first time in 1944.[37] This experiment required the making of enough curium to form a large enough target for the high energy helium ions to stand a chance of hitting its atomic nuclei and thereby combining to form a new element of higher atomic number. Curium is intensely radioactive and it took the team three years to collect enough of it for the test, and even so only a few micrograms were used. Their experiment produced around 5,000 atoms of californium, but this was enough to identify that it was a new element.

Economic element

Californium is produced in milligram quantities in nuclear reactors as the isotopes californium-249 and californium-252 where it is formed by the bombardment of plutonium-239 with neutrons. The former has a half-life of 350 years, the latter 2.645

---

[37] The equation which sums up this process is $^{242}Cm + {}^{4}He \rightarrow {}^{245}Cf$ ($t_{\frac{1}{2}} = 44\,\text{min.}$) + ^{1}n.

years. Californium-252 is a strong neutron emitter and used in portable neutron sources, with suitable precautions. It is employed in moisture gauges, in core analysis in drilling oil wells, and in on-the-spot activation analysis in gold prospecting. (Bombarding mineral samples with neutrons produces other radioactive elements which are easily identified by the radiation they emit.) A microgram of this element emits around 2 million neutrons per second.

Californium-252 is also used in cancer therapy—see below—an application that was encouraged by the US Nuclear Regulatory Commission who set its price at a relatively low $10 per microgram. The amount of californium in the world could now be tens of grams as some reactors have produced this element at the rate of 0.2 grams per year.

Medical element

Californium-252 neutron therapy was first investigated at the University of Kentucky in 1972 as a possible treatment for cervical and brain cancers, and when it proved successful, clinical trials began in 1976. Women with cervical cancer were the ones to benefit most and it was possible to increase their life span of 80 percent of those with the disease by 10 years, provided it was diagnosed early on. Even a third of those with late-diagnosed cancers had a chance of surviving 10 years.

Treatment required the implanting of a tiny amount of californium in the tumour in a relatively short operation. Within days the cancer began to shrink in size under the devastating impact of neutrons bombardment from the element. The result was usually an end to the pain and bleeding associated with the disease.

Environmental element

There are five californium isotopes that occur naturally on planet Earth—californium-249 to californium-253—but only in minute trace amounts and only in uranium-bearing deposits. A process of neutron capture followed by beta decay of uranium and plutonium isotopes will produce isotopes of americium, curium, and berkelium and it is from these that californium is produced.[38]

Californium was also produced for hundreds of thousands of years in natural nuclear reactors. These started up about 1.8 billion years ago and came about as a consequence of natural geological processes which concentrated uranium to an extent that it would start and sustain a nuclear reactor. This happened at 16 locations in what is now Oklo, Gabon, in Africa. These reactors functioned continuously and by neutron capture and beta decay processes they produced the transuranium elements from atomic numbers 83 (neptunium) to 100 (fermium) including californium.

---

[38] Curium-249 beta decays to berkelium-249 which undergoes further beta decay to produce californium-249. Berkelium-249 captures neutrons to become berkelium-250; this undergoes beta decay to produce californium-250, which through successive neutron capture reactions can produce californium-251, -252, and -253. These californium isotopes will be quite rare because a high uranium concentration and many neutron-captures are required to produce them.

Chemical element

Data file	
Chemical symbol	Cf
Atomic number	98
Atomic weight	252.1 (isotope Cf-252)
Melting point	900°C (est.)
Boiling point	1470°C (est.)
Density	15 g cm^{-3} (est.)
Oxide	Cf_2O_3

Californium is a radioactive metal and a member of the actinide group (row 5f) of the periodic table. In 1960, 3 micrograms of californium oxychloride (CfOCl) were made, enough to be visible with the naked eye. A sample of the metal itself was produced by reacting californium-249 oxide with lanthanum metal. The metal is expected to be readily attacked by air, steam, and acids, but not alkalis. The preferred oxidation states are +2 ($CfBr_2$ has been made), +3 (the most common state as shown by the green salts CfF_3 and $CfCl_3$), and +4 (again as a green compound CfF_4). In solution californium exists only as the Cf^{3+} ion.

There are 20 known isotopes of californium with mass numbers 237 to 256, and one known isomer. The longest-lived isotope is californium-251 with a half-life of 898 years. The next longest-lived ones are californium-249 (half-life 350 years) and californium-250 (13.08 years).

Element of surprises

1. Californium-252's neutrons can be used to examine sealed metal containers to see if they contain dangerous contents such as explosives. Neutrons are neutrally charged and so can pass unhindered through solid matter because they are not deflected by nor attracted to the electrons and nuclei of other atoms. However, they can collide with a nucleus whereupon they are absorbed and make that atom radioactive. As the new isotope decays it emits specific gamma rays by which it can be identified. For this reason californium-252 is also used to identify the type of oil that is being located at the bottom of a deep well.

2. Californium-251 is fissile and a small amount of around 5 kg would produce a chain-reaction explosion. It was considered possible to make so-called 'pocket nukes', that is, nuclear weapons of small size and low yield. However, due to the fact that large amounts of this isotope are hard to produce, such weapons have remained on the drawing board.

Carbon

Pronounced kar-bon, the name is derived from *carbo* the Latin name for charcoal.
French *carbone*; German *Kohlenstoff*; Italian *carbonio*; Spanish *carbono*; Portuguese *carbono*.

Essential element

Carbon stands supreme as having the chemical properties on which all life depends, and on supplies of which all modern living depends. Its compounds constitute organic chemistry, the most important brance of that science. Even pure carbon itself is being turned into new forms with unique features.

Cosmic element

Carbon is formed within stars by the coming together of three alpha particles, which are helium nuclei.[39] Carbon is one of the major elements to be found in the interstellar void and it is emitted by dying stars. A particularly carbon-rich star is CW Leonis (known to astronomers as IRC+10216) which is surrounded by a haze of carbon, making it appear very faint. By analysing the radio and micro-waves which this cloud of cosmic carbon emits, it has been possible to identify many carbon molecules including some with large numbers of carbon atoms. It was during laboratory experiments to replicate the formation of such molecules that the carbon spheres, the fullerenes, were discovered by Robert Curl Jr., Harry Kroto, and Richard Smalley, and this work was recognised by the 1996 Nobel Prize for chemistry.

Examination by advanced analytical techniques of the Murchison meteorite which landed in Australia in 1969 has reveals a vast number of different carbon molecules within its core, and these are not there as a result of contamination by terrestrial sources of carbon such as microbes. How and where in the universe they were formed remains a tantalising mystery.

In 2010 astronomers discovered that a planet (called WASP-12b) which was orbiting a star 1,200 light years from Earth contained a lot of carbon. It orbits close to the star, so much so that its surface temperature is more than 2,000°C. Most of its carbon is assumed to be in the diamond-form and located in its interior.

---

[39] In terms of protons this is 3×2 (He) = 6 (C).

Element of life

Carbon in the human body*	
BLOOD	varies minute by minute
BONE	0.8%
TISSUE	67% (based on dry matter)
TOTAL AMOUNT IN BODY	16 kg.

*This refers to the total weight of carbon in all its myriad forms, and as such has little meaning except on a comparative basis with other elements.

No element is more essential to life than carbon, and the reason is that only carbon forms bonds to itself that are strong enough and stable enough to resist chemical attack under ambient conditions. This gives carbon the ability to form long chains and rings of atoms, and these are the structural basis for many compounds that comprise the living cell, of which the most important is DNA.

Food element

Almost all that we eat—carbohydrates, fats, proteins, and fibre—is made up of compounds of carbon, and provides us with a total carbon intake of 300 g a day. Digestion consists of breaking these compounds down into simpler molecules that can be absorbed through the walls of the stomach or intestines. There they are transported by the blood to sites where they are utilised, some to supply the energy for warmth and movement, some to repair or replace damaged cells, some to be turned into body messengers, or to make the thousands of different chemicals that our body needs. Most carbohydrates are oxidised to release the energy they contain and then we breathe out the carbon as CO_2. This joins the other carbon dioxide in the atmosphere, from where it will again be extracted by plants and become part of the carbon cycle of nature—see below.

Dangerous element

In its elemental forms carbon is non-toxic, but some simple compounds can be very toxic, such as carbon monoxide (CO) and hydrogen cyanide (HCN). Both can kill within minutes, the former by blocking the haemoglobin of the blood so it cannot transfer oxygen to where it is needed, the latter by blocking the enzyme which the heart muscle uses to convert glucose to energy. Yet despite these dangers, the human body itself produces and uses both of these molecules, albeit in only safe amounts and as required. Carbon monoxide possesses the ability to lower blood pressure and dilate blood vessels, and cyanide can be involved in one form of vitamin B_{12}.

Many other molecules containing carbon, both natural and man-made, can be highly toxic and indeed their study forms a major part of the science of toxicology. Even carbon itself can pose a risk. Carbon black can be an irritant dust and there are fears that nanotubes particles—see below—could lodge in the lungs, as do tiny asbestos fibres, and likewise pose a long-term health risk.

Element of history

Carbon occurs naturally as anthracite (a type of coal), graphite, and diamond; these were known to the ancient civilisations. More readily available carbon was that produced by fire, such as soot or charcoal. Charcoal burning, which consists of heating wood in a limited supply of air, evolved to become an important trade supplying the industries that manufactured metals, glass, and pottery. Charcoal burning is still carried on in parts of the world to this day.

It was with the dawn of chemistry that these various forms of carbon were recognised as the same element. Many of the great chemists of the 1700s wrestled with the problem. Not surprisingly it was diamond that posed the greatest difficulty of identification. Naturalist Giuseppe Averani (1662–1738) and medic Cipriano Targioni (1672–1748) of Florence were the first to discover that diamonds could be destroyed by heating. In 1694 they focussed sunlight on to a diamond using a large magnifying glass and the gem eventually disappeared.

The experiment was repeated in July 1771 by the French chemist Pierre-Joseph Macquer (1718–84) and Godefroy de Villetaneuse (c.1735–c.1810) and they observed that the heated diamond burnt completely away leaving no ash. But it was not until 1796 that the English chemist, Smithson Tennant (1761–1813), proved that diamond was entirely a form of carbon and that it burned to form only CO_2.

Others argued that if diamond was carbon then it should be possible to convert other forms of carbon into the gem. However, early experimenters were unaware of the high temperatures and intense pressures that were needed to achieve this transformation, so we know that early claims to have made diamonds were false. Even the eminent chemist Henri Moissan (1852–1907), who claimed to have done this in 1893 and exhibited a stone 0.7 mm in size, was deluded. That diamond had been surreptitiously slipped into his experiment by one of his laboratory assistants.

Element of war

Gunpowder is a mixture of charcoal, sulfur, and saltpetre (potassium nitrate, KNO_3), and was for many centuries the explosive used in guns and cannon. Today it is no longer used in firearms and weapons, although it is still the basis of many fireworks. A fuller account of gunpowder is given under sulfur.

Economic element

The scale on which carbon in its various forms is extracted from the Earth's crust is far greater than that of any other element. Technically all the mining of limestone, dolomite and marble is the mining of carbon as carbonate, its most oxidised form. Similarly, the extraction of fossil reserves, coal, oil and gas, is equivalent to mining carbon, in this case reduced carbon.

World production of diamond is measured in carats, a strange unit of weight that originally corresponded to the weight of 24 carob seeds.[40] Today, it is defined as 200 milligrams. The total amount of diamond produced is 170 million carats (34 tonnes) and the major producing countries are, in order of importance, Russia, Botswana, Congo, Canada, and South Africa which together account for almost 80 percent of the total.

Today most diamonds are manufactured, although these are suitable only for industrial purposes. The first synthetic diamonds were made by the Swedish company ASEA in 1953 using temperatures of 3,000°C and pressures of 90,000 atmospheres, but they did not publish their results. The General Electric Company in the USA made diamonds using similar conditions and they announced their success in 1955. They went on to develop the process to the extent of fabricating gem-quality diamonds, although these are very costly to make. Today, Russia is the main producer of synthetic diamonds.

Diamond films can be made by a process known as chemical deposition in which a simple carbon-containing molecule such as methane is decomposed to its atoms and these will crystallise out as diamond onto the surface to be protected, such as a razor blade. The trick is to get the film to grow into a perfect array of carbon atoms with the same chemical bonding between them as in diamond and this is achieved by adding hydrogen gas to the methane before it is decomposed. The resulting film of carbon is invisible to the naked eye, but it gives the surface the same quality of strength and hardness of real diamond.

Less well known is the extraction of another relatively pure form of carbon: graphite. This is mined mainly in China, Brazil, Canada, and Mexico, with other countries such as Sri Lanka, Madagascar, and Russia producing small but significant amounts. Madagascar produces a grade of graphite that is suitable for making crucibles that can hold molten metal. World production of graphite is around 1 million tonnes a year of which China produces 70 percent. Graphite finds use as brushes in electrical motors, electrodes in fuel cells, furnace linings, extrusion dies, and even the so-called 'lead' of pencils.

Carbon itself is mainly used in its amorphous forms: as coke in steel making to reduce iron oxide to iron, as carbon black in printing and as a filler for tyres, and as activated charcoal in sugar refining, water treatment, and in respirators. Carbon black is made by burning natural gas in a limited supply of air, whereas coke is made by heating coal in the absence of air. Activated charcoal is produced by heating coconut shells, and this, in granulated form, it is used as an absorbent to filter out pollutants from gases and water, and is to be found in respirators, and in extractor hoods for kitchens.

Carbon fibre is produced by the controlled heating of acrylic polymer fibre until it chars to a carbonised material which is stronger than steel. This can be woven into fabrics which can absorb poisonous gases and these are used for protective clothing and fire-hoods. Carbon fibre is used to make laminates that combine great strength with low weight and are used in rockets and aircraft, and to reinforce plastics for sports equipment such as skis, fishing rods, and racquets.

---

[40] In Roman times these were used to check the standard gold coin which, when pure, weighed the same as 24 carob seeds. This is why we still refer to pure gold as 24 carat gold.

Carbides such as iron carbide (aka cementite) and tungsten carbide are very hard and they find use in applications such as abrasives and cutting tools.

In the 10 years since the first edition of this book there has been remarkable progress in two kinds of carbon: nanotubes and graphene. Both have potentially useful applications in nanotechnology and in electronics. Both are forms with the chemical structure of graphite in which every carbon is attached to three other carbons—see *Chemical element* below. Nanotubes are minute tubes of rolled-up graphite having single, double, or multiple walls, whereas graphene consists of flat single or multiple layers as in graphite itself, and these can be extracted from graphite layer by layer. Both nanotubes and graphene are good conductors of electricity and indeed graphene offers an alternative to indium tin oxide as the transparent but conducting material needed for screens and solar panels. Methods of producing graphene on an industrial scale, and even attaching it to flexible materials, are now available. Nanotubes can be cross-linked to form thin films sometimes referred to as 'black paper' and these too are likely to play a part in future high-tech electronics such as touchscreen displays and e-books.

Environmental element

Carbon in the environment	
EARTH'S CRUST	480 p.p.m.
	Carbon is the 15th most abundant element.
SEAWATER	28 p.p.m.
ATMOSPHERE	as CO_2 395 p.p.m.; as CH_4 1.9 p.p.m.; and as CO 0.1 p.p.m. These are global averages and vary depending on location.

The carbon of the Earth comes in several forms, as fully oxidised in the form of carbonate, CO_3^{2-} in rocks, of which there are 100 million trillion tonnes (10^{20}) or as hydrogen carbonate, HCO_3^- in the seas, or as reduced carbon in fossil deposits. In the last of these it is present as hydrocarbons as in natural gas, which is CH_4, or oil which approximates to CH_2, or coal which has a carbon-to-hydrogen ratio of about 1:1 although this is a particularly complex material. Known reserves of these fossil fuels amount to a vast 825 billion tonnes of coal, 170 billion tonnes of oil,[41] and 185 trillion cubic metres of gas; known reserves continue to increase slightly as more reserves were discovered than are being used up, although this is not likely to continue for much longer. Humans extract and use about 10 billion tonnes of fossil carbon each year.

Most of the reduced carbon in the Earth's crust is too widely dispersed to be economically exploitable and is known as kerogen, but altogether there is an incredible 375,000 billion tonnes of it, only a few percent will ever be accessible as fossil fuel.

---

[41] Oil is more usually measured in barrels of which there are 7.33 barrels in a tonne. A barrel has 159 litres so a tonne of oil corresponds to 1,165 litres.

Carbon

Carbon is probably the most important element from an environmental point of view. The Earth's early atmosphere may have contained a lot of carbon dioxide and methane but once life evolved this began to change.[42] Today, there is very little of these gases and a lot of oxygen instead, thanks chiefly to the action of plants which convert CO_2 and H_2O into carbohydrate and oxygen by photosynthesis. The Earth's atmosphere contains an increasing concentration of carbon dioxide and carbon monoxide, from fossil fuel burning, and of methane, from paddy fields and cows. Carbon dioxide also has a controlling effect on the acidity of rainwater, since this gas dissolves to form carbonic acid which leads to a pH of 5.7.

Marine species, especially algae, are thought to have been responsible for deposits of oil and gas, while plants on land laid down the layers of rotting vegetation that first turned into peat, then lignite, then bituminous coal, and finally anthracite coal which is almost pure carbon. The time to complete this geological sequence of events is around 250 million years.

The Carbon Cycle is the term used to describe the way in which this element moves between various parts of the terrestrial ecosphere (so-called 'reservoirs'). There are large scale transfers between these which rule the tempo of life on Earth, and 200 billion tonnes of carbon move from one reservoir to another every year. The amounts or carbon in the various 'reservoirs' are as follows:

Atmosphere: 724 billion tonnes.

Living things on land: 2,000 billion tonnes.

Oceans: 39,000 billion tonnes (mainly as dissolved hydrogen carbonate).

Living things in the seas: 40 billion tonnes.

Perhaps more important than movement between these reservoirs is the movement which occurs when carbon becomes part of the land-based and marine-based biological cycles. The first step in these cycles is the absorption of carbon dioxide by plants, which then provide food for animals. In this way carbon is passed up the various food chains, with each recipient releasing some of what it receives as CO_2, until most carbon is back where it started. Similar food chains exist among the species in the oceans, although there some of the carbon ends up as carbonate and settles down to the bottom sediments where it will remain for millions of years.

Most of the fossil carbon extracted every year is burnt mainly to generate energy in the form of heat or motion, although a few percent is used by the chemical industry to make things like plastics and fibres. In a sustainable future this source of carbon would not be used and all the carbon we require would have to come from biomass, and mainly grown as crops. However, Nature produces around 200 million tonnes of biomass a year of which carbon constitutes 40 billon tonnes. All it needs are chemists to convert some of it to useful products and we could become sustainable.[43]

---

[42] Other planets also have a lot of carbon in their atmosphere: on Venus the atmosphere is almost all carbon dioxide, while on Jupiter and Saturn it is methane.

[43] This is the theme of my book *A Healthy, Wealthy, Sustainable World*, published by the Royal Society of Chemistry, London in 2010.

The misuse of carbon is invoked as the main threat to humanity and indeed most of that which is extracted from the Earth's crust is burnt in wasteful ways. However, this contentious topic is beyond the scope of this book, being more to do with politics than science.

Chemical element

Data file	
CHEMICAL SYMBOL	C
ATOMIC NUMBER	6
ATOMIC WEIGHT	12.0107
MELTING POINT (DIAMOND)	3,550°C; sublimes at 4,800°C
DENSITY	graphite 2.3 g/cm³, diamond 3.5 g/cm³
OXIDES	CO and CO_2

The chemistry of carbon compounds is referred to as organic chemistry and its study constitutes the bulk of the subject, with more than 30 million different organic molecules having been made and recorded.

Carbon comes at the head of group 14 (row 2p) of the periodic table of the elements. When pure it occurs in various forms: graphite, diamond, fullerenes (of which buckminsterfullerene C_{60} is the best known), lonsdalite (see *Element of surprises*), carbon nanotubes, and graphene. Rather oddly, diamond is less stable than graphite, but the rate of conversion is immeasurably slow. More mundane forms of carbon, such as carbon black, are amorphous, that is, they have no well defined structure. Carbon displays formal oxidation states ranging from −4 (as in CH_4) to +4 (as in CF_4 and CO_2) although this concept has little relevance for carbon.

Elemental carbon is an element of extremes. Thus diamond is hard, transparent, an electrical insulator, and highly heat conducting, while graphite is soft to the point of being a lubricant, black, electrically conducting, and thermally insulating. Diamond consists of carbon atoms bonded to each other in a three dimensional array, each carbon forming four bonds to neighbouring carbons, and that's what gives it strength, rigidity and hardness. Graphite consists of carbon atoms in sheets; each carbon bonds to three neighbours, with the sheets stacked one on top of the other, in a rather weak array. This is what makes graphite easy to cleave, why it feels slippery, and why it is used as a lubricant. It also makes graphite able to conduct electricity in the direction of the sheets. Fullerenes have similar chemical bonding to graphite but instead of the sheets being flat they are rolled up into hollow balls or tiny tubes called nanotubes. They are less stable and on exposure to the air they are slowly oxidised.

There are 15 known isotopes of carbon with mass numbers ranging from 8 to 22, and no known isomers. There are two main isotopes of carbon: carbon-12, which makes up 99 percent of all carbon, and carbon-13, which accounts for 1 percent. These are non-radioactive isotopes, but a third natural form, carbon-14, is radioactive with a half-life of 5,730 years, and there is also a trace of this in the atmosphere—see box.

All atomic weights are defined with respect to the carbon-12 isotope whose mass is defined as exactly 12.00000...

Carbon dating

The American scientist Willard F. Libby developed a method of dating ancient relics by analysing the organic matter they contained for the radioactive carbon-14 content. He received the Nobel Prize for chemistry in 1960 for his work, which has proved invaluable not only for archaeologists but for detecting forgeries. Even fake wine has been uncovered this way, by its *lack* of radioactivity.

Carbon-14 is produced by cosmic rays bombarding nitrogen in the upper atmosphere and it is radioactive. About 7 kilograms of carbon-14 are produced per year which may not sound very much but it is almost a trillion trillion atoms, which is enough to ensure that it becomes incorporated into all living plants. When the plant dies the amount is fixed and no more carbon-14 is added. If it is a tree, and used in building or furniture, or if it is a fibrous plant, and used to make paper or textiles, then these can be dated by the amount of residual carbon-14 they contain.

The half-life of carbon-14 is 5,730 years, which means that objects from the ancient civilisations can be dated to within a hundred years or so. It also enables forgeries to be recognised and in this way the Round Table of Arthur, which is in Winchester Castle, was shown to be a forgery, as was the Turin Shroud.

Fake wine and spirits can be easily made from industrial alcohol (aka ethanol, CH_3CH_2OH) which is produced by the chemical industry. Because the carbon in this has been so long underground all its carbon-14 has disappeared and so it produces a drink whose alcohol also lacks radioactivity. Showing this to be so proved that the drink is phoney—although it will be just as potent. Today, alcohol is produced from maize and sugarcane on a massive scale for use a biofuel. No doubt some of this could be diverted to illegal use and this would not be detectable by its lack of carbon-14.

Element of surprises

1. Diamonds are called 'ice', not because they look like ice, but because when a large diamond is pressed against the lips it feels cold. The reason is its high thermal conductivity, which quickly drains heat to itself.

   Not all diamonds are crystal clear. In fact they can be anything from pale brown ('champagne') to green, the rarer orange, and the rarest of all, violet, which gets its colour from hydrogen atoms. Only about one diamond in a thousand is naturally coloured and they command high prices. Coloured diamonds can be produced by natural radiation in the Earth, when they can take on light green or blue colour, or they can be artificially irradiated in a nuclear reactor, when they take on a spectrum of hues.

2. The town of Nördlingen in Bavaria is built of stone that contains millions of minute diamonds that were formed when a 1 kilometre meteor crashed there about 15 million years ago creating what is known as the Reis crater. Such was the force of the impact that it turned a local deposit of graphite to diamond, creating an estimated 72,000 tonnes of them, but all less than 0.2 mm across. However, when a meteor containing graphite hits the Earth, as it did in Canyon Diablo, Arizona, then the force of the impact converts its graphite into a new brown mineral called lonsdalite which is like diamond but which retains the hexagonal carbon rings of its parent graphite although these are no longer flat but bent. Lonsdalite is not as hard as diamond. It has also been found at other meteor impact sites.

Cerium

Pronounced seer-iuhm, it was named after the asteroid, Ceres, which had been discovered two years prior to the discovery of the element in 1802, and which was itself named after *Ceres,* the Roman goddess of agriculture.

French, *cérium*; German, *Cer*; Italian, *cerio*; Spanish *cerio*; Portuguese, *cério*.

Cerium is one of 15 chemically similar elements referred to as the lanthanoid or rare earth elements, which extend from element atomic number 57 (lanthanum) to element atomic number 71 (lutetium). The term rare earth is a misnomer because some are not rare. The minerals from which they are extracted, and the properties and uses they have in common, are discussed under Lanthanoids, page 266.

Essential element

Cerium sulfide is the preferred red pigment for plastics, and cerium is also used in flatscreens and long-life, low-energy light bulbs.

Element of life

Cerium in the human body	
BLOOD	*approx.* 1 p.p.b.
BONE	*approx.* 3 p.p.m.
TISSUE	*approx.* 0.3 p.p.m.
TOTAL AMOUNT IN BODY	40 mg.

In 1880, the Italian Alfonso Cossa (1833–1902) reasoned that because cerium was often found in calcium phosphate ores, then it might also be found in bone, which is mainly calcium phosphate. He analysed bones and found traces of cerium. Cerium can mimic calcium and it was soon found in measurable amounts in barley, beech wood, and tobacco plants. This metal is not generally taken up by plant roots so little gets into the main food chain. The small amount that does is not considered a heath risk. Like all the lanthanoids, cerium is regarded as non-toxic.

As yet, no one has monitored dietary intake for cerium content, so it is difficult to judge how much is consumed. It is probably less than a milligram per day, but it is present beause urine tests positive for this element. Cerium has no known biological role, but it has been noted that cerium salts stimulate metabolism.

Dangerous element

Cerium is not toxic when taken by mouth but animals that were injected with large doses of cerium suffered heart attacks and died. It is more of a threat to aquatic species because it damages cell membranes, but thankfully there is almost no cerium in natural waters or seawater.

Medical element

In 1854, an eminent medic at the Department of Medicine and Midwifery at the University of Edinburgh, Professor Sir James Simpson (1811–70), reported that cerium nitrate prevented vomiting and especially that associated with the morning sickness of early pregnancy. He recommended a dose of one grain (65 mg) three times a day. He claimed it was also good for treating stomach pains and other gastrointestinal disorders.

Soon others were trying this new medicament and reporting excellent results, while some preferred to prescribe cerium oxalate which, being insoluble, could be given at 10 times Simpson's dose. This was particularly recommended for sea-sickness and by the end of the century medical pharmacopoeias were even suggesting doses of a gram at a time. Tests on animals showed that even higher levels were harmless although some research suggested that extremely high doses caused the very symptoms it was prescribed to alleviate, that is, nausea, vomiting, and diarrhoea. Meanwhile some doctors found it acted as a cough suppressant for people with tuberculosis.

Commercial preparations of cerium salts were on sale as Novonaurin and Cerocol tablets which continued to be manufactured well into the last century. Eventually it fell out of favour and looked likely never to return, until in 1995 a weak solution of cerium nitrate was found to be an effective first treatment for bathing the skin of people suffering extensive third degree burns. This is now standard procedure in some specialist burns units.

Element of history

Cerium was first identified by the Swedish chemist Jöns Jacob Berzelius (1779–1848) and the geologist Wilhelm Hisinger (1766–1852) in the winter of 1804. The German chemist Martin Klaproth (1743–1817) independently discovered it around the same time, and he wanted to call it cererium, again linking its name to the asteroid Ceres.

Although cerium is one of the lanthanoid elements it was discovered independently of them because there are some minerals that are almost exclusively cerium salts. Cerite, which is mainly cerium silicate, comes into this category. A lump of this reddish-brown mineral had been found in 1751 by the Swedish chemist Axel Cronstedt (1722–65) at a mine in Vestmanland, Sweden. He thought it was an iron ore, albeit strangely heavy, and when tungsten was reported in 1783 it seemed likely that it was an ore of that metal. Hisinger, whose aristocratic father owned the estate on which the mine stood, first sent a sample to Carl Scheele (1742–86) the famous chemist, to analyse it but he failed to find any tungsten. Hisinger remained curious so, in 1803, he and Berzelius worked on it and proved that it was the ore of a new element. Shortly thereafter, a black ore from Greenland, called allanite (an aluminium iron silicate) was also found to contain a lot of cerium.

There were several attempts to get a sample of the pure metal itself, but the usual procedures did not work, even heating cerium chloride with potassium failed. This produced an impure brown powder which probably contained some metallic cerium. It was not until 1875 that two American chemists, William Hillebrand and Thomas Norton, first obtained a pure specimen of the metal by passing an electric current through molten cerium chloride.

Economic element

Cerium comes mainly from rare earth ores but some is obtained from perovskite, a titanium mineral, and from the aluminium mineral allanite, both of which can have enough cerium to make them viable sources.

Cerium oxide, CeO_2, is the most important outlet for cerium. It is produced by heating bastnäsite ore, which oxidises the cerium to its insoluble oxide and it remains behind when the other metals are dissolved out using hydrochloric acid. The insoluble residue is known as cerium concentrate and production amounts to about 63,000 tonnes a year, but this is likely to increase this century as more uses are found for cerium, and existing uses expand.

During most of the 1800s, cerium and its compounds were considered to be of little interest outside the chemistry laboratory, but early the last century it finally became important in the production of incandescent mantles for gas lighting—see box.

The gentle glow of the gas-lit era

Karl Auer (1858–1929) discovered that thorium oxide would glow with a brilliant light when heated. Early supplies of town gas (coal gas) relied on traces of benzene to give a bright luminous flame, but once benzene was removed to be sold to chemical works as a raw material, an alternative way of boosting the luminosity of gaslight had to be found.

Auer came up with the gas mantle which did just that. It was made from a closely woven fabric impregnated with a solution of thorium nitrate and then baked to decompose it to the oxide. When the mantle was then heated in a gas flame it emitted a broad band of light centred on a wavelength of 500 nm. This is towards the blue-green end of the spectrum and the light was considered rather harsh. It was softened by adding 1 percent cerium nitrate to the thorium solution, and the cerium oxide this formed produced not only a more pleasing light, but it acted as a catalyst ensuring more complete combustion of the gas.

Cerium oxalate, when heated, decomposes to cerium oxide, and this is the way a pure form is made industrially. Cerium oxide (1 to 2 percent) protects glass against damage by radiation, X-rays, and cathode rays, and it will also filter out UV light. Most damage caused by light is due to UV radiation of wavelengths below 400 nm, but these are strongly absorbed by cerium. Optical glass for lenses also includes cerium, along with lanthanum and yttrium, to provide a high refraction index and low colour dispersion. Cerium oxide concentrate, as a slurry in water, is used in place of ferric oxide rouge for polishing high-grade and optical glass, giving a better finish more quickly to lenses and screens.

Cerium is important as a catalyst and used by the chemical industry in the production of alcohols, phenols and ketones. Cerium oxide is part of catalytic convertors which clean up vehicle exhausts and it enhances the performance of the other metals present. These anti-pollution devices consist of a ceramic or metal substrate, a coating of aluminium and cerium oxides, and then a layer of a finely dispersed metal such as platinum or rhodium which is the active surface. However, the cerium oxide plays more than just a support role, and when its catalytic effects were realised the amount of was increased. Researchers at the University of Trieste, Italy, have shown that cerium oxide works its magic by having sites on its surface where two oxide ions are missing leaving behind an active sits that consists of Ce^{4+} ions surrounding Ce^{3+} ions.

Minor uses of cerium oxide include self-cleaning ovens where it is incorporated into the walls and there catalyses the oxidation of cooking resides, which are mainly carbon. Cerium is also used to clean up the emissions from diesel engines. These emit so-called particulates, which are micron-sized carbon particles.

Until recently, cadmium red was the preferred choice of red pigment for containers, toys, household wares, and crates, and it replaced pigments derived from toxic heavy metals, such as lead and mercury, but cadmium itself is now considered environmentally undesirable and its pigments are being phased out in favour of cerium sulfide, Ce_2S_3. This gives a rich red colour, stable up to 350°C, and is completely non-toxic. It is made by heating cerium metal vapour in an atmosphere of sulfur. By adding traces of other rare-earth metals it is possible to produce a range of colours from deep maroons, through brilliant reds, to bright orange.

Cerium is used in many other ways, such as in flat screen TVs, low-energy light bulbs, magnetic-optic CD disks, and it is part of the core material in carbon-arc electrodes for film studio lights and for flood lighting. Cerium is also used in chromium plating, where a little is added to the electrolyte solution of the plating baths and assists in the plating process by preventing the formation of the Cr^{3+} ions which are difficult to reduce electrolytically to the metal.

There is only a small demand for cerium metal itself, which can be obtained by heating cerium fluoride with calcium, or by passing an electric current through molten cerium oxide. Cerium metal is added to aluminium to improve its corrosion resistance.

Cerium

Environmental element

Cerium in the environment	
EARTH'S CRUST	68 p.p.m.
	Cerium is the 25th most abundant element.
SOILS	approx. 50 p.p.m., range 2–150 p.p.m.
SEAWATER	1.5 p.p.t.
ATMOSPHERE	virtually nil.

As far as we know, cerium poses no threat to the environment. Indeed it is prov-ing essential in protecting it because it has three environmentally friendly virtues as indicated above: the oxide helps clean up vehicle exhausts, the red sulfide reduces the need to use toxic heavy metal pigments, and cerium is an essential is part of long-life, low-energy light bulbs.

Chemical element

Data file	
CHEMICAL SYMBOL	Ce
ATOMIC NUMBER	58
ATOMIC WEIGHT	140.116
MELTING POINT	799°C
BOILING POINT	3,426°C
DENSITY	8.2 g/cm^3
OXIDES	Ce_2O_3 and CeO_2

Cerium is a reactive, grey metal and a member of the lanthanoid group (row 4f) of the periodic table. It tarnishes in air, burns if scratched with a knife, reacts rapidly with water, and dissolves in acids. The metal is usually stored in light mineral oil to protect it against oxidation. Its preferred oxidation state is +3. Most of its salts are soluble but not cerium fluoride and cerium oxalate.

Cerium stands out from the other lanthanoids in having a stable higher oxidation state of +4 and its salts are generally yellow, orange, or red. In solution it exists as yellow Ce^{4+} ions. Ce^{4+} reagents have played an important part in analytical chemistry as powerful and stable oxidising solutions.

There are 39 known isotopes of cerium of atomic mass range 119 to 157, and 10 known isomers. There are four naturally occurring cerium isotopes in ores: cerium-140, which comprises 88.5 percent, cerium-142 (11 percent), cerium-138 (0.3 percent) and cerium-136 (0.2 percent). Although previously thought to be stable, it now turns out that only isotope-140 is stable; the others are radioactive albeit with exceedingly long half-lives: that of cerium-142 is 5×10^{16} years (equivalent to 50 million billion years), of cerium 138 is 1.5×10^{15} years (equivalent to 1.5 million billion years), and of cerium-136 is 0.7×10^{14} years (equivalent to 70 trillion years). There are also minute traces of other naturally occurring radioactive cerium isotopes produced in uranium-bearing deposits as fission fragments.

The next longest-lived radioactive isomer after those referred to above is cerium-143 with a half-life of 33 hours.

Element of surprises

Although cerium is one of the so-called rare earth metals it is the most abundant of them and indeed it refutes the lanthanoid elements to be so called. In the Earth's crust cerium is almost as common as zinc, four times more abundant than lead, and 30 times more abundant than tin.

Chlorine

Pronounced klor-een, the name is derived from the Greek, *chloros* meaning greenish yellow.

French *chlore*; German *Chlor*; Italian *cloro*; Spanish *cloro*; Portuguese *cloro*.

Chlorine and chloride: the former is the name of the element which exists as the molecule Cl_2, whereas chloride, as in sodium chloride, is a chlorine atom that has acquired a negative electron and is written Cl⁻. Cl_2 is reactive and dangerous, while Cl⁻ is stable and relatively safe.

Essential element

This is the element we have a love/hate relationship with. We enjoy its disease-preventing benefits and its remarkable PVC plastic, but worry about its drawbacks such as salt in the diet and the persistence of organochlorine compounds in the environment.

Element of life

Chlorine (as chloride) in the human body	
BLOOD	*av.* 0.3%.
BONE	900 p.p.m.
TISSUE	*approx.* 0.2–0.5%.
TOTAL AMOUNT IN BODY	95 g.

Chloride, Cl⁻, is essential to many species, including humans. Plants contain varying amounts and it is an essential micronutrient for higher plants where it concentrates in the chloroplasts. Cereals tend to absorb little chloride, having around 10–20 p.p.m. (dry weight) while potatoes can have levels of up to 5,000 p.p.m. Growth suffers if the amount of chloride in the soil falls below 2 p.p.m. but this rarely happens. The upper limit of tolerance varies according to the crop, with beans and apple trees suffering if levels exceed 700 p.p.m. while tobacco, tomatoes, cotton, and beet can cope with levels above 3,000 p.p.m.

Food element

The daily intake of dietary chloride is mostly in the form of common salt (sodium chloride, NaCl) and is on average around 6 grams, which is double the amount we really need. However, it is not the chloride component but the sodium which is seen as potentially dangerous, and especially for those prone to high blood pressure and heart disease—see page 502. Nevertheless food manufacturers have reduced the salt level in many processed foods and many people now try to avoid salt, to such an extent that doctors have warned that people who undertake vigorous exercise and/or drink excessive amounts of water, need to *increase* their salt intake to stay healthy.

Dangerous element[44]

Despite hydrochloric acid (HCl) being classed officially as a dangerous chemical, it is produced at relatively high concentrations in the human stomach to help break down food and destroy potentially dangerous bacteria. The element itself, chlorine gas (Cl_2), is very toxic. It immediately affects eyes and lungs at a concentration of only 3 p.p.m. in air. At 50 p.p.m. it is dangerous even for a short time; and breathing air with 500 p.p.m. for five minutes would almost certainly be fatal—as its use in World War I as a weapon to attack troops showed—see below. Symptoms of breathing chlorine gas are intense pain in the respiratory system, stinging of the eyes, and vomiting. Even when it is not fatal, within a short time it will cause permanent damage to the lungs.

Somewhat surprisingly, there is evidence that the white blood cells that defend the body against infection may use chlorine gas to do this; a paper in the *Journal of Clinical Investigation* in 1996 reported that these cells were capable of producing this deadly gas.

One of the worst peacetime disasters involving chlorine happened when a rail tanker of the liquid gas was ruptured in a rail crash at Graniteville, South Carolina, on 6 January 2005. The train has been diverted on to the wrong track and hit another train. Nine people died and 250 needed hospital treatment, while 5,400 people who lived within a mile of the accident were evacuated.

Medical element

The bleaching action of chlorine is best achieved as a solution of sodium hypochlorite (aka bleach, NaOCl) and this is an essential component of the worldwide battle against water borne diseases. Hypochlorite was first used to disinfect tap water at Maidstone, England in 1897 during an outbreak of typhoid; the epidemic was brought under control. Eventually this became the method of purifying drinking water throughout Britain and most of the developed world. Hypochlorite is the way in which it is used and this is made by bubbling chlorine up a column down which trickles a solution of the alkali sodium hydroxide forming NaOCl. This is a strong oxidising agent, safe to handle and stable for several months provided it is not exposed to heat, sunlight, or metals. Viruses and bacteria are extremely sensitive to it and it quickly destroys them even when very dilute. Because it is persistent, hypochlorite will keep water free of germs at very low concentrations for a long time.

[44] A full account of chlorine's deadly effects can be read in Hasok Chang and Catherine Jackson (Eds.), *An Element of Controversy: The Life of Chlorine in Science, Medicine, Technology, and War*, published by the British Society for the History of Science, 2007.

The chlorination of drinking water has been common practice for almost a century. It has virtually eliminated the water borne diseases such as typhoid, cholera and meningitis which were once common in overcrowded cites where water supplies were easily contaminated by sewage. Chlorination is cheap and, at low doses, is highly effective at ridding water of disease pathogens.

Element of history

Hydrochloric acid was known to the alchemists under names such as *acidum salis* or *spiritus salis*, while the early chemists called it muriatic acid, a name that persisted in some industries well into the last century, and indeed potassium chloride is still referred to as muriate of potash (MoP) when used as a fertilizer. The gaseous element itself was first produced in 1774 by 32-year-old Carl Wilhelm Scheele (1742–86) at Uppsala, Sweden, by heating hydrochloric acid with a powdered mineral he knew as brunsten (today it is known as pyrolusite and is native manganese dioxide, MnO_2). A dense, greenish-yellow gas was evolved which he recorded as having a choking smell and which dissolved in water to give an acid solution. He noted that it bleached litmus paper, and the leaves and flowers of plants. It attacked almost all metals which were exposed to the gas.

Scheele called it dephlogisticated muriatic acid, and it was known by that name for more than 30 years, until Humphry Davy (1778–1829) started to investigate it in 1807 and eventually concluded not only that it was a simple substance, but that it was truly an element. He announced this conclusion to the Royal Society in London in November 1810. Nevertheless, it took another 10 years for some chemists finally to accept that chlorine was an element.

Bleaching with chlorine dissolved in water was first demonstrated by James Watt (1736–1819) of Birmingham, England, in 1786. Within a few years this had become the standard method of bleaching both linen and cotton, both of which had hitherto been laid in strips in fields to be bleached by sunlight, a process that took several weeks and depended on the weather. A better bleaching solution was made by dissolving the gas in sodium hydroxide solution to form sodium hypochlorite (NaOCl) solution, and this became the general reagent used. It is still commonly used in homes for domestic hygiene and cleaning purposes. Industry preferred to use bleaching powder as a more convenient reagent, not least because it was easier to transport, and this was made by absorbing bleach on to slaked lime, calcium hydroxide, $Ca(OH)_2$.

Element of war

Chlorine was used as an offensive weapon in World War I in Flanders. It was first deployed on the morning of 22 April 1915 when the German army released the gas from hundreds of cylinders. The prevailing breeze carried the dense gas across no-man's land and into the trenches of the British troops. There it caused chaos, as 5,000 men died in agony and 15,000 were disabled by it. The threat was eventually countered by issuing gas masks, after which chlorine gas was little used.

Economic element

Halite (sodium chloride) is the main mineral mined for chlorine, but there are other chloride minerals such as carnallite (magnesium potassium chloride) and sylvite (potassium chloride). There are vast salt deposits in the USA, Poland, Russia, Germany, China, India, and Australia. World production of salt is 250 million tonnes annually, and it is produced in almost all countries. The major producers are China (60 million tonnes), USA (40 million tonnes), India (20 million tonnes), Canada (12 million tonnes), Brazil (11 million tonnes), and Australia (11 million tonnes). Reserves are almost limitless. The sea contains trillions of tonnes of chloride although most that is used commercially is extracted from deposits of rock salt, and it is from this that chlorine is made.

Chlorine gas is manufactured on a large scale (40 million tonnes annually) by the electrolysis of brine, which produces not only the gas but the equally useful chemical, sodium hydroxide (also known as caustic soda, NaOH) in about the same tonnage. Chlorine was made in mercury amalgam cells although most of these have been phased out because of the environmental threat this metal poses. Newer electrolysis cells employ either an asbestos diaphragm, or an ion-exchange membrane made from a fluorinated polymer, to keep the chlorine and sodium hydroxide separate. In energy terms the cost of producing chlorine is very expensive because it is an electrochemical process. For example, the UK's largest chlorine producer near Runcorn in Cheshire uses almost as much electricity per year (250 MW) as the nearby city of Liverpool.[45] The plant uses the membrane process.

Chlorine is involved in the manufacture of hundreds of products, directly or indirectly. The primary uses for chlorine are as follows: chemicals for industry (30 percent); PVC[46] manufacture (25 percent); water purification (20 percent); solvents (15 percent); and bleaches (10 percent), although these figures are only approximate, and are changing all the time.

PVC is a remarkable plastic with numerous uses ranging from rigid double glazing window frames to flexible tubing and even blood bags. It is long lasting and fire resistant so is ideal for electric wiring. It can be recycled many times going through a variety of uses such as garden furniture, fences, street signs, vinyl flooring, containers, and even handbags. PVC has the chemical formula $(CH_2CHCl)_n$ in which there are long chains of carbon atoms with a chlorine atom attached to every alternate one. It is made by polymerising vinyl chloride, hence its name, although this chemical is now more correctly called chloroethene.

Indirectly chlorine is needed to make the chemicals that are turned into healing drugs (85 percent of pharmaceuticals rely on chlorine in one form of another for their manufacture), silicones and polymers, but most of these no longer contain chlorine by the time they reach market. Some manufactured products do contain chlorine as an essential part of their makeup, such as flame retardants and pesticides.

---

[45] The plant produces a quarter of its energy needs by burning 750,000 tonnes of municipal waste to generate electricity.

[46] Poly(vinyl chloride).

Chloroform and carbon tetrachloride were once important chlorine-containing chemicals, the former as an anaesthetic, the latter as a dry-cleaning solvent. They are now strictly controlled because exposure to them causes liver damage. Other, more benign, organochlorines were substituted as solvents and dry-cleaning agents but these are now disapproved of because, as volatile organic chemicals (VOCs), they act as greenhouse gases when they leak into the atmosphere.

Environmental element

Chlorine in the environment	
EARTH'S CRUST	130 p.p.m.
	chlorine is the 20th most abundant element.
SOILS	very variable, range 50–2,000 p.p.m.
SEAWATER	1.8%
ATMOSPHERE	mainly as traces of organochlorine compounds.

Chloride in soils varies according to the distance from the sea. The average in top soils is about 100 p.p.m.

To many environmentalists of the last century, chlorine became to be regarded as *the* polluting chemical, and the reason being that so many products which contained it turned out to have adverse side-effects. In the first half of the 1900s many new organochlorine products were marketed offering great benefits, such as the dry-cleaning fluid carbon tetrachloride, insecticides such as dieldrin and DDT, herbicides such as 2,4-D and 2,4,5,-T, engineering oils such as the polychlorinated biphenyls (PCBs), and the refrigerant and aerosol gases known as CFCs (see Fluorine). All are now banned or strictly controlled because either they are regarded as potentially damaging to human health, or are too persistent in the environment and so pose a threat to wildlife. The CFCs in particular are not only greenhouse gases but their chlorine atoms can destroy the protective ozone layer of the upper atmosphere.

Organochlorines contaminate natural waters. The US Environmental Protection Agency and other authorities have set a limit of 100 p.p.b. for chloroform-type chemicals in drinking water. Some organochlorine compounds have caused cancer among those heavily exposed to them in industry, but at the low levels they are present in chlorinated water the health risk is negligible. Hypochlorite will react with organic matter in the water, such as leaf debris, to form traces of organochlorine compounds, which some people consider damaging to health and the environment, although this is unlikely at these concentrations.

For several years organochlorines were thought to be present in the environment solely as the products of the chemical industry, but it is now known that 2,400 organochlorine compounds occur naturally, and that chloromethane (CH_3Cl) is produced by plankton, trees, algae and fungi. Indeed it now seems likely that more than 75 percent of chlorinated substances at large in the environment come from natural causes or volcanic eruptions. Nevertheless, there are heavily contaminated sites due to man-made organochlorines and decontamination of these by natural microbes is long and slow, and not always complete. In 1997 a new bacterium was isolated from sewage that was capable of fully dechlorinating even the solvents tetrachloroethane and trichloroethene (aka trichlorethylene) which have been widely used as cleaning fluids.

In 2002 it was discovered that organochlorine compounds are common in uncontaminated soils and that microbes themselves produce them from chloride ions.

Chemical element

Data file	
Chemical symbol	Cl
Atomic number	17
Atomic weight	35.4527
Melting point	−101°C
Boiling point	−34°C
Density of gas	3.2 g per litre
Oxides	Cl_2O, ClO_2, Cl_2O_7

Chlorine is a greenish-yellow, dense, sharp-smelling gas that is a member of group 17 (row 3p), the halogen group, of the periodic table. The gas is soluble in water to the extent of 3 litres of gas dissolving in a litre of water at 10°C. Chlorine is extremely reactive and will form compounds with all elements except the noble gases helium, neon, argon and krypton. In some cases elements will form more than one stable chloride.

The most common oxidation state of chlorine is −1 as the chloride ion Cl^-. However, chlorine can exhibit positive oxidation states as high as +7 (in perchloric acid $HClO_4$ and perchlorate, ClO_4^-). Examples of other oxidation states are +1 (sodium hypochlorite, NaOCl), +3 (sodium chlorite, $NaClO_2$), +4 (chlorine dioxide, ClO_2), and +5 (sodium chlorate, $NaClO_3$).

There are 24 known isotopes of chlorine with mass numbers 28 to 51, and two known isomers. Naturally occurring chlorine is composed of two stable isotopes, chlorine-35 which makes up 76 percent, and chlorine-37 (24 percent), and traces of one radioactive one, chlorine-36. This is produced in the upper atmosphere from argon-36 and cosmic rays, and by neutron-capture by chlorine-35 in the Earth's crust. It has a half-life of 308,000 years. This isotope can be man-made for use in research. After chlorine-36, the next longest-lived radioactive isotope is chlorine-39 with a half-life of 56 minutes.

Element of surprises

Despite the benefits of chlorination, there were concerns expressed over the organochlorines that formed from dissolved organic matter when water was chlorinated, because some of them were shown to be carcinogenic when tested on rats, albeit at extremely high doses. This was widely reported by the environmental group

Greenpeace which had branded chlorine as 'the Devil's element' but this was to lead to an environmental disaster on a massive scale. Unfortunately, the public authorities in Peru were persuaded that the health threat from organochlorines outweighed the benefits of chlorination, so in 1991 they discontinued chlorinating the public water supplies in many areas. The result was a massive outbreak of cholera with over a million cases being reported throughout the 1990s, of whom 15,000 died. Chlorination was soon reinstated but controlling the disease took many years. The incident proved, if proof were needed, that human health greatly benefits if chlorine is properly used.

Chromium

Pronounced kroh-mi-uhm, the name is derived from the Greek *chroma* meaning colour, because its salts were strikingly coloured.

French *chrome*; German *Chrom*; Italian *cromo*; Spanish *cromo*; Portuguese *crómio*.

Essential element

This most glamorous and essential of elements harbours a secret which limits its usefulness.

Cosmic element

In 1817 André Laugier (1770–1832) detected chromium in the famous Pallas meteorite which fell near Krasnoyarsk, Siberia, in 1749 and which weighed 1,500 pounds (700 kilograms).

Element of life

Chromium in the human body	
BLOOD	6–100 p.p.b.
BONE	100–300 p.p.b.
TISSUE	25–800 p.p.b.
TOTAL AMOUNT IN BODY	1–2 mg, but can be as much as 12 mg.

Chromium is essential to some species, including humans. The daily intake varies according to the diet; it can be as high as 1 mg, but is more likely to be in the range 0.02–0.1 mg. Chromium is an essential element because it is needed to help the body utilise glucose, and its presence in RNA may indicate a second role. The organ with the highest level of chromium is the placenta.

It has been shown that animals which lack chromium have an impaired ability to use glucose, suffer mild diabetes, and have reduced cholesterol levels. The same might well be true of humans, but chromium supplements are rarely needed, although there are a few cases on record of people suffering chromium deficiency.

It has been observed in Americans that there is a steady fall in chromium in the body with age, but what this means is not known. If a lack of chromium were to be shown to be deleterious, then diets could be supplemented with high-chromium foods such as brewer's yeast, molasses, or wheat germ. (It would not be wise to supplement the diet with inorganic chromium salts.)

Food element

The foods which contain most chromium, and have more than 30 micrograms per 100 g, (300 p.p.b.) are oysters, calf's liver, egg yolk, peanuts, grape juice and black pepper. Even commonly eaten foods have some chromium, such as potatoes with 18 p.p.b., beans 9 p.p.b., carrots 18 p.p.b., and apples 8 p.p.b. Such tiny amounts may be all that is needed.

Dangerous element

Like other metals that are essential for life, chromium can be quite toxic in excess. However, chromium has two common oxidation states one of which, chromium(III) as Cr^{3+}, is safe, whereas the higher oxidation state, chromium(VI) as chromate CrO_4^{2-} is much more insidious and very toxic and carcinogenic. Rather unexpectedly, research has shown that chromium(III) could be oxidised by peroxides in the body and these would form chromium (V) and even chromium(VI). For this reason the addition of chromium(III) compounds to food supplements, most of which contain around 25 micrograms, cannot be assumed to be entirely beneficial.

Those working with chromium compounds are vulnerable to an industrial disease known as chrome ulcers, which were first reported in 1827 among workers in Glasgow, Scotland. It was also common among workers involved with chrome-plating, dyeing, French polishing, calico printing, and tanning, all of which could involve chromates. The symptoms were the sudden appearance of holes in the skin that exposed raw flesh and which itched unbearably. Exposure to chromates also leads to stomach ulcers.

The danger of polluted ground water coming from the dumping of chromate waste was highlighted in the 2002 film *Erin Brockovich* which featured Julia Roberts as the young campaigner who brought the issue to public attention and secured compensation for local people. Further research since then has appeared to show that the chromate in the local water supply (which is still present) may be of natural origin and formed within the local aquifer.

Element of history

Chromium was discovered and isolated in 1798 by the French chemist Nicholas Louis Vauquelin (1763–1829) in Paris. He was intrigued by a bright red mineral that had been discovered in a Siberian gold mine in 1766 and was referred to as Siberian red lead (it is now known as crocoite and is a form of lead chromate). Vauquelin analysed a sample in 1797 and confirmed that it did contain lead. Then he precipitated this out of a solution that he had made of the mineral, filtered this off, and focused his attention on the resulting solution. A year later he succeeded in isolating chromium, and intrigued by the range of colours that it could produce in solution he named it chromium. The same year he analysed an emerald and found its green colour was due to small amount of chromium.[47]

---

[47] The red of a ruby was also eventually proved to be due to this metal.

Economic element

The first commercially exploitable deposits of chromium were in Maryland, Pennsylvania, and Virginia; these supplied all that was needed for 30 years. Then, in 1848, large deposits were found in Turkey, and this became the main source of the metal for a long time. There it was mined as the black iron-chromium mineral, chromite ($FeCr_2O_4$), which is the main chromium ore.

A total of 24 million tonnes of chromite ore are currently extracted annually and the main producers are South Africa (10 million tonnes), India (5 million tonnes) Kazakhstan (3.5 million tonnes), and Turkey (1.5 million tonnes) although it is mined in many other countries as well. Global reserves are estimated to be of the order of 800 million tonnes. Chromite ore is used to make refractory bricks that can withstand the extremely high temperatures encountered in furnaces.

World production of chromium metal itself is around 20,000 tonnes per year. Chromium can be polished to a high shine and it resists oxidation in air because of a protective oxide layer which forms on its surface. Its main uses are in alloys such as stainless steel (which may contain up to 15 percent), and for chrome plating, and in metal ceramics. Chromium plating was once widely used to give steel a polished silvery mirror coating and all that was needed was a 1-microgram thick layer to achieve this effect, although it needs to be a hundred times this thickness to resist weathering. It was widely used in the last century to adorn cars, motorcycles, and even trucks. The chromium layer is deposited on top of a nickel layer which is electroplated on to the base metal such as iron. It is even possible to chrome-plate plastics.

Other uses for chromium are as an industrial catalyst, as potassium chromate ($K_2Cr_2O_7$) in dyeing, as chromium sulfate ($Cr_2(SO_4)_3$) in leather tanning, and as chromium oxide (Cr_2O_3) which is the most stable of green pigments. Today, about 90 percent of leather is chrome tanned, a process that was introduced in the 1860s. The trouble with chrome tanning is the effluent it produces which generally contains around 5 p.p.m. of chromium, and it is for this reason that tanners are searching for alternative tanning agents. Chromium copper arsenate is an excellent wood preservative although it is no longer allowed for wood that is used in the home and other environments that would bring it into contact with people.

Chromate pigments are bright yellow and these were once widely used to colour plastics, rubber, ceramic tiles, and floor coverings. Their insolubility made them relatively safe with respect to the chromate content. The most common were lead chromate (aka chrome yellow) and barium chromate (aka lemon chrome and ultramarine yellow) and while the workers who produced these pigments were prone to suffer from chrome ulcers, the users of the products which they coloured appear not to have been affected. Chrome yellow was particularly popular with artists and designers because of it brilliance, but its lead content resulted in its being withdrawn.

Ruby lasers rely on chromium atoms to produce the red light they emit.

Environmental element

Chromium in the environment	
EARTH'S CRUST	100 p.p.m.
	Chromium is 21st in order of abundance of the elements.
SOILS	varies widely between 1–450 p.p.m.; average 50 p.p.m.
SEAWATER	0.24 p.p.b.
ATMOSPHERE	barely detectable.

Chromium is not seen as a major environmental pollutant although it has caused problems in rivers taking untreated industrial waste, especially that from tanneries. Soluble chromate in soils gradually turns into insoluble chromium(III) salts and then becomes unavailable to plants. In this way the food chain is protected against excess chromium which cannot pass the soil-to-plant root barrier, so that even chromate-rich sewage from industrial areas does not pose a threat, though some has been shown to have levels as high as 4,000 p.p.m. (0.4 percent).

The manganese mineral birnessite can oxidise chromium(III) to chromium(IV) and thereby make it more mobile in the environment. The difficulty with this form of chromium is that it does not precipitate and so is carried to aquifers.

Chemical element

Data file	
CHEMICAL SYMBOL	Cr
ATOMIC NUMBER	24
ATOMIC WEIGHT	51.9961
MELTING POINT	1,860°C
BOILING POINT	2,672°C
DENSITY	7.2 g/cm³
OXIDE	Cr_2O_3

Chromium is a hard silvery metal with a blue tinge, and is a member of group 6 (row 3d) of the periodic table. It will dissolve in hydrochloric and sulfuric acids, but not phosphoric acid due to reaction and formation of a protective layer on the surface. Chromium exhibits oxidation states +1 to +6, but mainly +2 (as in the green salt $CrCl_2$), +3 (as in violet $CrCl_3$), and +6 (as in yellow K_2CrO_4, or orange $K_2Cr_2O_7$ aka dichromate).

There are 26 known isotopes of chromium with mass numbers 42 to 67, and two known isomers. Naturally occurring chromium consists of four isotopes: chromium-52, comprising 84 percent of the total, chromium-53 (9.5 percent), chromium-50 (4 percent) and chromium-54 (2.5 percent). Of these chromium-50 is radioactive although with an incredibly long half-life of 1.8×10^{17} years (which is millions of trillions of years). The next longest-lived radioactive isotope is chromium-51 with a half-life of 27.7 days.

Element of surprises

1. The precious gem alexandrite is the mineral chrysoberyl ($BeAl_2O_4$) with a small amount of chromium which provides its tantalising colour. Alexandrite is the July birth stone and it was named after Tsar Alexander I of Russia who reigned from 1801–25. It was prized not only for the beautiful blue/green colour it exhibited in daylight, but for the way this changed to a deep red when seen at night by artificial light.

2. The first stainless steel was discovered by accident in 1913 by Harry Brearley, a metallurgist working for a Sheffield steel maker. The previous year he had been researching better steels for rifle barrels and had made many alloys by adding chromium to steel but none performed as he had hoped, so he dumped all his samples in the mill yard where they were exposed to the acid rain and pollution for which that city was notorious. Several months later he noticed that one of the samples had been completely unaffected by the aggressive climate and that contained 12 percent chromium. The fact that steel could be made stainless had been recognised, not that this particular alloy was much use because it was too brittle.

Cobalt

Pronounced koh-bolt, the name comes from *kobald* the German word for goblin. It has been suggested that both words derive from the Greek *cobalos,* meaning a mine.

French *cobalt*; German *Cobalt*; Italian *cobalto*; Spanish cobalto; Portuguese *cobalto*.

Essential element

Famous for colouring glass, this element lies at the heart of an essential vitamin and bonds in an unexpected manner.

Cosmic element

Extraterrestrial cobalt was first detected in a meteorite found at the Cape of Good Hope in 1819.

Element of life

Cobalt in the human body	
BLOOD	0.2–20 p.p.b.
BONE	1–4 p.p.b.
TISSUE	*approx.* 0.3 p.p.b.
TOTAL AMOUNT IN BODY	1–2 mg.

The daily intake of cobalt is variable and may be as much as 1 mg, but almost all will pass through the body unabsorbed, except that which is incorporated into vitamin B_{12}. Cobalt is at the heart of this vitamin and is directly bonded to a methyl group, one of the rare examples of a metal-to-carbon bond in nature. Lack of this vitamin causes pernicious anaemia, a condition in which the body cannot produce enough red blood cells to transport all the oxygen that it needs. Vitamin B_{12} also helps maintain nerve tissue and is needed for the release of energy from carbohydrates and fats.

It has been said that nothing that grows out of the ground needs vitamin B_{12} but that every creature that walks, flies, or swims must have it. Microbes like bacteria and fungi also need it. Herbivores, such as cows and sheep, make enough B_{12} to satisfy their own needs provided they graze on land that is not deficient in this trace metal. Most other animal species can make what they require themselves, but not humans. Although there are bacteria living in our colon that produce it, we don't absorb it very effectively, hence the need to supplement it from our diet. This should not pose a problem because we require only 1.5 micrograms per day and the average intake is more like 5 micrograms.

Food element

The cobalt level of plant foodstuffs can be high, as in sweet corn, lettuce and cabbage (up to 100 p.p.m. dry weight), and low, as in fruits such as oranges and apples (8 p.p.m. dry weight). Cobalt enrichment of the soil finds its way into plants and they can also absorb the metal through their leaves if it is applied as a spray. Several species of plant are able to concentrate cobalt, such as legumes, borage, myrtle, and violets.

Foods particularly rich in B_{12} are clams, sardines, salmon, herring, liver, and eggs. The vitamin is depleted in foods exposed to light, and canning also has an adverse effect although the presence of vitamin C acts to protect it. The 1–2 milligrams of B_{12} we store in our body will last us for a long time, maybe two years or more, and it rarely needs supplementation. Vegan vegetarians must take B_{12} as an essential vitamin supplement.

Toxic element

For the average adult a large dose of a soluble cobalt salt could be life-threatening but this is not likely to occur by accident. Cobalt compounds are generally thought to be of low toxicity but they may affect the thyroid and damage the heart, and cobalt is a suspected carcinogen. Contact with cobalt salts can lead to dermatitis.

Metal workers using cobalt to make alloys were prone to a form of pneumoconiosis caused by it. In the 1960s, an epidemic of beer drinkers' cardiomyopathy (heart disease) was traced to the use of cobalt salts as foam stabilizers in the beer.

Element of history

Among the treasures discovered in the tomb of Pharaoh Tutankhamen, who ruled from 1361–52 BC, was a small glass object that was coloured deep blue with cobalt, and it is clear that using a cobalt mineral to achieve this rich colour was not unknown to the ancient world. Pieces of synthetic lapis lazuli[48] dating from 1400 BC have been found at Nippur, the religious centre of the Sumerians, and these were also coloured with it. Cobalt glass has been recovered from a shipwreck of around this time. Cobalt blue was known in China long before this era however, and it was used for pottery glazes, but it was always a rare pigment because cobalt minerals themselves are rare.

Later generations regarded cobalt ores as more of a curse than a colouring agent. When silver miners in Saxony in the 1500s tried to smelt what they believed was a silver ore they were disappointed to find that toxic fumes of arsenic were given off and no silver was to be won from it. This is not surprising, because they were dealing

---

[48] The blue colour of natural lapis lazuli, which is the alumino-silicate mineral sodalite ($Na_8Cl_2Al_6Si_6O_{24}$), is due to sulfur atoms replacing some of the chloride ions.

with an ore now known as smaltite, which is cobalt arsenide (CoAs$_2$). The miners cursed the ore, saying it had been bewitched by goblins and referred to it by the German name for these evil spirits, *kobald*. (See also the origin of the name nickel, page 348.)

Others, however, saw cobalt's potential as a deep-blue pigment and it soon began to colour glass and pottery. The material used was called zaffer or smalt, and was made by roasting smaltite with varying amounts of sand. The Venetian glass-makers were particularly fond of it to make the fine blue glass for which they were famous. Low grade zaffer was sold as a blue starch for laundry purposes and so-called 'dolly blue' is still used in some parts of the world to make white clothes appear whiter.

In 1730, chemist Georg Brandt (1694–1768) of Stockholm, Sweden, became interested in a dark blue ore from some local copper workings in Vestmanland and he eventually proved that this contained a hitherto unrecognised metal, cobalt, and he gave it the name by which its ore was known in Germany. He published his results in 1739. For many years his claim to have uncovered a new metal was disputed by other chemists who said his 'cobalt' was merely a compound of iron and arsenic, but eventually it was recognised as an element in its own right.

Element of war

Cobalt steel with high magnetic properties was first patented in Japan in 1917, and put to use at the start of World War II by Nazi Germany in the making of magnetic mines. These were anchored underwater and detected the presence of a ship by the magnetic field it generated which then triggered the mine to explode. At one time it appeared that magnetic mines would successfully sink all ships arriving or leaving the British Isles, but the problem was quickly solved by placing a 'degaussing' coil round ships that effectively neutralised the ship's magnetic field. (Once it was realised that these mines were ineffective, they were no longer employed against shipping, and the unused ones were turned into bombs and dropped on British cities instead, becoming the much feared 'land mines' of the Blitz.)

Economic element

Various cobalt arsenic minerals are known, such as cobaltite (CoAsS). World production of cobalt minerals is 62,000 tonnes a year coming mainly from Congo (25,000 tonnes), Canada (8,000 tonnes), Australia (6,000 tonnes), Zambia (4,000 tonnes), Brazil (4,000 tonnes), and Russia (3,500 tonnes). Production of the metal itself amounts to 53,000 tonnes with most being smelted in China (13,000 tonnes), Finland (9,000 tonnes), and Canada (6,000 tonnes). A lot of cobalt comes as a by-product of nickel production. Cobalt is used in alloys for magnets, in ceramics, in catalysts, and in paints. Certain types of stainless steel contain it, including that used to make razor blades. It is also used in alloys for jet engines. One alloy of cobalt, chromium, and tungsten is known as Stellite and is used for heavy duty, high temperature cutting tools. Some cobalt alloys are used in body parts such as knee replacements and in dentistry.

Cobalt can be magnetised, like iron, although it has only two thirds the magnetic strength of iron. However, it has the advantage of maintaining its magnetism to much higher temperatures; iron loses its magnetism at 770°C whereas for cobalt the temperature of its so-called Curie point is 1,130°C. Alnico, an alloy of mainly of aluminium, nickel and cobalt, is used for permanent magnets that are very powerful; 25 percent of all cobalt produced goes into making them. However, in several applications alnico has been superseded by neodymium-based magnets.

Cobalt blue is employed by artists as an important part of their palette of colours, and by craft workers in porcelain, pottery, stained glass, tiles and enamel jewellery. Its rich blue colour was also known as Sèvres blue and Thénard blue. Cobalt was particularly popular with glass bottle makers in the 1800s.

The radioactive isotope, cobalt-60, is made by bombarding cobalt-59 with neutrons in a nuclear reactor. It has a half-life of 5.3 years and undergoes decay with the emission of intense gamma-radiation (with energy of 1.3 million electron volts) which is very penetrating. This isotope is used in medical treatment and also to irradiate food and to sterilise medical supplies. It preserves and protects by destroying the organisms that cause decay, and the harmful bacteria that cause disease. (It is a popular misconception that irradiating food with cobalt-60 will cause the food to become radioactive, but for this reason this method of food preservation has been little used in some countries.) Cobalt-60 is used to check for cracks in large metal objects such as containers, pipes, and the hulls of ships. Irradiating diamonds with cobalt-60 colours them blue.

Cobalt catalysts are used in the chemical industry to convert syngas, which is a mixture of carbon monoxide and hydrogen, into liquid fuels. Cobalt phosphate catalysts have been made which can break water down to generate hydrogen gas under the effect of sunlight. These are thought to operate via cobalt in the rare oxidation state of +4. As yet, it has not led to a commercial application for making hydrogen gas.

Environmental element

Cobalt in the environment	
EARTH'S CRUST	20 p.p.m.
	Cobalt is 32nd in order of abundance of the elements.
SOILS	8 p.p.m.
SEAWATER	less than 1 p.p.t.
ATMOSPHERE	virtually nil.

Most of the Earth's cobalt is in its core and it is of relatively low abundance in the crust and in natural waters, from which it is precipitated as the highly insoluble cobalt sulfide (CoS). Although the average level of cobalt in soils is 8 p.p.m., there are soils with as little as 0.1 p.p.m. and others with as much as 70 p.p.m. Some soils have so little cobalt in them that it is necessary to supplement fertilizers with the element if they are to be used to grow crops for animal nutrition. However, a lack of cobalt does not retard the growth of plants. In the marine environment cobalt is needed by blue-green algae and other nitrogen-fixing organisms.

Chemical element

Data file	
CHEMICAL SYMBOL	Co
ATOMIC NUMBER	27
ATOMIC WEIGHT	58.933200
MELTING POINT	1,495 °C
BOILING POINT	2,870°C
DENSITY	8.9 g/cm³
COMMON OXIDES	CoO, Co$_2$O$_3$, and Co$_3$O$_4$

Cobalt is a lustrous, silvery-blue, hard metal which is also ferromagnetic. It is a member of group 9 (row 3d) of the periodic table. It is stable in air, unaffected by water, but slowly attacked by dilute acids. Its common oxidation state is +2 as in CoCl$_2$ (blue) and in solution as the ion Co^{2+} which is pink except in the presence of chloride ions when they form CoCl$_4^{2-}$ which is blue. Oxidation state +3 is also known for some compounds such as CoF$_3$.

There are 30 known isotopes of cobalt with mass numbers 47 to 76, and 11 known isomers. There is only one naturally occurring isotope, cobalt-59, and it is not radioactive. In the years after World War II, when nuclear weapons were tested above ground, then cobalt-60 contaminated the planet. Its half-life is 5.27 years and this is the longest-lived radioactive isotope of cobalt. The radioactive isotope cobalt-57, with a half-life of 272 days, has been used in medical investigation into the uptake and function of vitamin B$_{12}$.

Element of surprises

1. Cobalt was once popular for making invisible ink. This remains unseen until it is warmed. Exactly who first came across it is unclear, but it was used for sending secret messages in the 1600s, and even after Jean Hellot revealed its existence around 1700, it continued to be used in espionage.

   The ink was made by dissolving cobalt ore in aqua regia from which cobalt chloride crystals were produced. These were dissolved in water with a little glycerine (aka glycerol) to give an almost colourless (very pale pink) solution, which was then used to write with. Once this ink had dried another innocent letter in visible ink was written at right angles across the paper. When the letter reached its destination the recipient simply heated it and the hidden writing appeared. The heating drives off the water and the glycerine molecules that surround the cobalt, and chloride ions move in to take their place, giving a dark blue colour.

2. Cobalt chloride was popular in the 1800s for decorating artificial flowers that responded to changes in the weather. Coating their white petals with a solution of this salt dyed them pink, and they would remain this colour while the weather was wet or the atmosphere was humid. When the weather became sunny and dry they would turn violet, and when the day was very dry they turned blue.

Copernicium

Pronounced ko-per-nee-see-um, it is named after Nicolaus Copernicus (1473–1543) a Polish astronomer who published an influential book on the structure of the solar system.

French *copernicium*; German *Kopernicium*; Italian *copernicio;* Spanish *copernicio*; Portuguese *copernício.*

Element of history

A great deal of effort went into making atoms of element 112—see *Element of doubt* below, and the group which first succeeded and was officially recognised as such was led by Sigurd Hofmann at the Gesellschaft fur Schwerionenforschung (GSI) in Darmstadt, Germany. In 1996, they synthesised isotope 277 by bombarding lead-208 for two weeks with zinc-70 ions travelling at 30,000 km per second. In a two-week long experiment they observed the formation of a few atoms of element 112. These had a half-life of 0.24 milliseconds, decaying by a sequence of alpha-emissions via elements 110 (darmstadtium), 108 (hassium), 106 (seaborgium), etc. The GSI work was repeated at the Japanese Institute of Physical and Chemical Research, (RIKEN) in 2004 when two more atoms of copernicium were synthesised.

Since the original work at the GSI, other isotopes of copernicium have been made. In 1998, isotope-285 was observed as part of the decay sequence resulting from the discovery of element 114 at the Joint Institute for Nuclear Research (JINR) at Dubna, Russia. It was there that, two years later, isotope-284 was first observed as an alpha-decay product from their synthesising element 116.

Finally, in 2009, the claim of being the first to synthesise element 112 was recognised by IUPAC and this then gave the German team the right to name it. They proposed the name copernicium (symbol Cp) in July 2009. Within a few weeks, they changed the proposed symbol to Cn after they were informed that Cp was not permitted under IUPAC rules because it is the currently-used symbol for the molecule cyclopentadienyl.[49] Nor is Cn original because this had previously been used as the symbol for the element carolinium by Charles Baskerville who claimed, wrongly, to have separated this new element from thorium in 1900.

Chemical element

Data file	
CHEMICAL SYMBOL	Cn
ATOMIC NUMBER	112

---

[49] In fact Cp had also been the symbol for lutetium's original name of casseopeium.

ATOMIC WEIGHT	285 (longest-lived isotope yet known)
Properties such as melting point, boiling point, and density, are likely never to be measured because only the occasional atom has been made.	

From its position in the periodic table, in group 12 (row 6d) below mercury, this element should have the physical properties of a heavy metal. By analogy with mercury, copernicium might even be slightly volatile. Were its atoms long-enough lived for its chemistry to be investigated then it is expected to display oxidation states of +1 and +2, with the latter more stable. Even if only a few atoms were made then it might be possible to show that it forms the oxide CnO.

Copernicium has five known isotopes with mass numbers 277 and 282 to 285, with no known isomers. The half-life of the longest-lived isotope, copernicium-285 is 34 seconds. There have been claims for Cn-272 with a half-life of 47 days, but see below.

Two atoms of copernicium were made in 2007 by firing calcium-48 at plutonium-249. The nuclei fused and decayed to form copernicium with a half life of several seconds. During their brief existence a few properties were inferred such as that copernicium would be unreactive and indeed more like a noble gas than a metal.

Element of doubt

The first attempt to make element 112 occurred in 1971 and was made by Amnon Marinov, working at CERN, the European centre for nuclear research. He bombarded a tungsten target composed of isotopes tungsten-184 and tungsten-186, with strontium-86 and strontium-88. This was the first use of the so-called 'cold fusion' method. The experiment used 24 GeV protons to deform target and projectile nuclei, thereby making them easier to fuse. Marinov claimed to have made about 500 atoms of element 112, mass number 272, with a half-life of 47 days. What motivated Marinov was the discovery in the 1960s of long-lived isomers of isotopes that otherwise had short half-lives.

His method also produced mercury in the target material by nuclear reactions brought about by the proton bombardment. He used chemical extraction techniques to separate this from the target materials reasoning that element 112 would also be extracted because of its chemical similarity to mercury. He then identified element 112 by characteristic decay products from its spontaneous fission into fragments whose atomic numbers added up to 112. He submitted his claim to IUPAC in the 1980s, after exhaustive tests confirmed the methods of production and detection, but although three committees considered his work their members were not convinced.

Marinov's work has been essentially disregarded and his experiment has never been mentioned in any of the literature that discusses the history of element research and discovery—nor has there ever been an exact and accurate re-run of this experiment, although in 1990, the GSI physicists bombarded a tungsten target with strontium-88 in a similar experiment and detected fission fragments that probably came from element 112, mass number 271. They too considered the results not to be fully convincing, although this work is revealing because the tentative results seem to support Marinov's 1971 claim.

Copper

Pronounced kop-er, the name is derived from the Old English name *coper*, which came from the Latin *cuprum* the Roman name for Cyprus. The chemical symbol, Cu, derives from the Latin name. In fact Cyprus had been the main exporter of the metal long before it became part of the Roman Empire. It has even been suggested that the island, which was called *Kupros* by the Greeks, took its name from the metal rather than the other way round.

French *cuivre*; German *Kupfer*; Italian *rame*; Spanish *cobre*; Portuguese *cobre*.

Essential element

Copper has been an indispensible metal throughout history and it still continues to be important today, primarily as electrical cable. Copper is also the element of creatures with blue-blood.

Element of life

Copper in the human body	
BLOOD	1 p.p.m
BONE	1–25 p.p.m.
TISSUE	2–10 p.p.m. (highest in liver)
TOTAL AMOUNT IN BODY	70 mg.

Copper is essential to all species. There are more than 10 copper-containing enzymes. One such, *cyctochrome c oxidase*, is required by all cells to produce energy. Other important enzymes are *superoxide dismutase*, which protect against free radicals, and *tyrosinase*, which is needed for the synthesis of the pigment melanin. Yet others are involved in making hormones, removing dangerous amines, and repairing connective tissue. Certain bacteria use the enzyme *methane monooxygenase* to convert methane to methanol at ambient temperatures and this enzyme has two copper atoms at its active site.

Food element

There is little danger that our diet does not provide enough copper because it is abundant in certain foods. It may also come with drinking water if this is from a soft water area and supplied through copper pipes. It has even been suggested that we get too much copper and that this works against the iron and zinc in our bodies because copper can displace these metals from their active sites.

Copper

An adult needs to ingest around 1.2 mg of copper per day, and breast-feeding women need around 1.5 mg, and it has been suggested that 2 mg per day would be more beneficial for everyone. Volunteers on low-copper diets were found to have increased levels of cholesterol, high blood pressure and an impaired ability to digest glucose.

The foods with most copper are sea-foods such as oysters, crab, and lobster, while among meats it is lamb, duck, pork, and beef, which have the most—albeit less than the sea-foods—and especially the liver and kidneys of lamb and beef. The plant foods with most copper are almonds, walnuts, Brazil nuts, sunflower seeds, mushrooms, and bran. The amount of copper in a person's diet can be as much as 6 mg a day, depending on how much of these foods are eaten.

The copper in plants varies markedly: apples and cabbage have only 0.03 p.p.m., carrots 0.4 p.p.m., tomatoes 0.7 p.p.m., and sweet corn 1 p.p.m (all based on fresh weight). Lettuce normally has 0.1 p.p.m. but grown on contaminated land this can rise tenfold. Some microorganisms can take in remarkable amounts of copper, *Penicillium* for example can have 20 000 p.p.m. dry weight (that is, 2 percent).

Plants absorb copper but most tends to remain in their roots. That which is absorbed is linked to protein and it plays a key role in photosynthesis, carbohydrate distribution, cell wall metabolism, and especially in the production of DNA and RNA. It also helps the plant to resist disease. Crops grown on copper-deficient soils, that is, with less than 2 p.p.m., produce fewer seeds and it is believed that this condition may be more widespread than imagined because continued harvesting removes about 30 grams per hectare per year of nutrient copper.

Dangerous element

Essential though it be, copper can be toxic in excess. Normally vomiting soon starts if an excess of a copper compound is swallowed, and this generally acts to prevent acute toxic effects. A child who ate the sample of copper sulfate in a toy chemistry set died as a result and a few people have managed to commit suicide by drinking copper sulfate solution. A gram of this salt has been estimated to be a fatal amount, although would-be suicides have generally consumed much more than this and those that have survived have suffered major organ damage especially to the kidneys and brain.

In 2003, at a school in Canada, three teenage girls sought to play a rather malicious trick on some of their friends by adding copper sulfate to the Slush Puppy drink of the same blue colour. Several girls sampled the drink and were made very ill but thankfully none died. The three girls admitted in court to criminal negligence and were sentenced to 60 days in youth jail.

Medical element

The genetic illnesses known as Wilson's Disease and Menke's Disease are caused by the body's inability to utilise copper properly. In the former condition it builds up in the brain and in the latter disease there is a lack of copper in the body because the gene carrying the instruction for making the copper-transporting protein is missing. The result is retarded growth and an early death in infancy. Copper histidine therapy, which was introduced in the 1970s as a treatment, is not particularly effective in prolonging life and most victims die before they are six.

Copper compounds, such as copper acetate and even powdered copper metal, were used by the Egyptians as long ago as 3000 BC to treat diseases of the eye. In more recent times a lot of research went into potential medicaments based on copper for use as anti-ulcer, anti-inflammatory, anti-convulsant, and anti-cancer drugs. Few, if any, were successful. Some copper compounds, however, proved successful at reversing damage to those accidentally exposed to massive doses of ionising radiation.

The health benefits of wearing a copper bracelet, although popular, still remain to be proved and this alternative therapy is almost certainly useless. Nevertheless, research has shown that microbes cannot survive for long on a copper surface and this includes the MRSA bacteria which can be deadly for those with an impaired immune system. The microbes died within 90 minutes of coming into contact with copper. For this reason some hospitals have resorted to replacing metal fittings such as door handles, bath taps, toilet handles, rails, etc. with copper versions in an effort to reduce deaths from MRSA infection.

Element of history

Copper beads have been excavated in northern Iraq which were more than 10,000 years old and were presumably made from uncombined copper, nuggets of which can sometimes be found on the Earth's surface. (The biggest single mass of uncombined copper, weighing 420 tonnes, was found in Minnesota in 1857.) Smelting copper ores began around 5000 BC and all cultures made use of this metal for weapons, tools and jewelry, the copper generally being extracted from the easily identified green ore malachite, which is basic copper carbonate, $Cu_2CO_3(OH)_2$.

Later it was discovered that copper could be greatly strengthened and be honed to a fine cutting edge by adding one part tin to two parts copper to produce the alloy known as bronze. This was much more suitable for knives and swords, and cultures which advanced to this technology are distinguished as having moved out of the Stone Age to the so-called Bronze Age, which generally started around 3000 BC and lasted to around 1000 BC (these times varied considerably from one ancient civilisation to another). How bronze was discovered we do not know, but the peoples of Egypt, Mesopotamia, and the Indus valley were among the first to become familiar with it. Soon it spread to other areas and was smelted on a large scale. The prehistoric workings at Mitterberg in the Austrian Alps suggest that during the Bronze Age they yielded around 20,000 tonnes of copper.

Some of the ancient uses of bronze were spectacular. The Colossus of Rhodes was a 35-metre high bronze statue of the Sun-god Apollo erected at the harbour entrance around 280 BC. It fell down during an earthquake about 50 years later and was then sold as scrap. As the Roman Empire expanded so did the demand for everyday objects made of bronze, such as cutlery, needles, containers, coins, knives, razors, tools, jewellery, and musical instruments. Other cultures used bronze for making bells and the art of bell founding can be traced back to 1000 BC in China though it only became important in the West from around 1000 AD.

As it weathers and ages, copper becomes covered with a green layer, or patina, known as verdigris. In ancient Greece and Rome a similar green copper compound was produced as a pigment and a medicament. It was made by hanging strips of copper over vats of vinegar or the lees of wine; the copper became covered with a green layer of basic copper acetate. This method of production survived as a home-based industry around Montpellier in the south of France for more than 700 years. The women of that region were skilled in its production for which they used copper from Sweden and wine lees from local vintners. The verdigris they produced was sold all over Europe and used for both domestic and artists' paints. The actual recipe for making it was handed down from mother to daughter and never written down, and is now lost in the mists of time.

Copper is an excellent conductor of heat and so was used to make pans, kettles and boilers.[50] A lot of the metal was used in coins. Copper is traditionally known as one of the coinage metals, along with silver and gold, but it is the most common and therefore least valued of this group. The development of printing created a demand for copper plates for engravings, the first of which were produced in the 1430s. Later, copper sheeting was used to sheath the hulls of wooden ships which kept them relatively free of barnacles and other marine growth, thereby improving their speed.

Copper is not difficult to extract from its ores, but mineable deposits were not to be found easily. Some which were, such as the great copper mine at Falun, Sweden, date from the 1200s, were the source of great wealth. Fairly early on it was discovered that one way to extract the metal was to roast the sulfide ore then leach out the copper sulfate that formed, with water. This was then trickled over scrap iron on the surface of which the copper deposited, forming a flaky layer that was easily removed.

Economic element

In 1860 the Welsh town of Swansea smelted 90 percent of the world's copper from ores obtained from Cornwall and Anglesey. Today copper ore is mined, generally in open pit mines, in more than 50 countries to the extent of 16 million tonnes a year of which the major producers are Chile (5.5 million tonnes), USA (1.2 million tonnes), Peru (1.2 million tonnes), and China (1 million tonnes). Silver and gold are produced as by-products. Easily exploitable reserves are around 300 million tonnes, and more than three times this amount may also be recoverable, although there is more than a million times this amount in the Earth's crust.

The main ore is a yellow copper-iron sulfide called chalcopyrite ($CuFeS_2$) from which the copper is extracted first by roasting it to form copper sulfide and iron oxide. This is then heated with limestone and sand to form a molten slag from which the sulfide separates as a lower layer. This is tapped off, heated and blown with air, which converts it to copper metal, while the sulfur dioxide that forms is converted to sulfuric acid. The impure copper is first refined by having methane gas blown through it to remove any remaining oxygen and sulfur, before finally being cast into large anodes for electro-refining which gives a product that is 99.99 percent pure copper.

---

[50] By tradition, whisky stills are always made of copper.

The better known copper ore, green malachite, is used as such for polished slabs, tables and columns and it is mined in various countries. Other ores are cuprite (copper oxide) and azurite (copper carbonate).

About 2 million tonnes of copper a year are reclaimed by recycling and another source of copper is the spoil heaps of mines which can yield yet more copper by using microbes, such as *Thiobacillus ferrooxidans*. These break down iron(II) sulfide and oxidise the iron to Fe^{3+}, which in this oxidation state can oxidise any copper to soluble copper(II) which trickles from the spoil heap and on to scrap iron.

In Chile, at Mansa Mina, 20,000 tonnes per year of copper is extracted directly from low-grade ore by heat-loving bacteria such as *Sulfolobus thermofiles* and *Leptospirillium ferrooxidans* which digest the crushed rock in large open tanks kept at 65–85°C. Overall the global production of refined copper from all sources is 21 million tonnes: China produces 3.5 million tonnes, Mexico 3 million tonnes, Chile 3 million tonnes, Japan 1.5 million tonnes, and the USA 1.5 million tonnes. The UK, which was once the world's main source of copper now produces none.

Globally, most copper is used for electrical equipment (60 percent); for construction, such as roofing and plumbing, (20 percent); for industrial machinery, such as heat exchangers, (15 percent); and for alloys (5 percent), although these percentages vary widely in different parts of the world. For example, construction uses dominate in the USA, transport uses in the EU, and electrical uses in Asia. There are several long-established copper alloys apart from bronze, such as brass (copper/zinc) which combines hardness with a gold-like lustre. The alloy of copper with tin and zinc, which was strong enough to make guns and canons, was known as gun metal.[51]

The mixture of copper and nickel, known as cupronickel, is the preferred metal for low-denomination coins which are still referred to as copper. However, the US five-cent piece is called a nickel because of its colour, although it is an alloy consisting of 75 percent copper and 25 percent nickel.

The alloy consisting of 90 percent copper and 10 percent nickel finds use because it resists corrosion by sea water, and is important for marine pumps, propellers, and in desalination plants which produce drinkable water from sea water. Aluminium bronzes (copper with 7 percent aluminium) have a pleasing gold colour and polish well and are used as a decorative substitute for gold.

Copper is ideal for electrical wiring because it is easily worked and can be drawn into fine wire, and it has a high electrical conductivity. It appears in many guises as well as wiring, such as electrical appliances, piping, screws, hinges, and locks.

Copper in the form of Bordeaux mixture (a blue gelatinous suspension of copper sulfate and lime in water) was one of the first generation agrochemical pesticides, developed to control downy mildew on vines. This, and other variants using sodium carbonate or ammonium carbonate, continue to be used as a fungicide by organic farmers. Copper in small amounts is added to water to prevent algal growths.

---

[51] The copper content of this alloy was 85-90%.

Environmental element

Copper in the environment	
EARTH'S CRUST	50 p.p.m.
	Copper is the 26th most abundant element.
SOILS	20 p.p.m. on average.
SEAWATER	0.2 p.p.b.
ATMOSPHERE	small but significant.

Copper is rather immobile in soils and concentrates in the top layers where it is tightly bound to inorganic particles and organic matter. The loss of copper from soils is partly replaced by that deposited from the atmosphere, which in some industrial countries may exceed 200 g per hectare per year. It is also replenished by the use of copper-based agrochemicals, or even by using municipal sewage. Copper contaminated soils can have as much as 3,500 p.p.m. of copper in industrial regions and 1,500 p.p.m. on farmed land.

Exploitation of land in some areas has led to copper-deficient soils and reduced them to marginal status. These have been brought back into production by fertilizing them with copper sulfate, sometimes requiring as much as 23 kg of copper sulfate per hectare, although most required only a fraction of this amount to make them productive again.

Chemical element

Data file	
CHEMICAL SYMBOL	Cu
ATOMIC NUMBER	29
ATOMIC WEIGHT	63.546
MELTING POINT	1,084°C
BOILING POINT	2,567°C
DENSITY	9.0 g/cm³
OXIDE	Cu_2O and CuO

Copper is a golden/orange metal that is a member of group 11 (row 3d), also known as the coinage metals. It is malleable and ductile, with an electrical conductivity second only to that of silver. Copper is resistant to air and water but attacked by acids. It displays two main oxidation states: +1 as in Cu_2Cl_2, and +2 as in $CuCl_2$ and most other salts. In solution the Cu^{2+} state is the more stable. Higher oxidation states are known as in K_3CuF_6 (+3) and Cs_2CuF_6 (+4).

There are 31 known isotopes of copper with mass numbers 52 to 82, and 7 known isotopes. Copper has two naturally occurring isotopes: copper-63 which comprises

69 percent, and copper-64 (31 percent). Neither is radioactive.[52] This ratio produces a relative atomic weight of 63.5, a numerical value which puzzled early chemists—see page 639. The longest-lived radioactive isotope is copper-67 with a half-life of 62 hours.

Element of surprises

1. In 1847, Emil Harless (1820–62) discovered that the blood of the octopus was blue. The same is true of other creatures such as snails, spiders, and oysters. The reason is that these creatures rely on a blue copper compound to carry oxygen around the body, whereas most animals rely on the deep red iron compound at the heart of haemoglobin. In blue-blooded creatures the copper atom is at the centre of a molecule called haemocyanin, which does the job equally well.

2. A gold-coloured US dollar coin was issued in November 1999, made of copper (88.5 percent) alloyed with zinc (6 percent), manganese (3.5 percent) and nickel (2 percent). Each coin costs 13 cents to make and more than 20 *billion* of them were minted, yet few are in circulation because people in the USA preferred to hoard them. Starting in 2005, a billion new coins are to be issued annually and each year these will depict four former Presidents. Again most of these are expected to be hoarded.

3. The strong jaws of the blood-worm *Glycera dibranchiata* are made of the tough copper mineral atacamite, formula $Cu_2(OH)_3Cl$. This mineral is named after the Atacama Desert of Chile where it is to be found. The worm uses its hard-tipped jaws to bite and inject venom into its prey which it hunts for in marine sediments.

4. Ötzi the Iceman[53] was murdered around 3200 BC and his body was found in 1991 frozen and preserved high in the Alps on the Austria/Italian border. He had among his possessions an axe whose head was almost pure (99.7 percent) copper.

---

[52] In addition there some radioactive isotopes produced by fission in uranium-bearing deposits.
[53] He was named by the researchers who studied him and they took his name from the Otztal Valley glacier in which he had been found.

Curium

Pronounced kyuhr-ee-um, it was named in honour of Pierre and Marie Curie. The chemical symbol of Cm was chosen because m was Marie Curie's initial, or so it is said.[54]

French *curium*; German *Curium*; Italian *curio*; Spanish *curio*; Portuguese *curio*.

Essential element

The radioactivity of this man-made element provides power for systems in isolated situations, including the human body.

Element of life

Curium is rarely encountered outside nuclear facilities or research laboratories. It is dangerous because of its radioactivity, and tests on animals showed that if it gets into the body it concentrates in the bone marrow where its intense radiation destroys red blood cells. It also collects in the liver, but from there it is rapidly excreted.

Element of history

Curium was first made by the team of Glenn T. Seaborg (1912–99), Ralph A. James, and Albert Ghiorso (1915–) in the summer of 1944. Using the cyclotron at Berkeley, California, they bombarded a piece of the newly discovered element plutonium (isotope 239) with alpha-particles (helium nuclei) having energies of 32 million electronvolts. The irradiated plutonium was then sent to the Metallurgical Laboratory at the University of Chicago where, after much difficulty, a sample of curium was separated and identified.

The discovery was kept secret until the end of World War II and was first revealed by Seaborg when he appeared as the guest scientist on a children's radio show 'Quiz Kids' on 11 November 1945—see also Americium. The secrecy surrounding curium had been lifted a few days previously but the new element was not officially announced until a week later.

Economic element

Today, curium-242 is produced in nuclear reactors by bombarding plutonium with neutrons. This isotope is used as power source for pace-makers, navigational buoys, and on space missions because it gives off three watts of heat energy per gram of metal. Because it is an alpha-emitter its radiation can easily be shielded against.

---

[54] It could hardly have been otherwise because Cu is copper, Cr is chromium, and Ci is a unit of radioactivity (equivalent to 37 billion radioactive decays per second).

Environmental element

The environment is not entirely free of this element because a little curium was produced in above-ground nuclear-weapon tests which were carried out from 1945 to
1980. The amount is, however, negligible and its radiation contributes only a tiny
fraction of the Earth's background radiation.

There are eight curium isotopes that occur naturally on this planet and these have
mass numbers 242 to 249. They occur in minute traces in uranium-bearing deposits
where they are produced by a series of neutron-capture and beta-decay processes.
These produce americium isotopes which beta-decay to curium.[55]

Curium was also produced for hundreds of thousands of years in natural nuclear
reactors. Starting about 1.8 billion years ago, natural geological processes concentrated uranium to start sustained nuclear reactors at 16 locations in what is now Oklo,
Gabon, in Africa. These functioned continuously and by neutron capture and beta
decay processes they produced the transuranium elements from atomic numbers 83
(neptunium) to 100 (fermium).

Chemical element

Data file	
CHEMICAL SYMBOL	Cm
ATOMIC NUMBER	96
ATOMIC WEIGHT	247.1 (isotope curium-247)
MELTING POINT	1,340°C
BOILING POINT	not known
DENSITY	13.3 g/cm³
OXIDES	Cm_2O_3 and CmO_2

Curium is a member of the actinide group (row 5f) of the periodic table. The first
sample of a curium compound to be visible to the naked eye was of the isotope
curium-242, of which 30 micrograms was made in 1947 by Louis. B. Werner (1921–
2007) and Isadore Perlman (1915–91) at the University of California, Berkeley. The
metal itself was produced in 1951 by heating curium fluoride and barium at 1,300°C.

Curium metal is silver in colour although it tarnishes even in dry nitrogen, as well
as being rapidly attacked by oxygen, steam, and acids, but not alkalis. Its preferred

---

[55] Naturally occurring primordial plutoniun-244 beta decays to americium-244 which beta-decays to produce
curium-244. Naturally occurring americium-242, 244, and 245 beta decay to produce curium-243, 244, and
245. Curium-242 can capture neutrons to become curium-243. Curium-246 to 249 are produced by successive
neutron-capture reactions. These isotopes have long half-lives, so they last long enough in the uranium mineral
for the neutron-capture process to produce heavier isotopes. Curium-245 has a half-life of 8,560 years, curium-
246 has a half-life of 4,760 years, curium-247 has a half-life of 15.6 million years, and curium-248 has a half-life
of 349,000 years. Neutron-capture reactions in a highly-concentrated uranium-bearing deposit can eventually
produce curium-249 which undergoes beta decay to produce berkelium-249.

oxidation state is +3 as in CmF_3, $CmCl_3$, $CmBr_3$, and CmI_3, which are pale yellow. Oxidation state +4 is known as CmF_4. Attempts to study its solution chemistry are hampered because the heat given off, due to the radioactivity of the element, quickly evaporates the water. However, the preferred state in solution is Cm^{3+}.

There are 21 known isotopes of curium with mass numbers 232 to 252, and four known isomers. The longest-lived isotope is curium-247 with a half-life of 15.6 million years. The man-made isotopes produced in kilogram quantities are curium-242 (half-life 163 days) and curium-244 (18 years).

Element of surprises

Curium is being produced on a relatively large scale, possibly in tonne quantities, by bombarding plutonium with neutrons in nuclear reactors. This is seen as a way of reducing the accumulating world stockpile of plutonium, by turning it into something which is not only relatively short-lived, but much more useful.

Darmstadtium

Pronounced darm-stad-ee-um, it is named after the German city of Darmstadt which is where the Gesellschaft für Schwerionenforschung (GSI) is located and is where it was first conclusively proved to have been made.

French *darmstadtium*; German *Darmstadtium*; Italian *damrstadtio*; Spanish *darmstadtio*; Portuguese *darmstádio*.

Element of achievement

There had been a history of failed attempts to synthesise element 110 before it was finally achieved in 1991. In 1986, physicists at the Russian Joint Institute for Nuclear Research (JINR) at Dubna made several unsuccessful attempts. In these they had bombarded lead-208 with nickel-64, bismuth-209 with cobalt-59, thorium-232 with calcium-48, and thorium-232 with calcium-44. That same year the researchers at the German GSI bombarded lead-208 with nickel-64, also to no avail. The following year the JINR made four further unsuccessful attempts: bombarding thorium-232 with calcium-44; uranium-238 with argon-40; uranium-236 with argon-40, and uranium-235 with argon-40. In 1990, the GSI made two more attempts, bombarding both uranium-235 and uranium-233 with argon-40, but again no atom of element 110 was observed.

Finally in 1991, a team headed by Albert Ghiorso (1915–) at the Lawrence Berkeley National Laboratory (LBNL) at Berkeley, California, was successful and by bombarding bismuth-209 with cobalt-59 they produced element 110 isotope 267. After exhaustive analysing of the decay chain, which originally was not clear because of a brief drop-out of time in the detector during the experiment, they announced their results in mid-1994.

Other successful reports then followed. In September of 1994, a team headed by Yuri Oganessian and Vladimir Utyonkov at the JINR made element 110 isotope-273 by bombarding plutonium-244 with sulfur-34. Its half-life was 0.3 milliseconds and it decayed via a similar cascade of α-emissions. In November 1994, a team headed by Peter Armbruster and Gottfried Munzenberg at the GSI bombarded lead-208 with nickel-62 and synthesised isotope-269. Using nickel-64 instead of nickel-62 they got isotope-271.

Darmstadtium was also detected as a decay product of even heavier elements. Isotope-273, with a half-life of 110 microseconds, was detected as part of the α-decay chain of elements of element 112. In 1998, the JINR detected isotope-281 as part of the α-decay chain from its discovery of element 114 which has a half-life of 2.6 seconds.

Chemical element

Data file	
CHEMICAL SYMBOL	Ds
ATOMIC NUMBER	110
ATOMIC WEIGHT	278 (longest-lived known isotope)

Properties such as melting point, boiling point, and density, are likely never to be measured because there will never be more than the occasional atom to be observed.

Darmstadtium is a member of group 10 (row 6d) of the periodic table, coming below platinum. So far not enough of this element has been made at one time to be visible by the naked eye, but predictions are that it would be a silvery metal susceptible to attack by air, steam, and acids. Its preferred oxidation states are likely to be +2, +4, and +6. From its position below platinum, this element should have physical properties similar to that metal and, were it long-enough lived, it should be possible to make a wide variety of compounds, including oxides MO, MO_2, and MO_3.

Darmstadtium has five known isotopes of mass numbers 267, 269, 270, 271 and 273, and two known isomers. The longest-lived isotope is darmstadtium-278 with a half-life of 10 seconds, although darmstadtium-280 is calculated to have a half-life of 11 seconds.

Element of surprises

The laboratory which first made this element was not allowed to give it its name. In fact no name for element 110 was proposed until 2001 when discovery priority was awarded to the GSI by the Joint Working Party (JWP) set up by the International Union of Pure and Applied Chemistry (IUPAC) and the International Union of Pure and Applied Physics (IUPAP). The committee had been set up in 1998 specifically to study claims for elements 110, 111, and 112, and later extended to 118. The GSI's claim to priority was confirmed by re-runs at other laboratories which showed them to be complete and credible, whereas the data from the LBNL lacked complete decay-chain data because of a fraction of a second drop-out in their detector, and the data of the Russian physicists was not complete. Nevertheless, both of these institutions had succeeded in synthesising element 110 earlier than the GSI. The name darmstadtium, chosen by the GSI, was approved by IUPAC in 2003.

Dubnium

Pronounced dub-nee-um, the element is named after Dubna, the Russian town 120 km north of Moscow where it was first made.

French *dubnium*; German *Dubnium*; Italian *dubnio*; Spanish *dubnio*; Portuguese *dúbnio*.

Element of creation and contention

In 1968, a team led by Georgy Flerov (1913–90) at the Joint Institute for Nuclear Research (JINR) bombarded americium-243 with neon-22 and made an isotope of element 105. The element was not immediately given a name, as they waited to do more research to confirm their findings, but in 1970 they proposed calling it neilsbohrium (symbol Ns) in tribute to Niels Bohr (1885–1962), the Danish physicist and winner of the 1922 Nobel Prize for physics.

In 1970, a team led by Albert Ghiorso (1915–) at the Lawrence Berkeley Laboratory (LBL) bombarded californium-249 with neon-15 and obtained isotope 261. They disputed the claim of the JINR and proposed the name hahnium (symbol Ha) naming it after the German radiochemist Otto Hahn (1879–1968).

Both names were used quite extensively in the literature with hahnium becoming quite popular. However, the controversy over the naming of element 105 became part of a larger issue regarding the discoveries and names of other transfermium elements—see Transfermium elements for more details.

The International Union of Pure and Applied Chemistry (IUPAC) set up the Transfermium Working Group (TWG) to resolve the conflicts over claims and names. It suggested that the name hahnium should be moved to element 108 and that element 105 should be called joliotium (symbol Jl) in honour of the French nuclear physicist Jean-Frédéric Joliot (1900–58). This pleased no one. Eventually the name neilsbohrium, originally suggested by the JINR, was given to element 107, but shortened to bohrium. Finally, in 1997, the name dubnium was approved for element 105 in acknowledgement of the JINR's contributions to element research and discovery.

Chemical element

Data file	
CHEMICAL SYMBOL	Db
ATOMIC NUMBER	105
ATOMIC WEIGHT	268
MELTING POINT	not known
BOILING POINT	not known

Dubnium

DENSITY	not known
OXIDE	likely to be Db_2O_5

Dubnium is a radioactive element and a member of group 5 (row 6d) of the periodic table, coming below tantalum. Not enough of this element has been made at one time to be visible by the naked eye, but predictions are that it would be a silvery metal susceptible to attack by air, steam, and acids. A first attempt at assessing dubnium's chemical behaviour was made on isotope-262, produced by bombarding berkelium-249 with oxygen-18. Dubnium's chemical properties were first analysed in 1993 and oxidation state +5 was identified in the molecules $DbCl_5$, $DbBr_5$, $DbOCl_3$, and $DbOBr_3$. Other likely compounds would be DbF_5 and even the complex ion DbF_6^-. Lower oxidation states, namely +3 and +4, might also be expected by comparison with other members of group 5.

Dubnium has 12 known isotopes of mass numbers 256–263, 266–268, and 270, and one known isomer. The longest-lived isotope is dubnium-268 with a half-life of 29 hours. (No isotope has been theoretically calculated to have a longer half-life.) The next longest-lived is dubnium-269 with a half-life of three hours.

66
Dy

Dysprosium

Pronounced dis-pro-zee-um, the name is derived from the Greek word *dysprositos* meaning hard to get, for reasons explained below.

French *dysprosium*; German *Dysprosium*; Italian *disprosio*; Spanish *disprosio*; Portuguese *disprósio*.

Dysprosium is one of 15 chemically similar elements referred to as the lanthanoid or rare earth elements, which extend from element atomic number 57 (lanthanum) to element atomic number 71 (lutetium). The term rare earth is a misnomer because some are not rare. The minerals from which they are extracted, and the properties and uses they have in common, are discussed under Lanthanoids, on page 266.

Essential element

This not-so-rare rare earth metal helps to create extremely strong magnets and extremely strong beams of light.

Element of life

The amount of dysprosium in the average person is tiny and so far no one has monitored diet for dysprosium content. The metal has no biological role and, judging by the data of other lanthanoids, the levels will be highest in bone, with smaller amounts being present in the liver and kidneys. Dysprosium is not taken up by plant roots so little gets into the main food chain. It is difficult to assess how much we take in every day but it is probably less than a microgram.

Soluble dysprosium salts, such as chloride and nitrate, are mildly toxic by ingestion, but insoluble salts are non-toxic. Based on the toxicity of dysprosium chloride to mice it can be calculated that a dose of 500 g or more would be needed to put a person's life at risk.

Element of history

Dysprosium was discovered in 1886 by Paul-Émile Lecoq de Boisbaudran (1838–1912) in Paris. The steps which led to this are part of a sequence of events which began with the discovery of impure yttrium oxide from which erbium and terbium were later to be separated—see page 618. In 1878–79, erbium was also found to be harbouring two other rare earth oxides: those of holmium and thulium. But this was not the end of the story because de Boisbaudran, working on holmium oxide, separated from it yet another rare earth oxide, that of dysprosium. His procedure involved dissolving the oxide in acid and then adding ammonia to precipitate the hydroxide. He repeated this sequence 32 times, and followed them with 26 precipitations of the insoluble salt terbium oxalate from solution. This tedious operation eventually yielded the new rare earth dysprosium. He named it thus because it really had been *dysprositos* (difficult to get). He later revealed that the separation had been carried out on the marble slab of his fireplace at home. Perhaps, because of the obstacles in securing the first sample of dysprosium, this was one of the few elements whose discovery was not challenged.

Pure samples of dysprosium were not available until Frank Spedding (1902–84) and co-workers at Iowa State University developed the technique of ion-exchange chromatography around 1950. From then on it was possible to separate the rare earth elements in a reliable and efficient manner.

Economic element

Dysprosium is found in minerals that include all the rare earth elements. The most important ores are monazite and bastnäsite. Although it is not a major component of either, it is present in extractable amounts. (China is currently the major producer of this element.) Dysprosium is also found in several other minerals such as xenotime, fergusonite, gadolinite, euxenite, and polycrase. World production is around 2000 tonnes per year, and it is available as dysprosium metal itself and dysprosium oxide.

The metal is produced by heating dysprosium fluoride with calcium. It can be cut with a knife and it machines without sparking. As a pure metal, dysprosium is useless, because it rapidly corrodes.

Dysprosium finds itself in several applications along with other lanthanoid elements. Some is used in making alloys for the best kinds of permanent magnets, such as the alloy with neodymium, iron, and boron. Dysprosium (and its neighbour holmium) have the highest magnetic moments of any elements.

Dysprosium's best known use is in halide discharge lamps where dysprosium iodide (DyI_3) is used to get very intense light. The iodide dissociates into dysprosium atoms in the lamps hot centre and they absorb energy and re-emit it in the form of visible light in the spectral regions 470–500 nm and 570–600 nm, making it appear almost like white light.

Dysprosium is used in nuclear reactors as a cermet,[56] which is there to absorb neutrons and so 'cools' the chain reaction that is providing the energy. It makes a good cermet because it neither swells not contracts under prolonged neutron bombardment. Another use in the field of radioactivity is in dosimeters for monitoring exposure to ionizing radiation. When dysprosium-doped crystals of calcium sulfate or calcium fluoride are exposed to damaging radiation this excites the dysprosium atoms which then luminesce and this indicates the degree of exposure the dosimeter has been subjected to.

Dysprosium oxide is used as a dopant in special ceramics, such as barium titanium oxide ($BaTiO_3$), which are used to produce high-capacitance, small-size capacitors for

---

[56] Cermets are a composite material made of ceramic and sintered metal. They are highly resistant to high temperatures, corrosion and abrasion.

electronic applications. Other uses of dysprosium are in erasable optical laser-read discs, and in temperature-compensating capacitors.

Environmental element

Dysprosium in the environment	
EARTH'S CRUST	6 p.p.m.
	Dysprosium is the 42nd most abundant element.
SOILS	av. 5 p.p.m., range 1–12 p.p.m.
SEAWATER	1 p.p.t.
ATMOSPHERE	virtually nil.

Dysprosium is one of the more abundant lanthanoid elements and is more than twice as abundant as tin. It poses no environmental threat to plants or animals.

Chemical element

Data file	
CHEMICAL SYMBOL	Dy
ATOMIC NUMBER	66
ATOMIC WEIGHT	162.50
MELTING POINT	1,412°C
BOILING POINT	2,560°C
DENSITY	8.6 g/cm^3
OXIDE	Dy_2O_3

Dysprosium is a bright, silvery metal and a member of the lanthanoid group (row 4f) of the periodic table of the elements. It is slowly oxidised by oxygen, reacts with cold water, and rapidly dissolves in acids. Its preferred oxidation state is +3 as in its salts such as DyF_3, which is green, and in solution it exists as the yellow Dy^{3+} ion, around which cluster as many as nine water molecules.

There are 36 known isotopes of dysprosium with mass numbers 138 to 173, and 12 known isomers. In mineral deposits there are seven naturally occurring isotopes: dysprosium-164 constitutes 28 percent, dysprosium-162 (25.5 percent), dysprosium-163 (25 percent), dysprosium-161 (19 percent), dysprosium-160 (2.3 percent), dysprosium-158 (0.1 percent), and dysprosium-156 (0.06 percent).[57] All are stable except dysprosium-156 which is radioactive with a half-life of 10^{18} years (a *billion* billion years). The next longest lived radioactive isotope is dysprosium-154 with a half-life of 3 million years.

---

[57] There are also traces of dysprosium-170 produced by the double-beta-plus decay of erbium-170.

Element of surprises

1. A most unusual property is possessed by one type of dysprosium alloy, known as Terfenol-D. This visibly lengthens and shortens when exposed to a magnetic field. For more information about Terfenol-D see terbium.

2. Dysprosium might be the element with the highest stable atomic number: 66. Nuclear theory suggests that all elements above this are not stable and that for these heavier elements, where we regard them as stable, we only do so because their radioactive half-lives are so incredibly long.

Einsteinium

Pronounced iyn-sty-nee-um it was named after Albert Einstein (1879–1955) who won the Nobel Prize for physics in 1921.

French *einsteinium*; German *Einsteinium*; Italian *einsteinio*; Spanish *einstenio*; Portuguese *einstenio*.

Essential element

This element has an interesting past but, as yet, no future.

Element of life

Einsteinium has no role to play in living things, and it is never encountered outside nuclear facilities or research laboratories. It is highly dangerous because of the radiation it emits.

Element of history

A large number of scientists contributed to the discovery of einsteinium, prominent among them were Gregory R. Choppin (1927–), Stanley G. Thompson (1894–1953), Albert Ghiorso (1915–), and Bernard G. Harvey. Einsteinium was discovered in the debris of 'Mike', the code name for the first thermonuclear explosion which took place on Eniwetok, a Pacific atoll, on 1 November 1952. Mike was the precursor to the development of the hydrogen bomb.

Fall-out material, gathered from a neighbouring atoll, was sent to Berkeley, California, for analysis. There, tonnes of radioactive coral were sifted and examined, and a month later the new element was identified. However, it was not announced for another three years because the work was kept secret for security reasons.

In fact, fewer than 200 atoms had been found in the debris, but it was enough to show that the uranium-238 which had been used to provide the heat necessary for triggering a thermonuclear explosion, had been exposed to such a high flux of neutrons that some of its atoms had captured several of these particles and through a series of neutron-capture and beta-decay steps had formed einsteinium-253, which has a half-life of 20.5 days. (The process of neutron-capture, followed by beta decay, had produced elements of atomic numbers 93 to 100.)

By 1961, enough einsteinium had been collected together to be visible to the naked eye, and was even weighed on a specially constructed balance, although it amounted to a mere hundredth of a milligram (0.01 mg).

Economic element

Einsteinium is available for purchase in microgram quantities. Milligram quantities are produced by bombarding plutonium-239 with a high flux of neutrons inside a nuclear reactor for several months. The sequence of transformation is one of alternately accepting neutrons (n) and emitting beta-particles (β^-), and it goes via the elements americium, curium, berkelium, and californium, as follows:

Pu-239 + 2n $\rightarrow$ Pu-241 $-$ β^- $\rightarrow$ Am-241 + n $\rightarrow$ Am-242 $-$ β^- $\rightarrow$ Cm-242 + 7n $\rightarrow$ Cm-249 $-$ β^- $\rightarrow$ Bk-249 + n $\rightarrow$ Bk-250 $-$ β^- $\rightarrow$ Cf-250 + 3n $\rightarrow$ Cf-253 $-$ β^- $\rightarrow$ Es-253

If isotope einsteinium-253 then accepts a further neutron it becomes the isotope einsteinium-254 which then emits a beta particle and decays to fermium-254.

There are, as yet, no known applications of einsteinium.

Environmental element

Einsteinium occurred naturally on planet Earth a long time ago. It was produced for hundreds of thousands of years in natural nuclear reactors. Starting about 1.8 billion years ago, geological processes concentrated uranium to start sustained nuclear reactors at 16 locations in what is now Oklo, Gabon, in Africa. These functioned continuously and by neutron capture and beta decay processes they produced the transuranium elements neptunium (93) through to fermium (100).

Chemical element

Data file	
CHEMICAL SYMBOL	Es
ATOMIC NUMBER	99
ATOMIC WEIGHT	252.1 (isotope einsteinium-252)
MELTING POINT	about 860°C
BOILING POINT	not known
DENSITY	8.84 g/cm³
OXIDE	Es_2O_3

Einsteinium is a radioactive metal element and a member of the actinide group (row 5f) of the periodic table. It reacts with oxygen, steam, and acids, but not with alkalis. Its preferred oxidation state is +3 as in the compounds EsF_3, $EsCl_3$, and $EsBr_3$. A lower +2 oxidation state is known as the compound EsI_2.

There are 17 known isotopes of einsteinium with mass numbers from 241 to 257, and three known isomers. All einsteinium is radioactive, and the longest-lived isotope is einsteinium-252 with a half-life of 472 days. The next longest-lived is einsteinium-254 with a half-life of 276 days.

The symbol originally proposed for einsteinium was just E, but when IUPAC approved the name in 1957 they changed the symbol to Es to conform with their new rule that said all newly name elements should have symbols with two letters.

Element of surprises

The first thermonuclear explosion may have produced the first einsteinium but sadly it cost the life of US pilot, First Lieutenant Jimmy Robinson. His job was to fly through the cloud formed by the explosion, collecting samples of dust on filter papers. He waited too long before deciding to return to base, ran out of fuel, ditched his plane in the sea, and died.

Erbium

Pronounced erb-ee-um, it is named after Ytterby, Sweden—see 'The village of the four elements' under yttrium, page 618.

French *erbium*; German *Erbium*; Italian *erbio*; Spanish *erbio*; Portuguese *érbio*.

Erbium is one of 15 chemically similar elements referred to as the lanthanoid or rare earth elements, which extend from element atomic number 57 (lanthanum) to element atomic number 71 (lutetium). The term rare earth is a misnomer because some are not rare. The minerals from which they are extracted, and the properties and uses they have in common, are discussed under Lanthanoids, page 266.

Essential element

Think pink, think erbium—at least when it comes to art glass and goggles for glass-blowers.

Element of life

Erbium has no biological role, but it has been noted that erbium salts stimulate metabolism. It is difficult to separate out the various amounts of the different lanthanoids in the human body, but they are present, and the levels are highest in bone, with smaller amounts being present in the liver and kidneys.

There is no estimate of the amount of erbium in an average adult, and no one has monitored diet for erbium content, so it is difficult to judge how much we take in every day but it is probably only a milligram or so per year. Erbium is not taken up by plant roots so little gets into the main food chain.

Erbium can be mildly toxic by ingestion, but insoluble salts are non-toxic.

Medical element

The erbium-nickel alloy, of composition Er_3Ni, has a very high ability to absorb heat and this helps to ensure the economical operation of MRI (magnetic resonance imaging) body scanners, which work at liquid helium temperatures of around −270°C. Replacing some of the erbium with ytterbium makes them work even better.

Erbium:YAG lasers (YAG stands for yttrium aluminium garnet) are safe to use in dentistry and in dermatology and they rely on the erbium ion's emission at 2.9 microns wavelength. This is strongly absorbed by surface water molecules and boosts their temperature sufficient to strip off discolouring tooth stains, and it enables delicate cosmetic laser surgery to be performed without damage to underlying tissues.

The isotope erbium-169 (half-life 9.5 days) is a high-energy beta-particle emitter which is currently being studied for use in radiation therapy treatments.

Element of history

Erbium was discovered at Stockholm in 1843 by the Swedish chemist Carl Gustav Mosander (1797–1858). Erbium, as its oxide, was one of the first lanthanoid elements to be discovered, but the picture is clouded because early samples must have contained other rare-earths which were isolated from erbium oxide at a later date. In 1843, Mosander extracted two new oxides from yttrium oxide. These were erbium oxide, which was pink; and terbium oxide, which was yellow. He originally named these new oxides the other way round but later workers who confirmed his discovery got them confused and gave them the names by which we know them today.

Mosander's discovery was not the end of the matter, and in 1878 the Swiss chemist Jean-Charles Galissard de Marignac (1817–94), working at the University of Geneva, extracted another element from erbium oxide and called it ytterbium. This too was impure and scandium was extracted from it a year later.

A sample of pure erbium metal was not produced until 1934, when Wilhelm Klemm (1896–1985) and Heinrich Bommer achieved it by heating purified erbium chloride with potassium.

Economic element

Erbium is found in minerals that include all the rare earth elements. The chief mining area is China, and the most important ores are monazite and bastnäsite. Although erbium is not a major component of either ore, it is present in extractable amounts. Better sources of the element are xenotime (mainly yttrium phosphate) and euxerite (a complex ore of many metals). World production is 500 tonnes per year, mainly in the form of erbium oxide. The metal is produced by heating erbium chloride with calcium under vacuum and is available as ingots, lumps, or powder.

Erbium finds little use as the metal itself because it slowly tarnishes in air and it is attacked by water. However, it is more corrosion-resistant than the other lanthanoid elements. Sometimes erbium is alloyed with metals such as vanadium because it lowers their hardness making them more workable.

Erbium oxide is important in several situations that involve visible and infrared light. For example, it is added to the glass of special safety spectacles for workers, such as welders and glass-blowers, whose eyes are exposed to intense infrared light, because it absorbs these rays. Erbium is incorporated into phosphors that can convert infrared light into visible light giving a green image. When erbium is added to glass it produces a pleasing pink colour and this is used for some sunglasses and imitation gem stones.

Erbium is used to dope optical fibres at regular intervals to amplify signals; it does this by converting other wavelengths that are pumped down the fibre to the wavelength of the information-carrying signal.

The isotope erbium-167 is highly neutron absorbing and is used in the making of special nuclear fuel rods that have extended life-spans, enabling pressurised water reactors to work in research submersibles for periods of up to two years.

Environmental element

Erbium in the environment	
EARTH'S CRUST	4 p.p.m.
	Erbium is the 44th most abundant element.
SOILS	av. 1.6 p.p.m.
SEAWATER	0.8 p.p.t.
ATMOSPHERE	virtually nil.

Erbium is one of the more abundant rare earth elements and is almost twice as abundant as tin. It poses no environmental threat to plants or animals.

Chemical element

Data file	
CHEMICAL SYMBOL	Er
ATOMIC NUMBER	68
ATOMIC WEIGHT	167.26
MELTING POINT	1,529°C
BOILING POINT	860°C
DENSITY	9.1 g/cm³
OXIDE	Er_2O_3

Erbium is a bright silvery metal which is a member of the lanthanoid group (row 4f) of the periodic table. It reacts with oxygen and water, in both cases only very slowly, and it dissolves in acids. The preferred oxidation state is +3 as in compounds such as ErF_3 and $ErCl_3$, and in solution it exists as yellow Er^{3+} ions, surrounded by nine water molecules.

There are 35 known isotopes of erbium with mass numbers 143 to 177, and 13 known isomers. Erbium in minerals consists of six naturally occurring isotopes: erbium-166, which comprises 33.5 percent; erbium-168 (27 percent), and erbium-167 (23 percent), erbium-170 (15 percent), erbium-164 (1.5 percent) and erbium-162 (0.1 percent). Of these erbium-170 is radioactive with a half-life of 10^{17} years (equivalent to 100 million billion years), and so is erbium-162 with a half life of 10^{14} years (equivalent to 0.1 million billion years).[58] The next longest-lived radioactive isotope is erbium-169 with a half-life of 9.4 days.

---

[58] In fact nuclear theory suggests that all isotopes of erbium are radioactive and those we now think of as stable are so regarded because they have incredibly long half-lives.

Element of surprises

1. About the only way to tint glass pink is to add erbium oxide to it. Up to 5 percent of the pigment can be added and the result is an attractive colour that is used in sun glasses, crystal glassware, and jewellery. Ceramics can also be coloured pink this way. Erbium has this ability because it has a narrow absorption band at 530 nm in the green part of the spectrum, which means its salts appear pink. For the same reason when an erbium-impregnated gas mantle is heated it emits an emerald-green light.

2. Putting energy into something causes its temperature to rise—but not always. It can sometimes cause the temperature to drop. In 2006 it was reported from the University of the Basque Country at Bilbao, Spain, that adding small amounts of erbium ions, Er^{3+}, to a glass made of potassium lead chloride, KPb_2Cl_5, saw it get cooler when a laser beam was focussed on it. The drop in temperature was about 0.5°C.

Europium

Pronounced yoo-roh-pee-um, it is named after Europe.

French europium; German Europium; Italian europio; Spanish europio; Portuguese európio.

Europium is one of 15 chemically similar elements referred to as the lanthanoid or rare earth elements, which extend from element atomic number 57 (lanthanum) to element atomic number 71 (lutetium). The term rare earth is a misnomer because some are not rare. The minerals from which they are extracted, and the properties and uses they have in common, are discussed under Lanthanoids, page 266.

Cosmic element

Europium has been identified from its spectral line as being present in the sun and in certain stars.

Essential element

Europium makes low energy light bulbs possible, and it thwarts forgers who would seek to print euro banknotes.

Element of life

Europium is not easily absorbed by plant roots, so little gets into the main food chain. However, it is present in some plants but only in the 30–130 p.p.b. range (dry weight) and vegetables have generally much less than this with some having as little as 0.04 p.p.b. The amount of europium in humans is quite small and the metal has no biological role. It is difficult to single out europium from among the various lanthanoids in the human body, but it is present and is assumed to follow the same metabolic route as calcium and strontium. There is no estimate of the amount of europium in an average adult, and no one has monitored diet for europium content, so it is difficult to judge how much we take in every day but it is probably only a milligram or so per year.

Europium salts could be mildly toxic by ingestion, but insoluble salts are completely non-toxic.

Medical element

There is no known medicament which contains europium. However, europium tetracycline (aka EuTc) is used to screen for various medical conditions such as asthma, osteoporosis, Down's syndrome, and degenerative diseases of the brain and nervous system. Conditions such as these involve enzymes that control hydrogen peroxide, some releasing this chemical, others eliminating it, and their role can be highlighted by EuTc because this chemical reagent fluoresces when in contact with samples containing peroxides.

Element of history

Europium's story began with that of cerium which was discovered in 1803 and named after the then recently observed dwarf planet Ceres in the asteroid belt. Then in 1839 Carl Gustav Mosander (1797–1858) separated two other elements from cerium: lanthanum and one he called didymium. Didymium was recognised as an element for the next 40 years until in 1879 Karl Auer (1858–1929) showed it to be a mixture mainly of praseodymium and neodymium.

However, didymium also harboured another rarer metal, samarium, which was separated that same year by Paul-Émile Lecoq de Boisbaudran (1838–1912). But even his samarium was impure, and in 1886 Jean Charles Galissard de Marignac (1817–94) extracted another rare earth, gadolinium, from it. Despite all this activity, samples of samarium still had unexplained lines in its atomic spectrum, indicating the presence of yet another element. Eventually in 1901 the French chemist Eugène-Anatole Demarçay (1852–1904) carried out a painstaking sequence of crystallisations of samarium magnesium nitrate, and separated another new element, which he named europium.

Economic element

Europium as Eu^{2+} can replace calcium and strontium in certain minerals, such as the calcium feldspars (calcium aluminium silicates) and these can have unexpectedly high levels of europium. These are not the commercial source of the metal because it occurs at much higher levels in the ores monazite and bastnäsite, which contain all the lanthanoids, and while europium is not a major component, it is present in extractable amounts and it is one of the more abundant lanthanoids. (Monazite and bastnäsite are discussed in more detail together under lanthanum—see page 274.)

World production of rare earth ores is 120,000 tonnes per year, mostly in China. Pure europium metal production is around 270 tonnes a year, and this is obtained by heating europium oxide with lanthanum metal in a tantalum crucible under vacuum conditions. It can also be obtained by the electrolysis of a mix of molten europium chloride ($EuCl_3$) and sodium chloride (NaCl) with graphite electrodes.

Europium phosphors can be stimulated to give a bright red colour, thanks to its emission line at 610 nm. Indeed, it was this property which alerted early pioneers of television to the possibility of developing colour. The phosphor commonly used is yttrium oxysulfide doped with europium. Europium is also used to print euro banknotes to validate them. When these are exposed to UV light parts should glow red because of europium. Banknote counting machines can automatically detect a forgery by the lack of this element.

Low-energy light bulbs, which are referred to technically as trichromatic lamps, produce a soft light like that of incandescent light bulbs and not as harsh as the cold glare of fluorescent tubes. Europium provides not only the red component of the spectrum, but its lower oxidation state, europium(II), emits in the blue. Commercial blue phosphors rely on europium-doped materials, such that of strontium aluminate. For powerful street lighting a little europium is added to mercury vapour lamps to give a more natural light.

Other uses of europium are in thin-film superconductor alloys, and in lasers. A europium-doped barium fluoride bromide crystal will absorb X-rays and, when subsequently stimulated by a laser, will emit visible light.

Environmental element

Europium in the environment	
EARTH'S CRUST	2 p.p.m.
	Europium is the 50th most abundant element.
SOILS	*approx.* 1.2 p.p.m.
SEAWATER	0.2 p.p.t.
ATMOSPHERE	virtually nil.

Europium is one of the more abundant rare earth elements and is almost twice as abundant as tin. It poses no environmental threat to plants or animals.

Chemical element

Data file	
CHEMICAL SYMBOL	Eu
ATOMIC NUMBER	63
ATOMIC WEIGHT	151.964
MELTING POINT	822°C
BOILING POINT	1,597°C
DENSITY	5.2 g/cm^3
OXIDE	EuO, Eu$_2$O$_3$

Europium is a soft, silvery metal rather like lead, and a member of the lanthanoid group (4f) of the periodic table. It is the most reactive of the lanthanoid metals, reacting quickly with oxygen and water. It will ignite spontaneously when heated to around 180°C, and it burns in air. Europium is unusual among the rare earth elements in also having a stable +2 oxidation state which it exhibits in the form of salts such as EuSO$_4$ and the Eu^{2+} ions in solution. The preferred oxidation state is +3 as in EuCl$_3$, and Eu^{3+} ions in solution.

There are 38 known isotopes of europium with mass numbers ranging from 130 to 167, and 12 known isomers. There are two naturally occurring isotopes of europium: europium-151 which accounts for 48 percent, and europium-153 (52 percent). The former is now known to be radioactive with a half-life of 5 billion billion years

(5×10^{18}), which means that in 1 gram of europium only one atom a day will undergo radioactive decay. There are also minute traces of some radioactive europium isotopes in uranium ores, where they have been formed by nuclear fission. The second longest lived radioactive isotope is europium-150 which has a half-life 36.9 years.

Element of surprises

1. Had the early investigators of the rare earth elements known a little more about the chemistry of europium, they might have isolated it much earlier. Unlike the other lanthanoids, europium has a stable reduced form, europium(II), and this could have been the key to separating it easily from the other lanthanoid elements. The insoluble salt, europium(II) sulfate $(EuSO_4)$ will precipitate from solution leaving the other lanthanoids behind.

2. Europium was found in much higher concentrations than expected in samples of rock brought back from the Moon's surface. It was present as the reduced form of Eu^{2+}, rather than the more normal Eu^{3+}, suggesting that the minerals of the Moon were formed under highly reducing conditions. This unnaturally high abundance of europium suggests that the Earth and the Moon were not derived from a common source of cosmic material, and as yet there is no agreed theory about how the Moon was formed.[59]

3. Blue John is a rare transparent form of the mineral calcium fluoride which was discovered in Derbyshire, England. For inexplicable reasons crystals of Blue John glowed with an eerie light when exposed to UV. (The mineral gave its name to the phenomenon of fluorescence.) The reason it behaved this way was eventually traced to the presence of europium atoms in the crystals.

---

[59] The most likely explanation for the origin of the Moon is that a very large object hit the early Earth sending a lot of it into near-Earth space where it eventually coalesced to form the Moon. This then began collecting other debris orbiting in space that was gravitationally attracted to it.

Fermium

Pronounced ferm-ee-um, it is named after the nuclear physicist Enrico Fermi (1901–54) who won the Nobel Prize for physics in 1938, and who, in 1942 at the University of Chicago, initiated the first self-sustaining nuclear reaction to be produced by human endeavour.
French *fermium*; German *Fermium*; Italian *fermio*; Spanish *fermio*; Portuguese *fermio*.

Essential element

Too elusive to be captured *en masse*, yet tantalising in what it might one day provide.

Medical element

The isotope fermium-255, an alpha-particle emitter with a half-life of 20 hours, is currently being researched for possible use as a radiopharmaceutical for cancer treatment.

Element of history

A large number of scientists contributed to the discovery of fermium, prominent among them were Gregory R. Choppin (1927–), Stanley G. Thompson (1894–1953), Albert Ghiorso (1915–), and Bernard G. Harvey. It was discovered in 1953 in a sample of the debris of the first thermonuclear explosion (aka hydrogen bomb) which took place on the Pacific atoll of Eniwetok on 1 November 1952. Fall-out debris was gathered on a neighbouring atoll, and found to contain two new elements, einsteinium and fermium. There were in fact only about 200 atoms remaining of the fermium which had been formed during the explosion, in which a uranium-238 bomb (aka atomic bomb) was used to provide the heat necessary to trigger a thermonuclear explosion. The uranium-238 had been exposed to such a flux of neutrons that some of its atoms had captured several of them. Through a process of neutron-capture, followed by beta-decay, many elements were formed, including those of atomic numbers 93 to 100, among which was fermium-255.

Because of the Cold War, no announcement was made and no name suggested for element 101, the fear being that it might have military applications. News of its discovery was kept secret until 1955, by which time a group of physicists at the Nobel Institute in Stockholm had independently made a few atoms of it. In 1953 and 1954, they bombarded uranium-238 with oxygen-16 nuclei to produce fermium-250 with a half-life of 30 minutes. They suggested calling it centurium (chemical symbol Ct) based on its atomic number of 100, but unbeknownst to them it had already been discovered. The US group proposed the name fermium which was approved by IUPAC in 1957.

Economic element

Fermium-253 can be formed by the bombardment of plutonium-239 with neutrons in nuclear reactors; it has a half-life of three days. There are no commercial reasons for fermium to be produced, but it might one day have some uses in medicine. Annual world production of fermium is estimated to be only a few micrograms per year.

Environmental element

Fermium existed on planet Earth for hundreds of thousands of years, starting about 1.8 billion years ago when natural geological processes concentrated uranium at 16 locations in what is now Oklo, Gabon, in Africa. Thus began a series of nuclear reactors and these functioned continuously for hundreds of thousands of years producing transuranium elements of atomic numbers 93 (neptunium) to 100 (fermium), and in the latter case mainly as isotopes 254 to 257.

Chemical element

Data file	
CHEMICAL SYMBOL	Fm
ATOMIC NUMBER	100
ATOMIC WEIGHT	257 (longest-lived known isotope)
MELTING POINT	*ca.* 1,500°C (estimated)
BOILING POINT	not known
DENSITY	not known
OXIDE	none yet made, but likely to be Fm_2O_3

Fermium is a member of the actinide group (row 5f) of the periodic table. So far not enough fermium has been made at one time to be visible by the naked eye, but predictions are that it would be a silvery metal susceptible to attack by air, steam, and acids. Its preferred oxidation state appears to be +3 and this is the form in which it exists in solution as Fm^{3+}.

Fermium has 20 known isotopes with mass numbers ranging from 241 to 260, and two known isomers. The longest-lived is fermium-257 with a half-life of 100 days, and the next longest-lived is fermium-253 with a half-life of 3 days.

Element of surprises

Fermium is the element of highest atomic number that can be produced in a nuclear reactor. However, making it is an almost futile task because no sooner does an atom form than it is gone. The problem lies with fermium-257. Though this is the longest-lived isotope, it is also good at capturing neutrons. It can easily absorb another neutron to form fermium-258 which has a half-life of only 0.37 milliseconds. Nevertheless, the little fermium-257 that can be collected—less than a picogram at a time (a million millionth of a gram, which is 10^{-12} g)—has enabled some investigation of its properties to be carried out.

Fluorine

Pronounced flew-or-een, the name is derived from the Latin *fluere* meaning to flow, because the common fluoride mineral fluorspar (calcium fluoride, CaF_2) melts when heated in a flame.

French *fluor*; German *Fluor*; Italian *fluoro*; Spanish *flúor*; Portuguese *fluor*.

Fluorine and fluoride: the former is the name of the element which exists as the molecule F_2, whereas fluoride, as in sodium fluoride, is a fluorine atom that has acquired a negative electron and is written F^-. F_2 is dangerously reactive, while F^- is stable and is the form encountered in nature.

Essential element

Fluorine is difficult to tame but once it is under control then it can deliver some remarkable benefits as drugs and plastics.

Element of life

Fluorine (as fluoride) in the human body	
BLOOD	0.5 mg per litre
BONE	varies between 2,000 and 12,000 p.p.m.
TISSUE	*approx.* 0.05 p.p.m.
TOTAL AMOUNT IN BODY	3–6 g.

As long ago as 1802, fluoride was detected in ivory, bones, and teeth. By the mid 1800s it was found to be present in blood, seawater, eggs, urine, saliva, and hair. Although it appeared to be in all living things, this did not *prove* that it was an essential element. That had to wait until tests on animals showed that if it was excluded from their diet they failed to grow properly, or were anaemic and infertile. Today it is also regarded as essential for humans, but only in minute doses.

Too much fluoride, whether taken in from the soil by roots of plants, or absorbed from the atmosphere by their leaves, retards growth and reduces crop yields. However, some crops are resistant to fluoride, such as asparagus, beans, cabbage, and carrots, while those most affected are maize (corn) and apricots.

Fluoride has a powerful effect on enzymes, effectively blocking their activity, and this is made use of in biochemical research and, more practically, in cleaning fermentation vats in breweries and distilleries.

Five naturally occurring organo-fluorine compounds have been identified and the enzyme which enables these to be made was extracted from the soil bacterium *Streptomyces cattleya* in 2002. It has now been genetically engineered into a deep-sea microbe with the object of using it to manufacture organofluorine anti-cancer drugs.

Food element

The average person takes in between 0.3 and 3 mg of fluoride a day in their diet. Fluoridated tap water may provide some of this, but most comes from foods such as chicken, pork, eggs, potatoes, cheese, and tea (a cup of strong tea provides 0.4 mg). Mackerel, sardines, salmon, and sea salt are particularly rich in fluoride because they come from seawater which contains 1 p.p.m. Mackerel has 27 p.p.m. (fresh weight). Vegetables generally have much less than this, ranging from 3 to 20 p.p.m. (dry weight).

Medical and dental element

In humans, most of the fluoride accumulates in the bones and teeth. It strengthens both of these by converting the calcium phosphate of which they are made into fluoroapatite, a harder material. This, in tooth enamel, can better resist the corrosive effect of acids produced by oral bacteria. For this reason it is added to water supplies and toothpastes.

As long ago as 1901 it was noticed that children who had mottled teeth were less likely to suffer from the cavities caused by dental caries. We now know that this discoloration was due to a high intake of fluoride. Eventually there began a campaign to have water fluoridated and this was supported by comparing the average number of missing, decayed, and cavities in the teeth of children who drank such water, and that average was 2.8, with the number in children whose water was not fluoridated, and that average was 4.25. Clearly fluoridation is beneficial to teeth, but that has not prevented several groups opposing it on the basis that it might have adverse effects and the debate continues. In some countries fluoridation is mandatory, such as Ireland, in others it relies on local water authorities as in the UK and USA, while in Sweden it is banned.

About half the ingested fluoride ends up in bone and it has been found that those who live in areas where the water is fluoridated have bone levels of around 1 p.p.m compared to those in non-fluoridated regions where the levels are on average 0.6 p.p.m. Fluoride is prescribed in some countries as a preventative treatment for osteoporosis in the elderly.

Another medical role of fluorine comes in the form of organo-fluorine drugs, and these have brought heath benefits on a large scale. (About 20 percent of pharmaceuticals and 30 percent of agrochemicals are organofluorine compounds.) Several organofluorine drugs so potent that a single dose will clear up certain minor infections, for example, ofloxacin can cure gonorrhoea. Fluconazole is the drug of choice for attacking fungus diseases which can threaten the life of transplant patients whose immune system is deliberately suppressed, as well as curing more mundane infections such as vaginal thrush. These drugs, known collectively as fluoroquinolones, are easily absorbed and penetrate to parts of the body that conventional antibiotics find difficult to reach, for example the bladder. They work by interfering with microbial enzymes, such as that responsible for coiling bacterial DNA, which then prevents the disease pathogens from multiplying.

Antidepressants of the SSRI type (aka selective serotonin reuptake inhibitors) are generally organo-fluorine molecules such as fluoxetine (better known as Prozac). Many anaesthetics are fluorocarbon compounds such as desfluane.

The isotope fluorine-18 is used in medical diagnosis through the technique known as positron emission tomography (PET). This short-lived isotope has a half-life of only 110 minutes and it decays by emitting a positron. This immediately collides with its negative counterpart, an electron, and the two annihilate each other in a burst of intense energy in the form of gamma rays which can be used to monitor the body's vital organs. Fluorine-18 is made by bombarding oxygen-18, as water, with protons using a cyclotron or linear accelerator and it is then attached to a glucose molecule which is injected. More glucose is taken up by cells that are replicating rapidly such as cancer cells thereby enabling them to be highlighted by fluorine-18.

Dangerous element

Both fluorine and fluoride are dangerous. Humans, as well as animals, can be affected by too much fluoride and suffer from fluorosis, the first signs of which are mottled teeth.[60] Later, there may be a hardening of the bones, which can lead to a deformed skeleton. In certain parts of India, such as the Punjab, the condition is endemic, especially where villagers drink water from wells with high levels of fluoride, up to 15 p.p.m. About 25 million Indians suffer a mild form of fluorosis, with many thousands showing skeletal deformities. According to UNICEF skeletal fluorosis is endemic in 25 countries, most notably in India and China, in the former due to groundwater and in the latter due to burning fluoride-rich coal.

That fluoride concentrates in the skeleton is shown by its level in bone ash. Normal bone ash has up to 1,000 p.p.m. (0.1 percent) whereas the bone ash of those suffering serious fluorosis has around 8,000 p.p.m. (0.8 percent), while those of intermediate levels will have experienced such symptoms as chronic joint pain akin to arthritis.

Fluorine gas is extremely toxic and breathing it at concentrations of 0.1 percent for only a few minutes will kill. Luckily the gas is so pungent that it acts as its own early warning system, but in any case stringent legal precautions surround its use. Hydrofluoric acid is also dangerous, much more so than the related acid, hydrochloric acid, because it can penetrate skin and cause deep and painful burns which take a long time to heal. Fluoride itself is less toxic although it too is highly poisonous. As little as 0.25 g of fluoride will provoke a toxic reaction, while 5 g could prove fatal and 10 g would definitely kill the average adult. Because it is so toxic, fluoride has been used as an effective insecticide for cockroaches and ants.

It was just such a powder which led to a terrible accident at a hospital in the USA in 1943. Patients were served scrambled eggs to which sodium fluoride had been added instead of sodium chloride (common salt): 163 were taken ill and 47 died. Deliberate murders have been committed with fluoride and in 1949 this was to cause

---

[60] The current maximum allowable fluoride in drinking water in the USA is 4 p.p.m. above which it is likely to cause mottling.

the sudden death of Mamie Furr of Bogalusa, Louisiana. She was poisoned by Cola Leming, a neighbour who was having an affair with Mamie's husband. Cola had put sodium fluoride in Mamie's coffee which killed her within hours. She was found guilty and sentenced to life imprisonment.

When a new fire extinguisher was being tested in Italy in 1970 the two men who were researching its safety tested it on themselves by breathing the fluorocarbon vapour on which it was based and which would smother a fire. The first man passed out within minutes of breathing it and he eventually recovered. However the second man who went to his aid also fainted but he could not be revived and died almost immediately.

Element of war

Fluorine played a key role in the separation of uranium isotopes which made possible the first atomic bomb, the story of which is discussed under uranium.

The Nazis were also aware of the importance of fluorine, but for a different reason. They built a factory to produce chlorine trifluoride (ClF_3) as an incendiary agent for flame-throwers. It began operating in January 1945, but had to close almost immediately due to advancing Russian armies.

Today fluorine still has a role in warfare, but it is used to protect. Combat uniforms have a surface layer of a polymer to which perfluoro (C_8F_{17}) groups are attached. This not only repels water but also repels droplets containing chemical and biological weapons which also cannot adhere to the fabric.

Element of history

In 1529, Georgius Agricola (1494–1555) described the use of fluorspar (calcium fluoride, CaF_2) as a flux, pointing out that this mineral would melt and flow when heated in a fire, and that it was added to furnaces for smelting metals because the melt became more fluid and easier to work with. A second benefit was discovered in 1670 by Heinrick Schwandhard who found that if fluorspar was dissolved in acid the solution could be used to etch glass. The first glass to be so affected was reputed to be that of his spectacles.

The early chemists were aware that fluorides contained an unidentified element, fluorine, and that its compounds greatly resembled those of chlorine, but they could not isolate it. The French scientist, André Ampère (1775–1836), coined the name fluorine in 1812 but later changed his mind in favour of phthorine, derived from the Greek *phthoros*, meaning destructive. Thankfully the name fluorine prevailed. Humphry Davy (1778–1829) and his contemporaries were unable to isolate the element, and, like several of them, Davy became ill by trying to isolate it from its acid, hydrofluoric acid. Some chemists, such as the French chemist Jerome Nicklès of Nancy, died from breathing in hydrogen fluoride (HF) vapour.

The British chemist George Gore in 1869 passed an electric current through water-free hydrofluoric acid (HF) and found that the gas which was liberated at the anode reacted violently with his apparatus. This was fluorine, but he was unable to collect it and prove it to be so. It was not until 1886 that the French chemist Henri Moissan

(1852–1907) in Paris, obtained it by the electrolysis of a mixture of potassium bifluo-ride (KHF$_2$) dissolved in anhydrous HF. He was awarded the 1906 Nobel Prize for chemistry for this work (and for the development of the electric furnace).

Economic element

Annual world production of the mineral fluorite (aka fluorspar) is around 6 million tonnes. Most is mined in China which accounts for more than half of the world's production, followed by Mexico, Mongolia, and South Africa. Fluorspar deposits are to be found on every continent and world reserves amount to 240 millions tonnes with South Africa holding 40 million of them, followed by Mexico (32 million) and China (21 million).

Most fluorspar (35 percent) is used in the smelting and refining of metals espe-cially in steel making and aluminium production. A lot goes into the manufacture of hydrofluoric acid (400,000 tonnes per year), which in turn is used to make fluorine gas (15,000 tonnes per year). The hydrogen fluoride from which fluorine gas is made comes as a by-product of phosphoric acid manufacture because the apatite rock used as the source of this contains a significant amount of fluoride. Fluorine gas is still produced by the method devised by Moissan, although whereas he worked his appa-ratus at −23°C to reduce the gas's reactivity, modern electrolytic cells work at 167°C and are made of Monel metal, the nickel alloy which resists attack by F$_2$.

Fluorine gas is generally produced close to where it is needed, and the main uses are in making uranium hexafluoride (UF$_6$) for the nuclear industry, sulfur hexafluo-ride (SF$_6$) for the electricity supply industry, and chlorine trifluoride (ClF$_3$) for the chemical industry. UF$_6$ is needed to separate the nuclear fuel, uranium-235, from its more common isotope uranium-238. SF$_6$ is the inert insulating gas which surrounds high-power electricity transformers although it is a powerful greenhouse gas and it lingers long in the atmosphere. ClF$_3$ is used to make fluorocarbons. Fluorine gas is used for plasma etching in semiconductor manufacture and flat screen displays. Nitrogen fluoride (NF$_3$) gas is used to clean chambers and as an etching gas in the manufacture of semiconductors.

Fluorine gas can now be handled in bulk and even transported safely as a liquid in containers that are protected by liquid air. Diluted with nitrogen, fluorine is much safer to handle and this can be used to treat polyethene (polyethylene) containers to give them a coating of organo-fluoro polymer, so making then impermeable to sol-vents. Vehicle fuel tanks made of such treated plastic are less likely to rupture in a crash and so less likely to cause a fire.

Fluorocarbons can be liquids, based on short chains of carbon atoms, or polymers, which have long chains of carbon atoms. Both have lots of uses. Fully fluorinated short-chain carbons make excellent solvents with boiling ranges from 25 to 250°C, depending on the length of the chain. They are inert, heat-stable, non-flammable, non-toxic, and of low viscosity, but their escape into the environment poses a real threat because they are so stable and are greenhouse gases.

The fully-fluorinated polymer, popularly known as Teflon—more correctly known as poly(tetrafluoroethene) or PTFE for short—has many uses. It makes excellent cable insulation, it is an ideal lining for vessels holding corrosive liquids, and it is used for hoses and tubing, for fabric roofing, and plumber's tape for sealing joints in water pipes. It is also the basis of Goretex, the weather-proof fabric first made in 1969 by Bob Gore (1937–), who found a way of expanding PTFE by heating and stretching the polymer to form a membrane which had invisible pores that were small enough to keep water droplets out, but big enough to allow water molecules of sweat to escape. Goretex is also used for artificial veins and arteries in the treatment of cardio-vascular disorders. Scrap from the PTFE industry is re-used by grinding it to a microfine powder and adding it to printer's ink to make it flow more smoothly.

PTFE was discovered by Roy Plunkett at the DuPont research laboratories at Deepwater, New Jersey, on 6 April 1938. That morning he opened a cylinder of tetrafluoroethene which was supposed to hold exactly 1,000 g of the gas but only 990 g came out. The missing 10 g had turned into a curious white powder which was a plastic with some remarkable properties. It would not dissolve in anything, was not affected by heat, would not burn and was not attacked by hot corrosive acids.

More unusual was its ability to remain flexible down to *minus* 240°C, something that no other plastic was capable of doing. It was this property which made it so important for use in space and on the Moon where such temperatures prevail. In addition, it had a slippery feel, the reason being that nothing would stick to it. This was to be the secret of its commercial success. Combinations of tetrafluoroethylene with other polymers produces elastomeric materials such as Viton and these are idea for seals, gaskets, and hoses that have to withstand contact with strong acids, bases, oils, and solvents at all kinds of temperatures, and they are an essential part of modern car engines.

Environmental element

Fluorine in the environment	
EARTH'S CRUST	950 p.p.m.
	Fluorine is the 13th most abundant element.
SOILS	*approx.* 330 p.p.m. with a usual range of 150–400 p.p.m. although some soils can have as much as 1,000 p.p.m. and heavily contaminated soils have been found with 3,500 p.p.m.
SEAWATER	1.3 p.p.m.
ATMOSPHERE	0.6 p.p.b. mainly as CFCs. Up to 50 p.p.b. has been recorded in city environments.

It was known for centuries that cattle which grazed where volcanic dust had settled would become ill and lame; the reason was their high fluoride intake. Research in Iceland in 1970 showed that grass affected in this way could have as much as 4,300 p.p.m. fluoride (dry weight), although this fell to 30 p.p.m. within 40 days. Other activities can contaminate soil with fluoride, such as close proximity to an aluminium smelter, or a coal-fired power plant, or brick works, or a fertilizer factory—all industries that emit fluoride. Grass grown on such land can have as much as 5,400 p.p.m. (dry weight).

Just as replacing a hydrogen atom with a fluorine atom can turn a molecule into a powerful drug, so there are worries that other fluorinated molecules might well interfere with other bodily processes. Some of those used by industry have escaped into the environment and perfluoroocatanoic acid (aka PFOA, $C_7F_{15}CO_2H$) is one such molecule which can now be detected in all human beings, albeit in incredibly small amounts. This also forms from the degradation of the PTFE polymers used as coatings on things like carpets and paper. Nearly all humans contain traces—4 p.p.b. on average—of long-chain perfluoroalkyl compounds of which an estimated 7,000 tonnes have been lost to the environment. These are now to be found in all ecological niches, even those far removed from the centres of production. The companies manufacturing these products are now committed to changing their processes so that there will be no further release after 2015.

Volatile chlorofluorocarbons (CFCs), were once used in aerosols, air conditioning units, and refrigerators. Those manufactured on a large scale were CFC-11 (chemical formula CCl_3F), CFC-12 (CCl_2F_2), and CFC-13 (CCl_2FCClF_2) and while they performed well in the applications for which they were designed, nevertheless they were environmentally damaging to the ozone layer which protects the Earth from harsh ultraviolet radiation from the Sun. Their manufacture, and even the re-using of existing CFCs, are now banned. Their replacements, the hydrofluorocarbons (HFCs), do not damage the ozone layer because they are not persistent in the atmosphere but they are nevertheless potent greenhouse gases.

Chemical element

Data file

CHEMICAL SYMBOL	F
ATOMIC NUMBER	9
ATOMIC WEIGHT	18.9984032
MELTING POINT	−220°C
BOILING POINT	−188°C
OXIDES	OF_2 and O_2F_2

Fluorine is a pale yellow gas and the first member of group 17 (row 2p), also known as the halogen group, of the periodic table. It is the most reactive of all the elements. Only the noble gases helium, neon and argon are unaffected by it. Moreover when it combines with another element the bonds it forms are generally very strong. The most stable oxidation state for fluorine is −1, that is, as the fluoride ion F^-.

There are 18 known isotopes of fluorine with mass numbers 14 to 31, and one known isomer. Fluorine in nature has only one isotope, fluorine-19, which is not radioactive. However, there are minute traces of fluorine-23 of fleeting existence. This is produced by the decay of naturally occurring protactinium-231.[61] The half-life of this isotope is 2.23 seconds.

---

[61] This isotope undergoes a rare decay mode known as cluster decay which produces the fluorine-23.

The longest-lived radioactive isotope is fluorine-18 which has a half-life of 110 minutes. All the other isotopes have half-lives measured in seconds or less.

Element of surprises

Almost everything that fluorine does seems surprising as the above sections show. However, there are a couple of intriguing items that justify inclusion in this section.

1. If nothing will stick to Teflon, how does it stick to metal cookware? That little problem was solved in the 1950s by Louis Hartmann who treated a metal like aluminium with acid to form microscopic pits on its surface and then applied the polymer as an emulsion, followed by baking it at 400°C for a few minutes. The polymer melts, forms a film over the surface, and runs into the tiny etched pits, filling them and thereby becoming wedged fast and so gripping the rest of the Teflon film tightly to the surface.

2. The strongest commercially available acid is triflic acid (short for trifluoromethanesulfonic acid, CF_3SO_3H) which is many times stronger than even the most concentrated sulfuric acid. It is used as a catalyst.

Francium

Pronounced fran-see-um, the element is named after France.[62]
French *francium*; German *Francium*; Italian *francio*; Spanish *francio*; Portuguese *frâncio*.

Essential element

The only role of francium has been an historical one in that it completed group 1 of the periodic table, although one day it is likely to be joined by element 119—see unnamed elements.

Element of life

Francium is never encountered outside nuclear facilities or research laboratories, being potentially dangerous because of its intense radioactivity.

When francium was injected into rats it concentrated in the bladder, gut, kidneys, saliva, and liver. In those rats with cancer it also concentrated in the tumour, raising hopes that francium might have medical uses in treating this disease. For reasons given below, this was a forlorn hope.

Element of history

Mendeleyev said there should be an element like caesium waiting to be discovered because there was an unoccupied slot for it in his periodic table. The long form of the periodic table, which appeared in the early 1900s, confirmed there had to be such an element waiting to be found because heavier elements were already known like thorium and uranium. All who searched for it were fated to be disappointed because all its isotopes are radioactive and extremely short-lived.

Nevertheless, there were several attempts to discover it leading to claims, denials, and counterclaims with the disputes culminating in a controversy that was interrupted by World War II.

In 1925, Soviet chemist D.K. Dobroserdov observed weak radioactivity in potassium samples that he was analysing. He claimed this to be element 87, proposing the name russium (based on his native land Russia). The radioactivity was quickly shown to be from the radioactive isotope potassium-40, not from element 87.

In 1926, English chemists Gerald Druce (1894–1950) and Frederick H. Loring announced the results of their 1924 work in which they had analysed X-ray photos of manganese sulfate, claiming to have detected element 87. They proposed the name alkalinium (symbol Ak) because element 87 would be the ultimate alkali metal.

---

[62] France has the distinction of having two elements named after it. The name of gallium is derived from the Latin *Gallia*, the Roman name for that country.

In 1929, the American physicist Fred Allison (1882–1974) used a magneto-optical machine to analyse the minerals pollucite and lepidolite, claiming discovery of element 87 and proposing the name virginium (symbol Vi or Vm) after his home state of Virginia. Other researchers showed that his equipment and method of detection could not discover new elements.

In 1936, Romanian physicist Horia Hulubei (1896–1972) and French physicist Yvette Cauchois (1908–99) analysed samples of the mineral pollucite using a high-resolution X-ray apparatus and observed weak lines which they said were from element 87. They proposed the name moldavium (symbol Ml) from the Romanian province where they worked. The claim was criticised because it was already being said that element 87 did not occur in Nature; that was untrue because it does, albeit only in trace amounts.

Element 87 was finally discovered in 1939 by French physicist Marguerite Perey (1909–75) at the Curie Institute in Paris. She had purified a sample of actinium free of all its radioactive decay products and yet she observed that the beta-ray emission which it was giving out was more intense that it should have been. She knew that this meant there must be another element present, one that had previously escaped discovery, and which she deduced must be the missing element 87. It was.

This led to a dispute between those claiming to have discovered element 87, with Perey denigrating the Hulubei and Cauchois claim. However, Nobel prize-winner Jean-Babtiste Perrin (Physics, 1926) endorsed the Hulubei and Cauchois claim over that of Perey. World War II intervened and the issue was put on hold. Then in 1946, Perey renewed her claim and proposed the name catium because she said that element 87 was the most electro-positive cation of all the elements. This name was criticised so Perey changed it to francium (symbol Fa) naming it after her native country, France. At the time she chose the symbol Fa because the obvious choice of Fr had already been proposed for florentium, element 61, which had supposedly been discovered in 1924. Also at the time, no one seems to have supported the claim of Hulubei and Cauchois, although there was no indication that it was incorrect. However, since Romania had been an ally of Nazi Germany during the war, a name based on its territory probably seemed unacceptable and so may have played a role in Perey's claim being preferred to that of Hulubei and Cauchois. In any case, three years later, IUPAC approved the name francium, but changed the symbol to Fr which was now available because the claim for element florentium had foundered.

Economic element

Francium is an intensely radioactive metal, of which there are minute traces in uranium ores, but that required for research purposes is made either in a nuclear reactor by the neutron bombardment of radium, or in a cyclotron by the proton bombardment of thorium. Although francium can be made, it is difficult to investigate because the longest-lived atom is isotope-223 with a half-life of only 22 minutes. Not surprisingly, no use has been found for what little can be produced, and indeed no weighable quantity of francium has ever been made.

Environmental element

Francium occurs naturally in uranium minerals, but at any one time there is only one francium atom per *billion* billion atoms (10^{18}) of uranium. These few atoms of francium-223 are there as the radioactive decay product of actinium-227, which in turn is part of a uranium-235 decay series. The short half-life of francium-223 means that there is probably less than 30 g of francium present in the Earth's crust at any one time.

Chemical element

Data file	
CHEMICAL SYMBOL	Fr
ATOMIC NUMBER	87
ATOMIC WEIGHT	223.0 (isotope Fr-223)
MELTING POINT	*approx.* 27°C (*est.*)
BOILING POINT	*approx.* 660°C (*est.*)
DENSITY	not known
OXIDE	would be Fr_2O, if it were ever to be made

Francium is a radioactive metal element and a member of group 1 (row 7s), also known as the alkali metal group, of the periodic table. It has been studied by radio-chemical techniques, which show that its most stable state is +1 as in the ion Fr^+. Its chemistry would be like the element above it in the periodic table, caesium, and it was observed to co-precipitate along with the insoluble salt caesium perchlorate. Not that this enabled it to be separated from its sister element.

There are 34 known isotopes of francium with mass numbers 199 to 232, and seven known isomers. There are two naturally occurring isotopes, francium-221 and -223, which are produced in the decay chain series from uranium and neptunium. The longest-lived isotopes are francium-223 (22 minutes) followed by francium-212 (20 minutes).

The largest amount of francium ever assembled was a group of 10,000 atoms at the Stony Brook facility of the State University of New York in 1997. These were produced by firing a beam of oxygen-18 atoms at a target of gold-197 heated to near its melting point. The francium atoms were corralled into a group using laser beams. Among them was some francium-210 (half-life 3.2 minutes). There were enough atoms produced to observe the fluorescence caused by their radioactivity.

Element of surprises

Because all francium isotopes are transient beings with very short half-lives, it seems all the more remarkable that the element was discovered or at least detected before the nuclear age. Francium was the last element to be discovered in Nature.

Gadolinium

Pronounced gad-oh-lin-ee-um, it is named in honour of Johan Gadolin, who was born in 1760 at Åbo, then a part of Sweden, now a part of Finland; he died in 1852. Gadolin investigated the first rare earth mineral, which was later named gadolinite after him,[63] which was found at Ytterby, Sweden, in 1780.

French *gadolinium*; German *Gadolinium*; Italian *gadolinio*; Spanish *gadolinio*; Portuguese *gadolíneo*.

Gadolinium is one of 15 chemically similar elements referred to as the lanthanoid or rare earth elements, which extend from element atomic number 57 (lanthanum) to element atomic number 71 (lutetium). The term rare earth is a misnomer because some are not rare. The minerals from which they are extracted, and the properties and uses they have in common, are discussed under Lanthanoids.

Essential element

Gadolinium has the highest ability to capture neutrons and for this reason alone it is important in nuclear facilities. It also has microwave applications.

Element of life

The amount of gadolinium in the human body is not known, but it is tiny. Although gadolinium has no biological role, it has been noted that gadolinium salts stimulate metabolism. It is difficult to differentiate the various amounts of the different lanthanoids in the human body, but they are present, and the levels are highest in bone, with smaller amounts being present in the liver and kidneys.

No one has monitored diet for gadolinium content, so it is difficult to judge how much we take in every day but it is probably only a milligram or so per year. Gadolinium is not taken up by plant roots in any great amount—the amount detected in vegetables is less than 2 p.p.b. (dry weight)—so little gets into the human food chain. Some species, such as lichen, can have as much as 500 p.p.b. (dry weight) of gadolinium.

Medical element

Gadolinium compounds are used as contrast agents in magnetic resonance imaging for medical diagnosis and are particularly effective because they can be tracked in the human body by means of gadolinium's magnetic properties. These contrast agents

---

[63] Gadolinite is a beryllium iron silicate of formula $Be_2Fe(X)_2Si_2O_{10}$, where X can be a variety of lanthanoid elements and/or yttrium.

are produced commercially under names such as Magnevist and Ominscan. (A recent development has incorporated the gadolinium into nanotubes of carbon which are said to be particularly effective contrast agents.) Gadolinium can even help kill the deadly brain tumours which it locates. It does this because it is absorbed by the cancer cells and once inside them it travels to the cell's nucleus where it disrupts the DNA.

Dangerous element

Soluble gadolinium salts, such as the chloride and nitrate, are mildly toxic by ingestion, but insoluble salts are non-toxic. Gadolinium salts irritate the skin and eyes.

Element of history

Gadolinium was discovered in 1880 by Charles Galissard de Marignac (1817–94) at Geneva, Switzerland. He had long suspected that the didymium that Carl Gustav Mosander (1797–1858) had claimed as a new element was in fact not a pure substance. His suspicions were confirmed when Marc Delafontaine (1837–1911) and Paul-Emile Lecoq de Boisbaudran (1838–1912) in Paris reported that its spectral lines varied according to the source from which it came. Indeed, in 1879 they had already separated another rare earth oxide from a sample of didymium oxide which had been extracted from the mineral samarskite,[64] found in the Urals in Russia. This they had named samarium. The following year, 1880, Marignac extracted another new rare earth from didymium oxide which he tentatively referred to as alpha-yttrium. This was also obtained by Boisbaudran in 1886 and it was he who suggested the name gadolinium, to which Marignac agreed.

Economic element

Gadolinium is found in minerals that include all the rare earth elements and these are discussed in more detail together under lanthanum and the lanthanoids. The most important ores are monazite and bastnäsite. Although gadolinium is not a major constituent of either ore, it is present in extractable amounts. World production of pure gadolinium is about 7,500 tonnes per year, and it is separated by ion-exchange techniques. China is the main mining area although there are large deposits of rare earth ores in the USA, Brazil, India, Australia, Greenland, and Tanzania, and global reserves of gadolinium are estimated to be well in excess of a million tonnes. Gadolinium metal can be produced by heating gadolinium fluoride with calcium.

Gadolinium is rarely employed as the metal itself, but its alloys are needed to make magnets and electronic components. They are also used for magnetic data storage disks that can be permanent records. These consist of ultra-thin layers of various metals deposited on glass or plastic and gadolinium is a key component, along with other rare earth metals, such as terbium and dysprosium, plus the more common magnetic metals iron and cobalt. Gadolinium is magnetic like iron, but with a low

---

[64] Samarskite is a mixed oxide ore of several uncommon metals including yttrium, cerium, uranium, niobium, and tantalum.

Curie temperature of 20°C. (This is the temperature at which a magnet loses its magnetism.) It has been suggested that this could be utilised in so-called magnetic refrigeration, because as a magnetic field is applied its temperature drops slightly.

Gadolinium is unique in having the greatest neutron capturing ability of any known element. Its affinity for neutrons is used to regulate the neutron activity in the reactor core of nuclear power plants. It is ideally suited for this purpose by having two isotopes with superb neutron-absorbing capabilities: gadoniumium-155 and gadoniumium-157, the latter being four times more absorbing than the former and more than 300 times more absorbing than boron, the other element that is used to absorb neutrons. Around 5 percent of gadolinium oxide is intimately mixed with the uranium-235 oxide pellets used in nuclear reactors, and this makes for more efficient use of the nuclear fuel.

Gadolinium's neutron-capturing ability is put to use in the technique known as neutron radiography. This is used to monitor what is happening inside equipment such as engines, gear boxes, and generators when these are in motion. A screen made of gadolinium is used to trap neutrons, thereby becoming slightly radioactive which is then recorded on a film that is in contact with the screen.

Gadolinium yttrium garnets have microwave applications—see page 619. Adding gadolinium to metals like iron, chromium, and magnesium improves their workability and resistance to corrosion. Magnesium alloy is particularly improved this way.

Environmental element

Gadolinium in the environment	
EARTH'S CRUST	8 p.p.m.
	Gadolinium is the 41st most abundant element.
SOILS	*approx.* 6 p.p.m., range 1–15 p.p.m.
SEAWATER	1.2 p.p.t.
ATMOSPHERE	virtually nil.

Gadolinium is one of the more abundant rare earth elements and is more abundant than tin. It poses no environmental threat to plants or animals.

Chemical element

Data file	
Chemical symbol	Gd
Atomic number	64
Atomic weight	157.25
Melting point	1,313°C
Boiling point	3,270°C
Density	7.90 g/cm³
Oxide	Gd_2O_3

Gadolinium is a soft, shiny, silvery metal and a member of the lanthanoid group (row 4f) of the periodic table. It corrodes in air due to the formation of a flaky oxide layer which separates off, exposing more fresh metal to attack. Gadolinium reacts slowly with water and dissolves in acids. Its preferred oxidation state is +3 as exhibited by compounds like GdF_3 and $GdCl_3$. In solution it exists as the Gd^{3+} ion which is surrounded by as many as nine water molecules.

There are 36 known isotopes of gadolinium with mass numbers 134 to 169, and 10 known isomers. There are eight naturally occurring isotopes in ores, five of which are stable and three radioactive. The most abundant is gadolinium-158, accounting for 25 percent, gadolinium-160 (22 percent), gadolinium-156 (20 percent), gadolinium-157 (16 percent), gadolinium-155 (15 percent), gadolinium-154 (2 percent), gadolinium-152 (0.2 percent), and gadolinium-150, of which there are only traces. The radioactive ones are gadolinium-160 with a half-life of 1.03×10^{21} years,[65] gadolinium-152 with a half-life of 1.08×10^{14} years, and gadolinium-150 with a half-life of 1.79×10^6 years. This last isotope is produced by the alpha-decay of dysprosium-164. There are also minute traces of some radioactive isotopes in uranium ores, where they have been formed by nuclear fission.

Element of surprises

Unlike the other rare earth elements, which are used to produce the vivid colours in television screens, gadolinium is colourless because its spectrum has no absorption bands in the visible region (its compounds also appear white for this reason). Because of this property it is sometimes used as a support for other rare earth phosphors that luminesce, in the knowledge that it won't cause 'quenching' of their intense colours.

[65] 10^{21} is a billion trillion years.

Gallium

Pronounced gal-ee-um, the name is derived from *Gallia* the Latin name for France.[66]
French *gallium*; German *Gallium*; Italian *gallio*; Spanish *galio*; Portuguese *gálio*.

Essential element

This element has emerged from relative obscurity to the full glare of blu-ray technology and it is also a key component of modern LEDs (Light-Emitting Diodes).

Element of life

Gallium in the human body	
BLOOD	less than 0.1 mg per litre
BONE	data not available
TISSUE	av. 1 p.p.b.
TOTAL AMOUNT IN BODY	less than 1 mg.

Gallium has no biological role but is known to stimulate metabolism. It presents few heath risks because its salts generally have low toxicity, although rats fed 10 mg of gallium a day in their diet displayed some toxic symptoms.

Plants show a slight selective uptake of gallium depending on the level of this in the soil in which they grow. Land plants have 0.01–5 p.p.m. (dry weight) with edible vegetables 0.03–2 p.p.m. (again, dry weight).

Element of history

Gallium was discovered in Paris at the private laboratory of Paul-Émile Lecoq de Boisbaudran (1838–1912) at 3 a.m. on 25 August 1875. It was then that he saw a new pale violet line at 416 nanometres in the atomic spectrum of some zinc he had extracted from a sample of zinc blende ore (ZnS) from the Pyrenees. He knew that this new line meant there had to be a new element present in the zinc.

What Boisbaudran didn't know was that its existence, and properties, had been predicted six years earlier by the Russian chemist Dmitri Mendeleyev (1834–1907). He had devised his periodic table of the elements in 1869 and this showed there was a gap below aluminium which was yet to be occupied. He forecast that the missing element's atomic weight would be around 68 and its density would be 5.9 g/cm³.

---

[66] France has the distinction of having two elements named after it. The other one is francium.

By November 1875, Boisbaudran had isolated a sample of the new element, and even purified it by electrolysis of a potassium hydroxide solution in which it was soluble. He identified it as an element that was a metal like aluminium. On 6 December 1875 he presented a sample to the French Academy of Sciences and christened it gallium.

The choice of name, gallium, flattered the French Academy, because it appeared to be derived from Gallia, the Roman name for France, but it was suggested that Boisbaudran named the element after himself, although he denied that this was so. The idea probably arose because his middle name of Lecoq, is like '*le coq*' which is the French word for a cock, and the Latin name for that is *gallus*.

Medical element

Gallium in oxidation state +3 behaves like iron in this oxidation state and it will interact in the same way at sites in the body where iron(III) is involved. For this reason, and because iron(III) is important in many processes, gallium has been the subject of medical research looking for new drugs based on it. In 1998 one of its compounds was shown to be effective in attacking strains of malaria that have become resistant to conventional anti-malarial drugs such as chloroquine. As yet no gallium-based drug has appeared.

Gallium nitrate, $Ga(NO_3)_3$, aka Ganite, has been injected to treat bone cancers in an effort to control the osteoclasts on the surface of the bone, which is where calcium is lost.

The radioactive isotope gallium-67, which has a half-life of 78 hours, is used to locate and treat melanomas because it concentrates in these cancerous tissues.

Element of war

Gallium nitride—see below—makes possible radar detectors which do not need a rotating scanner; this means that they are less likely to reveal their location to an enemy. Potentially, and just as important in a major global conflict, they would not be affected by a burst of radiation from a nuclear weapon.

Economic element

Gallium minerals are rare; gallite is one such and its formula is $CuGaS_2$. Several ores, such as the aluminium ore bauxite, contain small amounts of gallium, and the same is true for coal, which may have relatively high gallium content. The flue dust from a coal-burning plant may have as much as 1.5 percent of the element. No ore is mined specifically for gallium and all that is produced is recovered as a by-product of zinc and copper refining. Annual world production is around 100 tonnes, half of which comes from mining and half from recycled scrap. The gallium needed for manufacturing semiconductors has to be at least 99.9999 percent pure.

The US Geological Survey estimates there are more than a million tonnes of gallium that could be extracted mainly from aluminium with some coming from zinc ores. Others have estimated that accessible gallium is more in the region of 1,500 tonnes and they have warned that these supplies will run out by 2020.

Gallium is a key component of many III/V semiconductors such as gallium arsenide (GaAs).[67] This can convert electricity to light and is used in light-emitting diodes (LEDs). Aluminium gallium arsenide was the characteristically red LED of numerical displays that were introduced in the 1960s, some of which are still functioning. However, it is its semi-conductivity which has brought gallium to prominence because it generates less heat than silicon and is therefore more suitable for use in laptop computers and mobile phones. The crystals are grown layer-by-layer using a process known as molecular beam epitaxy. In this, a mixture of volatile trimethyl gallium, $Ga(CH_3)_3$, and the gas arsine, AsH_3, are heated to strip them of the other atoms which make these elements volatile and the atoms then deposit in a regular array on the surface of a wafer-thin crystal substrate to give crystals of pure GaAs. Tiny amounts (p.p.m.) of other elements, such as tin or zinc, are included as dopants to adjust the semi-conductivity to that required.

Gallium arsenide has been used for solar panels in space, some of which were used for the machines sent to explore the surface of Mars. Chemists and physicists are also intrigued by gallium arsenide because it can form so-called 'quantum dots' on the surface of crystals and these are able to trap free electrons, which are normally the most mobile component of matter, and confine them to a small area. However, it is gallium nitride which has stolen the show—see box.

Gallium nitride GaN

This material is also a III/V semiconductor, and a rather remarkable one for three reasons: it has a band gap of 3 eV; it does not have to be made in a perfectly crystalline form as does silicon; and it can cope with much higher power. The band gap is the energy required to free an electron in a material from the control of its parent atom so that it can move freely and thereby carry electric current. Metals have low band gaps—less than 0.1 of an electron volt (eV)—which is why they make good electrical conductors—while insulators have high band gaps of 4 eV or more—which means that it is almost impossible for their electrons to escape the pull of the parent atoms. Semi-conductors have band gaps intermediate between the two such as gallium arsenide which is 1.4 eV and this enables it to emit low energy light at the red end of the spectrum as it returns to its ground state. Other semiconductors with higher band gaps made other colours possible but for a long time it seemed that a semiconductor with a relatively high band gap would not be found, until gallium nitride came along. Its band gap of 3 eV enables it to emit light at the high energy end of the visible spectrum and even into the UV region, thus making possible the full range of colours.

---

[67] The III/V semiconductors are named after the two groups of the older form of the periodic table in which the elements occur. That in which gallium occurs was called group III (now group 13) and the group headed by nitrogen, and in which arsenic is to be found, was group V (now group 15).

Blu-ray technology is a daily tribute to the effectiveness of GaN. It enables green and blue LEDs to be made for use in illuminated displays and general purpose lighting. It is stable at much higher temperatures than other semiconductor materials and can be used in sensors that have to operate under extreme conditions, such as engines. Moreover, it can handle more power and at higher frequencies making mobile phones and broadband more effective; and it can act as a sensitive pressure sensor for touch switches.

GaN is made by heating trimethyl gallium vapour, $Ga(CH_3)_3$, and ammonia, NH_3. GaN has widened the colour range of light-emitting diodes (LEDs) and eventually these will provide most of the illumination for indoors which was formerly obtained from incandescent light bulbs (which converted only 5 percent of the electricity they used into light) and fluorescent strip lights (which converted only 15 percent). LEDs have efficiencies in excess of 30 percent. White light LEDs are being used to back-light liquid crystal displays in flat screen TVs, monitors, and laptops.

Environmental element

Gallium in the environment

EARTH'S CRUST	18 p.p.m.
	Gallium is the 34th most abundant element.
SOILS	av. 28 p.p.m., range 1–70 p.p.m.
SEAWATER	30 p.p.t.
ATMOSPHERE	virtually nil.

Gallium is more abundant than lead but much less accessible because there are no geological processes that have selectively concentrated it into minerals, so it tends to be widely dispersed. It poses no environmental threat to plants or animals.

Chemical element

Data file

CHEMICAL SYMBOL	Ga
ATOMIC NUMBER	31
ATOMIC WEIGHT	69.723
MELTING POINT	29.7646°C*
BOILING POINT	2,403°C
DENSITY	solid 5.9 g/cm³, liquid 6.1 g/cm³
OXIDE	Ga_2O_3

* This melting point is one of the standard reference points on the International Temperature Scale.

Gallium is a silvery-white metal and a member of group 13 (row 4p) of the periodic table. It is soft enough to be cut with a knife. It is stable in air and with water, but it reacts with and dissolves in acids and alkalis. Gallium's preferred oxidation state is +3 as in GaF_3 and Ga_2Cl_6, and similar compounds, and in solution it is Ga^{3+}. However, there is a lower oxidation state +1 which is found as GaCl in some mixed chlorides.

There are 31 known isotopes of gallium with numbers 56–86, and 3 known isomers. Natural gallium consists of two stable isotopes: gallium-69 which comprises 60 percent and gallium-71, 40 percent.[68] The longest-lived radioactive isotope is gallium-67 which has a half-life of 3.26 days.

Element of surprises

1. Gallium will melt when held in the palm of the hand and on so doing it also shrinks in volume, albeit by only 3 percent, which is contrary to the way other metals behave. (The semi-metal antimony behaves likewise.) When alloyed with indium and tin it remains liquid down to −19°C and at one time it was seen as a replacement for mercury in thermometers. The alloy is known as Gallinstan and the proportions of the three elements are 69:21:10 (Ga:In:Sn).

2. Gallium has been used to detect neutrinos (these are elementary particle with zero charge and virtually zero mass) emitted by the Sun. More than 50 tonnes of molten gallium were used by Russian researchers to try and trap neutrinos by means of their reaction with the isotope gallium-71. In a more successful experiment, organised by the Laboratori Nazionali del Gran Sasso (LNGS) of Italy, only the gallium-71 isotope was used, in the form of 100 tonnes of gallium chloride solution, and neutrinos were detected. When neutrinos react with gallium-71 they are absorbed into the nucleus which then emits an electron and so is converted to germanium-71 and thereby are detected by its radioactivity (half-life 11.43 days). It has been estimated that at any one time there are only about a dozen atoms of germanium-71 present among the 100 *billion billion* billion atoms (10^{29}) of gallium in the tank holding the solution.

[68] In addition there are several radioactive isotopes produced as fission products in uranium-bearing deposits.

Germanium

Pronounced jer-may-nee-um, the name is derived from *Germania*, the Latin name for Germany.

French *germanium*; German *Germanium*; Italian *germanio*; Spanish *germanio*; Portuguese *germânio*.

Essential element

Although wrongly credited with special health benefits, this element has been rightly credited with special semiconductor properties and it is used in fibre optics.

Cosmic element

Germanium is produced within stars and has been detected in some very distant ones. It is also present in the atmosphere of the planet Jupiter.

Element of life

Germanium in the human body	
BLOOD	av. 0.4 mg per litre
BONE	data not available
TISSUE	0.1 p.p.m.
TOTAL AMOUNT IN BODY	5 mg.

There is no biological role for germanium, although it acts to stimulate metabolism. Its salts generally have low toxicity for mammals, but they are deadly to certain bacteria although no practical application of this has resulted.

Some plants, including those eaten as food by humans, can absorb germanium from the soil in which they grow, probably taking it in as the oxide. Average levels of the element of around 0.05 p.p.m. (fresh weight) were measured in grains and vegetables, but not all plants were found to contain it.

Food element/health food supplement

The estimated daily intake is around 1 mg, and there have been claims that germanium could be beneficial to health, although this has never been proved scientifically. Those claims began in 1980 when Kazuhiko Asai wrote a book *Miracle Cure: Organic Germanium*, which was published in Japan. In this he claimed that all kinds of health benefits would result from taking various germanium medicaments, which he referred to as *organic* germanium, by which he meant they were germanium compounds to which carbon-containing groups were attached. The principal compounds were GE-123, his name for a complex oxide also known as carboxyethylgermanium sesquioxide, and sanugerman, which was germanium citrate lactate. His theory was given wider publicity by Sandra Goodman in her book *Germanium: the health and life enhancer,* published in 1988.

A high intake of germanium was supposed to improve the immune system, boost the body's oxygen supply, make a person feel more alive, and destroy damaging free radicals. In addition it was said to protect the user against radiation. Expensive germanium food supplements appeared on the shelves of alternative health stores, and they were also labelled as 'organic' a term which is generally taken to mean the opposite of 'chemical' and therefore preferred by those who frequent such emporia. However, the craze was quickly brought to an end in the UK in 1989 when the Government's Department of Health warned against them, saying that germanium supplements had no nutritional or medical value and that taking them constituted a risk to health, rather than a benefit.

Element of history

Germanium was discovered by Clemens A. Winkler (1838–1902) at Freiberg, Germany, in 1886, 15 years after its existence had been predicted by Mendeleyev who saw that there had to be an element below silicon and above tin in group IV of his periodic table. He forecast some of its properties, predicting its atomic weight would be about 71 and that its density would be about 5.5 g/cm^3. Seven years earlier, a London chemist John Newlands had suggested that silicon and tin were part of a 'triad', as found with other elements, such as phosphorus/arsenic/antimony and sulfur/selenium/tellurium. If this were so, he said, then there should be another element linking silicon and tin.

In September 1885 a miner working in the deep level (460 metres underground) of the Himmelsfürst silver mine, at St Michaelis near Freiberg, came across an unusual ore. He gave it to the mine manager who sent it to Albin Weisbach at the nearby Mining Academy. He certified it was a new mineral, gave it the name argyrodite, and passed it to his colleague Winkler to analyse; he found it to be 75 percent silver and 18 percent sulfur. But 7 percent of its weight could not be accounted for.

Winkler spent the next few months working to separate and identify the unknown component and, by February 1886, he believed that it was a new metal-like element and he had produced a sample by heating its sulfide in hydrogen gas. After he had reported his findings, there began a debate as to which of the elements missing from Mendeleyev's periodic table it might be. Mendeleyev favoured a metal in the zinc/cadmium group, while Winkler thought it might be in group V and a neighbour of antimony.

As germanium's properties were revealed, it became clear that it was the missing element between silicon and tin because its chemical nature was much as Mendeleyev had predicted. Winkler named the element after his native land, Germany, and announced that argyrodite, from which it had come, had the formula Ag_8GeS_6.

Economic element

Germanium ores, like argyrodite, are rare. The least rare is germanite, a copper iron germanium sulfide with 8 percent of the element. No ore is mined as a commercial source of this metal. Germanium is widely distributed in the ores of other metals, such as zinc, and that which is required for manufacturing purposes is recovered as a by-product from the flue-dusts of zinc smelters. In Russia and China some coal deposits contain large concentrations of germanium and as such are used as a source of this element.

World production is about 140 tonnes per year and the metal is first separated by treatment with chlorine which converts the germanium to germanium tetrachloride ($GeCl_4$), a volatile liquid which boils at 86°C and which can be purified by distillation. The element itself is produced by reacting the $GeCl_4$ with hydrogen gas and the germanium this produces is then further purified by a technique known as zone refining until the level of impurities falls to less than one such atom in 10 million atoms of germanium. There are estimated to be stockpiles totaling around 500 tonnes of germanium in various countries around the world, with total world reserves being of the order of 500,000 tonnes.

Germanium is a semiconductor and, doped with other elements such as arsenic or gallium, it was used on a large scale for many years. Today it has been superseded by other elements in electronic devices, and most now finds use as the oxide, GeO_2, in special glass for fibre optics, wide-angle camera lenses (on account of its high refractive index) and for infrared devices (since it will transmit these rays). A little germanium is used in alloys. Silver is a metal which easily tarnishes but it can be made tarnish-proof by adding around 1 percent of germanium to the metal. The alloy is known as argentium.

In 1993, it was reported that doping silicon chips with germanium made them work almost as well as gallium arsenide, but little came from this discovery at the time. Silicon germanium as a semiconductor is now finding many applications previously served by silicon alone, or by gallium arsenide. Solar panels based on germanium are likely to be a major outlet for this element in future, having proved their worth in satellites. Germanium has the advantages of faster switching speeds, lower power consumption, and its semiconductors are bendable, and can be processed at lower temperatures. It should be possible to use germanium wafers to produce electricity from infrared heat and from the waste heat of many industries.

Germanium is used in radiation screening detectors in airports.

Environmental element

Germanium in the environment	
EARTH'S CRUST	2 p.p.m.
	Germanium is the 52nd most abundant element.
SOILS	av. 1 p.p.m., range 0.5–2 p.p.m.
SEAWATER	0.5 p.p.t., with more in the Pacific than the Atlantic.
ATMOSPHERE	virtually nil.

Germanium is less abundant than either tin or lead, which are the heavier metals of group 14, and it is less easily accessed because geological processes have concentrated only small amounts of it into minerals, so that it tends to be widely dispersed. It poses no environmental threat to plants or animals.

Organo-germanium compounds have been detected in seawater, suggesting that the element might have an effect on some micro-organisms, but these have not been identified.

Chemical element

Data file

CHEMICAL SYMBOL	Ge
ATOMIC NUMBER	32
ATOMIC WEIGHT	72.61
MELTING POINT	938°C
BOILING POINT	2,830°C
DENSITY	5.3 g/cm³
OXIDE	GeO$_2$

Ultrapure germanium is a silvery-white, brittle, semi-metal element that comes in group 14 (row 4p) of the periodic table. It is stable in air and water, is unaffected by alkalis and acids, except HNO$_3$. It can exhibit a range of oxidation states, such as −4 in GeH$_4$, +2 in a lower oxide GeO, and +4, which is the most common state, as in GeO$_2$ and GeF$_4$. In solution it does not exist as simple Ge^{4+} ions but as germanates GeO$_3^{2-}$.

There are 32 known isotopes of germanium with numbers 58 to 90, and 13 known isomers. There are five naturally occurring stable isotopes: germanium-74 (abundance 36.5 percent), germanium-72 (27 percent), germanium-70 (20.5 percent), germanium-73 (8 percent), and germanium-76 (8 percent). The last of these is radioactive with an incredibly long half-life of 1.78×10^{21} years (= 1.78 billion trillion years).[69] The next longest-lived radioactive isotope is germanium-68 with a half-life of 271 days.

Element of surprise

Germanium was the first element to be used in transistors. Work on it began in 1942 as part of the US war effort. What had attracted attention to germanium was its potential as a rectifier of electrical current, because it had the ability to conduct electricity in one direction but not in the opposite direction. The problem the US Government faced was that the two known sources of germanium were held by Nazi Germany. The answer was to extract the element from the waste of the Eagle-Picher

---

[69] In addition, there are several radioactive isotopes produced in trace amounts in uranium-bearing deposits.

Industries zinc smelter in Oklahoma, where an ore with a high germanium content had been processed for many years. A method of extracting it from this waste was developed and provided all the germanium that was required for the war effort. The benefits of germanium were then to find peaceful outlets. By 1948, the first transistor radios were on sale to the public, with germanium transistors replacing diode valves.

Gold

Pronounced gowld, the name is the Anglo-Saxon word *gold,* which may have come from the word *geolo* meaning yellow; the chemical symbol for the element is Au which derives from the Latin, *aurum,* meaning 'glow of sunrise'.

French *or*; German *Gold*; Italian *oro*; Spanish *oro*; Portuguese *ouro*.

Essential element

Gold had been used for coins, jewellery, teeth, and gilded domes because it resists corrosion. Today it has a role in medication, and an increasingly important one as a catalyst in the chemical industry.

Cosmic element

It is possible that there is gold on Mars. At least it is present in meteorites which were created by impact events on that planet and which eventually reached Earth. The gold levels were on average 3–4 p.p.m.

Element of life

Gold in the human body	
BLOOD	1–40 p.p.b.
BONE	16 p.p.b.
TISSUE	0.4 p.p.b. in liver
TOTAL AMOUNT IN BODY	less than 0.2 mg.

Gold has no biological role and both the metal and gold salts generally have low toxicity because they are poorly absorbed by the body.

Plants can absorb gold, and those plants that metabolise cyanide absorb most gold, because this helps to solubilise it. The Douglas fir can have 20 p.p.b. (dry weight), and honeysuckle 15 p.p.b. (dry weight). The Indian mustard plant, *Brassica juncea,* is particularly good at doing this, especially if the soil is treated with ammonium thiocyanate which aids gold solubilisation and uptake. Whether it could be grown as a gold-gathering crop seems unlikely, although technically it could be done at a profit.

Medical (and dental) element

Gold is part of the medical armoury against rheumatoid arthritis, and treatment with it even merits a special name, chrysotherapy, from the Greek word for gold: *chrysos*. It is prescribed when non-steroid anti-inflammatory drugs fail to give relief. Gold can take up to 10 weeks to work, but there is a limit to how long it can be given, and even if a patient's symptoms are alleviated, the doctor will have to phase the treatment out after a few years because of the side effects associated with a build-up of gold in the body, which may amount to as much as 500 mg.

Gold therapy began in 1927 when gold sodium thiomalate (known as myocrisin) was used to treat TB, unsuccessfully as it happened, but a beneficial side-effect was noted: patients with arthritis had this condition held in check. That the effect was real was proved in 1956 when the Empire Rheumatism Council conducted double-blind tests, with some patients reporting an improvement up to a year after treatment ended.

Myocrisin had to be injected, so chemists produced an oral version, auranofin (trade name Ridaura). After successful clinical trials, this was approved by the US Government's Food and Drugs Administration in 1985. Moreover auranofin has fewer side effects. The recommended dose is 6 mg per day which delivers 2 mg of gold, of which about 0.5 mg is absorbed into the body.

Other gold-containing drugs are showing promise against parasitic diseases and cancers. In the latter case this involves injecting the radioactive isotope, gold-198, which has a half-life of 2.7 days. Alternatively, gold nanorods can be grown from a dilute solution of auric acid ($HAuCl_4$) by seeding the solution with gold nanoparticles. These can then be incorporated into certain monoclonal antibodies which bind preferentially to the surface of cancer cells. When exposed to near-infrared radiation the nanorods then melt and eject their atoms at high velocity thereby destroying the cancer cell.

Dentists have been using gold since the time of the Etruscans, who lived in Italy from 1000–400 BC, where excavated graves have revealed skulls with false teeth held in place by gold bridges. Today more than 60 tonnes a year of gold are used in dentistry. Gold fillings, crowns, and teeth are particularly popular in Germany and Japan. Dental gold is alloyed with silver and palladium, plus a trace of zinc (0.2 percent) to harden it, with the proportion of gold being around 75 percent.

Element of history

Gold has been known since prehistoric times and was one of the first metals to be worked, mainly because it was to be found as nuggets or as particles in the beds of streams from which the gold could easily be panned. The Nile was a fruitful source of such alluvial gold, but by 2000 BC the Egyptians were already beginning to mine gold in Nubia (southern Egypt). Their use of gold could be extravagant, witness the magnificent death mask of Tutankhamen which was discovered in 1922 and which contained more than 100 kg of the metal. It shone as bright as the day when he was buried 3,500 years earlier. The royal graves of ancient Ur (modern Iraq), which flourished 5,000 years ago, contained gold jewellery that had been skilfully worked.

Gold

Gold was easy to refine and reclaim from its alloys; all that was needed was to heat it in a furnace with salt and brick dust which either volatilised or absorbed the other metals leaving the gold unaffected. The minting of gold coins began around 2,650 years ago in the Lydian Kingdom (modern Turkey) using electrum, a native alloy of gold and silver. The first pure gold coins were minted in the reign of King Croesus, who ruled from 561–547 BC.[70]

Archimedes, the first forensic chemist?

When King Hieron II of Syracuse commissioned a new gold crown he was pleased with what the goldsmith produced but suspected that the metal had been debased with silver. He asked a local scientist, the famous Archimedes (257–212 BC), whether it would be possible to prove this without damaging the crown. The answer came as Archimedes was taking a bath and realised that the volume of water he displaced was equal to the volume of his body. *Eureka!*

The same would be true for the crown, and knowing its volume and its weight it would then be easy to compare it with the weight of a similar volume of pure gold. Today we would talk in terms of density, which is weight divided by volume, and say that gold has a density of $19.3\,g/cm^3$, whereas that of silver is $10.5\,g/cm^3$. Alloying gold with silver would reduce its density significantly. As the king suspected, his crown was not pure gold, and the hapless goldsmith's crime was revealed.

All civilisations have valued gold and developed sophisticated techniques for working with it, including weaving gold thread into garments and casting works of art by the lost-wax technique. The lure of gold drove some to seek ways of actually making it, and the alchemists expended much time fruitlessly searching for the Philosopher's Stone, the material that would transmute base metals like lead into gold. In late Roman times, the worry was that they might succeed.

The emperor Diocletian, who ruled from 285–300 AD, was afraid that his currency reforms might be undermined, so he ordered the destruction of all alchemical texts, thereby consigning to the flames the vast amount of chemical knowledge that had been acquired over the centuries in ancient Egypt. Even so, it did not stem the alchemists' quest, although it explained why they worked under a cloud of secrecy thereafter. In the Middle Ages, alchemy became respectable once again, and by the 1600s quite eminent individuals, such as the great scientist, Isaac Newton (1643–1727), and his king, Charles II, were engaging in it.

Not all alchemists' time was wasted. In the 1200s they found that gold would dissolve in a mixture of concentrated nitric and hydrochloric acids called *aqua regia* ('king of waters'). When this solution was diluted with oil of rosemary the gold stayed

[70] He was defeated by the Persian King Cyrus II.

soluble and this potion, called *aurum potabile*, was prescribed as a cure-all. Colloidal gold can be precipitated when pure tin is added to the solution. Slowly, a brilliant purple precipitate forms and is known as Purple of Cassius named after Andreas Cassius of Potsdam Germany, who first published the method of making it in 1685. The pigment was used to colour glass, to which it gave a ruby tint, or to decorate porcelain, such as that of Meissen and Sevres, on which it retained its original purple colour. It is still used to this day for colouring high quality tableware.[71]

Economic element

Gold is not the rarest of metals, nor the most expensive; platinum, rhodium, osmium and iridium are rarer and more costly. How much gold has been produced over thousands of years is impossible to estimate, but it has been calculated that in the past 500 years around 160,000 tonnes have been mined. In volume terms that would be a cube of side 20 metres. Gold can sometimes be found as huge nuggets. The largest one on record weighed 112 kg (4,000 ounces) and was found in Australia. The famous gold rushes of the 1800s were driven by the discovery of easily accessible alluvial gold.

Traditionally gold was garnered from river sediment using mercury to dissolve it out of the mud. The mercury was then distilled off leaving the gold. Artisan gold mining of this kind is still thought to account for a quarter of all gold produced. However, most gold is mined and comes from quartz veins, or is associated with pyrites deposits. The amount of gold can be as little as 1 p.p.m. and still be economic to extract from rocks. More than 80 countries have gold mines of which eight produce more than 100 tonnes a year and they are, in order of output in recent years: China, South Africa, Australia, USA, Peru, Russia, Indonesia, and Canada. In South Africa gold is mined at depths of 3,000 metres. Even Greenland produces 2 tonnes a year. Despite all this activity the amount of gold produced declines year on year by about 1 percent. It amounted to 2,350 tonnes in 2009. Gold reserves are estimated to be around 85,000 tonnes. Even Western Europe annually produces 30 tonnes of gold, mainly as a by-product of zinc and copper refining, but some is mined in France and Scandinavia.

Mined gold is extracted using sodium cyanide which forms a water-soluble ion $Au(CN)_4^-$. This is then reacted with zinc to precipitate the gold. A new method of extracting gold employs bacteria to liberate the gold from sulfide ores such as iron pyrites (FeS_2) and arsenopyrite (FeAsS) which bind the metal to such an extent that the cyanide process can only extract half of it at best, and only a quarter in many cases. The heat-loving bacterium *Sulfolobus acidocalderius* releases all the gold by digesting the rock to get at the sulfur which it uses as a source of energy. This digestion takes about three days at 50°C and gold from the Youanmi deposit in Western Australia is extracted this way.

---

[71] In the centuries following its discovery, chemists tried to explain Purple of Cassius but it was just pure gold. The answer came with the work of the Viennese chemist and Nobel Prizewinner, Richard Zsigmondy (1865–1929) who researched the nature of colloids and was thus able to explain how differently sized particles could produce different colours by the way they interacted with light.

Gold is used as bullion, in jewellery, glass and electronics. Traditionally gold has been measured in carats, with pure gold being 24 carats. The word derives from the word *keratia* used by the Byzantine goldsmiths. The other grades of gold are 22 carat, which is 92 percent, 18 carat (75 percent), 14 carat (58 percent), and 9 carat (38 percent). This last kind tarnishes with age because of the high amount of alloying metals. Most gold is alloyed with silver or copper, plus 0.2 percent zinc to harden it.

Jewellery consumes around 75 percent of all gold produced. Goldsmiths are still basically a craft industry although the making of some items, such as gold chains is now done mechanically. Pforzheim in Germany in a major centre for jewellery, with 25,000 people there being employed in the trade. India is the biggest importer of gold and takes in about 800 tonnes a year (25 percent) of world production.

Gold for jewellery can be given a range of hues depending on the metal it is alloyed with. There is white gold (90 percent by weight of gold, 10 percent of nickel); red gold (50 percent gold, 50 percent copper); blue gold (46 percent gold, 54 percent indium); purple gold (80 percent gold, 20 percent aluminium); green gold (73 percent gold, 27 percent silver); and even black gold (75 percent gold, 25 percent cobalt). Green gold was known to the Lydians as long ago as 860 BC, and they called it electrum; it occurred naturally as a native alloy of gold and silver, and was more of a pale lemon colour. Red gold was produced in Peru in 1300 AD. The most recent alloy for jewellery is known as '990' gold and consists of 99 percent gold, 1 percent titanium (23.76 carat). This alloy is much stronger than pure gold and is ideal also for coins and medals.

Thin coatings of gold are used as lubricants in aerospace industries for environments where there is a high temperature, high vacuum, or high radiation exposure, as in space applications. Metallic gold reflects infrared rays and has been applied to the windows of some large buildings to reflect the heat of the sun. Gold electroplating can produce a film microns thick and is used to coat electrical connectors and printed circuit boards, to protect their copper components and improve solderability. Gold is ideal for coating electrical contacts because it never corrodes, and thereby breaking the contact, which other metals are likely to do. Gold plating is also applied to the gears of watches, for artificial limb joints, as well as cheap jewellery.

That gold could act as a catalyst was first demonstrated in the 1980s by Masatake Haruta in Japan who observed that carbon monoxide (CO) was oxidised to carbon dioxide (CO_2) at low temperatures. Gold catalysts are now being used industrially in the form of nanoparticles. They are particularly good at removing traces of CO in gases by oxidising it to CO_2 and they are used in new type safety masks. They can remove all traces of CO from the hydrogen gas used in fuel cells and this has to be done because CO ruins the electrodes. There is even a possibility that gold will become part of catalytic convertors partly to replace the more expensive platinum.

Gold atoms deposited on a support of titanium dioxide (the two are co-precipitated from solution) is also a powerful catalyst although the support atoms play an active part in the process as well. Other catalysts have palladium atoms on gold atoms and they too are proving to be remarkably efficient. It has been shown that size is important and that particles of 1.4 nm size were best and these corresponded to gold clusters of about 55 atoms. These are able to act as catalysts in their own right and without the need to be deposited on a support medium. One of the most successful uses of gold catalysts is the production of vinyl acetate monomer, a key starting material for adhesives, fibres, paints, and resins.

Gold has the highest malleability and ductility of any element, and this manifests itself in the way the metal can be beaten into paper-thin films that have been used in ornamental architecture for millennia, and to coat space satellites. A mere gram of gold, which is the size of a grain of rice, can be beaten to a gold film covering one square metre when it then appears translucent and transmits greenish light. Applied to glass this gold can be used to de-ice windows by passing electrical current through it.

Environmental element

Gold in the environment	
EARTH'S CRUST	1 p.p.b.
	Gold is the 73rd most abundant element.
SOIL	*approx.* 1 p.p.b., range 0.05–8 p.p.b.
SEAWATER	10 p.p.t.
ATMOSPHERE	virtually nil.

Gold occurs mainly as the metal, occasionally as crystals, but more generally as grains, sheets and flakes in other rocks. There are some gold ores, such as sylvanite which is a combination of gold, silver, and tellurium. The continuously erupting volcano Mount Erebus, in Antarctica, spews forth gold dust and is unique among volcanoes in this respect.

Chemical element

Data file	
CHEMICAL SYMBOL	Au
ATOMIC NUMBER	79
ATOMIC WEIGHT	196.96655
MELTING POINT	1,064°C*
BOILING POINT	2,807°C
DENSITY	19.3 g/cm³
OXIDE	Au_2O_3 (unstable)

*The melting point of pure gold used to be taken as 1,063.0 °C and was a fixed reference point on the scale of temperatures. That was until in 1968, when it was measured again and found to be too low. The exact value is now 1,064.43°C.

Gold is a soft metal with a characteristic shiny yellow colour and a member of group 11 (row 5d) of the periodic table. Gold is unaffected by air, water, alkalis, and acids, except *aqua regia* and selenic acid, H_2SeO_4. Supercritical water will also dissolve gold—see under hydrogen page 238. Although gold is known for being chemically unreactive many gold compounds are known, mainly in the lower oxidation states +1, as with AuCl, but also +3, as with AuF_3. There is even gold in oxidation state +5, as with AuF_5. And gold atoms can form cages, witness Au_{16}.

There are 37 known isotopes of gold with mass numbers 169 to 205, and 34 known isomers. Natural gold has only one isotope, gold-197, which is not radioactive. The longest-lived radioactive isotope is gold-195 with a half-life of 186 days. The next longest-lived is gold-196 with a half-life of 6.18 days. (Nuclear theory suggests that even gold-197 should be radioactive but so far this has not been observed on account of its having an incredibly long half-life.)

Hidden gold

It has been said that Niels Bohr, the Nobel Prize winner, dissolved his gold medal in aqua regia in 1943 to prevent it falling into the hands of the Nazis before he fled his native Denmark, and left it unobtrusively in his laboratory. When he returned in 1945 the bottle of acid was still there and so he reclaimed the gold, and a new medal was struck.

The reason for gold's imperviousness to oxidation and corrosion lies in the remarkable grip it exerts on its lone outermost electron.

Element of surprises

1. Even though the concentration of gold in the sea is only 10 p.p.t. it means that more than 13 million tonnes of the metal are in the oceans of the world. Many have been tantalised by the idea of claiming some of this bounty, and even some eminent chemists have tried to extract it, including the 1918 Nobel Prize winner for chemistry, Fritz Haber (1868–1934)—see nitrogen. On the basis of early analyses he mistakenly believed that the level of gold in seawater was 10 p.p.b. (a thousand times greater than it really is) and he argued that it would be possible to extract enough gold from the sea to repay the punitive reparations of 20 billion marks imposed on Germany by the Allies after World War I. Despite much research over several years, it proved impossible to do it economically.

2. Wedding rings do slowly wear away. A typical 18-carat wedding ring weighs about 6 grams and will lose about 6 mg of gold a year, or about 0.12 mg per week. Vigorous clapping can double this weekly loss. If there are one billion wedding rings being worn around the world then the loss per annum amounts to six tonnes of gold.

3. Gold is sometimes used as food decoration (it is coded E175) and there are alcoholic drinks which contain flakes of gold. While earlier civilisations believed that consuming gold could only confer benefits, we now know that it will simply pass through the body without being absorbed.

4. When gold is in contact with the skin of certain people it leaves a black mark. In medieval times this was linked to witchcraft, while today it is regarded by some as a sure sign of anaemia. In fact it is neither, and is probably indicative of the gold being of poor quality with the alloying metal being responsible for the black mark.

5. One way of proving that a supposedly ancient artefact made of gold is really a fake is to measure the helium it contains. This gas increases with time and is generated by traces of uranium and thorium in the gold. Newly refined gold will have levels of helium around 0.2 p.p.m. whereas gold that is hundreds of thousands of years old will have levels of helium around 60 p.p.m.

Hafnium

Pronounced haf-nee-um, it is derived from *Hafnia* the Latin name for Copenhagen.
French *hafnium*; German *Hafnium*; Italian *afnio*; Spanish *hafnio*; Portuguese *háfnio*.

Essential element

This element is needed to ensure tiny microchips and nuclear reactors work securely.

Element of life

The amount of hafnium in the human body is not known, but is tiny. It has no known biological role, and its salts generally have low toxicity. The element is poorly absorbed by the body and poisoning by hafnium compounds is unheard of. Plants take up small amounts of hafnium from the soil in which they grow and levels of 0.01–0.4 p.p.m. (dry weight) have been recorded.

Element of history

In 1911, Georges Urbain (1872–1938) reported the discovery of a 'missing' element, the one that came below zirconium in group 4 of the periodic table, and he called it celtium. However, further research, done in collaboration with a young Englishman, Henry Moseley (1887–1915), showed he was mistaken.

Moseley, born in Weymouth, England, in 1887, was in his early 20s and was already regarded as a scientific genius, following his discovery in 1913 of the fundamental property of atomic number.[72] This he deduced from the X-ray spectra of elements and as a result he was able to announce with confidence that elements with atomic numbers 43, 61 and 75 still remained to be discovered.

The same was also true of element 72, but Moseley had originally been misled into thinking this was the recently announced celtium. Indeed he never lived to see the discovery of the missing element because he was killed by a sniper's bullet during the ill-fated Dardanelles campaign in World War I. His death, at 27, is regarded as one of the great tragedies of science.

With celtium discredited, the search was on once again for the missing element. It was finally tracked down in 1923 by the Hungarian chemist George Charles de Hevesy[73] (1889–1966) working with Dirk Coster (1889–1950) at the University of Copenhagen. It turned up, perhaps not surprisingly, in a zirconium ore. When they

---

[72] This is the number of protons in an atom's nucleus, and it determines which element it is.
[73] De Hevesey won the Nobel Prize in Chemistry in 1943 for his work on isotopes.

examined a sample of Norwegian zircon, using Moseley's method of X-ray analysis, they observed lines that showed that another element was present. This proved to be hafnium, but it also proved very difficult to separate it from the zirconium. Although neither of the discoverers was Danish they named the new element after the city in which it was found.

Historic specimens of zirconium minerals from museums were examined by Hevesy, and some were found to contain as much as 5 percent hafnium. That it had remained undiscovered for so long is not so surprising because the two elements share many features in common for reasons explained in the *Chemical element* section below. These findings necessitated the revision of the atomic weight of zirconium once samples free of hafnium had been produced.

The first pure sample of hafnium was made in 1925 by decomposing hafnium tetra-iodide (HfI_4) over a hot tungsten wire.

Economic element

Hafnium ores are rare, but two are known: hafnon, which is hafnium silicate, $HfSiO_4$, and alvite, which is a mixed hafnium, thorium, zirconium silicate, $(Hf,Th,Zr)SiO_4$. Zircon ores contain hafnium and they are found in Brazil, Malawi, Western Australia, and New South Wales. World resources of recoverable hafnium are estimated to be less than 5,000 tonnes and global production of the metal itself is around 100 tonnes per year. The main producing countries are France and the USA which produce an estimated 30 tonnes and 50 tonnes respectively. The biggest users of hafnium are the nuclear industry (60 percent) while around 30 percent goes into metal alloys.

Although hafnium and zirconium are so alike chemically that they are extremely difficult to separate, when it comes to their nuclear behaviour they could not be more different. Zirconium is transparent to neutrons, while hafnium absorbs them, and is 500 times better at doing so than is zirconium. The nuclear power industry uses both of them, the zirconium for cladding fuel rods in reactors and the hafnium, and its alloys, for control rods in reactors. Nuclear submarines particularly need hafnium for this purpose, and it also has a very high melting point and is corrosion resistant. However, it is expensive and the 50 or so control rods needed for a large nuclear reactor can cost as much as $1 million.

Hafnium has to be separated from zirconium, whose ores contain around 5 percent of hafnium. There is no simple chemical process for doing this and it is done using a liquid-liquid extraction process which relies on small differences in solubility of hafnium and zirconium thiocyanates, $Hf(SCN)_4$ and $Zr(SCN)_4$, in the solvent methyl isobutyl ketone. The separated hafnium is then reacted to form hafnium chloride and this is reduced to the metal by reaction with magnesium at 1,100°C. To make ultrapure hafnium, the metal is converted to hafnium iodide, HfI_4, and this is decomposed on a tungsten filament at 1,700°C.

Hafnium is used in high temperature alloys and ceramics. Hafnium carbide and nitride are some of the most refractory materials known, in other words they will not melt except under the most extreme temperatures, which in the case of hafnium nitride is 3,310°C. The mixed carbide of hafnium and tantalum, $HfTa_4C_5$ has the highest melting point of any know material: 4,215°C.

In transistors, the current flows between two terminals, one of which is called the source and the other is called the drain, and it is controlled by the voltage at a third terminal called the gate. For a transistor to switch efficiently, the gate has to be isolated from the source and the drain by an insulating layer, and this used to be a film of silicon dioxide but this was made thinner and thinner, as microchips got smaller and smaller, until it was on the verge of being unacceptable on account of current leakage. Hafnium oxide is now being used as the gate insulator and it performs much better. The result is referred to as 45 nanometre technology, referring to the size of the features on each chip. The full stop at the end of this sentence corresponds to the area taken up by two million of these new chips, they are so tiny. More recent microchips are even smaller than this.

Hafnium catalysts have been introduced for the polymerisation of the olefins 1-octene and ethylene, and in the case of co-polymerisations of a mixture of olefins the resulting polymers have more regular features leading to higher melting points and other desirable traits.

Plasma welding torches do not require fuel to generate high temperatures but rely instead on a flow of plasma from the spark which arcs between two electrodes. The one which releases the electrons to generate the plasma has to be hafnium because only this metal remains unaffected, in other words it neither melts nor corrodes, at the high temperatures at which the arc operates.

Environmental element

Hafnium in the environment	
EARTH'S CRUST	5.3 p.p.m.
	Hafnium is the 45th most abundant element.
SOILS	av. 5 p.p.m., with a range of 2–20 p.p.m.
SEAWATER	0.007 p.p.t.
ATMOSPHERE	virtually nil.

Hafnium was the next to the last of the non-radioactive elements to be discovered—rhenium came two years later—but whereas rhenium is extremely rare, hafnium is relatively abundant. It poses no environmental threat to plants or animals.

Chemical element

Data file	
CHEMICAL SYMBOL	Hf
ATOMIC NUMBER	72
ATOMIC WEIGHT	178.49
MELTING POINT	2,230°C
DENSITY	13.3 g/cm³
OXIDE	HfO_2

Hafnium is a lustrous, silvery, ductile metal and a member of group 4 (row 5d) of the periodic table of the elements. It resists corrosion due to a tough, impenetrable oxide film on its surface. However, powdered hafnium will burn in air. The metal is unaffected by alkalis and acids, except hydrofluoric acid. Its preferred oxidation state is +4 as compounds such as HfF_4 and $HfCl_4$. Hafnium will rapidly absorb hydrogen gas when heated to 700°C and forms a compound of approximate composition Hf_5H_9. Hafnium in the form of a complex with cyclopentadienyl[74] has the ability to break some of the strongest bonds in chemistry such as those between the two nitrogen atoms of N_2 and between the carbon and oxygen of CO—and it can perform this action at normal temperatures.

Hafnium is difficult to separate from its group 4 partner, zirconium, because the two elements have atoms that are the same size. Normally as you go down a group of the periodic table the atoms get larger, but for hafnium and zirconium this general rule does not hold. The extra electrons in hafnium occupy an inner orbit. With atoms of the two metals being the same size, and with a similar arrangement of outermost electrons they are almost indistinguishable chemically.

There are 36 known isotopes of hafnium with mass numbers 153 to 188, and 27 known isomers. Naturally occurring hafnium mainly consists of five stable isotopes, hafnium-180 (which accounts for 35.1 percent), hafnium-178 (27.3 percent), hafnium-177 (18.6 percent), hafnium-179 (13.6 percent), hafnium-176 (5.3 percent), and one radioactive one, hafnium-174 (0.16 percent) which has an incredibly long half-life of 2×10^{15} years (equivalent to 2 million billion years) and which is a weak alpha-emitter. It is this isotope that gives hafnium its high neutron-capturing ability. The next longest-lived radioactive isotope is hafnium-182 which has a half-life of 9.6 million years, and this occurs in trace amounts.

Element of surprises

1. Hafnium might well have been discovered much earlier than it was. In the late 1800s an American, Edgar Smith, was investigating monazite sand, the mineral that contains the rare earth elements (see lanthanoids, page 266). Other metals are also present in this ore and one of these was zirconium which he removed by converting it to the sulfate. Smith retained the remaining material, telling his co-workers that he was convinced there was another element concealed within it, and that he would find this when he had the leisure to do it. Smith was right in his belief—there was hafnium present—but he was wrong in thinking he would be given the time to discover it.

2. The isomer hafnium-178m2, with its half-life of 31 years, has a particularly high energy capacity. This energy can be pumped into the atom using low-energy X-rays, but when the atom decays it releases 60 times as much energy in the form of deadly gamma rays. One gram of the isotope contains around 1,300 MJ of energy which is the same as that contained in 300 kg of TNT. Hafnium-178m2 is

---

[74] This consists of a ring of five carbon atoms which incorporates two double bonds.

made by bombarding tantalum with protons but thankfully only tiny amounts can be made at any one time. Thankfully, because it might otherwise be possible to deploy it via miniature missiles which would kill everything within a wide radius by the emission of powerful gamma rays, but without the radioactive fallout associated with nuclear bombs.

Hassium

Pronounced hass-ee-um, the element is named after the German state of Hesse, whose Latin name was Hassia.

French *hassium*; German *Hassium*; Italian *hassio*; Spanish *hassio*; Portuguese *hássio*.

Essential element

Essentially unknown, except to physicists, but despite only making a few atoms they have proved its basic chemical similarity to osmium.[75]

Element of history

Interest in this element began in the 1960s when it was simply referred to as element 108. Theory was that it might occur on Earth in trace amounts as long-lived deformed isomers. This came out of research to explain the extreme radiation damage in certain minerals that could not have been produced by energies emitted by any known radioactive isotope, but might be emitted by heavy elements—see unbihexium page 592 for further details. The mineral molybdenite was analysed by Victor Cherdyntsev who even claimed to have detected element 108 in 1963. He suggested the name sergenium for it and this was based on an area of Kazakhstan where mineral samples were collected which supposedly contained this element.

The first attempt to synthesise element 108 took place in 1978 at Russia's Joint Institute for Nuclear Research (JINR) in Dubna. A team headed by Yuri Oganessian and Vladimir Utyonkov bombarded radium-226 with calcium-48 to get element 108 isotope 270, and by bombarding lead 208 with iron-58 they appeared to get isotope 264. They were cautious, because the data was less than convincing. (Re-runs of the lead-iron experiment in 1984 indicated that element 108 had probably been produced.)

In 1983, the JINR carried out a series of experiments in which they bombarded bismuth-209 with manganese-55 and got element 108 isotope 263. They also bombarded californium-249 with neon-22 and got isotope 270, and by bombarding lead-207 and lead-208 with iron-58 they obtained isotope 264. A re-run in 1984 confirmed the last of these.

In 1984, at Germany's Gesellschaft für Schwerionenforschung (GSI) in Darmstadt, a team headed by Peter Armbruster and Gottfried Münzenberg bombarded lead-208 with iron-58 and synthesised isotope 265.

To whom should the accolade for discovery of element 108 be awarded? That was the weighty issue that faced the Transfermium Working Group (TWG), a joint

---

[75] There may be long-lived isomers of hassium we do not know about and which could conceivably have some practical use some day.

IUPAC/IUPAP committee that had been set up in 1985 to study claims for all trans-fermium elements. It chose to acknowledge the GSI. In a strangely-worded decision, it said that the GSI claim was 'more detailed' and 'carried conviction' although all the data from both laboratories taken together supported the JINR's claim of 1983 and the TWG even acknowledged there was little doubt that 108 had first been produced there.

The physicists at the GSI proposed naming the new element after the German state of Hesse in which Darmstadt is located and suggested hassium. The IUPAC committee thought otherwise, and said that the German state of Hesse did not merit having an element named in its honour. Instead it suggested hahnium (symbol Hn) for element 108, naming it after the German radiochemist Otto Hahn (1897–1968), but the GSI insisted on their choice. The American Chemical Society preferred the name hassium and began using this in its publications in 1995. IUPAC dropped their objection and approved this name in 1997, along with the names of elements 104–109.

Environmental element

Based on several claims that hassium might occur naturally in trace amounts, the JINR in 2004 began a search for naturally occurring hassium in minerals as part of the SHIN experiment (Search for Heavy Elements in Nature). SHIN is located at the LSM *Laboratoire Souterrain de Modane* (Modane Subterranean Laboratory) in France. These element searches are done deep below the Earth's surface to eliminate the masking effects of cosmic rays, which can obscure the infrequent decay emission of a heavy element. Cosmic rays cannot penetrate to the underground laboratory, so an occasional decay signal of a trace element can be detected.

Chemical element

Data file	
CHEMICAL SYMBOL	Hs
ATOMIC NUMBER	108
ATOMIC WEIGHT	269 (longest-lived known isotope)
Properties such as melting point, boiling point, and density, are likely never to be measured because there will never be more than a few atoms produced at any one time.	
oxide	not known but likely to be HsO_4

Hassium is a radioactive element and a member of group 8 (row 6d) of the transition metals of the periodic table, coming below osmium. So far not enough of this element has been made at one time to be visible by the naked eye, but predictions are that it would be a silvery metal susceptible to attack by the oxygen of the air. By analogy with other members of group 8 its preferred oxidation state should be +8 although lower oxidation states, namely +3, +4, and +5, might also be stable.

More than 100 hassium atoms have been synthesised although none now exists. In 2001, the GSI studied the chemical properties of hassium, and using only seven atoms of hassium the made hassium tetroxide (HsO_4) which was similar to osmium tetroxide (OsO_4) indicating its chemical relationship to osmium which comes above it in the periodic table, and showing its preference for the oxidation state +8.

In 2006, four atoms of a new isotope, hassium-270 were produced for the first time at the GSI in Germany by firing magnesium at curium. These had a half-life of 22 seconds. This particular isotope has 114 protons and 162 neutrons which theory suggests should also confer stability on the nucleus. What was particularly noteworthy was that they were able to confirm hassium tetroxide, HsO_4.

There are 2 known isotopes of mass number 263 to 271, 273, 275, one and 277, and three known isomer. The longest-lived known isotope is hassium-269 with a half-life of 9.7 seconds. (Hassium-276 is calculated to have a half-life of one hour.)

Element of surprises

Despite the brevity of its existence, a chemical reaction has been carried out on hassium tetroxide. This was with sodium hydroxide and the reaction product was the salt disodium dihydroxytetraoxohassate (VIII), chemical formula $Na_2[HsO_4(OH)_2]$.

Helium

Pronounced heel-ee-um, the name comes from the Greek *helios* meaning the Sun. Helium is unlike the other noble gases, neon, argon, krypton, xenon, and radon, in that its name does not end in '-on' for reasons explained below.

French *hélium*; German *Helium*; Italian *elio*; Spanish *helio*; Portuguese *hélio*.

Essential element

This element, which will react with nothing, has a myriad of uses, the most important of which, as liquid helium, is to cool the powerful magnets of MRI scanners.

Cosmic element

Helium had the distinction of being discovered on the Sun before it was found on Earth.

It was one of the elements produced during the big bang, along with hydrogen and lithium, and its presence throughout the Universe supports the theory that such an event occurred. (Traces of beryllium and boron were produced from these first three elements shortly thereafter by radioactive decay and nuclear reactions.) Helium is the second most abundant element in the Universe, making up 24 percent of its mass (hydrogen is the first with 75 percent), although in terms of the numbers of atoms there are 13 times as many hydrogen atoms as helium atoms. Together these two elements account for 99 percent of the elemental matter in the Universe.[76]

Stars like the Sun produce helium by the fusion of hydrogen nuclei, and in so doing release massive amounts of energy. The fusion of two helium nuclei to form beryllium, is not energetically favourable, however, but the fusion of three heliums to form carbon does release energy. A helium nucleus is also known as an alpha particle (α-particle) and this process of forming carbon is referred to as the triple alpha reaction.

Element of life

There are traces of helium in the human body, mainly dissolved from the air into the blood stream as it passes through the lungs. A little is also produced from the radioactive decay of uranium atoms which are mainly lodged in the skeleton. Helium has no biological role, and although it is a harmless gas, it could asphyxiate if were to exclude oxygen from the lungs, although only two people are reported to have died this way in the past 10 years.

[76] Elemental matter comprises only about 4% of the Universe. The other 96% consists of so-called 'dark matter' which comprises 23%, and 'dark energy' which accounts for 73%. Neither of these 'dark' materials has yet been identified.

Medical element

Helium is needed for medical diagnosis and about a quarter of all helium that is gathered is used this way. Liquid helium, at a temperature of −269°C, cools the superconductors which produce the intense magnetic fields needed for magnetic resonance imaging (MRI). When a human is placed in such a magnetic field and exposed to radio waves, these interact with the hydrogen atoms of water and other molecules in the body and an image is produced of internal organs, enabling possible cancers to be diagnosed.

Element of history

When the light from a hot atom is observed through a prism it presents a pattern of coloured lines, known as its spectrum. The position of these lines can be accurately measured and will identify the element producing them. This technique of atomic spectroscopy led to the discovery of several elements in the 1800s – see for example caesium, lithium, rubidium, and thallium. It was even possible to identify elements such as sodium in the light from the Sun, but such was the intensity of sunlight that faint lines from other elements could not be seen. However, at the moment of a total eclipse most of this would be blocked out and analysis of the spectrum of the corona might well reveal more information. So thought Pierre J.C. Janssen (1824–1907) who travelled to the city of Guntur in India to study a total eclipse forecast for 18 August 1868, and he was lucky in that the sky was clear and he could take his measurements. He observed a yellow line in the corona spectrum at a wavelength of 587.49 nm, which is near to the two sodium lines, but distinct from them, and he speculated that this was from a hitherto unknown element.

The British astronomer Joseph Norman Lockyer[77] (1836–1920) also travelled to India to the city of Vijaydurg where at the fort there he too observed helium in the spectrum of the Sun. He made no announcement, however, because he wanted to do further research to see if it was possible to detect helium during regular observations of the Sun and not just during an eclipse. Using a specially designed telescope, he and Edward Frankland (1825–99) made their observations of the Sun on 20 October 1868 through the smoky skies of an overcast London and, confirming the same line, agreed that there was a new element which they postulated could exist only on the Sun. Lockyer assumed it would be a metal and so he called it helium. His presumption to have discovered a new element on the Sun, and having the temerity to name it, were ridiculed by some at the time but in the end he had the last laugh.

Unbeknown to Janssen, Lockyer, Frankland and their contemporaries, helium existed on Earth but only in tiny amounts in the atmosphere, and in uranium minerals. Its discovery on planet Earth was made by the Italian physicist Luigi Palmieri (1807–96) in 1882 when he identified it by its spectral line while analysing lava from the volcano Mount Versuvius. When the US geologist William Hillebrand (1821–86)

---

[77] Lockyer was a civil servant from Wimbledon, London, who wrote the first book setting down the rules of golf as played at St. Andrews in Scotland. He was also the founder of the Science Museum in London.

dissolved the mineral uraninite (UO_2) in acid in 1889 he observed that bubbles of gas escaped from it. He could not identify the gas and did not pursue his observations further. However, when Per Teodor Cleve (1840–1905) and Nils Abraham Langer in Uppsala, Sweden, in 1895, made the same observations from a new mineral (which at the time was named cleveite after Cleve, although it later turned out to be uraninite) they were able to collect enough of the gas positively to identify it as the element helium and to measure its atomic weight accurately. Quite independently, that same year Sir William Ramsay (1852–1916) and Morris Travers (1872–1961) at University College, London, also collected a sample of the new gas which they too identified as helium. Ramsay even wrote to Lockyer on 29 April 1895 suggesting he change the name of the gas to helion to make it consistent with the other noble gases. Lockyer ignored his request and so helium it has remained.

In May 1903, a company drilling for oil near the town of Dexter, Kansas, struck a gas geyser. The pressure of the escaping gas caused a loud howling that could be heard for miles. Before the well was capped, it was decided to celebrate the company's good fortune by igniting some of the gas which was escaping at the rate of a quarter of a million cubic metres a day. Crowds assembled to witness the promised pillar of flame which was to be ignited by a burning bale of hay pushed towards a pipe that had been set up to draw off some of the escaping gas. Unexpectedly it not only failed to ignite the gas, but the burning bale was extinguished.

The Kansas state geologist, Professor Erasmus Howard, collected samples of the escaping gas and took them back to the University of Kansas at Lawrence for analysis. Helped by colleagues Hamily Cady and David McFarland, he found that the gas was 72 percent by volume nitrogen, 15 percent methane (not enough to make the gas combustible), 1 percent hydrogen, and 12 percent of something else which refused to be identified.

To remove all the other components they used activated charcoal which they knew would absorb all gases, except hydrogen, helium, and neon, when cooled to liquid air temperatures (–190°C). They then tested the gas that was left spectroscopically and found the yellow line of helium, the gas that had only been reported a few years previously and which was still regarded as a rarity. Far from being rare it was present in vast quantities in the gas fields beneath Kansas, Texas, and Oklahoma.

In 1907, Ernest Rutherford (1871–1937) and Thomas Royds (1884–1955) showed that alpha-particles were really helium nuclei by allowing them to penetrate a thin layer of glass into an evacuated tube where they picked up the electrons needed to form helium atoms. When an electric discharge was passed through the tube it glowed with the characteristic spectrum of helium.

Economic element

Although helium is present in the atmosphere, it is at present uneconomical to extract it from liquid air. The commercial source of helium is natural gas which may contain as much as 7 percent. It is collected mainly in the USA, although Poland, Algeria, and Qatar also produce it. Europe's helium gas comes from Algeria where there is a reserve which produces 17 million cubic metres per year. Helium is continually being formed within the Earth's crust by the many naturally occurring radioactive elements that undergo alpha-decay, emitting alpha-particles, such as uranium, thorium, radium, radon, and polonium.

Radioactivity is the source of the 175 billion litres of this gas which is extracted from gas wells annually, and it has been estimated that 3,000 tonnes of helium are generated in the Earth's crust every year.

Helium is needed for low-temperature instruments, such as the superconducting magnets mentioned above, for providing inert atmospheres for industrial uses such as welding air-sensitive metals, for detecting minute leaks in equipment, for deep-sea diving, for rocket launches (the valves that release the hydrogen and oxygen for the rockets are controlled by compressed helium), for weather balloons and airships, and for gas lasers. Helium-neon gas lasers are used in supermarket check-outs to scan bar codes. Another major use is in the production of optical fibres which are made under an atmosphere of helium which suffuses the molten glass and then promotes rapid cooling as it escapes from the extruded fibres. Helium can provide the inert atmosphere needed for growing ultrapure crystals for silicon wafers.

Helium is used in balloons and airships. Indeed, this was the earliest use of helium, and the first use of large volumes of the gas. It was first used to lift an airship in 1921 when it raised the US Navy's 198-foot airship, the non-rigid C-7 blimp. Helium has about the same lifting power as hydrogen gas. A cubic metre of helium gas can lift a load of 1 kilogram. To lift a ton would require a thousand cubic metres, or a balloon of around 12 metres diameter.

In 1925 the US Government voted to set up a strategic National Helium Reserve at Amarillo in Texas, thinking that it would be needed for military airships of the future. In the 1920s and early 1930s, the US Navy explored rigid-design airships, one of which was a zeppelin obtained as war reparation from Germany, and three were built in the USA. All three US-built airships had crashed by 1935, due to bad weather or design flaws. The German-built zeppelin, which flew under the name Los Angeles, was retired shortly thereafter as the Navy cancelled the rigid-airship programme. Helium-filled non-rigid airships (aka blimps) continued in service until 1962, used for anti-submarine reconnaissance, and being very successful during World War II in this role—see also hydrogen, page 230.

The National Helium Reserve was even expanded when the importance of helium to the space race was apparent; it was required to keep the liquid hydrogen and oxygen cool that powered the Apollo space vehicles. The helium was collected from gas wells and piped to an underground storage facility, which was an exhausted gas well. By 1995, a billion cubic metres of the gas had been collected and the project had debts of $1.4 billion. Congress voted to end the scheme in 1996 with the gas being sold off to repay the loans, although some will be retained. The known reserves of helium in the USA might be exhausted early this century unless more effort is made to trap it out of natural gas and to recapture it from industries which use it.

Helium is part of the air supplied to deep sea divers. This can prevent too much oxygen from entering the brain and prevent high-pressure nervous syndrome. At depths below 600 metres, even helium cannot always prevent this happening and then a mixture of hydrogen, helium, and oxygen, known as 'hydreliox' is used.

New uses for helium continue to be found. The first ever helium-ion microscope was installed in 2007 at the US National Institute of Standards and Technology. This type of microscopy gives better resolution than scanning electron microscopy, which is currently used to examine semiconductor surfaces. Helium ions have shorter wavelengths than electrons so can be focussed into a narrower beam and the particles interact more strongly with materials thereby producing clearer images.

The Large Hadron Collider (aka LHC) at CERN uses 96 tonnes of liquid helium to keep the temperature at 1.9 Kelvins (−271.3 °C). This machine is the world's largest particle accelerator with a circuit 27 km in diameter and is able to produce energies of up to 14 TeV (teraelectron volts, 14×10^{12}). It is designed to probe elementary particles of matter.

Environmental element

Helium in the environment	
EARTH'S CRUST	8 p.p.b.
	Helium is the 71st most abundant element.
SEAWATER	4 p.p.t.
ATMOSPHERE	5 p.p.m. by volume.

Air contains helium, but unlike most other gases in the atmosphere, it can escape into space. Even though there are hundreds of millions of tonnes of the gas circling the Earth, it is not worth extracting from this source at present. Helium is less soluble in water than any other gas which means there is very little in the sea.

The age of thorium and uranium minerals can be calculated from the amount of helium trapped within them.

Chemical element

Data file	
CHEMICAL SYMBOL	He
ELEMENT NUMBER	2
ATOMIC WEIGHT	4.002602
MELTING POINT	−272°C (0.95 K) at 26 atmos. pressure*
BOILING POINT	−269°C (4.2K)
DENSITY OF THE GAS	0.18 g per litre
OXIDE	none known, nor ever likely to be

*Solid helium was first made in 1926.

Helium is one of the so-called noble gases, and is generally assigned to group 18 (row 1) of the periodic table—but see page 650. It is a colourless, odourless gas which is totally unreactive chemically. However, some molecular ion species have been made such as VHe^{3+} and $HePtHe^{2+}$ but these are highly unstable. Theoretical calculations suggest that helium might form weak bonds to cyclic molecules such as cyclobutadiene but such compounds have yet to be shown to exist.

Helium has eight known isotopes with mass numbers ranging from 3 to 10, and no known isomers. There are two stable isotopes which occur naturally: helium-4 and helium-3. There are six radioactive isotopes with mass numbers ranging from 5 to 10. The longest-lived is helium-6 with a half-life of a mere 0.807 second.

Almost all the helium-3 which is present on Earth has been around since the planet formed although small amounts arrive from cosmic dust and from the decay of hydrogen-3 (aka tritium) which is produced in nuclear reactors and which has a half-life of 12.3 years, converting to helium-3. The amount of helium-3 in the Earth's atmosphere is a million times less than that of helium-4. Nevertheless supplies of man-made helium-3 are available and widely used in oil and gas drilling to measure the porosity of rock to assess whether a pocket in the Earth's crust is likely to contain gas, oil, or just water.

It can also be used to detect the presence of nuclear material. Supplies of helium-3 are available from governments with nuclear arsenals because it is a by-product of the radioactive decay of tritium (hydrogen-3) which is needed for hydrogen bombs. Helium-3 can detect neutrons by the amount of energy that is released when they collide and such detectors are used in the US to check for smuggled nuclear material.

In 2000 the USA alone had 200,000 litres of helium-3 and Russia was exporting 25,000 litres of the gas annually. Both continued to harvest helium-3, but production was declining while the gas was being used up, so that by 2010 the USA reserve had shrunk to less than 50,000 litres. Meanwhile Russia no longer exports helium-3.

Element of surprises

1. Breathing helium dramatically lightens the voice because sound travels three times faster in helium than in air, and while the wavelength of the sounds that the human voice box generates does not change, the frequency changes to compensate and the pitch rises.

2. When the temperature of liquid helium falls below 4K (−269°C), it is at first simply a colourless liquid, albeit extremely cold, but when the temperature drops below 2K something rather strange happens to it. It becomes a liquid with bizarre properties and is known as helium(II). Not only does it expand on cooling, but it is a million times more heat conducting than normal liquid helium and its viscosity is zero, in other words nothing can slow its flow. It will escape from any container that is not totally sealed; it does this by flowing up the sides and out of the top, apparently defying even gravity.

3. Helium can only be solidified at near absolute zero (−273°C) by applying a pressure of about 26 atmospheres. Solid helium is a frictionless material. It moves by the action of helium atoms filling holes in the lattice of the solid left by other helium atoms which have moved, and so it creeps effortlessly along.

Holmium

Pronounced hol-mee-um, it is derived from *Holmia*, the Latin name for Stockholm.

French *holmium*; German *Holmium*; Italian *olmio*; Spanish *holmio*; Portuguese *hólmio*.

Holmium is one of 15 chemically similar elements referred to as the lanthanoid or rare earth elements, which extend from element atomic number 57 (lanthanum) to element atomic number 71 (lutetium). The term rare earth is a misnomer because some are not rare. The minerals from which they are extracted, and the properties and uses they have in common, are discussed under Lanthanoids.

Essential element

When it comes to focussing powerful magnets, nothing beats holmium.

Element of life

The total amount of holmium in the human body is not known, but is tiny. Holmium has no biological role, but it has been noted that holmium salts stimulate metabolism. It is difficult to identify exactly how much of the various lanthanoids are present in the human body, but they are present and holmium is likely to be one of the least. No one has monitored diet for holmium content, so it is difficult to judge how much we take in every day but it is probably only a milligram or so per year. Holmium is not readily taken up by plant roots so very little gets into the human food chain, and vegetables that have been analysed for holmium have been found to have less than 0.1 p.p.b. dry weight.

Medical element

Holmium is used in lasers as Ho-YAG (holmium yttrium aluminum garnet) and because they emit light at 2.08 microns they are eye-safe. These lasers have been approved for various types of surgery, combining as they do the ability to be transmitted down optical fibres with good tissue cutting power.

Element of History

Holmium was discovered at Geneva in 1878 by the Swiss chemists Marc Delafontaine (1837–1911) and Louis Soret (1827–90), and independently by Per Teodor Cleve

(1840–1905), who was professor of chemistry at Uppsala, Sweden. The discovery came as a result of work on the element yttrium, which was contaminated with traces of other lanthanoid elements although this had not been realised at the time yttrium was discovered. Later erbium and terbium were separated from it, and then erbium was found to contain ytterbium. When, in 1874, Cleve looked more closely at the residual erbium left behind after the ytterbium had been removed, he realised it must contain yet other elements because he found that its atomic weight varied, according to how it was made. He separated holmium from erbium in 1878. However, Delafontaine and Soret had announced a few months earlier that erbium oxide must contain another element, which they referred to as 'element X', because its atomic spectrum gave bands that could not be explained. Eventually when the spectrum of 'element X' was compared to that of holmium they were found to be identical. Even so, we cannot be certain that Cleve had produced a *pure* sample of the new element because, within a few years, yet another rare earth, dysprosium, was to be extracted from holmium.

Economic element

Holmium is found in minerals such as monazite and bastnäsite, although it is only a minor component of such ores, for example, holmium is present to the extent of 0.5 percent in monazite. It is extracted from those ores that are processed for the yttrium they contain. The main mining area is currently China, although the USA, Brazil, India, Greenland, and Tanzania have exploitable deposits of rare earth ores. The reserves of holmium are estimated to be around 400,000 tonnes. World production of holmium metal itself is only around 10 tonnes per year. This is made by heating holmium chloride or holmium fluoride with calcium.

Holmium is also used in nuclear reactors as a 'burnable poison' which is one that burns up while keeping a chain reaction from running out of control.

Environmental element

Holmium in the environment	
EARTH'S CRUST	1.4 p.p.m.
	Holmium is the 56th most abundant element.
SOILS	1 p.p.m.
SEAWATER	0.4 p.p.t.
ATMOSPHERE	virtually nil.

Holmium is one of the rarer of the rare earth elements but is, nevertheless, 20 times more abundant than silver. It poses no environmental threat to plants or animals.

Chemical element

Data file	
CHEMICAL SYMBOL	Ho
ATOMIC NUMBER	67
ATOMIC WEIGHT	164.93032
MELTING POINT	1,474°C
DENSITY	8.8 g/cm³
OXIDE	Ho_2O_3

Holmium is a bright, soft, silvery metal and one of the lanthanoid group (row 4f) of the periodic table. It is slowly attacked by oxygen and water, and dissolves in acids. Its preferred oxidation state is +3 in compounds such as HoF_3 which is pink. In solution it exists as Ho^{3+} ions, surrounded by nine water molecules.

There are 35 known isotopes of holmium with mass numbers 140 to 174, and 28 known isomers. Naturally occurring holmium in minerals consists entirely of a single isotope, holmium-165, which is not radioactive. The longest-lived radioactive isotope is holmium-163 with a half-life of 4,570 years. Another long-lived isomer, holmium-166m, which has a half-life of 1,200 years, is used in the calibration of gamma ray spectrometers. This isotope produces a wide spectrum of such rays.

Element of surprises

1. Holmium alloys are used as a magnetic flux concentrator for high magnetic fields because it has the highest magnetic moment of any element: 10.6 Bohr magnetons. The pole pieces of the most powerful magnets are made of holmium.

2. Nuclear theory suggests that all isotopes of holmium are radioactive and that includes holmium-165 which is currently listed as the only stable isotope. Theory is that this has an incredibly long half-life which, as yet, has not been measured.

1
H

Hydrogen

Pronounced hy-dro-jen, the name is derived from the Greek words *hydro genes,* meaning water-forming. This is the only element that has a different name for its three main isotopes. Hydrogen-1, the most common isotope, is also called protium although that name is almost never used; hydrogen-2 is called deuterium, chemical symbol D; and that name is invariably used, as it tritium for hydrogen-3, chemical symbol T.

French *hydrogène*; German *Wasserstoff*; Italian *idrogeno*; Spanish *hidrógeno*; Portuguese *hidrogênio*.

Essential element

Hydrogen is the number one element in more ways than one: it was the first element of Creation; it is the first element of the periodic table; it is the most abundant element in the Universe; it is the element that fuels the Sun and stars; it is the lightest of all gases; it is the atom which confers acidity on a molecule; and it will link molecules together in its own unique fashion. This last point is particularly important for water and DNA and this unique kind of intermolecular linking is called hydrogen bonding. Replacing hydrogen with deuterium changes materials subtly and this has important scientific implications.

Cosmic element

Hydrogen accounts for 75 percent of the atoms in the universe, and was one of the three elements, along with helium and lithium, that were produced in the big bang. It continues to be the fuel of the stars. At the centre of a star like the Sun, the temperature is around 13 million degrees and the density about 200 kg per litre. Under such conditions, hydrogen will begin to 'burn', in other words protons, which are hydrogen nuclei, fuse together to form helium nuclei releasing vast amounts of energy and radiation as they do. Our star converts around 600 million tonnes of hydrogen per second into helium, and during this process around 5 million tonnes of matter are converted into energy.[78]

The fusing of two hydrogen nuclei first forms deuterium, as one proton converts to a neutron and releases a positron (the positive equivalent of an electron). The deuterium then reacts with another proton to form helium-3 and when two of these nuclei collide they form normal helium-4 while ejecting two protons to go through the process again. Because the first step in this process is slow to occur it means that stars like the Sun have enough fuel to burn for 10 billion years.

[78] In accord with Einstein's equation relating energy to mass: $E = mc^2$.

On Earth, hydrogen is light enough to reach the upper atmosphere and thence escape into space. Only a planet the size of Jupiter has the gravity to prevent this from happening. A probe in 1995 showed that planet was composed of 99.8 percent hydrogen and helium. Indeed, such is the pressure at the centre of Jupiter, that it is believed that the hydrogen there is converted to a metallic form of the element. Hydrogen can exist as a metal and this was proved in 1996 when a layer of liquid hydrogen 0.5 mm thick was subjected to a pressure of two million atmospheres. Under such conditions the element became electrically conducting, which is taken as evidence that it had become metallic.

In space there is also the molecular-ion H_3^+ which is formed by cosmic rays, and this too has been observed in the upper atmosphere of Jupiter.

Element of life

Hydrogen in the human body	
BLOOD	whole blood is 85% water, so consists of about 9.4% of hydrogen
BONE	about 5.2%
TISSUE	about 9.3%
TOTAL AMOUNT IN BODY	about 7 kg.

Water is *the* element of life in that nothing can live without it. It makes up around 65 percent of the weight of the human body, and because hydrogen constitutes 11 percent of the weight of water, so the total amount of hydrogen we contain comes to around 7 kg. The human body requires around 2.5 litres of water a day to keep itself healthy: half of this is taken in the form of drinks and the other half as food.

Hydrogen is also part of almost all of the other molecules which make up a living cell and it is an essential element for life because it is a constituent of DNA and as such is part of the genetic code.

Dangerous element

Hydrogen gas (H_2) is non-toxic, and so is its chief compound, water (H_2O), yet both can kill if they exclude air from the lungs. Water can also be fatal if too much is given to people who are badly dehydrated, such as those found dying of thirst. A sudden excess upsets the balance of sodium and potassium in the heart muscle resulting in a heart attack.

High acidity in the blood can pose a threat to health although this is usually a symptom of other conditions. The exchange of sodium ions (Na^+) and hydrogen ions (H^+) in heart tissue plays an important role in coronary heart disease. Yet the stomach can tolerate quite high acid levels necessary for digestion, although generally such concentrated acids are quite corrosive.

Were a person to drink D_2O instead of H_2O, they would not survive long because its chemical nature is sufficiently different to upset certain key cell processes.

Medical element

Pharmaceutical chemists have discovered that various existing drugs can be made more effective by replacing some of their hydrogen atoms with deuterium atoms. For some drugs this makes them last longer in the patient, in other cases it improves their effectiveness. The drugs most likely to benefit from this new approach are those used against HIV and those which treat kidney disease.

Element of history

Hydrogen gas was known long before its nature was realised. As far back as the 1500s the Swiss alchemist, Paracelsus (1493–1541) noted that the bubbles given off when iron filings were added to sulfuric acid were flammable. In 1671, Robert Boyle (1626–91) dissolved iron filings in dilute hydrochloric acid and made the same observation, saying that as the gas burned it gave off a little light and a lot of heat 'with more strength than one would easily suspect.'

The person credited with hydrogen's discovery, in 1766, was the eccentric chemist, Henry Cavendish (1731–1810). He was one of the richest men in England and lived in Soho in central London, but spent most of his time in his laboratory at his other home in the suburb of Clapham. It was there that he collected the bubbles rising from the chemical reaction of iron filings with dilute sulfuric acid, and showed that they were different from other gases. He was also the first person to demonstrate that when hydrogen burned it formed water, and this he did in 1781, thereby ending the ancient Greek theory that the world was composed of four elements of which water was one. The gas was named *hydrogène* (water-former) by the great French chemist, Antoine Lavoisier (1743–94), and the name stuck.

Until the 1900s, hydrogen had little commercial application except in theatres where it was used to produce limelight (see calcium) and in airships, and yet its uniqueness fascinated chemists. In 1898, James Dewar (1842–1923) invented the vacuum flask, and with it he was able to liquefy hydrogen gas for the first time; he produced solid hydrogen the following year.

In December 1931, Harold Urey (1893–1981) and his colleagues at Columbia University in the USA detected another form of hydrogen in the residue of a large amount of liquid hydrogen which they had allowed to evaporate. This heavier form has a neutron as well as a proton in its nucleus and so has twice the mass of normal hydrogen. They named it deuterium, chemical symbol D. The following year they showed that when water is decomposed to oxygen and hydrogen by electrolysis, deuterium-containing molecules were left behind; this became known as heavy water (D_2O).

Hydrogen provided humans with their first reliable form of air transport. This began in 1783 when the French scientist Jacques Charles invented the hydrogen-filled balloon and demonstrated it not long after the first hot-air balloon had successfully carried a passenger. Hydrogen balloons were used for sight-seeing flights throughout the 1800s.

The airship was invented in 1852 by Henri Giffard, and lifted by hydrogen. The rigid version was invented in 1896 by David Schwarz. This invention inspired the German

Count Ferdinand von Zeppelin to create a more sophisticated design. He promoted the idea of rigid airships which became known as zeppelins. The first zeppelin flew in 1900 and it became the standard design for all later rigid airships. These were all lifted by hydrogen gas. They began a regular air transport service in 1910, with flights to many major cities in Europe. By the outbreak of World War I, in August 1914, they had carried more than 35,000 passengers without a serious incident, demonstrating how safe hydrogen-filled airships could be—or at least appeared to be.

After that war, in 1919, a British airship, the R-34, made the first non-stop round-trip transatlantic crossing. In the 1920s and 1930s zeppelins returned to passenger service and it seemed as if they would be made completely safe when helium from gas wells (see page 221) was discovered. However, at that time, the US Government refused to sell the gas for this purpose, seeing it as a strategic resource, although they did offer it for use in the 1930s.

Despite this, flights continued without incident until the spectacular crash of the Hindenberg on 6 May 1937 as it came into land at an airfield in New Jersey, killing 35 and injuring dozens of others.[79] This was filmed, and broadcast, as it happened. Although a leak of hydrogen was assumed to be responsible for the crash, it was later conclusively shown that powdered aluminium in the fabric coating was the cause of the fire, and that this had been ignited by static electricity. (Powdered aluminium is used in fireworks and as fuel in solid-fuelled rockets such as the space shuttle boosters.)

Hydrogen lifted the largest aircraft ever to fly, the rigid airship Graf Zeppelin II which was 261 metres (857 feet) in length. It made its first flight in 1938. Powdered bronze now replaced the powdered aluminium in the hull-fabric after scientists at the airship-manufacturing company had determined the cause of the crash in July 1937. They never revealed their findings because of concerns over legal action. The bronze coating gave this airship a unique golden-yellow colour. The Graf Zeppelin II was scheduled to enter transatlantic service, but concerns over German Government policies caused the US Government to cancel the landing permit for the US Naval Air Station at Lakehurst, New Jersey. This military base was the only site available for landing large airships in the New York area. They also withdrew a promise to sell helium to the airship company. The privately-owned airline company was never prohibited from flying to the USA but without a New York landing site, it cancelled its scheduled flight. Seized by the German Government, the Graf Zeppelin II was used for reconnaissance flights in Europe and off the coast of the UK in 1939. Its last flight was on 30 August, four days before the start of World War II. Withdrawn from service, it was stored in a shed until early the following year when it was dismantled to reclaim the aluminium of its metal frame for aircraft production.

Hydrogen made possible the first flight to the North Pole which occurred on 12 May 1926 when the Italian semi-rigid airship Norge crossed the pole on its way to Alaska. Until the 1990s, the credit for this endeavour was accorded to the aviator Admiral Richard E. Byrd who claimed to have flown his aeroplane to the pole on 9 May

---

[79] Most of those who died did so because they jumped from the descending airship, whereas the 60 people who stayed on board were able to escape unharmed when the passenger section, known as the gondola, touched ground.

thus beating the Italians. However, it was later revealed that he has falsified his log book and indeed his plane did not carry enough fuel to make the journey, but he was held in such high esteem, because of his achievements during the early days of aviation, that no one doubted his claim for almost 70 years.

Element of war

Hydrogen balloons were used for military applications, in the form of observation and reconnaissance platforms, soon after they were demonstrated to be feasible. In World War I the Germans used zeppelins to bomb London, often with impunity because they could fly much higher than fighter aircraft, although some were brought down when incendiary bullets ignited them. In that same war the British used hydrogen-filled airships for anti-submarine reconnaissance.

Although it has never been used in war, the most fearful weapon of all is the hydrogen bomb (a thermonuclear explosion). At its heart is an atomic bomb which, when detonated, releases enough neutrons and energy to create a high enough temperature for hydrogen fusion to occur among the surrounding mass of lithium deuteride (LiD) which also contains some tritium, the third isotope of hydrogen. When a lithium atom absorbs a neutron, more tritium is formed (Li + n → He + T); this, together with the original tritium, then fuses with the deuterium to releases a vast fireball of explosive energy, equivalent to millions of tonnes of TNT, and large enough to obliterate a city of a million people.

Current research is aimed at developing a thermonuclear weapon that would not need a fission explosion (atomic bomb) to set it off. The technique being explored is called inertial confinement fusion (ICF), and this uses a powerfully focussed laser beam to condense hydrogen to a density and temperature that can ignite an explosive fusion reaction.

Economic element

Worldwide industrial production of hydrogen gas is 40 million tonnes a year (470 billion cubic metres). Known reserves of hydrogen are almost limitless because it can be made from water, and the oceans hold 140 *million* billion tonnes of the element. (The estimated 130 million tonnes of hydrogen gas in the Earth's atmosphere is too dilute to be reclaimed.) Most hydrogen is made in countries with large chemical industries and there is a network of pipelines for moving hydrogen around in North America and Europe. It goes into making ammonia (NH_3) for fertilizers, with sizeable amounts being used to make other chemicals like methanol (CH_3OH) for plastics, and some even goes into making margarine by converting plant oils into fats.

Currently most hydrogen is produced from natural gas (which is mainly CH_4) by heating it with steam at around 1,000°C and this forms a mixture of hydrogen and carbon monoxide which is called syngas. This can also be produced by treating biomass with steam, and a process has been developed in the USA called a plasma-enhanced melter (PEM) which not only converts solid wastes into syngas with high levels of hydrogen but produces enough energy to power the PEM itself. Even glycerol (aka glycerine) can be a source of hydrogen when reacted with steam at 600°C.

Coal-burning power stations can also be made more efficient if they convert the coal to syngas first and these plants, called integrated gasification combined cycle (IGCC) power stations, have efficiencies of more than 40 percent compared to the 30 percent efficiency of a typical coal-fired power station.

The hydrogen of syngas can be separated out with the aid of a silver-palladium membrane which allows hydrogen gas to pass through but not the carbon monoxide. The hydrogen so obtained is 99.99 percent pure. A quicker, if less effective, method of removing carbon monoxide is with a zeolite which can trap larger gases but not hydrogen.

There are other ways of making hydrogen. It can be produced by the electrolysis of water, and this has been suggested as a way of using the surplus electricity of nuclear power stations, or where there is plenty of hydroelectricity. It is still not an economical process despite improvements in efficiency such as the electrolysis of steam inside porous electrodes of zirconium oxide.

It has been known for more than 30 years that in the presence of titanium dioxide, ultraviolet light will decompose water into its constituent elements. To date yields of the gas have been tiny and at best they represent only 8 percent energy conversion which is still short of the 10 percent deemed necessary for commercially viable solar H_2 production. The iron oxide haematite (Fe_2O_3) is also capable of acting as a catalyst to split water, but it has to be made by chemical vapour deposition (using iron carbonyl as the volatile form of iron). Adding silicon also increases its ability to generate H_2.

Using sunlight and a metal catalyst to generate hydrogen from water has relied up to now on the UV part of the spectrum. However, Chinese chemists have found a catalyst which can produce hydrogen from water using visible light in the 420 nm (violet) region and with a quantum efficiency of more than 90 percent. The new catalyst was made from cadmium sulfide doped with palladium sulfide and platinum.

Another way of using sunlight to generate hydrogen is in operation at the Hydrosol II plant in Almeira, Spain. There, an array of mirrors focuses the sun's rays onto a honeycomb-like ceramic coated with iron and nickel oxides which reaches temperatures of around 1,000°C and under these conditions water is stripped of its oxygen and off comes hydrogen gas. The catalyst is then recycled by heating at 1,200°C to drive off the oxygen which can also be collected. The plant converts about 30 kg of water to 3 kilograms of hydrogen gas per hour, which in energy terms is equivalent to 100 kW.

Nature has evolved its own hydrogen-generating enzymes, the *hydrogenases*, and these offer a milder way of producing the gas from organic matter, and indeed it is so produced in small amounts in our own intestines. When conditions are just right, species like green algae and cyanobacteria will emit hydrogen. In 2004 a group of Japanese researchers discovered a new bacterium which can digest food waste to release hydrogen gas. One day it may be possible to ferment organic wastes with genetically engineered bacteria and release hydrogen in useable amounts. That day is probably still far into the future and the process will undoubtedly be slow, because the enzymes that generate H_2 have their active site buried deep within the enzyme to protect them against oxygen.

Some environmentalists and politicians have been advocating hydrogen gas as the clean fuel of the future, generated from water and returning to water when it is burnt, with almost no pollution; they speak of a 'hydrogen economy' but it is slow in coming—if it ever does come. Before the hydrogen economy becomes a reality there are four stages which have to be functioning: affordable hydrogen production; transporting this to where it is needed; storing it, especially if it is to be used as a transport fuel; and burning it in an efficient manner such as in a fuel cell.

The energy given out by burning hydrogen exceeds that of conventional fuels if we compare them on weight-for-weight: hydrogen releases 25 kilocalories (104 kilojoules) per gram compared to 10 kilocalories (42 kilojoules) per gram for natural methane gas. However hydrogen is less dense than methane and it requires three times the volume of gas to provide the same amount of heat. Yet, moving hydrogen through pipelines is a much more efficient ways of transporting energy than electric power lines; and it can easily be stored in subterranean caverns. The difficulty with hydrogen is using it for transport.

If a car is to travel 500 km (approximately 300 miles) on hydrogen as a fuel then it will need to burn around 10 kg of the gas, and must carry this in a convenient form, either as compressed gas, or liquid hydrogen, or absorbed into another material which occupies only a small volume. The liquid hydrogen car designed by BMW holds 120 litres of the fuel in a tank whose insulation consists of 70 thin layers of aluminium and glass fibre vacuum-packed into a 3 cm gap between two metal skins. Refuelling takes only three minutes and is done automatically with no loss of hydrogen.

All sorts of materials have shown promise as absorbents for hydrogen: metal alloys, zeolites, metal-organic frameworks, polymers, ammonia borane, graphite, and even Prussian blue, the dye of blue ink. So far, few have met the requirements of efficient storage capacity for vehicle use. Alloys have proved most promising as hydrogen receptors, such as those made of titanium and iron, or magnesium and nickel, and they can soak up hydrogen to such an extent it is as if they were storing it as liquid hydrogen. They will release the gas as required, but their major drawback is that they tend to lose their efficacy over time. This problem has been solved with an alloy of magnesium, nickel and aluminium which has been patented by the US-based company Ergenic and is called Hy-Stor. It can be charged and discharged more than 100,000 times.

Another substance in which hydrogen might be stored is one that consists of nanocubes. This type of material has open lattices with cube-like voids which are made by reacting zinc oxide with special acids. The empty cavities in this compound are so large that the material itself has the lowest density of any crystalline solid ever recorded at 0.21 g/cm^3.[80] Nor have these voids to be packed with molecules to support this open framework; it retains its shape even when its pores are empty, and it is robust enough to be heated to 400°C. Every void in such nanocubes could well hold more than 50 hydrogen (H_2) molecules. Other materials with large voids are zeolites. They are also possible as storage compounds and incorporating some magnesium into their structure increases their capacity.

Another storage material is lithium nitride (Li_3N) which absorbs hydrogen to form lithium amide $LiNH_2$. When this is heated it releases H_2 as required, and if the lithium nitride is in the form of nanofibres then it absorbs and releases hydrogen in minutes.

[80] Zinc oxide itself has a density of 5.6 g/cc which is 25 times heavier.

Storing hydrogen could be in the form of solids like ammonia borane (NH_3BH_3), which is 20 percent by weight of hydrogen and it can be released on heating to 150°C. Finally there have been proposals that hydrogen could even be transported by converting toluene ($C_6H_5CH_3$) to methylcyclohexane ($C_6H_{11}CH_3$) so that it could be shipped in bulk as this liquid to be converted back to toluene and hydrogen gas at its destination.

A tonne of H_2 occupies a volume of 11 million litres but only 14,000 litres as a liquid, which is how large quantities are transported and stored. The US space programme requires massive amounts of liquid hydrogen and uses road and rail tankers carrying 75,000 litres at a time. One storage tank at Cape Canaveral holds over 3 million litres. The space shuttle required 500, 000 litres of liquid oxygen and 1.5 million litres of liquid hydrogen for take-off, but the dangers were all too apparent as the Challenger disaster showed when it exploded in a giant fireball on 26 January 1986 during its 25th launch.

Buses running on hydrogen have been demonstrated in some European cities such as Berlin and experimental cars running on hydrogen have been produced by German and Japanese companies. In 1991 such a car travelled 300 kilometres on a 100-litre tankful of liquid hydrogen, held in an insulated aluminium container. Engineers at the Mazda car company have built a hydrogen car which stores its fuel in metal alloys. Hydrogen cannot compete with hydrocarbon fuels at the moment although it may one day be used to power cars, trains and even aircraft.

There are two ways in which hydrogen gas can be used to drive a car. It can simply be pumped into the engine along with air and sparked, as in a normal engine burning petrol vapour, or it can be turned into electric current in a fuel cell and used to drive an electric motor. In a fuel cell, hydrogen still reacts with the oxygen of the air, but it does so in a controlled way and giving up most of its energy to generate electricity, rather than as heat, although some heat is produced.

A fuel cell is an electrochemical device. The hydrogen molecules enter the anode compartment where a platinum catalyst breaks it into hydrogen atoms and these are separated from their electrons which move off to provide the current to drive an electric motor. The hydrogen nuclei, which are simply protons, then move through an electrolyte and a membrane towards a cathode where they meet up with the oxygen of compressed air to form water. The only emission is steam, and energy can even be extracted from this.

There are various kinds of fuel cell depending on the material used for the electrolyte. The first ones had an alkaline electrolyte and operated at temperatures in the range 50 to 250°C. Medium-scale power generation cells use phosphoric acid and work at 200°C. More popular are proton-exchange membrane fuel cells because these operate at lower temperatures (80°C), higher efficiency, and higher power density, and it is these that are seen as ideal for motor vehicles. Large-scale power generation would use much higher temperatures, such as molten carbonate at 600°C or solid oxides at 1,000°C.[81] These are commercially available and are used in hospitals, hotels and

---

[81] The companies most closely associated with developing these are International Fuel Cells (phosphoric acid type), Ballard Power Systems (proton exchange membrane), and Westinghouse (solid oxide and molten carbonate).

schools as backup energy supplies, or to provide energy in remote locations. These fuel cells use cheaper nickel-based catalysts, and have an energy efficiency of more than 50 percent. Moreover, the hot steam from such a power unit could be used to run a microturbine, thereby increasing the efficiency to more than 70 percent, and the steam finally emitted could even be used for heating.

The favoured fuel cell for cars is the proton electrolyte membrane and a stack of these comprising a 75kW power unit takes up about the same space as a conventional car engine. Hybrid cars are also being developed and these run on both a fuel cell and a small internal combustion engine, which kicks in when the vehicle is running at a constant speed. Such hybrid cars achieve an overall efficiency of 60 percent, double that of a normal car. What restricts their use is their need for platinum. A typical fuel cell requires about 50 grams of this rare metal which adds £1,250 (or $2,000) to the price of a car.

Hydrogen may be poised to be the fuel of the future but as far as transport for cars is concerned, it is currently caught in a vicious circle. Few vehicles run on hydrogen because there are not enough refilling stations, and there are not enough refuelling sites because there are two few hydrogen driven vehicles to make them profitable.

Hydrogen is involved in another form of energy generation: nuclear power. The deuterium form of water D_2O, known as heavy water, it is used mainly as a moderator in nuclear reactors; by slowing down neutrons it makes it easier for uranium atoms to capture them. The world could have unlimited power if physicists and engineers could harness the nuclear reactions which fuel the Sun, in other words to turn hydrogen into helium. When a mere four grams of hydrogen combines to form a little under four grams of helium they release a 2.5 billion kilojoules of energy. But making hydrogens combine to form helium in a controlled way has proved elusive because it requires temperatures of millions of degrees to start the reaction.

Environmental element

Hydrogen in the environment	
EARTH'S CRUST	1500 p.p.m. (0.15%)
	Hydrogen is the 10th most abundant element.
SOILS	major constituent, as water.
SEAWATER	major constituent as water, but also contains a little dissolved H_2 gas.
ATMOSPHERE	0.5 p.p.m. as H_2, up to 4% as water vapour.

The hydrogen of the Earth's crust is chiefly present as water. Hydrogen is also a major component of biomass, constituting about 14 percent by weight.

Without water vapour it has been estimated that the Earth would be much cooler than it is, even too cold to sustain life. Water vapour in the atmosphere acts as the principal greenhouse gas which warms the air by more than 30°C giving an average global temperature of +15°C instead of −20°C. The water of the oceans also acts as a great moderator of global temperature and indeed without liquid water, a planet would never even see life begin in its most primitive form.

Chemical element

Data file	
CHEMICAL SYMBOL	H
ATOMIC NUMBER	1
ATOMIC WEIGHT	1.00794
MELTING POINT	−259°C
BOILING POINT	−253°C
DENSITY	0.0899 g per litre
OXIDES	H_2O (water) and H_2O_2 (hydrogen peroxide)

Molecular hydrogen (H_2) is a colourless, odourless gas, almost insoluble in water. It burns in air, and mixtures with air are explosive. Hydrogen gas can exist in two forms: ortho hydrogen and para hydrogen, which are distinguished by the spins of the nucleus of the atoms. In ortho hydrogen both of the nuclei are spinning in the same direction, in para hydrogen they spin in opposite directions. At room temperature hydrogen is 75 percent ortho and 25 percent para.

Hydrogen is a component of most molecules albeit only a minor one. In most it forms just a single bond although in some compounds like diborane, B_2H_6, two of the hydrogen atoms form two bonds bridging the two boron atoms. In some compounds hydrogen accepts an electron to form the H⁻ ion as in sodium hydride, NaH. When this reacts with water, hydrogen gas is released.

Nothing could be simpler than a water molecule, yet nothing is as complex in its behaviour. For example, H_2O which is a very light molecule, might be expected to be a gas at room temperature, as is hydrogen sulfide (H_2S), which twice its weight; yet water is a liquid. Moreover, when it freezes at 0°C its solid form, ice, floats instead of sinking. The reason is that each water molecule in ice forms four so-called hydrogen-bonds to surrounding waters and in so doing arranges itself into a crystal with a very open lattice, making it less dense than liquid water, so it floats.

Hydrogen-bonds influence the structure of all living things. The double helix of DNA is held together by hydrogen-bonds. The same is true of all the protein molecules of our body, from the largest organs, such as our muscles and our brain, to the smallest components of cells, such as enzymes. Hydrogen-bonds are generally formed between molecules with oxygen and nitrogen atoms, and occur when a hydrogen attached to one of them is attracted to the other. The two molecules are drawn close together by means of a so-called 'hydrogen bond' which is designated as a row of dots, thus: ⋯. Although hydrogen-bonds are weak compared to normal bonds, having less than a quarter of the energy of these, there are many of them and quantity makes up for quality.

In living things the most important type of hydrogen bond is between nitrogen and oxygen, N-H⋯O, and these are the ones that hold together the strands of DNA. The genetic code of DNA is represented by the letters A, C, G and T, which stand for the bases adenine, cytosine, guanine and thymine. A and T form two complementary hydrogen-bonds, while C and G form three. Because these combinations are so well matched, when a strand of DNA separates in order to form a copy of itself, it does so exactly, because each A, C, G and T uses its hydrogen-bonds to seek out the right partner.

Some molecules can split off one or more of their hydrogen atoms to give positively charged hydrogen ions, H^+, which are protons, and as such they behave as acids when dissolved in water because these protons add to another water molecule to form the active molecular-ion H_3O^+. The best known acids are sulfuric, nitric, and hydrochloric acids, all of which are produced commercially. The higher the concentration of H_3O^+ the stronger the acid and acidity can vary so widely that a special scale, known as pH, is needed to measure it.

The pH is a logarithmic scale in which the concentration of H_3O^+ is expressed in terms of negative powers of 10. For example in neutral water the concentration of H_3O^+ is very low, only 10^{-7} molecular units per litre, so its pH is 7. The pH scale is an inverse scale as far as acids are concerned, in other words the lower the pH number, the stronger the acid. The range of normal acidities is from 1 (the strongest) to 7 (neutral water) and being a logarithmic scale this means that an acid of pH 1 has 10 times more H_3O^+ than one of pH 2, and this has 10 times more than pH 3, and so on. Thus an acid of pH1 has a million times more H_3O^+ than in neutral water. Water at pH 7 has equal amounts of H_3O^+ and its counterbalancing ion OH^-, hence its neutrality.

The pH scale can also be extended even further to cover alkaline conditions from 7 to 14 when OH^- predominates and the concentration of H_3O^+ falls by a further factor of a million. The span of pH from 1 to 14 represents a difference in acidity of a *million* million times. Nevertheless we can come across materials in our everyday life which encompass the full range:

The acids and alkalis of everyday experience

pH	Typical examples	Chemical composition
0*	chemical reagent	concentrated sulfuric acid
1	stomach acid	dilute hydrochloric acid
2	lemon juice	citric acid
3	vinegar	acetic acid
4	tomato juice	ascorbic acid (vitamin C)
5	beer, rain water	carbonic acid (H_2CO_3)
6	milk	lactic acid
7	blood	neutral conditions
8	seawater	dissolved calcium carbonate
9	bicarbonate solution	sodium hydrogen carbonate in water
10	milk of magnesia	magnesium hydroxide in water
11	household ammonia	ammonia solution in water
12	garden lime	calcium hydroxide solution
13	drain cleaner	sodium hydroxide solution
14	caustic soda	concentrated sodium hydroxide solution

* There are some very strong acids with negative pH values, but these are not met with in everyday life.

There are seven known isotopes of hydrogen with mass numbers 1 to 7, and no known isomers. There are three naturally occurring isotopes: hydrogen-1 which comprises over 99.9 percent; hydrogen-2 (deuterium) which accounts for 0.0156 percent; and hydrogen-3 (tritium) which is present only in trace amounts and is produced naturally in the upper atmosphere by cosmic rays. This last isotope is radioactive with a half-life of 12.5 years. The other isotopes have half-lives that are a billion trillionth of a second or less. The longest-lived is hydrogen-5 and its half-life is uncertain but thought to be around 1×10^{-21} second.

Element of surprises

1. When pure hydrogen burns, its flame is almost colourless, the reason being that the light which it emits is UV light.

2. When salt water in a test tube is exposed to radio waves then it emits a mixture of hydrogen and oxygen which can be ignited and will burn as it comes off—see demonstrations on YouTube.

3. Another process involving hydrogen releasing energy is via 'hydrinos' which are supposed to be hydrogen in a lower ground state than normal hydrogen. As an atom falls to this lower energy level it emits energy. This can supposedly be brought about by heating H_2 until it breaks into atoms and then putting these in contact with potassium atoms which act as a catalyst for the formation of these magical hydrinos, which then emit a vast excess of energy. Needless to say, the evidence for hydrinos lack scientific support, but they do have a number of adherents who believe in them.

4. Although water boils at 100°C, this is only strictly true at sea level. At the top of Mount Everest it would boil at about 75°C because of the reduced air pressure. If we increase the pressure, we can increase the boiling point up to 374°C, but to do so requires a pressure of 220 times atmospheric. Above this temperature water becomes a so-called supercritical fluid, in which it behaves both like a gas and a liquid. As such, it will dissolve almost anything, even oils, and when it does so the volume of fluid can suddenly shrink to a half or less. This happens because supercritical water tends to pack tightly around other molecules. More strangely still, organic materials will flame and 'burn' in it. Treatment with supercritical water has been suggested as a way of disposing of sewage sludge, which is converted to a crystal clear, odourless, germ-free solution. The trouble with supercritical water is that it is capable of corroding almost any metal, even gold.

5. The discovery of cosmic rays was made possible by hydrogen. In 1911, Victor Hess (1883–1964) detected them during a balloon ascent to investigate why electroscope reading of radiation changed with altitude.

6. In November 2010, researchers at the Large Hadron Collider at CERN, on the France-Switzerland border, were able to collect together 38 atoms of antihydrogen, which is the antimatter version of hydrogen, aka the positron. (When antimatter

meets matter they annihilate each other instantly.) What they did at CERN was trap the antihydrogen in a magnetic 'bottle' although not for long because within two tenths of second its antihydrogen had vanished. However, it was long enough to prove that it was possible, and now the search is for slower moving antimatter hydrogen atoms so that they can be trapped for longer and studied.

Indium

Pronounced in-di-uhm, its name comes from the Latin word *indicum* meaning violet or indigo which refers to the brightest line in its atomic spectrum.

French *indium*; German *Indium*; Italian *indio*; Spanish *indio*; Portuguese *índio*.

Essential element

Indium tin oxide is a vital part of almost all solar panels and flatscreens because it combines three essential features: it bonds strongly to glass; it is transparent; and it conducts electricity.

Element of life

Indium in the human body	
BLOOD	not known, but low
BONE	not known
TISSUE	*approx.* 15 p.p.b.
TOTAL AMOUNT IN BODY	about 0.4 mg.

There is no biological role for indium but in small doses its salts stimulate metabolism. However, if more than a few milligrams are consumed, they will provoke a toxic reaction and affect the liver, heart, and kidneys.

Medical element

The isotope indium-111 is radioactive, with a half-life of 2.8 days, and it emits gamma-rays. It is used in nuclear medicine to track the movement of white blood cells in the body. A sample of blood is take from the patient, its white blood cells are separated, and then labelled with indium-111 before being injected back into the blood stream. Where these cells then accumulate will reveal areas that are targeted by the white blood cells because that is where they are needed to fight infection.

Element of history

Indium was discovered in 1863 by Ferdinand Reich (1799–1882) and Hieronymous Richter (1824–98) at the Freiberg School of Mines in Germany. Reich was professor of physics and was investigating a sample of the mineral zinc blende (now known as sphalerite, ZnS) which he believed might contain the recently discovered element thallium. He produced a yellow precipitate which he thought was thallium sulfide, but his atomic spectroscope showed lines that were not those of thallium. However, because he was colour-blind he called in his chemistry colleague, Richter, to look at the spectrum, and what was immediately evident to him was a brilliant violet line. It was this, at 451 nm wavelength, that gave rise to the name of indium for the new element.

They realised that a significant amount of indium must be present in the zinc blende, and so set to work on it and were able to isolate a small sample of the new element. Together they published a paper announcing the discovery. Subsequently the two men fell out when Reich learned that Richter was laying claim to the discovery when he exhibited samples of the new metal at the Académie des Sciences, Paris, in April 1867.

Searches for minerals of the new element resulted in none being found until 1876, when indium-rich zinc blendes were discovered in Colorado and at Bergamo, Italy. Indium was detected in the pumice ejected from Mount Krakatoa when it erupted in 1883. Some zinc minerals have sufficient indium to give the characteristic indigo line without the need of chemical concentration.

Although people were interested in the element, little use was found for it, despite the efforts of a Mr. Murray who, in 1932, displayed a 3 kg ingot of pure indium to the members of the Rotary Club of Utica, New York state. Today, indium is very much in demand.

Economic element

Specimens of uncombined indium metal have been found in the Transbaikal region of Russia. An indium mineral, indite ($FeIn_2S_4$) has been found in Siberia, but it is rare. No one has yet found a deposit worth mining for indium alone, and the chances of this happening appear small because this is one of the least abundant elements on Earth. The indium which is produced for industry comes mainly from Canada and is a by-product of smelting zinc, tin, and lead ores, some of which can contain 1 percent indium. (The indium ion, In^{3+}, and the zinc ion, Zn^{2+}, are about the same size, which explains why indium can be accommodated in place of zinc in a zinc sulfide ore.) World production of indium is around 600 tonnes per year, 40 percent of which is new metal and 60 percent recycled. Canada has reserves of the metal estimated to be around 1,500 tonnes, while total world reserves may be as low as 6,000 tonnes.

Despite its rarity, indium has found important niches in the world economy. Most indium (70 percent) ends up as indium tin oxide (aka ITO) while 15 percent goes into indium semiconductors such as indium-arsenide and indium-antimonide. Another use is for low melting alloys for fire-sprinkler systems. Alloys of bismuth and indium will melt at temperatures just above 100 °C; and one alloy of 24 percent indium and 76 percent gallium is even a liquid at room temperature.

ITO is vital to the world economy because it has three properties that make it invaluable: it adheres strongly to glass, it is transparent, and it conducts electricity. Light can pass through it both when it acts as an electrode in a solar panel to generate electricity, and in the opposite direction when it is used to create electronic images in a flat panel display.

Indium metal will 'glue' itself to glass and when it is evaporated and allowed to deposit on glass it produces a mirror as good a quality as that of silver and more corrosion resistant. Indium can be used as a heat-reflecting coating to glass, acting as mirror towards infrared radiation (heat) and for this reason it is added to window glass in tall buildings and as the protective film on the outside of the goggles worn by welders and glass blowers to protect their eyes against heat radiation.

Indium nitride, indium phosphide, and indium antimonide are type III/V semi-conductors and computer chips based on indium phosphide can operate at 7 billion cycles per second, which is three times faster than other microchips.

First generation photovoltaic (PV) cells are made of crystalline silicon and they account for most of the solar electricity currently generated, but this material is not very efficient in turning sunlight to electricity. Second generation cells are made from alternative materials such as copper-indium-diselenide, $CuInSe_2$, (CIS). Metal-based semiconductors are potentially much better PV materials because they can more closely match the energy of the incoming light rays. At the US National Renewable Energy Laboratory in Golden, Colorado, a cell made of CIS has achieved an energy efficiency of almost 20 percent.

CIS is already being manufactured into thin film solar panels which can generate a peak power of 40 watts at 16 volts. The skill in making them starts with the production of semiconductor indium which has to be at least 99.9999 percent pure. Replacing about a quarter of the indium in CIS with gallium to give copper indium gallium diselenide (aka CIGS) not only reduces the amount of the more expensive metal, but also produces an alloy that is easier to manufacture, and appears even to improve the efficiency. Indium(III) phosphide is also used as a semiconductor, while indium(III) arsenide is used for infrared detectors, and indium(III) antimonide will even generate electricity when subjected to infrared radiation.

Indium foils are used to assess what is going on inside nuclear reactors. The isotope, indium-115, has a high neutron capturing ability and as it does so other radio-active isotopes are produced, and these indicate the activity inside the reactor.

Indium metal has a low coefficient of friction and was used to coat ball bearings in fighter aircraft engines in World War II because they could function without the need for a lubricant. Ball bearings in Formula 1 racing cars are also of this type.

Indium gallium nitride and indium gallium phosphide are used in light-emitting diodes.

Environmental element

Indium in the environment	
EARTH'S CRUST	0.1 p.p.m.
	Indium is the 69th most abundant element.
SOILS	0.01 p.p.m.
SEAWATER	0.1 p.p.t.
ATMOSPHERE	virtually nil.

Indium is not widely dispersed throughout the environment and it poses no threat to land or marine life. Cultivated soils are reported to be richer in indium than non-cultivated sites, some even have levels as high as 4 p.p.m. This might have a slight inhibiting effect on certain soil micro-organisms such as the nitrate-forming bacteria.

Chemical element

Data file

CHEMICAL SYMBOL	In
ATOMIC NUMBER	49
ATOMIC WEIGHT	114.818
MELTING POINT	156.5985 °C*
BOILING POINT	2,080°C
DENSITY	7.3 g/cm³
OXIDE	In_2O_3

* This is a defined point on the international temperature scale.

Indium is a soft, silvery metal that is a member of group 13 (row 5p) of the periodic table. It is stable in air and in water, but dissolves in acids. Its preferred oxidation state is +3 which manifests itself as the ion In^{3+}.

Indium has 39 known isotopes with mass numbers ranging from 97 to 135, and 47 known isomers. There are two stable isotopes which occur naturally: indium-115 which accounts for 96 percent, and indium-113 (4 percent). The former is weakly radioactive with a half-life of 4.4×10^{14} years (= 440 trillion years) and is a beta-emitter.[82] There are man-made, radioactive, isotopes, the longest-lived of which is the isomer indium-114m which has a half-life of 50 days.

Element of surprises

1. When a piece of indium metal is bent it is said to let out a high-pitched shriek, although in fact it is little more than a barely-audible squeak. That was the most notable feature of an otherwise uninteresting element in the hundred years following its discovery.

2. A molecule with a chain of six indium atoms has been produced at Imperial College London, the first time that indium has shown this capability.

---

[82] In addition there are several radioactive isotopes produced as fission products in uranium-bearing deposits.

Iodine

Pronounced iyo-deen, the name is derived from the Greek *iodes*, meaning violet.

French *iode*; German *Iod*; Italian *iodio*; Spanish *yodo*; Portuguese *iodo*.

Iodine and iodide: iodine is the name of the element itself which exists as the molecule I_2 with two iodine atoms joined together. This is the iodine of black crystals, purple vapour, and yellow antiseptic solution. When iodine acquires a negative electron, as it is inclined to do, it forms the more stable *iodide* ion, written I^-. The general term, iodine, is used collectively for all aspects of dietary, medical, and environmental iodine, when its actual molecular form is of secondary importance.

Essential element

This element has had a fascinating history, not least because of the uses to which it has been put. Although we need less than a tenth of a milligram of iodine a day, there are millions of people who don't get enough and millions of children have been doomed to a life of cretinism (very low IQ) because of their mothers' lack of this vital element.

Element of life

Iodine in the human body	
BLOOD	0.06 mg per litre
BONE	0.3 p.p.m.
TISSUE	varies widely in the range 0.05–0.7 p.p.m.
TOTAL AMOUNT IN BODY	10–20 mg, of which most is in the thyroid gland

Iodine is an essential element for humans and animals. The iodine in the human body is concentrated mainly in the thyroid gland in the form of two hormones: 3,5,3',5'-tetraiodothyronine (aka thyroxine or T_4) and 3,3',5-triiodothyronine (aka tri-iodothyronine or T_3). These regulate growth in the young and control body temperature throughout our lives. Other organs where iodine is found are the salivary glands, stomach, pituitary, ovaries, epidermis, and the immune system. The body can recycle iodine to some extent but a little is lost each day via the urine so it has to be part of the diet.

Iodine is not an essential element for plants, although these are capable of absorbing it from soil water through their roots, or from the atmosphere via their leaves.

Food element

We need a minimum of 70 micrograms (μg) of iodide per day and a pregnant and breastfeeding woman needs twice this amount, without which her developing baby's brain will be impaired. In the UK, the dietary reference value is set at 140 μg while the US Food and Nutrition Board and Institute of Medicine recommends 150 μg. In the USA in 2000, the intake of iodine of the US population was judged to be more than 200 μg per day for both men and women. In Japan it was more than 300 μg per day due to the level of seafood and seaweed in the diet. A safe upper limit is believed to be 1 mg (= 1000 μg) per day, above which there is a risk of adversely affecting the thyroid gland. Iodine deficiency is best assessed by urine analysis, and when its level is less than 50 μg per litre then iodine dietary deficiency (IDD) is indicated.

People living in developed societies get all the iodine they require from their diet. Milk is a major source and the level in this has increased in recent years due to its supplementation in cattle feed. Edible plants which take up most iodine are cabbages, onions, and mushrooms all of which can have 10 p.p.m. (dry weight). Foods richest in iodine are cod, oysters, shrimp, herring, lobster, sunflower seeds, seaweed, and mushrooms. Haddock is especially rich in iodine and a 150 gram (six ounce) portion of this fish contains 300 μg. Some foods, such as cassava, maize, bamboo shoots, and sweet potatoes can interfere with iodine uptake, but they only threaten people in regions where dietary iodine is already in short supply.

IDD was estimated to affect 750 million people in the developing world in 2000 and of these around 10 million were suffering cretinism. Those most at risk were in India and China, where centuries of intensive cultivation had depleted the level of iodine in the soil. The World Health Organisation (WHO) and other agencies, such as UNICEF, ran a successful campaign to have all edible salt iodised by the year 2000 and most of the countries where IDD was endemic have now made this a legal requirement.

Salt is iodised at a level of 15 p.p.m. so that a daily intake of around 5 g of salt will provide a basic 70 μg of iodide. It can be added as either potassium iodide (KI) or potassium iodate (KIO_3), the latter being more stable in hotter and humid climates. In fact it would require only 175 tonnes a year of iodine to supply all human needs worldwide. Yet in 2007, *The Lancet* noted that iodine deficiency still remains a worldwide problem, as in India where 500 million people suffer IDD and where there are 2 million with cretinism.

Dangerous element

Iodine vapour irritates the eyes and lungs, and the maximum permissible concentration in the workplace is 1 mg per cubic metre. The element itself is toxic if taken orally and as little as 3 g can kill. On the other hand, iodides are relatively safe, although even these can cause an adverse reaction if taken in excess.

Medical element

The use of iodine in medicine began around 180 years ago. Its role was eventually linked to deficiency diseases, of which the best known are goitre, which results in a swollen neck caused by an enlarged thyroid gland, and hypothyroidism, when a person becomes listless and feels cold all the time. Too much iodine, or its mal-function in the body, results in hyperthyroidism with its associated restlessness and hyperactivity.

In 1820, Dr Jean-Francois Coindet (1774–1834) introduced iodine into medicine in the form of a solution of iodine and potassium iodide dissolved in alcohol, known as tincture of iodine, and he advocated prescribing it to those suffering from goitre. He had noted that earlier treatments for this disease advised taking the ash of seaweed and he pointed out that as this was a rich source of iodine this element might well be the active ingredient. He was right, but when patients were given his tincture they suffered from severe stomach pains due to its irritant effects and the treatment did not catch on.

In 1829 a French physician, Jean Lugol (1786–1851), discovered that iodine would dissolve readily in a solution of potassium iodide in water and this became known as Lugol's iodine. It was much less painful when used on open wounds than tincture of iodine. Around this time it was observed that regions where goitre and cretin-ism were endemic, such as the Alps, had noticeably little iodide in the water supply. Attempts to cure goitre with iodide solutions like Lugol's were also abandoned when patients were adversely affected by too much iodide.

The relationship between iodine and the thyroid gland was uncovered in 1895 by a Dr Bauman who is reputed to have spilled some concentrated nitric acid on a sample of thyroid gland and seen purple fumes rising from the decomposing tissue. How-ever, it was not until 1916 that an American biologist, David Marine (1888–1976), of Ohio, showed that goitre could be treated and prevented with iodide provided that it was taken as a dietary supplement, and the best way to do this was to add it to common salt. Iodised salt was introduced in the USA in 1930 and in the following two decades goitre was virtually eliminated in that country.

Although iodine solution had failed as a remedy for goitre, it soon became the accepted treatment for open wounds even though it was not realised at the time that its effectiveness was due to its germ-killing properties. Although iodine tincture was much in vogue as an antiseptic and disinfectant, it was eventually replaced by much less painful chemicals. However, it is still used to disinfect the udders of cows after milking. Previously it had been used prior to milking, and some ended up in the milk thereby making this a source of dietary iodine.

The isotopes iodine-123 (half-life 13.2 hours) and iodine-125 (half-life 59.4 days) are used as radioactive tracers in the human body and especially to observe the action of the thyroid gland. Iodine-125 is a gamma-ray emitter, and it is also used to treat prostate cancer where it is implanted in the form of tiny titanium pellets. Radioactive iodide isotopes can be used to kill thyroid cancer cells because this is where the iodine accumulates so the level of the radioactivity does not affect other organs of the body.

Because it is a heavy element, iodine blocks X-rays, and so iodine compounds can be used as contrasting agents and given by intravenous injection. The radioactive isotope iodine-131 (half-life 8.0 days) has also been used in medical diagnosis, particularly for thyroid conditions, and for their treatment. Iodine-131 was the first radio nuclide to be used therapeutically in the form of a drug, and it became available after World War II. It has a half-life of eight days and emits beta and gamma rays as it converts to xenon gas.

Colposcopy is the medical term for screening the vagina and cervix and this is done with Lugol's iodine solution which detects abnormal, cancerous tissue by virtue of this not being stained brown because it lacks the glycogen of healthy tissue.

Iodine is no longer to be found in domestic medicine cabinets, although a new treatment known as Iodozyme has been introduced for long-standing leg ulcers. This involves applying a dressing consisting of two layers and the iodine is generated *in situ* by the oxidation of iodide by hydrogen peroxide.

Element of history

Iodine was discovered as an indirect result of the Napoleonic Wars of the early 1800s, during which the Royal Navy blockaded France in an effort to cut off the supply of saltpetre (potassium nitrate) that was needed to make gunpowder. Saltpetre was imported from the East where its crystals were collected from latrines and cesspits and where it forms as a result of bacteria acting on the nitrogenous components of human waste. In republican France a cottage industry sprang up to supply the demand for saltpetre and this was produced from heaps of rotting manure and animal offal mixed with humus-rich soil and ashes, these last providing the necessary potassium.

Bernard Courtois (1777–1838), a chemist engaged in saltpetre manufacture on the outskirts of Paris, used seaweed ash as a source of potassium. He boiled this with water and extracted potassium chloride by selective crystallisation. One day, in 1811, he added sulfuric acid to the residual liquor and was surprised to see purple fumes rising from the pan. When he repeated the reaction in a retort, the fumes condensed to beautiful black crystals with a metallic lustre. Courtois was fascinated and guessed he had discovered a new element. He even carried out a few simple experiments with it, such as adding the iodine to ammonia solution from which he got explosive nitrogen triiodide crystals.[83]

Courtois gave samples of his new chemical to two other French chemists that he knew, Charles-Bernard Desormes (1771–1862) and Nicolas Clément (1779–1842), and they carried out a more systematic investigation. On 29 November 1813 they exhibited a sample of iodine at the Institut Imperial de France. But was it really a new *element*? That was soon proved by fellow countryman Joseph Gay-Lussac (1778–1850) and confirmed by the Englishman Humphry Davy (1778–1829). This famous chemist

---

[83] Courtois' business went bankrupt following Napoleon's final defeat at the Battle of Waterloo in 1815, after which the naval blockade was lifted. This allowed cheap imports of saltpetre to flood into Europe from the East, just at a time when the need for gunpowder dramatically declined. Nor did Courtois profit from the growing market for iodine, despite being able to sell it at 6 francs per 10 grams, because he could make so little of it. Sadly, he died penniless in 1838.

had been given permission by Napoleon to visit Paris, even though the two countries were at war. Davy was given a sample of iodine and he worked on it in his hotel room using his travelling chemistry set. He announced his results on 11 December 1813 to the Institute Imperial, and sent back a report to the Royal Institution in London where it was read at a meeting on 20 January 1814. It was Gay-Lussac who named it *iode*.

The British, not knowing of Gay-Lussac's parallel proof, thought Davy was the first person to show that iodine was an element and a lingering disagreement over priority continued for the next century and right up to 1913 when the French celebrated the centenary of the announcement of iodine. Authentic samples of Courtois' iodine were exhibited at the event and his part in the discovery was reaffirmed.

Economic element

There are some iodine-containing minerals, such as lautarite, $Ca(IO_3)_2$ which is found in Chile, and iodargyrite, AgI, specimens of which are to be found in Colorado, Nevada and New Mexico. The Chilean nitrate deposits contain 5 percent sodium iodate ($NaIO_3$). Known reserves of easily accessible iodine amount to around two million tonnes. Iodine is not currently being produced from seaweed, which was the original source, although this method was employed from the 1820s to the 1950s and was based mainly on kelp. This seaweed contains 0.45 percent iodide (dry weight) and its ash is 1.5 percent iodide, permitting 15 kg of the element to be produced from a tonne of the ash.

Worldwide industrial production of iodine in 2003 was around 22,000 tonnes mostly in Chile (40 percent) and Japan (30 percent), with smaller amounts being produced in Russia and the USA. Iodine is also extracted from natural brines and oil well brines, which may have up to 100 p.p.m. of the element. The Minami Kanto gas field in Japan and the Anadarko Basin gas field in Oklahoma yield brines rich in iodide which is extracted by acidifying the brine with sulfuric acid and then bubbling chlorine and air through the solution. This releases the iodine which is then trapped out of the gas stream. (It can be made 99.5 percent pure by converting it to hydriodic acid (HI) and then back again to iodine.) The iodate of nitrate ores is extracted using sulfur dioxide to release the iodine, and after further treatment is converted to a solid form known as prill which consists of tiny iodine spheres that are dust-free and safe to handle.

The main outlets for the element are in pharmaceuticals (including iodine-based disinfectants) which account for 25 percent; animal feeds 15 percent; printing inks and dyes 15 percent; industrial catalysts 15 percent; with a variety of miscellaneous uses, including art photography, accounting for the remainder. In many of these the iodine is turned into iodides such as titanium iodide for catalytic use, zirconium iodide which is used to prepared the pure metal zirconium, and potassium iodide which is given as an animal feed supplement, and silver iodide for photography. The last of these was one of the first commercial uses of iodine.

In 1839, Louis Daguerre (1787–1851) invented a method of taking photographs and he called his images daguerreotypes. These were produced on a piece of metal that had been coated with a film of silver. This was then treated with iodine vapour in order to from a surface layer of light-sensitive silver iodide. It was then exposed and the stronger the light falling on it, the more the silver iodide was converted back to silver metal and iodine. The iodine could be washed away leaving a positive image, and this could be improved by exposing it to mercury vapour. In 1851, a wet plate alternative was introduced with the silver iodide film on glass. Daguerreotypes are now highly prized.

Environmental element

Iodine in the environment	
EARTH'S CRUST	0.14 p.p.m.
	Iodine is the 64th most abundant element.
SOILS	av. 3 p.p.m., with between 0.1 to 10 p.p.m. in common soils.
SEAWATER	0.06 p.p.m.
ATMOSPHERE	0.2-60 p.p.b., rainwater contains 0.7 µg per litre.

Seaweed is thought to accumulate iodide as a protection against UV damage. Its role is to act as an antioxidant and in so doing it converts iodide to iodine, which explains why I_2 is detectable in the atmosphere of coastal regions. About 400,000 tonnes of iodine escape from the oceans every year as elemental iodine, iodomethane (aka methyl iodide CH_3I), and diiodomethane (CH_2I_2) which are produced by marine organisms. Some of this is deposited on land where it may become part of the bio-cycle via soil micro-organisms such as fungi which are known to accumulate it. Iodine as iodide is continually leached from the land to rivers and returned to the oceans. Recently it has been shown that rice plants emit methyl iodide, CH_3I, and that this accounts for about 4 percent of that which is present in the atmosphere. Methyl iodide is now the preferred soil fumigant, having been approved as a replacement for methyl bromide which was volatile and stable enough to reach the ozone layer, and was being used at a rate of 50,000 tonnes a year. Methyl iodide molecules break down in the lower atmosphere by the action of UV, so it poses no threat to the ozone layer.

In some regions, such as the Baraba Steppe of Russia, the level of iodine in soils can be as high as 300 p.p.m. and soils along the coasts in Japan and Wales can have 150 p.p.m. However, where the soils have been exposed to glaciation, the amount can be very low, which is why goitre was prevalent in such regions as the Alps.

Radioactive iodine-131 has been released in large amounts following nuclear accidents such as that in Chernobyl in 1986. This isotope has a half-life of eight days but is picked up by grazing animals and may enter the human food chain rapidly as milk or in water supplies. To counter its effects, potassium iodide tablets can be issued to the population and especially to children, but even so this may come too late as it did in the Chernobyl region where the incidence of thyroid cancers has increased. If a large dose of potassium iodide is given immediately exposure is imminent it prevents the radioactive form from being absorbed, thereby limiting the damage it can cause.

Chemical element

Data file	
CHEMICAL SYMBOL	I
ATOMIC NUMBER	53
ATOMIC WEIGHT	126.90447
MELTING POINT	114°C
BOILING POINT	184°C (under pressure)
DENSITY	4.9 g/cm^3
OXIDES	I_2O_5 which is stable, and I_2O_4 and I_4O_9 which are less stable

Iodine is a black, shiny, non-metallic solid and a member of group 17 (row 5p), also known as the halogen group of the periodic table. It sublimes easily on heating to give a purple vapour. Elemental iodine is not very soluble in water and a litre will only dissolve 0.3 grams at 20°C, but it dissolves in some solvents like carbon tetrachloride to give a beautiful purple colour, while in other solvents such as ether and water it is brown. Its solubility in water is greatly enhanced by the addition of potassium iodide which forms the triiodide ion, I_3^-. Iodine can exhibit a range of oxidation states ranging from -1 to +7. Examples of these are −1 as the iodide ion I$^-$, +1 as in ICl (iodine monochloride), +3 as in IF$_3$, +4 as in the oxide I_2O_4, +5 as in HIO$_3$ (iodic acid), and +7 as in HIO$_4$ (meta-periodic acid) and H$_5$IO$_6$ (ortho-periodic acid).

There are 38 known iodine isotopes with mass numbers 108 to 145, and 14 known isomers. In nature, iodine occurs mainly as a single, stable isotope, iodine-127, which accounts for virtually 100 percent. There are also traces of a radioactive isotope-129, with a half-life of 15.7 million years, which is produced by cosmic ray interaction with xenon-130 in the atmosphere. Some of this isotope is also produced as a fission product in uranium-bearing deposits, along with other radioactive iodine isotopes produced in the same manner. The next longest-lived isotope is iodine-125 with a half-life of 59 days.

Element of surprises

1. Iodine was once an important part of chemical analysis where its dramatic blue colour with starch was used as to indicate an endpoint in analytical titrations. This chemical effect is still used to test banknotes. When a detection pen is wiped across a counterfeit note its Lugol solution will leave a blue streak as it reacts with the starch in the paper. All commercial paper contains starch, except that produced especially and exclusively for banknotes.

2. As well as gaining an electron to form the iodide ion, I$^-$, iodine can lose one to form a positive ion I$^+$. This form of the element may even be involved in the way in which the organo-iodine thyroxine is produced in the thyroid gland.

Iridium

Pronounced irid-ee-um, the name is derived from the Latin *Iris*, the goddess of the rainbow, and it was called this because of the palette of colours which its discoverer, Smithson Tennant, observed when he first dissolved the metal in concentrated hydrochloric acid.

French *iridium*; German *Iridium*; Italian *iridio*; Spanish *iridio*; Portuguese *irídio*.

Essential element

This metal resists corrosion like no other and because of this it has found applications as diverse as fountain pen nibs and aero engine parts.

Element of life

The amount of iridium in the human body is not accurately known but is very low, and in tissue it is about 20 p.p.t. The metal has no biological role. Iridium chloride is moderately toxic by ingestion, although most of its compounds are insoluble and if consumed they would not be absorbed by the body.

Medical element

The radioactive isotope, iridium-192, half-life 73.8 days, is a gamma-ray emitter and is used in a type of radiation therapy called brachytherapy. This is used against cancers of the prostrate, bile duct, and cervix. It involves implanting a radioactive source near the cancer and thereby slowly killlling it.

Element of history

Iridium and osmium were discovered together in 1803 by Smithson Tennant (1761–1813) in London. He had tried to dissolve some crude platinum in aqua regia but found that a black residue was left behind. He was not the first person to have observed this, but earlier reports had attributed the residue to graphite. Tennant was not convinced that it was graphite, and began working on the insoluble material. By treating it alternately with alkalis and acids he was able to separate it into two new elements. These he announced to the Royal Institution in London in the spring of 1804, naming one iridium, because its salts were so colourful and the other osmium because it had a curious odour (see osmium).

In July 1813, a chemist, William Allen (1779–1843), recorded in his diary that he had gone with Tennant, Humphry Davy (1778–1829), and about 30 other eminent scientists to the home of John Children (1777–1852), Secretary of the Royal Society, who lived in Tunbridge. There the assembled throng set up a massive battery of electrical cells consisting of copper and zinc plates with a mixture of acids as the electrolyte solution. Such was the current this generated that they were able to melt a sample of Tennant's iridium with it, to great acclaim.

Economic element

Iridium is found as the uncombined element, and also as two iridium-osmium alloys: osmiridium (which has more iridium than osmium) and iridosmine (*vice versa*), of which the former is the more common. Most iridium, however, is obtained as a by-product of nickel refining and most comes from Canada. Annual world production, expressed in the industrial units of troy ounces, is normally around 100,000 oz, which is around 3 tonnes. Iridium comes as a by-product of nickel and copper refining. Reserves have not been estimated but there are iridium-rich deposits in South Africa, Russia, and Canada.

Iridium is used by the chemical industry (25 percent ends up this way), the electrical industries (25 percent), and the electrochemical industry, and in particular the chlor-alkali industry, (again around 25 percent). The remaining 25 percent finds a variety of outlets such as for crucibles, spark plugs, and alloys, and especially for providing high mechanical strength. The crucibles are used for growing crystals of ultrapure semiconductors at high temperatures, and for making the single crystal sapphires needed for the lasers used in robot-controlled welding.

Iridium is the most corrosion-resistant metal known. Traditionally iridium was the coating of nibs for fountain pens and for compass bearings. The standard metre bar kept in Paris, France, is made of a platinum-iridium alloy (90 percent platinum and 10 percent iridium) but this was superseded as the basic unit of length in 1960 by a line in the atomic spectrum of krypton.

The aerospace industry relies on iridium for certain long-life engine parts, and an iridium-titanium alloy is used in corrosion resistant pipes for deep-water projects. The Cativa process for making acetic acid has an iridium catalyst. Indium black pigment is the metal finely divided and this is used to paint porcelain and gives an intense black.

Environmental element

Iridium in the environment	
EARTH'S CRUST	about 3 p.p.t.
	Iridium is the 84th most abundant element.*
SOILS	not recorded.
SEAWATER	about 0.13 p.p.t.
ATMOSPHERE	nil.

* Another way of putting this would be to say it is one of the *least* abundant of the metals.

Because of its attraction to iron, most of the iridium of this planet is thought to be in the Earth's molten core. Some volcanoes emit small amounts of the metal, but nowhere is it concentrated in mineable quantities in its own right. The level of iridium in land plants is below 20 p.p.b. (dry weight) and for all intents and purposes the element today has no impact on the environment or living things, although its presence in certain geological deposits can serve to indicate what might have caused past mass extinctions—see *Element of surprises*.

Chemical element

Data file

CHEMICAL SYMBOL	Ir
ATOMIC NUMBER	77
ATOMIC WEIGHT	192.217
MELTING POINT	2,410°C
BOILING POINT	4,130°C
DENSITY	22.6 g/cm³
OXIDE	IrO_2

Iridium is a hard, brittle, lustrous, silvery metal of group 9 (row 5d) of the periodic table. It is one of the the so-called platinum subgroup of elements. The solid metal has the second highest density known, and for a time it was thought to be equal to, if not slightly ahead, of osmium in this respect. It is unaffected by air, water, and acids, including aqua regia, and the only two substances that the metal will dissolve in are molten sodium cyanide and potassium cyanide. Its oxidation states range from −3 to +6 (IrF_6) but the most common are +3 (as in $IrCl_3$) and +4 (as in IrF_4).

There are 36 known isotopes of iridium of mass numbers 164 to 199, and 32 known isomers. There are two naturally occurring stable isotopes: iridium-191 which comprises 37 percent and iridium-193 (63 percent).[84] The longest-lived radioactive isotope is iridium-192 with a half-life of 74 days. However, there is also a long-lived isomer, irdium-192m1, which has a half-life of 241 years.

Element of surprises

1. It came as a surprise when geologists discovered a thin layer of iridium in the rocks that were deposited at the boundary layer between those of the Cretaceous and the Tertiary geological periods. This was the time, around 65 million years ago, when a traumatic event happened on Earth which led to the extinction of most of the species then in existence, most notably the dinosaurs.

   One theory is that this was a time of great volcanic activity with flood basalt eruptions in India that covered thousands of square km with hot molten lava and which released vast amounts of the poisonous gas sulfur dioxide into the atmosphere.

---

[84] There are also minute traces of iridium-194 which is produced by the beta decay of osmium-194 and that comes from the alpha decay of long-lived platnium-198, which has a half-life of 3.2×10^{14} years. (Iridium-194 has a half-life of 19 hours.)

Another theory blames the extinctions on the impact of a 10 km meteor which struck the region that is now the Yucatan peninsula, off the Gulf of Mexico, forming a crater 100 km across, and that dust from this impact caused a darkening of the sky for several seasons with the result that few creatures were able to survive.

The evidence for the meteor impact having global significance is the amount of iridium in the boundary layer. This metal can only have come from space, and indeed several meteorites have been found to be relatively rich in this rarest of metals. Similar meteoric impacts around 36 million years ago left craters at Popigai in northern Russia and Chesapeake Bay in the USA. These also left an imprint of iridium on the planet. However, they did not lead to mass extinctions and it is now thought that, while the Yucatan meteorite had a serious effect, it was not the main reason why the Cretaceous period came to an end. It is now thought that this might have been just one of several contributing factors.

2. Nuclear theory suggests that all isotopes of iridium should be radioactive, and that also includes the two isotopes iridium-191 and iridium-193 which are currently regarded as stable. If they are radioactive, then their half-lives must be incredibly long.

Iron

Pronounced iy-on, the name comes from the Anglo-Saxon *iren*, and the chemical symbol, Fe, from *ferrum*, the Latin for iron.

French *fer*; German *Eisen*; Italian *ferro*; Spanish *hierro*; Portuguese *ferro*.

Essential element

Historically we are still living in the Iron Age and this metal is still the most important one that we use and which is produced on a scale which dwarfs all others. And there are ways in which it might even transform life in the oceans of this planet.

Cosmic element

Iron is the heaviest element that can be made in the nuclear fusion furnace which fuels a typical star. The nucleus of an iron atom is the most tightly bound of all nuclei, which means that no more energy can be liberated by fusing it with other nuclei. In effect it is the 'ash' of the nuclear burning process and when the core of a star has become mainly iron, that star has run out of its primary source of energy.

Element of life

Iron in the human body	
BLOOD	av. 415 p.p.m. (range 380–450 p.p.m.)
BONE	varies 3–380 p.p.m.
TISSUE	av. 180 p.p.m. (range 20–1,400 p.p.m.)
TOTAL AMOUNT IN BODY	4 g

Iron is essential to almost all living things, from micro-organisms to humans. A person must have a regular intake of iron because we lose a little of this metal every day through the walls of our stomach and intestines. Even so, it is rare for normal people to be lacking iron, even though they may sometimes lose a large amount of blood, such as women when they are menstruating. Those who regularly donate blood should not require iron supplements if they eat a balanced diet, and the same is true of pregnant women, despite a growing foetus requiring about 700 mg of iron.

Green plants need iron for energy transformation processes and other uses. Fodder plants can have up to 1,000 p.p.m. (dry weight) of iron but food plants generally have much less. Beans are particularly rich in the metal having as much as 9 p.p.m. fresh weight (80 p.p.m. dry weight), while lettuce also has a lot with 5 p.p.m. fresh weight (130 p.p.m. dry weight).[85] At the other extreme, apples, carrots, cucumbers, oranges, and onions have relatively little iron.

Some bacteria use particles of iron in the form of magnetite, Fe_2O_3, which is a magnetic material and this acts as compass to give them a sense of direction. Molluscs also use this mineral to form their teeth, while the very strong teeth of limpets are made of the mineral goethite, $FeO(OH)$.

Food element

A man needs an average daily intake of 7 mg of iron, a woman 11 mg, and a normal diet will generally provide all that is needed. Of the iron in food, only a certain fraction is capable of being absorbed, maybe as little as 25 percent, but this is enough to replenish that which is lost each day. Iron-rich foods are liver, corned beef, iron-fortified breakfast cereals, baked beans, peanut butter, raisins, bread, eggs, curries, and cakes made with black treacle (molasses). The iron of red meat is particularly good at being absorbed. Extra vitamin C helps absorption, while foods rich in phosphate (such as processed cheese) and phytate (such as bran) hinder it.

The average western diet provides about 20 mg of iron per day, but it is estimated that around the world some 500 million people are anaemic through lack of dietary iron. This problem is likely to get worse because the high yield crops that have been developed do not provide much in the way of dietary iron.

Dangerous element

A dose of 5 grams of iron in an easily absorbed form, such as iron(II) sulfate (aka ferrous sulfate), would put the average adult's life at risk and indeed children have died as a result of consuming such tablets that have been prescribed as a cure for anaemia in adults. Iron in excess causes liver and kidney damage, as it overwhelms the transferrin protein which binds to it and whose job it is to move it and other essential metals around the body.

Some people retain too much iron and suffer iron overload. Individuals who have the genetic disorder haemochromatosis are afflicted this way and iron builds up in their pancreas, liver, spleen, and heart. Patients who need multiple blood transfusions for inherited anaemic diseases such thalassaemia, also accumulate too much and they need to take drugs that can help the body shed its burden of iron.

Working in an atmosphere of iron dust can lead to a form of pneumoconiosis known as welder's lung.

---

[85] As is popularly known, spinach contains a lot of iron, but what is less well known is that this is present in a form that the body can not assimilate. For most of the last century its reputation as a rich source of iron was grossly exaggerated, due mainly to a printing error in an official US Government booklet on the nutrient content of foods.

Medical element

Generally the problem for humans is iron deficiency which leads to anaemia. Iron has many roles in the body but its best known one is to carry oxygen from the lungs to where it is needed to generate energy, such as to the heart and brain, and then take the waste gas, carbon dioxide, back to the lungs where it can be exhaled and replaced by fresh oxygen.

The ancient Greek physician, Galen,[86] suggested taking iron filings as a medicament, recommending it as a laxative. (Today we know that iron is more likely to have the opposite effect.) In the 1700s, following a Dr Sydenham's suggestion that iron had healing powers, pharmacists began to stock it in the form of iron filings coated with sugar, although a more popular way to take the element was to dissolve the filings in an acidic wine and drink that. Today, if iron is recommended, it is given as a soluble iron(II) sulfate, $FeSO_4$.

Iron was found to be present in vegetable ash early in the 1700s, and in 1745 the Italian physician, Vincenzo Menghini (1704–59), demonstrated its presence in blood. He did this by feeding dogs iron preparations, bleeding them, drying and burning the blood, and then observing that particles from the ash were attracted to the blade of a magnetised knife. Despite this, and the work of others, the medical role of iron was not understood, and it was not until 1832 that a Dr Bland cured a case of anaemia with iron(II) sulfate. As the role of iron in the blood became clearer, then its deficiency was associated with anaemia and other symptoms such as listlessness and general fatigue. Iron tonics and pills became popular and were regarded as a way of keeping a person fit and healthy.

As well as being part of haemoglobin, iron plays a key role in various enzymes, such as those involved in the synthesis of DNA, those which scavenge free radicals and protect the body, and those which enable cells to release energy by using glucose. Normal brain function needs iron. There are regions of the brain that are rich in iron, which may explain why iron deficiency in infants and children has been associated with lowered mental development. Surplus iron is stored mainly in the liver and as much as a gram of iron may be held in reserve this way. Bone marrow is rich in iron too, because this is the site where haemoglobin in synthesised.

Inside the body, iron, as iron(III) is strongly bound by transferrin, a protein found in serum and other secretions, and it is this which transfers the metal between cells. Transferrin binds iron tightly, and because it does so, it acts as a powerful antibiotic simply by denying this essential metal to any invading bacteria which need iron to multiply. As soon as our body registers a bacterial invasion, it produces more transferrin to mop up any free iron in the blood stream and hide it in the liver. Breastmilk also contains a form of transferrin called lactotransferrin, and egg white contains ovotransferrin, both of which function in the same iron-binding way, and both of which provide anti-bacterial protection.

Nevertheless bacteria have their own molecules, called siderophores, which scavenge free iron and deliver it back to the microbe. Some bacteria go even further, and gonorrhoea produces a protein that can even wrest iron from transferrin.

---

[86] He practised medicine in the second half of the second century AD.

Element of history

No one discovered the element iron as such, although no doubt individuals often came across samples of the metal lying on the surface. We know this because iron articles have been found in Egypt dating from around 3500 BC. These were of meteoric origin because they contain about 7.5 percent nickel, which is the typical composition of meteorites known as siderites. These iron artefacts from Egyptian tombs were able to resist the ravages of rusting on account of this nickel.

The ancient Hittites of Asia Minor, today's Turkey, appear to have been the first to discover the smelting of iron from its ores around 1500 BC and this new metal gave them economic and political power, so they kept the process a closely guarded secret. Not that it saved them. An invasion of warriors known as the Sea People destroyed their civilisation around 1200 BC, after which the iron workers dispersed and took their craft with them. The Iron Age had begun.

Some iron workers ended up in western Iran where they discovered a superior kind of iron: steel. That this depended on its carbon content was not appreciated. Generally, the skills of iron-working made up for the deficiencies of the metal, and blacksmiths became respected craftsmen. Iron working was particularly well developed in India, witness the 7 m high iron pillar of Delhi. This remarkable object has stood for 1,500 years without rusting, the reason being a layer of iron hydrogen phosphate on its surface formed from the unusually high levels of phosphate in the iron itself.

Thales (c.624–c.546 BC), the Greek philosopher, reported in 585 BC that pieces of iron ore that came from Magnesia in Lydia, Asia Minor, had the strange power to attract iron filings. They were called magnets after the place they came from. The mineral is actually magnetite and pieces of it are referred to as lodestone. The ancient Greek potters also used iron as a pigment, having discovered that if items were fired under oxidising condition the colour would be red (due to the oxide Fe_2O_3) while under reducing conditions it would be black (due to Fe_3O_4).

Iron-making on a large scale was an essential part of the Industrial Revolution that began in England in the mid-1700s, and still there was no recognition that small amounts of carbon will transform iron into steel. The first person to suggest this was René Antoine Ferchault de Réaumur (1683–1757) who said that steel, wrought iron, and cast iron, were to be distinguished by the amount of a black combustible material they contained and that this came from the charcoal then used to smelt the metal.

Technological developments in smelting in the 1700s, such as the use of coke rather than charcoal, resulted in cheap wrought iron becoming available and so began an era in which it was widely used in transport and architecture. The first all-iron bridge was built in Shropshire, England, in 1778 and from which the town of Ironbridge took its name. The first iron ship, *Volcano*, was launched in 1818 and was in service for 50 years on the river Thames, but the biggest boost to the industry was the development of railways from the 1830s onwards.[87]

---

[87] Perhaps the greatest achievements of this era were the Forth Railway Bridge in Scotland, opened to great acclaim in 1890, and the Eiffel Tower in Paris, which was built in 1889. The latter is now a much revered monument, although at the time it was derided by the leading artists and designers.

Element of war

One undesirable side effect of the Iron Age was the appearance of warriors armed with iron swords against which there was limited defence and these weapons dominated warfare for more than 2,000 years. The Roman God of War was Mars and the element associated with him was iron. Although the Romans made extensive use of iron, the author Pliny had mixed feeling about the metal because of its role in warfare: 'with iron we construct houses, cleave rocks, and perform many other useful acts of life, but it is also with iron that wars, murders, and robberies take place, and not only hand-to-hand but from a distance...with winged weapons launched from engines.'

Yet this latter role for iron in warfare lay mainly in the future, first in the form of cannon balls, then later as grenades and bombs. In the World Wars of the last century the quantity of iron discharged in the form of bombs and shells reached colossal proportions, with large sections of all the major industrial economies devoted to turning the metal into instruments of death and destruction.

Economic element

Economically workable reserves of iron ores exceed 100 billion tonnes and the main ones are haematite (Fe_2O_3), which is frequently found as black sand along beaches, magnetite (Fe_3O_4), goethite ($FeO(OH)$), lepidocrocite ($FeO(OH)$) and siderite ($FeCO_3$). World production of iron ore is around 2.4 billion tonnes a year, and the main producers are China (700 million tonnes), Brazil (350 million tonnes), Australia (300 million tonnes), and Russia (100 million tonnes).

Many countries also have iron ore mines, and even more produce the basic metal as pig iron. Total production of this is 1 billion tonnes of which China smelts almost half. Iron is produced by heating its ore with coke (carbon) and limestone (calcium carbonate) in a blast furnace, and the result is so-called pig iron. Around 95 percent of iron is smelted this way. Pig iron has 3 percent carbon plus small amounts of sulfur, silicon, manganese and phosphorus; it is not used as such but is the raw material for making steel by removing the impurities and adjusting the residual amount of carbon. Most steel has around 1.7 percent carbon and this makes the iron more durable, less brittle and more resistant to corrosion. Production of crude steel amounts to 1.3 billion tonnes annually, and China (500 million tonnes), Japan (120 million tonnes), and the USA (100 million tonnes) are the major producers, with Germany, India, and South Korea producing more than 50 million tonnes per year. A great deal of ferro-alloys, incorporating metals like chromium, nickel, and manganese, are also produced.

Cast iron contains 3–5 percent carbon and is used for valves, pipes, and pumps. It is not as tough as steel but it is cheaper, and can be improved by the addition of small amounts of nickel-magnesium alloy. Then it is known as ductile cast iron which can be machined. Pure forms of iron, of 99.9 percent or better, are named after the process that produces them, such as ingot iron, electrolytic iron, reduced iron, and carbonyl iron, all of which have special uses.

Iron is a metal that is recycled on a large scale. It has more applications than any other metal and it still accounts for around 90 percent of all metal that is refined. From food cans to family cars, from screws to railway lines, from cargo ships to paper staples, iron has a part to play. Iron is the most versatile of all the metal elements: it can be cast, beaten, welded, mechanically drawn into wire, and forged—but it rusts. A great deal of effort is expended in trying to prevent this by painting, galvanizing with zinc, plating with cadmium, or coating with plastic.

Rusting can be prevented in various ways but other alloys of iron offer better resistance to rusting as well as making the metal more suitable for use. Nickel steel is best for bridges, electricity pylons and bicycle chains; tungsten and vanadium steel is used for cutting tools; manganese steel is used for rifle barrels and power shovels; while the ubiquitous stainless steel has high levels of chromium (18 percent) and nickel (8 percent). This is known as '18-8 stainless' and finds use in architecture, cutlery, and even jewellery.

Although it forms alloys with many metals, iron does not form an amalgam with mercury, and that is why iron containers have been used to hold this metal since ancient times.

Iron catalysts are used in two major industrial processes and they are the Haber Bosch process which converts nitrogen to ammonia and thence to fertilizers, and the Fischer Tropsch process which converts syngas, which is a mixture of carbon monoxide and hydrogen, into liquid fuels.

Environmental element

Iron in the environment	
EARTH'S CRUST	41,000 p.p.m. (4.1%)
	Iron is the 4th most abundant element.
SOILS	0.5–5%
SEAWATER	2.5 p.p.b. (greater in the Atlantic than Pacific).
ATMOSPHERE	traces.

Iron is really the Earth's most abundant element because the 7,000 km diameter core of the planet consists mainly of the molten metal, although the very centre may be a solid iron core about 2,500 km in diameter. In effect the planet is a large iron sphere, alloyed with nickel. Although iron is an abundant element at the Earth's surface, there are regions of the planet where it is so lacking that it is the limiting factor to life.

The oceans are often thought of as teeming with life, but this is true of only a few regions, and these tend to be over-fished as a consequence. More than 80 percent of the boundless ocean is empty. In the mid-1980s, John Martin of the Moss Landing Marine Laboratories, in California, put forward the theory that it was the lack of iron in the upper levels of the sea which prevented plankton from growing, and without plankton to feed on, other forms of marine life have no food supply to support them.

Ten years later Martin's idea was tested by a joint US-UK research team, which used a solution of iron sulfate to fertilize 60 km^2 of the Pacific Ocean, west of the Galapagos Islands. The results were dramatic. Within a week this barren span of ocean bloomed and turned green with plankton, proving that it was simply lack of this metal that was limiting their growth. Since then several other sea fertilization experiments have been carried out, and one in 2009 covered an area of 300 km^2 using 10 tonnes of iron(II) sulfate. Again there was a massive plankton bloom, and this was rapidly consumed by zooplankton. It is now realised that as far as the seas are concerned there has to be more than just the essential elements nitrogen and phosphorus in the surface waters, there has to be iron as well and the relative proportions of these elements needed to ensure maximum growth has now been established as 16 nitrogen, 1 phosphorus, and 0.001 iron.

It may be that the oceans will one day be fertilized, and that this could be done with ferrous sulfate, which is easily and cheaply made from rusty old iron and sulfuric acid. The seas could bloom again, as they must once have done when the world was young, and when cellular organisms first flourished and learned how to make use of this versatile metal.

Soils contain a lot of iron and it is easily transformed by organic matter into various types of iron oxide and other compounds. Iron in oxidation state +3 is the common environmental form, but it can be reduced to oxidation state +2 in waterlogged soils, and it is so reduced by plant roots to facilitate its absorption.

Chemical element

Data file

CHEMICAL SYMBOL	Fe
ATOMIC NUMBER	26
ATOMIC WEIGHT	55.845
MELTING POINT	1,535°C
BOILING POINT	2,750°C
DENSITY	7.9 g/cm^3
OXIDES	FeO, Fe$_2$O$_3$, Fe$_3$O$_4$

Iron, when absolutely pure, is lustrous, silvery and 'soft', meaning it is easily worked. The element is a member of group 8 (row 3d) of the periodic table. Iron rusts in damp air, but not in dry air. It dissolves readily in dilute acids. Its preferred oxidation states are +2 (aka ferrous iron) as in FeSO$_4$, and the Fe^{2+} ion in solution, and +3 (aka ferric iron) as in FeCl$_3$ and the Fe^{3+} ion in solution. Iron can exhibit other oxidation states such as +6 as in potassium ferrate (K$_2$FeO$_4$), and even mixed oxidation state compounds exist such as Fe$_7$(CN)$_{18}$, aka Prussian blue.

There are 30 known isotopes of iron with mass numbers 45 to 74, and six known isomers. Naturally occurring iron consists of four isotopes: iron-56, which comprises 92 percent; iron-54 (6 percent); iron-57 (2 percent); and iron-58 (0.3 percent). One of these is now known to be radioactive: iron-54 has a half-life of 3.1×10^{22} years, so long that perhaps it is not surprising that it was formerly considered to be stable.[88] The next longest-lived radioactive isotope is iron-60 with a half-life of 1.5 million years. This may well have been present on Earth when it originally formed but it has all now disappeared.

The isotope, iron-45, is extremely proton-rich and decays by the rare decay mode of two-proton emission. This mechanism, long considered theoretically possible, was first conclusively detected in 2002 in iron-45.

Element of surprises

1. The musty smell of iron that we may notice when handling iron objects is not due to the metal itself but to its reaction with a body oil on the surface of the skin. This reaction results in the formation of the chemical 1-octen-3-one and that is what we smell.

2. Rusty iron is used to clean up water supplies such as the drinking water that is drawn from deep wells in Bangladesh and parts of India, and which contains high levels of arsenic. This is removed as the water trickles over the rusty metal, being converted to insoluble iron arsenate.

3. Iron sulfide, FeS_2, as the mineral pyrites, shines like gold hence its popular name of fool's gold because it led some prospectors into thinking they had found what they were seeking.

[88] 10^{22} is 10 million trillion years.

Krypton

Pronounced krip-ton, the name is derived from the Greek *kryptos* meaning hidden.
French *krypton*; German *Krypton*; Italian *cripto*; Spanish *kriptón*; Portuguese *criptônio*.

Essential element

Krypton finds use in lighting, and the detection in the atmosphere of one of its radio-active isomers is a way of monitoring the build-up of nuclear arsenals around the world.

Cosmic element

Krypton gas has been detected in the atmosphere of Mars of which it accounts for 0.03 p.p.m.

Element of life

Krypton can have no biological role, and the gas is harmless. The isotope krypton-83 has the right magnetic properties for MRI imaging and it has been used medically to research the performance of the lungs.

Element of history

Having discovered helium, with an atomic weight of 4, and argon, with an atomic weight of 40, William Ramsay (1852–1916), who was based at University College, London, was convinced he had stumbled upon a new group of elements of the periodic table, and so he knew that there must be others wating to be found. Together with his assistant, Morris William Travers (1872–1961), he began to search for them but to no avail. He thought they might be released from minerals, which is how helium had been found, but none came to light.

Eventually they decided to examine the 15 litres of argon gas that they had extracted from air. This had been produced by removing all the oxygen, nitrogen and carbon dioxide by chemical means, which left the unreactive argon behind. They reasoned that perhaps other gases were also present, and by a process of liquefying the argon and allowing it slowly to evaporate they hoped it might leave behind a heavier component. They were right.

On the afternoon of 30 May 1898, their hopes were realised and they isolated about 25 cm³ of a residual gas. This they immediately tested in a spectrometer, and saw from its atomic spectrum that it was a new element, with orange and green lines that could not be attributed to any other gas. By midnight that day they had measured its density and this confirmed that they had indeed found a new element, which they appropriately named krypton. An alternative name was suggested by the eminent French chemist Marcellin Berthelot (1827–1907) who said it should be called eosium from the Greek word *eos* which is the colour of the dawn sky and which he said its spectrum resembled. His suggestion was not adopted.

Economic element

Krypton may be one of the rarest gases in the atmosphere, but in total there are more than 15 billion tonnes of it circulating the planet, of which only about 8 tonnes a year are extracted, via liquid air.

Krypton has some uses, such as in strip lighting where it produces a soft violet glow by itself, or it can be used to modify the colour of other 'neon' lights. Certain high-speed photographic flash lamps and strobe lights use krypton gas because it has an extremely fast response to an electric current. It is also used in research in the form of krypton fluoride lasers which produce intense, uniform beams of short wavelength.

Liquid krypton is sometimes used in special calorimeters, which measure heat, and one such piece of equipment at the European Organization for Nuclear Research, CERN, contains 27 tonnes of liquid krypton.

From 1960 to 1983, the fundamental standard of length was based on the sharp orange line in the spectrum of krypton-86, and one metre was defined as exactly 1,650,763.73 wavelengths of this line.[89]

Environmental element

Krypton in the environment	
EARTH'S CRUST	10 p.p.t.
	Krypton is the 83rd most abundant element.
SOILS	nil
SEAWATER	80 p.p.t.
ATMOSPHERE	1 p.p.m.

Krypton has no role to play in any part of the environment because it is almost totally inert chemically and of low solubility in water.

Chemical element

Data file	
CHEMICAL SYMBOL	Kr
ATOMIC NUMBER	36

[89] The standard changed in 1983 to one based on the speed of light in a vacuum, a metre being the distance light travelled in 1/299,792,458th of a second, as measured by a light beam from a helium-neon laser.

ATOMIC WEIGHT	83.80
MELTING POINT	−157°C
BOILING POINT	−52°C
DENSITY OF THE GAS	3.7 g per litre
OXIDE	none.

Krypton is a colourless, odourless gas and one of the noble gases that form group 18 (row 4p) of the periodic table. For many years it appeared to be chemically inert to everything but fluorine gas with which it forms only KrF_2, KrF^+ and $Kr_2F_3^+$, but in 1988 a molecular ion of formula $HCNKrF^+$ was produced. Unfortunately, this compound explodes above −50°C. Other compounds are also unstable, unless isolated in a matrix at very low temperatures, but under such conditions compounds with krypton-hydrogen, krypton-carbon and krypton-chlorine bonds have been made.

There are 33 known isotopes of krypton with mass numbers 69 to 101, and eight known isomers. There are five naturally occurring isotopes: krypton-84 accounts for 57 percent. krypton-86 (17 percent); krypton-82 (12 percent), krypton-83 (11 percent), and krypton-80 (2 percent). Also naturally occurring are three radioactive isotopes of which krypton-78 is the most abundant at 0.35 percent and which has an incredibly long half-life of 1.1×10^{20} years (equivalent to 110 *billion* billion years)—so long that until recently it too was regarded as stable. Otherwise the longest-lived radioactive isotope is krypton-81, which has a half-life of 229,000 years, and this too occurs on Earth, along with another radioactive isotope krypton-85, which has a half-life of 10.78 years. Both are formed naturally in the upper atmosphere from cosmic rays interacting with krypton-82 and -86 isotopes. Some krypton isotopes are also produced as the fission products in uranium-bearing deposits.

Element of surprises

Radioactive krypton-85 is given off by nuclear reactors and nuclear fuel reprocessing plants. It escapes to the atmosphere but is not considered to pose a threat to life because like all krypton it is chemically inert. The level of this gas in the atmosphere was carefully monitored during the Cold War years (1947–91) by the West because it revealed the extent to which the Soviet bloc was producing nuclear material. Knowing how much krypton-85 was coming from reactors in the USA and Europe, it was possible to deduce how much was being emitted by the USSR and its partners, and from that to know the amount of material they had available for nuclear weapons.

The Lanthanoids[90]

These are the elements with atomic numbers 57–70 and collectively they are named lanthanoids after the first member of the series, lanthanum. In fact the term can be extended to cover element 71 (lutetium) as well and on many periodic tables lanthanum is placed in group 3 with the elements cerium (element 58) to lutetium placed in the 4f block as the lanthanoid series. For more discussion on the lanthanoids and their position in the periodic table see page 651.

The alternative name for the lanthanoids is rare earth elements and this is how they have been viewed in the past but not all of them are rare. The chemists of the nineteenth century were mystified by the rare earths, and towards the end of the century, William Crookes (1832–1919), the eminent British chemist, summed them up in exasperation as follows:

[T]he rare earth elements perplex us in our researches, baffle us in our speculations, and haunt us in our very dreams. They stretch like an unknown sea before us, mocking, mystifying and murmuring strange revelations and possibilities.

It was hardly surprising that he viewed them thus because there were 15 of them which always occurred together in certain minerals, and extracting them was a tedious process. The story of the discovery of the lanthanoids began with that of yttrium which is not regarded as one of them, but is very similar, and often found with them. The yttrium mineral gadolinite was eventually to yield several rare earth elements: terbium, erbium,

Table 1: The abundances of the lanthanoids in the Earth's crust and in their main mined ores (averages)

At. no.	Element	Abundance (p.p.m.)	Monazite ore	Bastnäsite ore
57	Lanthanum	32	20%	33%
58	Cerium	68	43%	49%
59	Praseodymium	9.5	4.5%	4.3%
60	Neodymium	38	16%	12%
61	Promethium	trace	0	0
62	Samarium	8	2.5%	0.8%
63	Europium	2	0.1%	0.1%
64	Gadolinium	8	1.5%	0.2%
65	Terbium	1	500 p.p.m.	160 p.p.m.
66	Dysprosium	6	0.6%	300 p.p.m.

[90] Previously these were called the lanthanides, and before that the rare earths, a term that is still commonly used. Lanthanoid is the term approved by IUPAC.

67	Holmium	1.4	50 p.p.m.	50 p.p.m.
68	Erbium	4	0.2%	35 p.p.m.
69	Thulium	0.5	200 p.p.m.	10p.p.m.
70	Ytterbium	3	0.1%	5 p.p.m.
71	Lutetium*	0.5	200 p.p.m.	1 p.p.m.

* Technically speaking there can be only 14 lanthanoids occupying the spaces in the periodic table, although traditionally chemists have always included lutetium as one of the rare earths, and this is the next element beyond the lanthanoid group.

ytterbium, lutetium, holmium, thulium, and dysprosium. Other members of the series were extracted from another mineral, cerite, which first yielded cerium and then lanthanum, neodymium, praseodymium, samarium, gadolinium and europium.

As the table shows, the elements with even atomic numbers are more abundant than those adjacent to them with odd atomic numbers, and this is a common feature of all the elements. Monazite ores generally have higher concentrations of the heavier lanthanoids than bastnäsite ores. In line with its rarity, the most expensive lanthanoid is lutetium, followed by thulium.

What follows is an account of the lanthanoids as they impinge on terrestrial and human affairs. In this account no distinction has been made as to the part played by individual elements; here we are dealing with them together.

Elements of life

Lanthanoids were detected in plants in the 1870s and the amounts were related to the soil on which they grew. Generally, roots reject lanthanoids and the levels in plants are less than 0.01 percent of the level of these metals in the soil. However, some plants can concentrate them, such as the hickory tree, and in the eastern USA these trees have been found to have over 2,000 p.p.m. of lanthanoids, mainly lanthanum, cerium and neodymium, which are the most abundant of the rare-earth metals. Mosses and ferns also concentrate lanthanoids.

The amount of the various lanthanoids that become part of the human food chain can be judged from the following table. While the ratios in cow manure do not cor-

Table 2: Lanthanoids in cow manure and municipal sewage (p.p.m. dry weight)

Element	Abundance in cow manure (p.p.m.)	Abundance in human sewage (p.p.m.)
Lanthanum	23.7	12.7
Cerium	55.0	41.9
Praseodymium	10.7	4.3
Neodymium	2.5	2.5
Promethium	0	0
Samarium	5.2	3.5

Europium	0.7	3.7
Gadolinium	1.5	6.8
Terbium	0.3	1.4
Dysprosium	1.0	4.7
Holmium	0.36	0.32
Erbium	0.70	1.16
Thulium	0.14	0.39
Ytterbium	1.76	0.60
Lutetium	0.60	0.12

respond exactly with those in human sewage they are somewhat similar and indicate that the rare earth elements are part of our dietary intake and may come mainly from dairy products.

Lanthanoids do not move up the food chain easily, and those which do so move have little toxic effect. When animals have been fed insoluble compounds, such as lanthanoid oxides and sulfates, it has proved impossible to register any toxic effect no matter how much they were given. Some rats were fed the equivalent of one gram per kilogram body weight (which would be equivalent to a human eating 70 grams) without any effect. Even 10 times this amount had no effect when lanthanum oxide itself was used.

It is difficult to identify the various amounts of the different lanthanoids in humans but they are present and the levels are highest in bone, with much smaller amounts being present in the liver and kidney. In most human organs the levels are measured only at parts per billion although they have been found at levels of parts per million in the eyes, kidneys, and spleen. The amounts in the lens of the eyes increase when cataracts form.

Studies on workers in a smelter who were exposed to lanthanoids as part of the job showed twice the level of lanthanoids in their body than normal—though that was still not very much. When they moved on to other jobs, and were contacted years later, it was found that the level of lanthanum was unchanged, suggesting that once these metals have been absorbed by the body they stay put.

A lot of research has been carried out into the behaviour of lanthanoids in animals, generally by injecting a radioactive isotope and following the way it moves, finding where it ends up in the animal's body, and how long it takes to be excreted. Injections of the lighter lanthanoids into rats led to so-called 'rare-earth fatty liver' in which this organ become pale and enlarged with fatty deposits.

Dangerous elements

Broken skin is sensitive to lanthanoids and exposure results in ulcers which are very slow to heal—terbium will even irritate unbroken skin. Lanthanoid dusts irritate the eyes, neodymium especially so. Workers exposed to the fumes and dusts coming from arc lights, in which lanthanoids are used, have been known to develop a lung condition called rare earth pneumoconiosis.

Medical elements

Early clinical trials of lanthanoid salts for the treatment of TB, cholera, and leprosy were encouraging but eventually led nowhere. There was some concern that these heavy metals might be toxic, but fears were soon allayed when they proved to be harmless. Radioactive isotopes were tried in anti-cancer treatments following the discovery that cancer cells tended to absorb lanthanoids, but again this did not prove to be a successful form of treatment.

The lanthanoids do have some effect on human metabolism, lowering blood pressure and cholesterol levels, preventing blood coagulation, reducing appetite and, in animals, of preventing atherosclerosis. In Hungary a lanthanoid creme called Phlogodym is used to reduce inflammation, and cerium nitrate has proved successful as an antiseptic for burn wounds.

When lanthanoids are injected directly into the brain they have the same pain-killing ability as opiates such as morphine, but for reasons that are not understood.

Environmental elements

Lanthanoids are present in trace amounts in many ores due to the fact that their tri-positive ions Ln^{3+} are about the same size as those of calcium, Ca^{2+}, which they can partly replace in the mineral lattice. Consequently when calcium ores (of which there are many) break down under weathering they release lanthanoids into the environment, but these are soon rendered immobile by contact with carbonate and phosphate ions with which they form insoluble salts. This insolubility in water explains why the amounts of these elements in natural waters are invariably low.

Lanthanoids are to be found in massive rock formations such as basalts, granites, gneisses, shales, and silicate rocks—but only in low concentrations of 10–300 p.p.m. None is an exploitable source of the lanthanoids. There are over a hundred lanthanoid minerals, and although most of these are so rare as to be of little interest except to collectors, there are a few that are mined specifically as a source of rare earth metals. These rare earth minerals were formed deep in the Earth's crust, where they precipitated from superheated solutions or molten rock at very high pressures.

In parts of China, lanthanoid salts are added to fish farm tanks to prevent diseases of the scales, gills and intestines caused by microbes. It is even claimed that fish grow faster. The Chinese also produce a fertilizer with lanthanoid salts added, called Nong-le, which is said to enhance crop yields of sugar-cane, apples, wheat, and rice. Using the fertilizer spreads about 500 g of lanthanoid salts per hectare.

Economic elements

The most important mineral sources are monazite, which is a phosphate ore, and bastnäsite, which is a fluoride carbonate ore. Both contain the full spectrum of lanthanoid elements although cerium and lanthanum predominate—see Table 1 above. Monazite sands have built up over millions of years along the coasts of Southern India, Western Australia, Brazil, South Africa and Sri Lanka. Unfortunately, monazite also has high levels of the radioactive element thorium. The largest deposits of bast-näsite, which are free of this undesirable impurity, are in China and the USA. Other minerals, such as cerite, loparite, and samarskite are rich in lanthanoids but are relatively rare. The main lanthanoid-producing country is China in the province of Inner Mongolia. Indeed China has invested heavily in this area and now dominates lanthanoid production and they have the only facility which can separate them to the required level of purity which in some cases is 99.9999 percent. Deng Xiaoping, the leader of China in the early 1990s, said that the rare earths are to China as oil is to the Middle East and they invested accordingly.

World reserves of lanthanoid metals are estimated to exceed 110 million tonnes. The majority of the world's lanthanoids currently come from a mine in China near the city of Baotao in Inner Mongolia. It was the output of this mine and the price-cutting strategy of the early 2000s which made production uneconomic from mines in the US, Russia, India, and Malaysia which mostly closed down. The world's largest deposit of rare earth minerals are the rocks of a geological formation known as the Ilimaussaq Intrusion which is located on the barren plateau of southwest Greenland. This may well be mined later this century, as will a large deposit that has been discovered in Tanzania.

Today, China produces around 120,000 tonnes of lanthanoids per year, Brazil around 1,000 tonnes and Malaysia around 600 tonnes, with several other countries producing small amounts such as Indonesia, Kazakhstan, North Korea, Mozambique, Nigeria, Russia, and Vietnam, but not the USA which was once the primary source of these elements. However, the large rare earth mine at Mountain Pass, California, is due to restart production in 2012.

Processing monazite is done by treating the ore in alkali, which brings the thorium into solution but precipitates the lanthanoids as their insoluble hydroxides. Care has to be taken to remove all traces of the radioactive thorium. Some beach sands are dredged for their titanium and zirconium ores and monazite is also part of these. Bastnäsite is upgraded by leaching with hydrochloric acid and then heated to remove the carbonate as CO_2. After these primary digestion processes, the mixture of rare earth elements then has to be separated. The early chemists tried a slow method of seemingly endless sequential crystallisations, each one giving a slightly better separation, but such methods are not suitable for commercial processes.

A better method is ion-exchange in which a solution of the rare earth elements flows down a column packed with special resin which picks up the metals at different rates. This method, common about 50 years ago, has now been superseded by liquid-liquid extraction. Liquid-liquid extraction, which involves a dissolved compound moving between an oil and a water layer, can be very effective because the relative solubility varies from metal to metal and salt to salt. Even so, liquid-liquid extraction is a slow process requiring 60 or more stages.

It is difficult to produce the metals themselves because they have high melting points and are easily oxidised. Two methods are used: chemical reduction and electrolytic reduction. The former consists of heating the metal chloride, or better still the fluoride, with calcium in a tantalum crucible under an atmosphere of argon gas. The electrolytic method involves passing an electric current through a mixture of the molten metal chloride and sodium chloride in a graphite lined steel cell, which serves as the cathode, and a graphite rod which serves as the anode. The lighter and lower melting point lanthanoids, cerium, samarium and europium can be obtained by this method, and it is also used for ytterbium. Most lanthanoid metals are produced by the calcium process.

One lanthanoid is missing from lanthanoid ores and that is promethium, element number 61, of which there are no stable isotopes. It is, however, present on planet Earth in trace amounts in uranium ores where it is continuously being produced by nuclear fission. Its most common isotope in uranium minerals is promethium-147. Promethium has such a short half-life compared to the time taken for the Earth to form that it had decayed away before the planet was complete, although some was always going to be present due to the presence of uranium.

The lanthanoid metals have become essential to many products of modern living. They are to be found in mobile phones, laptop computers, liquid crystal displays, plasma screens, smart weapons, hybrid cars, disc drives, and wind generators. These last three uses rely on the remarkable magnetic properties of neodymium and dysprosium which can produce permanent magnets far more powerful than conventional ones and this means that the output of electricity from a generator can be considerably increased. Hybrid cars have batteries that need lanthanum, and metals like cerium are needed to make catalytic convertors more efficient.

Fluorescent lights rely on lanthanoid phosphors to convert the emission from a mercury arc, which has a UV wavelength of 254 nm, into visible light in the 400–700 nm spectrum. Such lamps are more correctly called trichromatic fluorescent bulbs because they rely on these phosphors to emit bands of light in the blue, green and red parts of the spectrum, that is, at 450, 550 and 610 nm respectively. The resulting visible radiation is perceived as white light by the eye.

In trichromatic lights, the blue emission band comes from europium, the green from a mixture of lanthanum, cerium and terbium, and the red from a mixture of yttrium and europium. Such light bulbs not only last for five years, or more, but an 18 watt bulb will give out as much illumination as a conventional 75 watt incandescent bulb.

Specialised high-intensity lighting using a high pressure mercury discharge (halogen lamps) also relies on lanthanoid elements in the form of metal halides. These dissociate in the arc column forming free metal atoms which emit visible light in many parts of the spectrum.

The rare earth elements need not be separated for some of their uses. Together they are converted into their chlorides and then melted and electrolysed to produce a mixture of metals. This is known as mischmetall, and it consists of about 50 percent cerium, 25 percent lanthanum, and 25 percent other rare earth metals. A form of mischmetall was originally patented in 1903 as flints for cigarette lighters.

Mischmetall is used in steel making where it acts to scavenge oxygen and sulfur, thereby improving the physical properties to give high-strength low-alloy steels. This use is declining, however, as modern steel-making technologies produce 'cleaner' steels, that is, those that have lower levels of oxygen and sulfur, but it is still used in China. Iron, aluminium and magnesium, with some mischmetall added, become stronger and more workable. Some goes to making cobalt and nickel alloys which are magnetic. Mischmetall has been used as a 'getter' in vacuum tubes, that is, it is ignited to remove final traces of air.

Chemical elements

All the lanthanoids are chemically similar, which is what made them so difficult to separate by the traditional methods available to chemists in the 1800s. The reason for this similarity is that as we go from one element to the next in the series, the additional electron is not added to the outer orbit of the atom, which would greatly affect its chemical behaviour, but to an inner orbit where its effect is muted. The result is that the atoms of the lanthanoid elements present the same three outer electrons to the world and consequently their chemistry is dominated by loss of these to form positively charged ions, Ln^{3+}. This is the form of these elements that is the most stable.

Rather unexpectedly, as the atomic number of the lanthanoids increases across the series, and the number of electrons in the atoms increases, they actually grow smaller. This phenomenon is known as the lanthanoid contraction and is due to the extra electrons going into an inner orbit within the atom, where they feel the pull of the nucleus more strongly.

Element of surprises

The lanthanoids played an important role in the discovery of nuclear fission. These elements were mysteriously appearing when uranium was bombarded with neutrons. Fission was first achieved by Enrico Fermi (1901–54) in early 1934 when he bombarded uranium with neutrons in an attempt to produce heavier elements beyond uranium in the periodic table. He was unsuccessful, producing fission products instead, although he did not realise that this was happening, and it was Ida Tacke-Noddack (1896–1978) who correctly pointed out that he had actually achieved nuclear fission in a paper published in September 1934.

As a result of this, in 1934 Irene Joliot-Curie (1897–1956) decided to bombard uranium with neutrons specifically to produce fission. She documented the presence of fission products by identifying elements in the middle of the periodic table. She published these results in two papers, in late 1937 and in 1938. She was cautious, stating that more research was required. At a lecture in November 1938, she made the illuminating comment that 'the entire periodic system arises from uranium plus neutron bombardment' clearly demonstrating her grasp of the reality of nuclear fission.

In late 1938, Otto Hahn (1897–1968), Fritz Strassmann (1902–80), Lise Meitner (1878–1968), and Otto Frisch (1904–79) set up an experiment to confirm the results of the earlier experiments that had produced fission of the uranium. The work was performed at the Kaiser Wilhelm Institute in Berlin. Most of the researchers were not totally convinced by the results they were getting. The scientist in charge, Otto Hahn, was puzzled to find that several lanthanoids were present in the bombarded sample.

We now know that these consisted of the radioactive isotopes lanthanum-140 (half-life 41 hours), cerium-141 (21 days), cerium-143 (4 hours), cerium-144 (284 days), praseodymium-143 (14 days), neodymium-147 (11 days), promethium-147 (2.6 years), samarium-151 (93 years) and europium-154 (16 years). Hahn wrote about his findings to Lise Meitner, who had fled to Sweden because the Nazi race laws had excluded her from working at the Institute. Meitner correctly explained the results: the uranium was undergoing nuclear fission, splitting into lighter elements.

Meitner often pointed out that she considered Irene Joliot-Curie to be the researcher who was the first to set up an experiment to specifically demonstrate nuclear fission and was successful in identifying this phenomenon. Meitner indicated that the experiment of which she was part merely confirmed what had already been done.

Thankfully the lanthanoids in the fall-out from nuclear fission bombs and nuclear accidents are generally of short half-lives and their insolubility means that they have not posed the same threat as some of the other radioactive elements released into the environment.

Lanthanum

Pronounced lan-than-um, the name comes from the Greek *lanthanein*, meaning to lie hidden.

French *lanthane*; German *Lanthan*; Italian *lantanio*; Spanish *lantano*; Portuguese *lantânio*.

Lanthanum is one of 15 chemically similar elements referred to as the lanthanoid or rare earth elements, which extend from element atomic number 57 (lanthanum) to element atomic number 71 (lutetium). The term rare earth is a misnomer because some are not rare. The minerals from which they are extracted, and the properties and uses they have in common, are discussed under Lanthanoids.

Essential element

Although in the two centuries following its discovery there were no commercial applications of this metal, today it plays an important role in special types of lighting and special kinds of glass, and this century its role could be even more important in the batteries of hybrid cars.

Element of life

The amount of lanthanum in humans is quite small, probably not more than 1 mg. In bone it has been detected at 80 p.p.t. and in tissue at 0.4 p.p.t. The metal has no biological role, and it has almost no impact on living things. There is no estimate of how much lanthanum we take in each day with our diet but it is part of the food chain because plants will absorb the metal. Some vegetables are reported to have as much as 0.2 percent (dry weight) and some plants have been found with 1.5 percent (dry weight).

When lanthanum chloride solution was injected into mice it ended up in the liver and the bone, whereas when dogs breathed a dust of this salt it was found to collect in the lungs and gut.

Medical element

Lanthanum salts, such as the chloride, are not regarded as toxic when taken by mouth, and injections of lanthanum chloride, $LaCl_3$, were once explored as a possible treatment for cancer. The side-effects of such treatment were somewhat unpleasant and included chills, fever, muscle pains, and stomach cramps.

People suffering from chronic kidney dysfunction generally have too much phosphate in their blood, an illness known as hyperphosphataemia. (The average diet contains much more phosphate that the body requires.) If this is left untreated it will result in osteodystrophy, a painful bone condition in which there are abnormalities in the structure of the skeleton. Renal dialysis can do little to reduce phosphate levels, and although patients with hyperphosphataemia can be put a low-phosphate diet, it is almost impossible to prevent the bones from deteriorating.

A new method of treatment relies on lanthanum, which binds to phosphate so strongly that it cannot be absorbed from the gut. Lanthanum carbonate is added to the food of hyperphosphataemia patients and it combines with any phosphate that is released as the food is digested in the stomach and intestines, and carries it out of the body.

Element of history

Lanthanum was discovered in January 1839 by the Swedish chemist, Carl Gustav Mosander (1797–1858) at the Karolinska Institute, Stockholm. Mosander was professor of chemistry and mineralogy, and it was there that he extracted a new 'earth' from cerium nitrate, the element which had been discovered in 1803. He heated cerium nitrate to decompose it to cerium oxide, and then stirred this with dilute nitric acid in which it was supposed to be insoluble. Mosander noticed that some did dissolve and when he filtered the solution, and evaporated it, he obtained crystals. These were not cerium nitrate, but the nitrate of a new element. He told his friend Jöns Jacob Berzelius (1779–1848) about his discovery and it was he who suggested the name lanthanum to which Mosander agreed. Word of the new element quickly spread but Mosander became strangely silent about it and did not even publish an account announcing its discovery.

That same year, Axel Erdmann, a student at the Karolinska Institute, discovered lanthanum in a new mineral from Låven island, Langesundfjord, Norway. He named this mosandrite, in honour of Mosander himself.

Mosander, meanwhile, was not to be hurried, as most of his time was taken up running a mineral water business. A year passed, and then two, while chemists all over Europe waited and speculated. Finally, Mosander explained the delay, saying that he had extracted yet another new earth from his lanthanum oxide, and this he called didymium, from the Greek word *didumos*, meaning twin. Didymium formed a rose-pink oxide and it explained why both cerium oxide and lanthanum oxide appeared pink, because they were contaminated with it. When pure they were white. It had taken Mosander two long years of patient extraction to separate didymium from his lanthanum. (Didymium was not to last, because it too was a mixture, and in 1885 it was separated into praseodymium and neodymium.)

Element of war

Lanthanum had an important role during World War II in helping to design atomic bombs, in what was known as the RaLa project (short for **ra**dioactive **la**nthanum). This was a series of 254 test-material detonations that used the radioactive lanthanum-140 isotope as a calibration substance in the development of implosion nuclear weapons. These tests began in 1944 and were designed to produce a plutonium bomb which cannot be used in the barrel-type technique of a uranium weapon. Lanthanum-140 emits gamma rays and these were used to calibrate the rate of density increase that would be needed to achieve the necessary density of plutonium-metal with the critical mass necessary for a nuclear explosion. The sphere of plutonium had to be compressed by the shock wave of conventional explosives in order for this to occur. The tests were conducted from 1944 to 1961 and helped to design more efficient nuclear bombs.

Economic element

There are no ores which contain only lanthanum as the metal component; it is found in minerals that include all the lanthanoid elements, and these are discussed in more detail in that section. Rare earth ores contain a lot of lanthanum: monazite has around 25 percent and bastnäsite around 38 percent. Today, world production of lanthanum oxide is more than 30,000 tonnes per year, and currently known reserves of lanthanum are around 6 million tonnes. Several lanthanum compounds are commercially available, such as the oxide, carbonate, chloride, fluoride, nitrate and sulfate. The metal itself is obtained by the reaction of lanthanum fluoride and calcium metal.

There are no commercial uses for pure lanthanum metal, but there are for its alloys. That of lanthanum and nickel, $LaNi_5$, is extremely good at absorbing hydrogen gas, and as a powder it can absorb as much as 400 times its own volume. It can act as a storage system for hydrogen. While few vehicles use hydrogen as a fuel, more are relying on lanthanum in the form of nickel metal hydride batteries which are used in hybrid cars and these rely on lanthanum as a vital part of the alloy of the anode. A typical hybrid car will have around 10 kg of lanthanum in this form.

Lanthanum is used as the core material in carbon-arc electrodes for film and photographic studio lights and for flood lighting. It increases the brightness of the arc lights and gives an emission spectrum almost identical to sunlight.

Lanthanum oxide is added to the glass for making lenses because it improves its refractive index and low colour dispersion. Up to 40 percent of lanthanum oxide can be incorporated into optical glass, along with yttrium oxide (up to 5 percent) and gadolinium oxide (up to 10 percent). Lanthanum-containing glass is also much more resistant to attack by alkalis. Lanthanum fluoride (LaF_3) is part of ZBLAN glass which consist of the fluorides of zirconium, barium, lanthanum, aluminium, and sodium and which give the glass its name (N stands for Na, that is, sodium). It will transmit infrared light and it is used in fibre optic systems because it can carry much more data than conventional fibre optics. This is the most stable fluoride glass known and is used in spectroscopy and for lasers.

Lanthanum salts are included in the zeolite catalysts used in petroleum refining because they stabilise the zeolite at high temperatures.

Environmental element

Lanthanum in the environment	
EARTH'S CRUST	32 p.p.m.
	Lanthanum is the 28th most abundant element.

SOILS	approx. 26 p.p.m., ranging from 1–120 p.p.m.
SEAWATER	5 p.p.t.
ATMOSPHERE	virtually nil.

Lanthanum is one of the more abundant rare earth elements, being as common as lead and tin together. It poses no environmental threat to plants or animals.

Chemical element

Data file

CHEMICAL SYMBOL	La
ATOMIC NUMBER	57
ATOMIC WEIGHT	138.9055
MELTING POINT	921°C
BOILING POINT	3,460°C
DENSITY	6.1 g/cm^3
OXIDE	La$_2$O$_3$

Lanthanum is a silvery-white metal and is the first member of the lanthanoid group (row 4f) of the periodic table. It is soft enough to be cut with a knife. It rapidly tarnishes in air and burns easily if ignited. It is one of the most reactive of the rare earth metals and even with water it reacts to release hydrogen gas. Lanthanum salts are often very insoluble, such as the oxide, hydroxide, carbonate, fluoride, phosphate, and oxalate. The chloride, bromide, and nitrate are readily soluble. The preferred oxidation state is +3, as in LaCl$_3$ and the La^{3+} ion in solution, although a few compounds with oxidation state +2 are also known, such as LaI$_2$.

There are 39 known isotopes of lanthanum with mass numbers 117 to 155, and 12 known isomers. Lanthanum in minerals consists essentially of two isotopes, the heavier of which is lanthanum-139 and this comprises 99.9 percent of the total. The remaining 0.1 percent is the lighter isotope, lanthanum-138, which is weakly radioactive with a half-life of 100 billion years (10^{11}). There are also minute traces of some radioactive lanthanum isotopes in uranium ores, such as lanthanum-140, half-life 1.68 days, where they have been formed by nuclear fission. The second longest-lived radioactive isotope, after lanthanum-138, is lanthanum-137 with a half-life of 60,000 years.

Element of surprises

Although lanthanum forms trivalent ions, that is, La^{3+}, whereas calcium forms only divalent ones, that is, Ca^{2+}, nevertheless the former is used by researchers as an easily monitored substitute for the latter in living things. The reason is that the two ions are virtually the same size, so that receptors that recognise calcium generally accept lanthanum as well.

Lawrencium

Pronounced law-ren-see-um, the element is named after Ernest O. Lawrence (1901–1958), the inventor in 1929 of a circular type of particle accelerator which he named the cyclotron. It was with this research instrument that several new elements were produced, and for which he won the Nobel Prize in physics in 1939.

French *lawrencium*; German *Lawrencium*; Italian *lawrentio*; Spanish *lawrencio*; Portuguese, *laurêncio*

Element of creation and contention

This element has a long controversial history of discovery. In 1958, the Lawrence Berkeley Laboratory (LBL)[91], bombarded curium-244[92] with nitrogen-14 and appeared to get isotope-257 of a new element of atomic number 103. No follow-up was done because of an incident which involved the destruction of the target. In 1960, the LBL tried a californiuim-252 target with boron-11 projectiles hoping to get isotope-259 but the results were inconclusive. In 1961, the LBL bombarded a mixed isotope curium target of four isotopes (249 to 252) with boron-10/11 mixed projectiles claiming to get isotope-257. They filed a discovery claim, proposing the name lawrencium and the symbol Lw, later changed to Lr.

In 1965, the Soviet Union's nuclear research centre, the Joint Institute for Nuclear Research (JINR), successfully bombarded an americium-243 target with oxygen-18 projectiles to get isotope-256. The also checked the LBL's work, but found that the isotope, energy, and half-life claimed were not accurate. The JINR filed a claim in 1967 proposing the name rutherfordium. The physicists at LBL proposed this same name for element 104. The LBL adjusted the information in its claim based on the JINR's findings and said their product must have been isotope-258, but they did no follow-up work to prove it.

In 1971, IUPAC awarded discovery to the LBL despite the LBL's lack of definitive or credible data, and approved the name lawrencium. (Under present IUPAC regulations a discovery has to be verified by other physicists.) However, lawrencium was now caught up in the naming controversies which surrounded elements 102, 103, 104, and 105—for more details see Transfermium elements.

In 1992, IUPAC set up the Transfermium Working Group (TWG) to deal with discovery claims and it said that while the JINR had made the first definitive production of element 103, the LBL had made important contributions. The TWG recommended that the discovery should be shared. Because of the long-term use of the name lawrencium, IUPAC reaffirmed the name.

---

[91] The name was later changed to the Lawrence Berkeley National Laboratory (LBNL).
[92] This also included 5% of the isotope curium-246.

Chemical element

Data file	
CHEMICAL SYMBOL	Lr (see *Element of surprises*)
ATOMIC NUMBER	103
ATOMIC WEIGHT	262 (isotope 262)
Properties such as melting point, boiling point, and density, are likely never to be measured because there will never be more than a few atoms produced at any one time.	
OXIDE	likely to be Lr_2O_3.

Lawrencium is a radioactive element and a member of the actinide group (row 5f) of the periodic table. To date, not enough has been made at any one time to be visible by the naked eye, but predictions are that it would be a silvery metal susceptible to attack by air, steam, and acids. Its preferred oxidation state, at least in solution, appears to be +3, and with chlorine gas a chloride appeared to be formed which was likely to have been $LrCl_3$. The workers at Berkeley were able to extract lawrencium atoms from an aqueous solution into an organic solvent and deduce that, as expected, its preferred oxidation state was +3, as Lr^{3+}, and there were indications that the lower oxidation state Lr^{2+} could also exist, but was much less stable.

Lawrencium has 11 known isotopes of mass numbers 252 to 262, and one known isomer. The half-life of the longest-lived known isotope lawrencium-262 is 3.6 hours. (Lawrencium-264 is theoretically-calculated to have a half-life of 10 hours.)

Element of surprises

The original chemical symbol for lawrencium was Lw, but this was changed in 1963 to Lr by the Committee on Nomenclature of Inorganic Chemistry, in order to comply with a new IUPAC ruling which said that the second letter of an element symbol should preferably be a consonant, and in multi-syllable names, it should be chosen from the first letter of the second syllable if possible.

Lead

Pronounced led, the name comes from the Anglo-Saxon word *lead*. The chemical symbol, Pb, derives from *plumbum*, the Latin for lead.

French *plomb*; German *Blei*; Italian *piombo*; Spanish *plomo*; Portuguese *chumbo*.

Essential element

Despite the odium which surrounds lead and the desire to minimise human contact with it, this metal continues to be mined on a large scale and finds a variety of uses ranging from car batteries to cable sheathing.

Cosmic element

Lead can be used to measure things on a geological time scale using the uranium-lead dating method. Lead is the end product of the radioactive decay of elements of higher atomic number, and in particular uranium and thorium. By comparing the amount of these metals to lead in a sample it is possible to calculate its age.

Element of life

Lead in the human body	
BLOOD	0.2 p.p.m.
BONE	3–30 p.p.m.
TISSUE	0.2–3 p.p.m.
TOTAL AMOUNT IN BODY	120 mg.

Lead has no biological role for any living species, although some microbes have learnt to live with it, such as the bacterium *Ralstonia metallidurans* which can survive in highly polluted environments of toxic metals. For humans, lead is a cumulative poison and was once an occupational hazard for painters, plumbers, and pottery workers, while for the majority of people in urban environments it was absorbed from the diet they ate and the dust in the air they breathed. When leaded petrol was the norm, the dust of city streets was heavily contaminated. Soluble lead can leach from old water pipes, badly glazed pottery and lead crystal decanters.[93] Dining off antique tableware can also add lead to a meal and it has been found that some dishes can contaminate food with lead to quite high levels. A likely source of

---

[93] Research in Canada showed that port wine stored in lead crystal decanters contained around 2,000 p.p.m. lead (0.2%) after being left for a year, while Scotch whisky had around 1,500 p.p.m. (0.15%).

contamination for young children can be their nibbling of flakes of paint from old woodwork.

In the modern world, the level of lead in bone is still about twice that found in bones taken from prehistoric burials, but it is many times less that the lead in bones of more recent generations. A tiny amount of lead gets into the food chain because all plants contain some lead, albeit very little. Vegetables contain only minute amounts, such as cucumber which has 24 p.p.b. (fresh weight), while sweet corn has 22 p.p.b., cabbage 16 p.p.b., tomatoes 2 p.p.b. and apples a mere 1 p.p.b. However, lettuces grown on soil near a lead processing plant were found to have up to 3,000 p.p.b. (3 p.p.m.) although even this would be unlikely to produce any symptoms of lead poisoning.

Most lead passes through the body without being digested and the average daily intake of around 1 mg results in only about 0.1 mg being absorbed. Exposure to lead in the diet depends on drinking water and the World Health Organisation's recommended upper limit for the level of lead in this is 10 p.p.b.

Medical element

The effects of lead were known to the Romans, and the architect and engineer Marcus Vitruvius, who lived in the first century BC, observed that labourers in lead smelters had pale complexions. The Greek physician Hippocrates (460–377 BC) described a severe attack of colic in a patient who was a lead miner. Neither writer linked what they saw to its true cause, which was exposure to lead.

Despite the health risks, lead was used by doctors for around 2,000 years to treat illnesses. The practice started with Tiberius Claudius Menecrates, physician to the Emperor Tiberius (ruled 14–37 AD), who invented diachylon plasters, which contained a paste made of lead oxide and olive oil. They were prescribed for skin complaints, although sometimes women sought to procure abortions by eating the paste.

In the 1700s, Thomas Goulard (1697–1784), a surgeon of Montpellier, France, produced a medicine by boiling lead oxide in vinegar which reacts to form soluble lead acetate, and he recommended this not only for skin diseases, such as abscesses, erysipelas, and ulcers, but also for piles and even for cancer, which it clearly could not cure. Victorians doctors went further and gave lead acetate with opium as a cure for diarrhoea—it was effective because the lead paralysed the gut, while the opium deadened the associated stomach pains. Internal bleeding and gonorrhoea were also treated with lead acetate, and doses as high as a gram a day were given. Lead medicine was likewise seen as a remedy for neuralgia, persistent coughs, and 'hysteria'.

Andrew Jackson (1767–1845), seventh President of the USA, was exposed to too much lead as indicated by samples of his hair cut in 1815 which were found to have 131 p.p.m., confirming those historians who believed that he suffered from plumbism, the technical name for chronic lead poisoning. Its source remains a mystery: did it come from medicines or alcoholic drinks? Ludwig van Beethoven (1770–1827) was also affected by lead as shown by his hair and parts of his skull. Samples of Beethoven's hair taken on his death bed show that he was treated with lead by his doctor several times in the months leading to his death. These treatments involved puncturing his abdominal cavity to reduce the build up of fluids and then sealing the wound with a lead-based poultice. Six strands of his hair were analysed at the US Department of Energy's Argonne National Laboratory in the year 2000 using a synchrotron and this revealed that Beethoven's hair had 60 p.p.m. of lead, a hundred times more than normal.[94] There could have been many sources of lead contaminating the food and drink of the great composer. The more likely ones would have been water taken from a lead cistern, wine, pewter drinking vessels, or lead glazed pottery. All these were known to deliver high doses of lead, and those afflicted suffered just as Beethoven did, with painful colic and constipation, damage to the nervous system, and maybe even his increasing deafness, which became total when he was about 50.

Lead medicaments are no longer prescribed because we know how dangerous this metal is once it gets into the body. However, metallic lead still finds use in hospitals, providing protective shielding for employees whose job it is to operate diagnostic imaging devices such as X-ray scanners and CAT (computerised axial tomography) equipment.

The body can eliminate lead, taking about 30 days to remove it from the blood, but taking about this number of years to remove it from the skeleton. Chelating agents such as dimercaprol or sodium calcium edentate, which can bind strongly to lead ions, are used as antidotes to lead poisoning.

Dangerous element

The symptoms of mild lead poisoning—headaches, stomach pains, and constipation—can easily be overlooked or attributed to other causes. A single dose of a lead compound is unlikely to kill; what is more insidious is its unsuspected absorption over a period of time. The lead which penetrates the body's defences tends to remain for a long time, locked away in the bones as lead phosphate.

Historically, lead was used to sweeten wine, which it did by reacting to form lead acetate (sugar of lead). In the Middle Ages, this often led to mysterious local outbreaks of 'colic' whose symptoms were stomach cramps, constipation, weariness, anaemia, insanity, and lingering death. There was Picton colic in France, Devon colic in England, and also dry gripes, which afflicted the West Indies and Massachusetts in the 1600s. All were caused by lead-contaminated drinks, in these cases of wine, cider, and rum respectively. In the 1700s, the English physician George Baker (1722–1809) proved the cause of Devon colic was the lead-lined apple presses used in cider making.

Lead and gout

In the 1700s and 1800s, many prominent people suffered from gout. This condition is caused by excess uric acid in the body crystallising in the joints, making movement very painful. Among those affected were Benjamin Franklin, Lord Tennyson the poet, Charles Darwin the biologist, and John Wesley the founder of Methodism. The popular belief

---

[94] The amount of mercury in his hair was normal, thus scotching a long-standing rumour that Beethoven's illness was due to mercury, and that this had been prescribed as a cure for syphilis.

that gout was a sign of too much rich living and drinking port wine may have had some foundation. Early last century, it was found that more than a third of those suffering from gout had high levels of lead in their blood, and port wine was invariably 'improved' with lead acetate, and kept in lead crystal decanters.

George Frederick Handel (1685–1759) was appointed musician to the court of the Elector of Hanover, and when the Elector became King George I of England, Handel moved with him and lived the rest of his life in London. He wrote a stream of operas and oratorios, his greatest work being the *Messiah* which was premiered in 1742. This masterpiece may well have been written while Handel was suffering from lead poisoning which, although it did little to hamper his creative spirit, did affect his health and he suffered from gout, minor strokes, and went blind. The lead undoubtedly came from his drink and he was particularly partial to port wine. On one of Handel's manuscripts he jotted a reminder to order 12 gallons of port from his wine merchant. Adding a little lead to port wine, to improve its flavour and keeping quality, was a practice that continued well into the 1800s; a piece of lead shot was added to each bottle and this slowly dissolved.

Lead poisoning may have destroyed a whole expedition. In May 1845, John Franklin set off to search for the supposed North West Passage around the top of Canada, which was seen as an alternative route from the Atlantic to the Pacific. His two ships were well provisioned with five year's supply of food for the 129 officers and crew. In August the ships were seen in Baffin Bay, but then they disappeared. Five years later a second expedition set out to find them, but all they ever discovered were three graves on Beechey Island. These were of crew members John Torrington, John Hartnell, and William Brain and they died in 1846.

In 1988, researchers at the University of Alberta, Canada, were allowed to exhume and analyse the perfectly preserved remains of these three men and this showed such high levels of lead that it seems they almost certainly died of lead poisoning, probably exacerbated by scurvy. Not only that, the researchers were able to prove that the lead came from the solder of the canned food that they ate, by analysing empty cans found nearby. The ratio of lead isotopes in the victims was the same as that of the lead solder, and quite different from the ratio of lead isotopes in local Inuit people.

Why is lead so dangerous to health? Lead is absorbed into the blood stream where it deactivates the enzymes that make haemoglobin and this results in the build up of the precursor molecule aminolevulinic acid (ALA). It is ALA that causes the various symptoms. It paralyses the gut, hence the stomach cramps and constipation; it results in excess fluid in the brain, causing pressure headaches and loss of sleep; it affects the reproductive system, leading to infertility and miscarriages. Prolonged exposure to lead eventually results in anaemia which exacerbates these ailments.

The most serious effect of all is on the brain where it weakens the walls of the blood vessels so that they leak and this causes a pressure of water to build up in the brain, and ALA itself can then permeate the brain. These symptoms may range from mild headaches, depression, and sleeplessness in a mild attack, but become hallucinations, insomnia, fits, blindness and coma in severe cases. The swelling of the brain, known as lead encephaly, is the danger most feared in children because it can result in permanent damage. Lead poisoning may in part be due to its replacing zinc at certain enzyme sites and it is known that it binds to these sufficiently differently to impair their operation, such as in the formation of proteins. Whereas zinc will bond to four sulfurs, lead will bond to only three because it is reluctant to involve two of its electrons in forming chemical bonds—the so-called inert pair effect.

Mass poisonings with lead are rare but they have occurred in China, British Columbia, and Western Australia, when water from lead mines has found its way into rivers from which drinking water is extracted. In the Western Australia incident it affected mainly children in the town of Port Esperance, not fatally in their case, but 4,000 birds in the area were not so lucky and they died.

The role of lead in violent criminal behaviour and hyperactivity in children appeared to be supported when comparisons are made between such individuals and those behaving normally. Lead levels in hair samples showed there to be four times as much lead as normal in those who are pre-disposed to violent behaviour.[95]

Element of history

Lead has been mined for more than 6,000 years, and it was certainly known to the Ancient Egyptians who used white and red lead in pigments and cosmetics as well as casting the metal itself into small figures. Cosmetics found in tombs of the second millennium BC were found to consist of galena (black lead sulfide), cerussite (white basic lead carbonate), laurionite (white basic lead chloride) and phosgenite (brown lead chloride carbonate), and since the last is rare in nature it seems likely that the Egyptians were able to make it chemically from lead oxide and salt. It has even been suggested that when used as mascara, the lead-based cosmetics also provided protection against eye infections.

The proverbial heaviness of lead is mentioned a few times in the *Bible*, for example, in Exodus (15:10) which described how the Red Sea engulfed the Egyptians: 'they sank as lead in the mighty waters.' That was written about 900 BC but referred to events that allegedly happened several centuries earlier.

The Greeks mined lead on a large scale from 650–350 BC when they exploited a large deposit at Laurion near Athens. Although their aim was to extract silver, of which the mine produced 7,000 tonnes, it had associated with it more than 2 million tonnes of lead. It was not long before they found uses for this by-product. The ancient Greek writer, Theophrasus (372–286 BC) described how it could be turned into white lead, which was used as a pigment for paint. Lead strips were exposed to vinegar fumes (acetic acid vapour) which formed a white deposit after about 10 days. This was scraped off to expose more fresh metal and the process repeated until all the lead had reacted. The white deposit was then ground to a powder, boiled in water, and left to settle to produce white lead.

The Romans were the first civilisation to employ lead on a large scale, mining it mainly in Spain and Britain. They used this easily worked metal for water pipes,

---

[95] The same was true of other metals like aluminium and cadmium.

coffins, pewter tableware, and as white lead in paint, as well as using red lead as colouring for interior decoration.[96] They also used lead to debase their silver coinage. According to one theory, lead was the cause of the low birth rate of the Romans, and indeed the population of their Empire stagnated at around 50 million, even though the level of hygiene was relatively high.

Small lead nuggets have been found in pre-Columbian Peru, Yucatan, and Guatemala.

The Dark Ages (c.500–1000 AD) saw lead mining come to an end in many places, but thereafter it began again and, in addition to the traditional uses, new outlets were found for it, such as pottery glazes, bullets, and printing type. A lot was used as roofing material.

Economic element

A total of 350 million tonnes of lead have been mined throughout human history. Galena (PbS) is the main lead ore and it is 87 percent lead by weight. There are also deposits of cerussite ($PbCO_3$) and anglesite ($PbSO_4$) which are mined. At various times, deposits of pyromorphite (a lead chloride phosphate) and boulangerite (lead antimony sulfide), have been mined, and the latter is still mined in France. Lead occasionally occurs as the native metal.

Newly mined lead amounts to around 4,000,000 tonnes per year, coming mainly from China (40 percent), Australia (20 percent), the USA (10 percent), and Peru (10 percent). The ores are crushed, mixed with coke and limestone, and roasted. Workable reserves of lead minerals are an estimated 140 million tonnes. Most lead is recycled lead from batteries and this accounts for around 5 million tonnes per year, and again most of this is processed in China although most developed countries have facilities for doing this.

Lead is refined to remove the silver it contains (as much as 1.2 kg per tonne). This is done by the Parkes process which involves adding zinc to the molten metal and allowing it to cool slowly until the zinc settles out as a separate layer carrying the silver with it. This is removed and the lead then heated under vacuum to drive off the remaining zinc giving a product which is 99.99 percent pure.

In many cultures white lead, $2PbCO_3.Pb(OH)_2$, was used as a cosmetic for the same reason it was used in paints, because of its brilliant whiteness and its covering power; a little went a long way. In the last century, many of the traditional uses of lead declined or were banned.[97] Nevertheless, some new ones were found such as the insecticide lead arsenate and the additive tetraethyl lead (TEL) which was used to raise the octane level of petrol (gasoline). They too have been phased out, as has the lead acetate in hair dye for men and which turns grey hairs dark brown.

Lead is still in demand, however, and most (80 percent) goes into lead-acid batteries, mainly for vehicles. In these the anode is spongy lead and the cathode a paste

---

[96] Red lead was made by heating white lead. Another colourful lead pigment is chrome yellow, which is lead chromate.

[97] Except in the UK where it is mandatory to restore the outside fabric of historic buildings (aka Grade 1) with the same materials that were used when they were constructed and that means painting with lead-based paints.

of lead oxide on a lead alloy metal grid. Other lead uses are in pigments, lead sheeting, ammunition (but no longer for shotgun pellets, at least in the USA), alloys, cable sheathing, weights for lifting, weight belts for diving, and protective clothing used to shield against dangerous radiation including neutrons. (Nor does it become radioactive when so exposed.) A little lead goes into making lead crystal glassware, and into tiling and plastic linings which are specially designed to block out noise. 'Chemical' grade lead, which contains small amounts of silver and copper, is used in the chemical industry because it resists the action of corrosive liquids. Lead-based semiconductors include lead selenide and lead telluride are used in infrared detectors.

Lead is still used in architecture as roof cladding and for stained glass windows. It provides protection that will last for centuries, even in industrial and coastal regions, nor does it cause discolouration of surrounding stone or brick work. The protective layer slowly changes over the centuries from lead oxide to basic lead carbonate, then to lead carbonate itself, and via lead sulfide to lead sulfate in urban or industrial environments. In all cases the lead compound that is formed is insoluble and, unlike rust, does not flake from the surface.

Environmental element

Lead in the environment	
EARTH'S CRUST	14 p.p.m.
	Lead is the 36th most abundant element.
SOILS	approx. 23 p.p.m., range 2–190 p.p.m.
SEAWATER	2 p.p.t.
ATMOSPHERE	trace.

More lead has been discharged into the environment as a result of human activity than any other metal. There are two components of the natural environment which ensure that most lead on the planet remains immobile: sulfur and phosphorus, the sulfur as sulfide, S^{2-}, and the phosphorus as phosphate, PO_4^{3-}. Both form lead compounds that are extremely insoluble.

Plants will tolerate quite high levels of lead in soil, up to 500 p.p.m., before it affects their growth. In parts of the UK that have been heavily industrialised for 250 years, the levels of lead in some soils exceeds 15,000 p.p.m. (1.5 percent) in urban gardens and more than 20,000 p.p.m. (2 percent) near old lead mines (both dry weight values). The danger of such high levels is mainly to micro-organisms which are a key part of the biological activity of soils. Farmed soil is generally protected by the application of phosphate fertilizers, and in non-farmed soils the formation of sulfides by some micro-organisms will also render lead inactive.

There are several environmental archives—polar ice, glaciers, peat bogs, lake, and marine sediments, and even trees—which preserve a record of lead deposition from the atmosphere, although some of these can be contaminated by local pollution. However, some of them, such as the ice of the poles, Greenland, and high Swiss glaciers, are free of interference and these not only preserve a reliable record but the ratio of lead isotopes (lead-206 compared to lead-207) can tell us where any increase in lead was coming from.

Lead in lake sediments has preserved a record of the contamination of the global environment over the ages. In layers that represent the times before the mining and use of lead, there are only faint traces of the metal which came from atmospheric dust and amounted to only 0.5 p.p.t. However, during the Roman era the level rose to 2 p.p.t., and it has been estimated that the Romans smelted around 80,000 tonnes per year of the metal. As the empire declined after 400 AD the level sank back to near pre-historic levels.

The lead analysis of a peat bog on a Swiss Mountain has allowed William Shotyk of Bern University to construct an atmospheric lead profile going back 14,500 years. He has measured not only the amount of lead in the various layers but the ratio of the two isotopes, lead-206 / lead-207. His results show that agriculture started in Europe around 4000 BC with forest clearing and soil tilling which put soil dust into the atmosphere, carrying lead with it. About 1000 BC, the isotope ratio changed when serious lead mining began, probably by the Phoenicians, because this increased the level of lead-207 compared to lead-206. In recent times, the ratio increased further when Australian lead ores were mined because these had even more lead-207.

In 1921, Thomas Midgley found that adding TEL to petrol boosted its performance, and by the 1960s all cars were running on leaded petrol. Lead in the snows of the Arctic reflected this and reached a maximum 300 p.p.t. by the late 1970s, although levels have since declined.

A widely publicised report of 1979 claimed that reduced intelligence and lack of concentration among city children could be attributed to environmental lead pollution. They were exposed to traffic fumes from leaded petrol, and from leaded paint in older homes. In the USA, inner-city children are automatically given a blood test for lead before starting school. In 1993, more than 400,000 New York State children had their blood lead measured and about 5 percent were found to have worryingly high levels. Linking blood lead levels to lower IQ was supported by an investigation at Cornell University in 2003.

In 1997, a study was carried out by the University Hospitals of Lund and Malmö, Sweden, on 38 men who had worked at a lead smelter for 10 years. The conclusion was that, despite having blood lead levels of around 0.4 p.p.m., which is twice the average, they behaved no differently in memory and behavioural tests when compared to a control group whose blood lead levels were a low average of 0.04 p.p.m.

A report by Columbia University in the USA in 1998 found that sediments in the lake of New York's Central Park showed highest airborne lead levels in the 1930s, before leaded petrol was introduced, and linked the cause to municipal waste incineration, saying that this was always likely to be the major source of lead pollution in cities.

Another environmental threat, at least to wildlife, were the lead weights used as sinkers by fishermen. These were responsible for the deaths of many swans who scooped up and swallowed lost sinkers as they fed in the mud along the bottom of rivers. These have now been replaced by non-toxic ones.

Chemical element

Data file	
CHEMICAL SYMBOL	Pb
ATOMIC NUMBER	82
ATOMIC WEIGHT	207.2
MELTING POINT	334°C
BOILING POINT	1,740°C
DENSITY	11.4 g/cm³
OXIDES	PbO, PbO_2, Pb_3O_4 (red lead)*

* This is an oxide of both oxidation states lead(II) and lead(IV).

Lead is a soft, weak, ductile, dull silver-grey metal that is a member of group 14 (row 6p) of the periodic table. It tarnishes in moist air but is stable to oxygen and water. It dissolves in nitric and acetic acids. Lead has two oxidation states: +2, as with $PbCl_2$ and Pb^{2+} in solution, and +4, as in $PbCl_4$ and lead dioxide (PbO_2) which is a powerful oxidising agent.

Lead has four main naturally occurring isotopes: lead-208 accounts for 52.5 percent; lead-206 (24 percent); lead-207 (22 percent); and lead-204 (1.5 percent). All are radioactive and lead-204 has a half-life of 1.4×10^{17} years, this being the shorter lived isotope, but still is around 100 million times the age of the universe. It means that only a few atoms decay each year.

Nuclear theory determines that because of the structure of the nucleus, all lead isotopes must be radioactive. The lead isotopes once considered stable were lead-206, lead-207, and lead-208 but these are now calculated to be alpha-emitters also with incredibly long half-lives. Lead-208 is calculated to have a half life of 2×10^{19} years (20 billion billion years) with lead-206 and lead-207 being similar. Research is being done to detect and measure the radioactivity of these isotopes.

There are 38 known isotopes of lead with mass numbers 178 to 215, and 36 known isomers. The isotopes lead-209 to lead-212, and lead-214 occur in trace amounts in thorium and uranium minerals where they are produced as part of the decay chain of those elements. Several are new isotopes which have recently been detected from the decay of uranium isotopes by the newly discovered exotic decay mode called cluster decay.

The longest-lived isotopes, other than those already mentioned, are lead-205 with a half-life of 15.3 million years, and lead-202 with a half-life of 52,500 years.

Element of surprises

1. The effect of using white lead in paintings can be seen in many ancient manuscripts. Faces that were originally painted pale pink are now deep black and this was caused by the hydrogen sulfide given off by coal and gas fires, which were popular in the 1800s. This reacts with the white lead oxide to form black lead sulfide.

2. Famous artists may have been affected by the lead of the pigments they used, such as white lead, chrome yellow (aka lead chromate, $PbCrO_4$), and red lead (Pb_3O_4). Goya is thought to have been affected, as was Van Gogh who was particularly fond of sucking the paint from his brushes. His erratic behaviour and mental state, which led to his committing suicide when he was only 37, are consistent with chronic lead poisoning.

3. A premature baby born to a Sikh woman in Adelaide, Australia, in 2001, had record high levels of lead in its body. The mother had been taking a traditional Indian remedy which contained 9 percent of lead compounds. Her baby's blood was found to have almost 250 µg per 100 ml (2.5 p.p.m.), and it was suffering from various symptoms of advanced lead poisoning. Treatment with antidotes eventually saved the baby's life.

4. Roman cooks used an ingredient called *sapa* as a sweetening agent and this was made by boiling down sour wine in lead pans. The popular Roman recipe book, *The Apician Cookbook*, had *sapa* in 85 of its recipes, and *sapa* was used by vintners to preserve wine, and especially Greek wines. *Sapa* tasted sweet because it was mainly lead acetate (aka sugar of lead). Old recipes for making *sapa* have been repeated in recent times, and analysed, showing that the syrup contained around 1,000 p.p.m. of lead (0.1 percent). A spoonful of *sapa* would deliver a dose of lead that would undoubtedly lead to symptoms of sterility, miscarriages, constipation, headaches, and insomnia—all of which afflicted the upper classes in Roman times.

Lithium

Pronounced lith-ee-um, and the name comes from the Greek word *lithos* meaning stone.
French *lithium*; German *Lithium*; Italian *litio*; Spanish *litio*; Portuguese *litio*.

Essential element

Previously this lightest of metals was of limited use, but it now has a key role in the rechargeable batteries of modern lightweight personal equipment.

Cosmic element

Lithium is rare in the Universe, although it was one of the three elements, along with hydrogen and helium, to be created by the big bang. However, young stars have many times the amount of lithium produced at that event, indicating that other processes can generate it. Nevertheless, lithium is destroyed in the interior of a star if the temperature exceeds 2.4 million degrees, which in most stars it does. Destruction occurs when a lithium and a hydrogen (proton) combine and are converted to two atoms of helium.

The presence of lithium has been used by astronomers to distinguish brown dwarf stars from red dwarf stars, both of which are much smaller than our Sun. Brown dwarfs are hot enough to destroy lithium, so if this metal shows up in the spectrum of light from a star, as indicated by a line at a wavelength of 670.7 nm, then it must be a red dwarf.

In 1992, an orange star was discovered, in the star system called V404 Cygni, whose spectrum revealed it to have enormous amounts of lithium. A possible explanation was the presence of a nearby black hole that was dragging the lithium from the centre of the star to its cooler surface. Other lithium-rich orange stars have also been found, one of which, Centaurus X-4, orbits a neutron star and that may be having the same gravitational pull on its partner.

In 2002, the red giant star V838 Monocerotis became ten thousand times brighter than normal, and then dimmed, became brighter again, then dimmed again, and did this three times. As this happened its spectrum changed and showed a lot of lithium to be present. What this star had in fact done was to consume three of its planets and as it vaporised these they released their lithium.

Element of life

Lithium in the human body	
BLOOD	4 p.p.b.
BONE	1.3 p.p.m.
TISSUE	av. 24 p.p.b..
TOTAL AMOUNT IN BODY	7 mg.

Lithium is easily absorbed by plants, and this can be a good guide to the lithium status of the soil. No plant species has yet been found for which the element is an essential nutrient, but for some it does appear to affect growth and development if levels reach around 200 p.p.m. While too much lithium can be toxic to some plants, this can easily be overcome by liming the soil, in other words adding calcium hydroxide, $Ca(OH)_2$, since calcium inhibits uptake of the lighter metal.

The amount of lithium in plants varies widely, with Solanaceae (the family to which potatoes belong) having most, and as much as 30 p.p.m. (dry weight) in some cases. Corn grains have around 0.05 p.p.b., oranges 0.2 p.p.m., lettuce 0.3 p.p.m., and cabbage 0.5 p.p.m. (all dry weight).

Goats who were not exposed to any lithium in their diet from birth were found to put on less weight than those who were given a normal diet. In this respect lithium may be an essential element, at least for one animal, but this has not been proved and no role within the body has been found in which it is involved. The human body appears to have no biological need for lithium, but because this element is widely dispersed in nature we take in some each day and absorb it, although most is excreted.

Dangerous element

Lithium can be toxic by ingestion but there are wide variations of tolerance. In the late 1940s in the USA, lithium chloride was marketed as a salt substitute, known as Westsal, and was used by those who needed to be on a salt-free diet because of heart disease. Then in 1949, several patients suddenly died and this was deduced to be lithium poisoning. The treatment was banned and the US Food and Drugs Administration (FDA) carried out research on lithium chloride which showed that it could adversely affect the kidneys when taken in large doses.

A blood level of 10 mg per litre registers as mild lithium poisoning, and when it reaches 15 mg per litre there can be side effects, such as confusion and slurred speech. Only if it were to exceed 20 mg per litre would there be the risk of death. Nevertheless, at the lower levels it is safe and a person can be kept on lithium medication for as long as five years without coming to any harm.

Medical element

It is believed that around half a million people in the developed world are using lithium medication as a treatment for bipolar depression. This used to be called manic depression, and the sufferer experiences uncontrollable mood swings from euphoria to misery, and it is a condition which does not respond merely to antidepressant medication. The oscillation between these two extremes can be moderated with lithium.[98] In 1871 a British doctor, William Hammond, recommended lithium bromide as a

---

[98] Lithium therapy was acknowledged in the theme of a song of that name by the successful US rock band Evanescence. Lithium is also the title of a song by an earlier, at that time little known US rock band, Nirvana, and again the theme is personal depression.

treatment for what was then described as 'acute mania' as did a Danish neurologist, Carl Lange, but these treatments were not taken up by the medical profession for another 100 years. Indeed lithium became linked to what was seen as a more disabling medical condition of the time: gout.

In the 1800s, lithium enjoyed a vogue as the treatment for gout, the painful condition that develops when sharp crystals of uric acid form between the joints, especially of the feet. Uric acid is not very soluble, so that once such crystals form they may take a long time to dissolve again. Because the lithium salt of uric acid is very soluble, it was thought that taking lithium-rich spa waters would aid recovery—a theory advanced by a Dr Ure in 1843. This belief held sway until in 1912 a Dr Pfeiffer showed that lithium actually slowed up elimination of uric acid from the body in patients with gout. In any case the lithium in spa waters, often taken as the treatment, was too dilute to have any therapeutic effect.

In 1949, an Australian doctor, John Cade (1912–80), senior medical officer in the Victoria Department of Mental Hygiene, was experimenting with guinea pigs and injecting them with urine taken from manic-depressive patients in the hospital in the hope of proving that their condition was caused by an excess of some body chemical. The animals died, and Cade thought this might have been due to excess uric acid in the urine. To test whether this was so he injected guinea pigs with a solution of the lithium salt of uric acid and noted that soon they became very lethargic, but recovered within a few hours.

He then tested the effect the lithium might be having by injecting them with a 0.5 percent solution of lithium carbonate. To his surprise these normally highly-strung animals became docile, and indeed were so calm that they could be turned on their backs and would lie placidly in that position for several hours. Cade then gave lithium carbonate to his most mentally disturbed patient, who had been admitted to a secure unit five years earlier. The man responded so well that within days he was transferred to a normal hospital ward and within two months was able to return home and take up his old job. Other manic depressives responded similarly. Cade's treatment was tried by other doctors with equally good results, and within 10 years lithium medication was used throughout Europe, and by 1970, throughout the USA as well.

With the right dose of lithium, generally given as 250 mg lithium carbonate tablets, a patient can be kept from either of the extremes of mania or depression. The level of lithium in the blood is monitored and should be between 4 and 8 mg per litre.

How lithium works is still not known for certain, but because only the brain is affected, the explanation seems to be that it interferes with an inositol phosphate chemical messenger which is being over-produced and causing the extreme mood swings. This messenger is made in the brain from glucose, and lithium appears to regulate its production to more normal levels. The lithium reduces the productivity of the enzyme which is generating too much of the chemical messenger.

According to one Japanese study, published in 2009, those communities whose water supply contains lithium have significantly lower suicide rates.

Element of history

Lithium was discovered in 1817 by Johan August Arfvedson (1792–1841) in Stockholm, Sweden, and the metal itself was first isolated in 1821 by William Thomas Brande (1788–1866).

The first lithium minerals to be discovered were petalite and spodumene, and these were found on the Swedish island of Utö by the Brazilian mineralogist, Jozé Bonifácio de Andralda e Silva (1763–1838), when he toured Europe in the 1790s. The curious nature of these new minerals was noted, in that they were difficult to analyse and that they gave an intense crimson flame when added to a fire; the colour is due to the molecule LiOH.

Then, in 1817, Arfvedson analysed petalite and realised that it contained a previously unknown metal, which he called lithium because it had come from a stone. He announced his discovery in 1818, correctly identifying lithium as a new alkali metal and a lighter version of the element sodium. Soon after, he found that spodumene and another new mineral, lepidolite, also contained the element.

Although it was an alkali metal, like sodium and potassium, Arfvedson had not been able to separate it by electrolysis, and while William Brande obtained a tiny amount of the metal this way, it was not enough on which to make measurements. Humphry Davy (1778–1829) also managed to isolate a small amount, but again was unable to made measurements of its physical properties. It was not until 1855 that the German chemist Robert Bunsen (1811–99), and the British chemist Augustus Matthiessen (1831–1870) independently succeeded in isolating enough of the metal to study it. They achieved this by the electrolysis of molten lithium chloride.[99]

Soon after its discovery, lithium was found in other minerals and even in spa waters at Karlsbad, Marienbad, and Vichy. It was detected in seawater by means of a spectroscope in 1859. In the years that followed, this same technique, which allowed lithium to be revealed by the characteristic red line it imparts to the spectrum, also resulted in the metal being found in grapes, seaweed, tobacco, vegetables, milk, blood, and human urine. (Patients on lithium therapy have the level in their blood easily monitored by the intensity of this red colour.)

Element of war

The hydrogen of hydrogen bombs is the compound lithium hydride, in which the lithium is the enriched lithium-6 isotope and the hydrogen is the hydrogen-2 isotope, also known as deuterium. This lithium deuteride is capable of releasing massive amounts of energy by nuclear fusion. This is achieved by packing it around the core of an atomic bomb which provides the heat necessary to initiate nuclear fusion reactions as well as providing the necessary neutron flux.

When the bomb detonates, it releases neutrons from the fission of its uranium-235, and these are absorbed by the nuclei of lithium-6 which immediately decays to form helium and hydrogen-3. The hydrogen-3 then fuses with the deuterium to form more

[99] It is rather ironic in view of what we now know of lithium's power to control unbalanced mental states, that Matthiessen committed suicide in 1870 when only 39 because of severe depression.

helium and this releases yet more neutrons. These are absorbed by the casing of the bomb, which is made of uranium-238, and they convert it to plutonium-239 which then adds a third explosion, again one of nuclear fission.

The consequence of all this is to release a *billion* billion joules of energy in a fraction of a second with the explosive force of millions of tonnes of TNT. By the end of the Cold War in 1991, the USA had more than 40,000 tonnes of lithium stored as the hydroxide and from which it had removed 75 percent of the lithium-6 isotope.

Economic element

The main ore is spodumene ($LiAlSi_2O_6$). The other common ores are petalite ($LiAlSi_4O_{10}$), lepidolite ($K(Li,Al)_3(Si,Al)_4O_{10}(F,OH)_2$) and amblygonite (($Li,Na$) $AlPO_4(F,OH)$). There is a large deposit of spodumene in South Dakota where a single crystal of the ore was once discovered which weighed over 10 tonnes and was 15 metres long. Lithium is also extracted from the brines of certain lakes, which may have as much as 1 g per litre in their waters, such as Searles Lake in California, as well as in Nevada and northern Chile.

The main producer countries are Chile, which produces around 12,000 tonnes, Australia (7,000 tonnes), China (3,500 tonnes), and Argentina (3,000 tonnes). The metal is also produced in significant amounts in other countries, such as Portugal, while a significant deposit of spodumene has been discovered in Finland. Most lithium currently comes from Chilean brines, which are first concentrated by solar evaporation, a process that crystallises out calcium sulfate, sodium chloride, and potassium magnesium chloride and leaves a solution that has 3 percent lithium from which lithium carbonate can be obtained by adding sodium carbonate.

The Chilean desert of Salar de Atacama has below its surface a 3,000 square kilometre lake of brine rich in lithium, potassium, and boron. The water from this lake is pumped into plastic-lined ponds on the surface and it emerges a bright blue colour although it slowly turns yellow over a period of several months as it evaporates and the concentration of lithium reaches 6 percent, after which it is taken in tankers to a refinery where it is precipitated to lithium carbonate and exported around the world. Bolivia has its Salar de Uyuni salt lake covering 4000 square miles (10,0000 km²) similarly rich in lithium with an estimated 5 million tonnes of the element.

Chile's reserves of this metal are around 3 million tonnes and those of China are 1 million. Total world reserves of lithium are of the order of 11 million tonnes, and some estimates put it much higher, with a possible total of 35 million tonnes.

Reported industrial production of new metal is about 25,000 tonnes a year, and this is produced by the electrolysis of molten lithium chloride and potassium chloride in steel cells at around 450°C. The production figures of lithium for nuclear weapons are not published, but may exceed 20,000 tonnes a year in the USA alone.

Lithium is used in several ways: lithium oxide accounts for about half the production and this goes primarily in to glass and glass ceramics. About a third goes into lithium chemicals, of which lithium carbonate for pharmaceuticals accounts for a significant amount. Lithium metal itself accounts for the rest, some of which goes into alloys with aluminium and magnesium, and it greatly improves their strength while making them lighter, so much so that the magnesium-lithium alloy is used for protective armour plating. Perhaps the single most important use in now in rechargeable batteries for items such as mobile phones, digital cameras, mp3 players, iPods, etc., as well as electric vehicles.

Lithium, being the lightest of all metals, produces alloys which are light. But this was not why the first commercial lithium alloy, called Bahnmetall, was produced. This alloy of lead, with 0.7 percent calcium, 0.6 percent sodium and 0.04 percent lithium, was developed by the Germans in World War I as a replacement for tin, the importing of which was blockaded. It is still used by the railways there as an anti-friction alloy for bearings.

The most important alloy is with aluminium, which can be used to reduce the weight of commercial aircraft where every kg of weight reduction saves over 125 litres of fuel per year. Aluminium-lithium alloys (with 2–3 percent lithium) are the ones that offer not only the greatest saving, but the lithium makes the metal stronger and less prone to fatigue. However, the alloy is more brittle and less ductile than aluminium. These drawbacks can be overcome by adding small amounts of other metals like zirconium, copper, or magnesium.

An aluminium-lithium alloy with small amounts of silver, magnesium, and copper is about 30 percent stronger than aluminium alloys now used, and it is used in more down-to-earth applications, such as bicycle frames and high-speed trains.

Lithium batteries (most encountered as the AA and AAA type) operate at 3-volts or more, and are used wherever reliability and lightness are important. They consists of lithium metal which releases electrons as the anode and manganese which accepts them as the cathode between which is an electrolyte consisting of lithium perchlorate in propylene carbonate and dimethoxyethane.

Lithium is also part of the batteries implanted to supply the electrical energy for heart pacemakers. They function with lithium as the anode and iodine as the solid electrolyte and have a lifespan of 10 years. This longevity has been extended to lithium batteries which work at the more common 1.5-volts used in toys, clocks, and various small gadgets. The lower voltage was achieved by replacing the usual manganese dioxide cathode with iron disulfide.

Lithium-ion batteries are one of the most commonly used types of rechargeable batteries and can deliver a higher voltage of 3.6 volts than conventional batteries such as the rechargeable nickel-cadmium battery's 1.2 volts. They have the enormous advantage of having a solid polymer gel across which the charge-carrying lithium-ions, Li^+, move, and by which they carry the current. The positive electrode is commonly made of lithium cobalt oxide and the negative electrode of carbon. When current is drawn from such a battery then Li^+ ions move from one electrode to the other and when the battery is recharged they are forced back again. They can be made very slim and are an ideal power source for many modern electronic devices such as laptops and personal equipment such as mobile phones, and also in medical devices. Lithium-ion batteries are also likely to be the preferred batteries for electric cars, when each would involve about 6.5 kg of the metal. Lithium-ion batteries are now the largest commercial outlet for this element, and indeed it is this use of lithium which puts this metal on the list of those elements likely to be in short supply later this century.

Lithium chloride is one of the most water-absorbing solids known and because of this ability, it is used to dry industrial gases and in air conditioning, such as that in submarines. See also *Element of surprises*.

Lithium stearate, made by reacting stearic acid with lithium hydroxide, is an all-purpose high-temperature grease and most greases contain it. It will even work well at temperatures as low as −60°C and has been used for vehicles in the Antarctic.

The ability of lithium salts to act as a flux, in other words to make something flow more easily, is put to use in aluminium smelting, where lithium carbonate is added to the baths of molten salts, and this is another major use of lithium. The carbonate is also put as a flux in baths used in dip-soldering and in welding, and in addition it is added to ceramics and glass, partly to make melts flow better. Lithium produces glass which is resistant to sudden heating or cooling because of its low thermal expansion coefficients. Lithium carbonate is also used to treat glass fibres and make then anti-static.

Lithium has a curious relationship with hydrogen. Together they will form lithium hydride, which is produced as a white powder when lithium and hydrogen react at 750°C. In contact with water this will release the hydrogen again and this is one way of storing the gas. A kg of lithium hydride will release 2,800 litres of the gas when treated with water.

Extracting lithium from seawater has been shown to be possible using a liquid membrane based on dichloromethane (CH_2Cl_2) in which there is present a molecule that selectively bonds to lithium in preference to sodium. While this latter element far exceeds the amount of lithium in seawater, the membrane is able to concentrate lithium from its normal 0.005 percent in seawater to 80 percent in the receiving phase. It can then be recovered by evaporation as lithium chloride (LiCl).

Lithium nitride, Li_3N, is lightweight and stable and is being seen as having two commercially exploitable roles: in rechargeable batteries and as a means of hydrogen storage. For the former it needs to be doped with traces of metals like copper, nickel, and cobalt, and then it become suitable as conducting anode material rather like graphite. In terms of hydrogen storage, then lithium nitride can absorb up to 9 percent by weight of this gas which forms a mixture of lithium hydride, LiH, and lithium amide, $LiNH_2$. However, the temperatures needed to absorb and release the hydrogen, not all of which is released, have to be of the order of 250°C.

Lithium also has a few curious niche uses. It has the highest specific heat capacity of any solid element and the liquid metal makes an excellent coolant, in theory, but is only used for special applications. Lithium fluoride crystals are used for optical devices because this material has the lowest refractive index of all common transparent materials. Molten lithium fluoride and beryllium fluoride are used as the medium for nuclear reactors which have thorium as the fuel—see thorium for more details.

Environmental element

Lithium in the environment	
EARTH'S CRUST	20 p.p.m.
	Lithium is the 31st most abundant element.
SOILS	av. 40 p.p.m., range 1–160 p.p.m.
SEAWATER	0.17 p.p.m.
ATMOSPHERE	traces.

Lithium is found in small amounts in nearly all rocks, and in many mineral spring waters. It poses little threat to plants or animals, on land or in the seas.

Chemical element

Data file	
CHEMICAL SYMBOL	Li
ATOMIC NUMBER	3
ATOMIC WEIGHT	6.941, but variable—see below.
MELTING POINT	181°C
BOILING POINT	1,347°C
DENSITY	0.53 g/cm³
OXIDE	Li_2O

Lithium is a soft, silvery-white, metal that heads group 1, aka the alkali metals, of the periodic table (row 2s). It reacts vigorously with water. Storing it is a problem; it cannot be kept under oil, as sodium can, because it is so light it floats, and will even float on such low density liquids like petrol (gasoline). It is stored by being coated with vaseline (petroleum jelly). Its preferred, indeed only oxidation state, is +1 as in LiCl and other salts, and as Li^+ in solution. Lithium carbonate, Li_2CO_3, is insoluble and this can be used to separate and purify this element. Lithium aluminium hydride, $LiAlH_4$, is a powerful reducing agent used in chemistry.

Somewhat unexpectedly, in view of the behaviour of other elements of group 1, it does not react with oxygen unless heated to 100°C, but it will react with nitrogen from the atmosphere to form a red-brown compound lithium nitride, Li_3N.

There are 10 known isotopes of lithium of mass numbers 4 to 13, and two known isomers. The metal is made up of two isotopes, lithium-6 (8 percent) and lithium-7 (92 percent), neither of which is radioactive. The longest-lived radioactive isotope is lithium-8 with a half-life of 0.84 seconds.

The atomic weight of lithium depends on the mineral from which it has been extracted. Because lithium ions are very similar to magnesium ions they can substitute for these in clay minerals and when that exchange occurs then lithium-6 is preferred over lithium-7 to a certain extent resulting in the enrichment of the lighter isotope.

Element of surprises

1. Lithium as its hydroxide, LiOH, was an important part of the Apollo moon missions and the Shuttle programme, and continues to be important for the International Space Station, on account of its ability to absorb the CO_2 breathed out by those on board. Weight for weight, lithium hydroxide absorbs more of this gas than other alkali metal hydroxides such as sodium hydroxide, which is why it is used.

2. Lithium's atomic weight varies between 6.94 and 6.99 because separation processes might accidentally, or more likely deliberately, influence the isotopic

composition of the element. Lithium, which has been processed to remove some of the lithium-6 for nuclear weapons, has been sold as depleted lithium without its origin being declared.

3. The ion LiO^- is the most basic substance known, in other words it is avid to combine with an acidic H^+. It was made by fragmenting lithium oxalate in a mass spectrometer and can only exist in such a rarified environment.

Lutetium

Pronounced loo-tee-shee-um, it is derived from Lutetia, the Latin name for Paris—but see *Element of surprises*. The spelling was originally lutecium, but this was changed to lutetium in 1949.

French *lutétium*; German *Lutetium*; Italian *lutezio*; Spanish *lutecio*; Portuguese *lutécio*.

Lutetium is is one of 15 chemically similar elements referred to as the lanthanoid or rare earth elements, which extend from element atomic number 57 (lanthanum) to element atomic number 71 (lutetium). The term rare earth is a misnomer because some are not rare. The minerals from which they are extracted, and the properties and uses they have in common, are discussed under Lanthanoids.

Essential element

Lutetium is the most costly metal in the world, despite being more abundant than silver, and having few uses, and its position in the periodic table is somewhat controversial.

Element of life

The amount of lutetium in humans is quite small and the metal has no biological role, but it has been noted that lutetium salts stimulate metabolism. It is difficult to separate out the various amounts of the different lanthanoids in the human body, but all are present, and the levels are highest in bone, with smaller amounts being present in the liver and kidneys.

No one has monitored diet for lutetium content, so it is difficult to judge how much we take in every day but it is probably only a few micrograms per year. Only tiny amounts of lutetium are taken up by plant roots so little gets into the human food chain—some vegetables have as little as 10 p.p.t. (dry weight)—and its low impact on humans can be judged by the analysis of sewage which contains less lutetium that any of the other lanthanoid elements.

Lutetium is mildly toxic by ingestion, but insoluble salts are non-toxic.

Medical element

The radioactive isotope, lutetium-177, is a high-energy beta-particle emitter with a half-life of 6.75 days. It has the potential to be used to diagnose and treat cancers.

Element of history

Lutetium was discovered in 1907 by the French chemist, Georges Urbain (1872–1938) at the Sorbonne in Paris.[100] This element was the final link in a sequence of rare earth discoveries which began with yttrium in 1794. That element was contaminated with traces of other lanthanoids. First erbium and terbium were extracted from it in 1843, and then erbium yielded holmium in 1878, thulium in 1879, and finally lutetium in 1907. This came as a result of painstaking work by Urbain, who called the element lutecium, later changed to lutetium by IUPAC. Meanwhile, two other chemists had also discovered lutetium and were on the verge of reporting it: the Austrian chemist Karl Auer (1858–1929) working in Germany and Charles James (1880–1928) in the USA.

Auer had isolated the element from yttrium oxide, and he called it cassiopeium after the constellation Cassiopeia, a name that was widely used in Germany for many years. Meanwhile James, who was Professor of Chemistry at the University of New Hampshire, had also separated lutetium and indeed had produced a lot of it. However, he was rather cautious about publishing his findings and so the credit went to Urbain. James, who had been born and educated in England, spent his whole working life researching the rare earth elements and devised the method of separating them that was used until the advent of the modern techniques of ion-exchange and liquid-liquid extraction. A sample of pure lutetium metal was not made until 1953.

Economic element

Lutetium is found in minerals that include all the lanthanoid elements, and these are discussed in more detail under Lanthanoids. The main mining area is China. There are also deposits of these minerals in other parts of the world such as the USA, Brazil, India, Australia, Greenland, and Tanzania. However, the amount of lutetium is tiny in all minerals, for example, monazite contains only 0.003 percent. Reserves of the element are estimated to be around 200,000 tonnes.

World production of lutetium is around 10 tonnes per year, as lutetium oxide. The metal is obtained by heating lutetium fluoride and calcium, but little use is made of it except for research purposes. One commercial application has been as a pure beta-emitter, and this is lutetium which has been exposed to neutron activation. As such it has been employed in the oil refining industries. What generally prevents lutetium being more widely used is its cost and the fact that it offers very little which cannot be obtained more cheaply from other lanthanoid elements.

A tiny amount of lutetium is added as a dopant to gadolinium gallium garnet (aka GGG, formula $Gd_3Ga_5O_{12}$) which is used in magnetic bubble memory devices. Lutetium aluminium garnet has been used experimentally as lens material.

---

[100] Urbain was also a noted sculptor and painter.

Environmental element

Lutetium in the environment	
EARTH'S CRUST	0.5 p.p.m.
	Lutetium is the 60th most abundant element.
SOILS	av. 0.3 p.p.m., dry weight.
SEAWATER	0.3 p.p.t.
ATMOSPHERE	virtually nil.

Lutetium is one of the rarer of the rare earth elements but is, nevertheless, seven times more common than silver. It poses no environmental threat to plants or animals.

Chemical element

Data file	
CHEMICAL SYMBOL	Lu
ATOMIC NUMBER	71
ATOMIC WEIGHT	174.967
MELTING POINT	1,663°C
BOILING POINT	3,400°C
DENSITY	9.8 g/cm^3
OXIDE	Lu_2O_3

Lutetium is the hardest and densest of the lanthanoids, and this is to be expected since it is regarded as the last in the series (row 4f). In fact it is best placed in group 3 of the periodic table and not as commonly shown at the end of the 4f row with lanthanum occupying the slot in group 3 (row 5d)—see the chapter on the periodic table for further discussion of where lutetium should be.

Lutetium is silvery-white and resists corrosion in air. Its preferred oxidation state is +3 as in the oxide and salts like the insoluble LuF_3 and the soluble $LuCl_3$ which are colourless. In solution it exists as the Lu^{3+} ion and is surrounded by 9 water molecules.

There are 35 known isotopes of lutetium with mass numbers 150 to 184, and 42 known isomers. Lutetium in minerals consists of two isotopes: lutetium-175 which comprises 97.4 percent, and lutetium-176 (2.6 percent). The heavier of these is a weakly radioactive beta-emitter with a half-life of 38.5 billion years. This isotope has also been used to date meteorites. The next longest lived radioactive isotope is lutetium-174 with a half-life of 3.3 years. (Nuclear theory suggests that there are no stable isotopes of lutetium and that lutetium-175 itself should be radioactive, albeit with an incredibly long half-life.)

Element of surprises

1. Although lutetium is supposed to be named after Paris, which was first settled by the Parisii tribe in the third century BC, the name really derives from Lutetia meaning 'boatyard on the river.'

2. Lutetium is the most expensive metal in the world costing $75 per gram, and well ahead of the highly valued metals like platinum and gold.

3. The isotope lutetium-176 undergoes beta-decay to hafnium-176 and comparing the ratio of these two isotopes in rocks was a way of dating them, at least in theory although it was almost impossible to carry out due to the difficulty of analysing the tiny amounts of hafnium that were present. Then, in 2001, researchers at the Institute of Mineralogy, University of Münster, Germany, found that this could be done using inductively coupled plasma mass spectrometry. The upshot was a re-evaluation of the date of formation of the continental crust from the Earth's mantle, which occurred around 4.3 billion years ago. This is 200 million years earlier that previously assumed. Indeed, parts of the crust formed when the planet was still only 300 million years old and some of it even managed to survive the intense meteorite bombardment to which the early Earth was subjected.

Magnesium

Pronounced mag-neez-ee-um, the element is named after Magnesia, a district of Eastern Thessaly in Greece, which in turn took its name from the ancient Magnetes tribe. They also colonised parts of Western Turkey which is likewise known as Magnesia.

French *magnésium*; German *Magnesium*; Italian *magnesio*; Spanish *magnesio*; Portuguese *magnésio*.

Essential element

Despite its reputation for burning fiercely and spectacularly, magnesium is used as a metal for all kinds of things including aircraft, bicycles, cameras, and luggage.

Cosmic element

Magnesium is abundant throughout the Universe where it has been formed in supernova stars by the combining of three alpha particles, each of which has two protons, and a carbon nucleus, which has six protons. Consequently it is the ninth most abundant element in the Universe and this is reflected on Earth where it is the seventh most abundant element in the crust.

Element of life

Magnesium in the human body	
BLOOD	38 p.p.m.
BONE	varies between 700 and 1,800 p.p.m.
TISSUE	*approx.* 900 p.p.m.
TOTAL AMOUNT IN BODY	25 g.

Magnesium is an essential element for almost all living things, except some insects. It is certainly an essential element for all plants because it is at the heart of the chlorophyll molecule, by means of which plants capture the energy of the sun in order to convert carbon dioxide and water into glucose, which then goes into making cellulose, starch, and many other molecules. Leaves appear green because magnesium-chlorophyll absorbs the blue and red components of sunlight but not the green. Plants take their magnesium from the soil, and humans get theirs by eating them, or eating the animals which feed on them.

Magnesium disperses throughout the body, with most (60 percent) going into the skeleton where it helps maintain bone structure and also act as a store. Magnesium has four main functions: it regulates movement through membranes; it is essential to the working of more than 300 enzymes, including those which release energy from food; it is used for building proteins, which means it is also important in foetal development; and it is involved in the replication of DNA.

Food element

Humans take in between 250 and 350 mg of magnesium every day and we need at least 200 mg, but the body deals very efficiently with this element, taking it from food when it can, or by recycling what it already has when it cannot. It is possible to suffer from magnesium deficiency caused by malnutrition, illness, or old age but a normal diet usually provides more than enough.

While cooking does not affect magnesium, it can leach it from vegetables into the water in which they are boiled, and they may lose as much as half of their magnesium this way. Phosphate in foods interferes with magnesium uptake because it forms insoluble magnesium phosphate which cannot be absorbed. Foods with high levels of magnesium are almonds, brazil nuts, cashew nuts, soybeans, parsnips, bran, chocolate, coca and brewer's yeast, all having more than 200 mg per 100 g. Green leafy vegetables contain a lot.

Some brands of beer contain a lot of magnesium, such as Webster's Yorkshire Bitter, and this may owe some of its unique taste to the high levels of magnesium sulfate in the water used to brew it. By themselves, magnesium salts have a bitter taste.

Medical element

The summer of 1618 saw England suffering a terrible drought, and yet as Henry Wicker walked across Epsom Common he was puzzled to see a pool of water being ignored by nearby cattle. He found that it tasted very bitter and the reason was explained by the crystals he obtained when he evaporated some of it. These became known as Epsom salts which were magnesium sulfate ($MgSO_4$) and soon they were being taken as a medicament, reputedly curing all kinds of ailments, but most noticeably constipation. Epsom salts became popular all over Europe and were made either from magnesium-rich brines or from sea water.

At one time magnesium bromide, $MgBr_2$, was prescribed as a sedative, but too much magnesium taken at one time acts as a muscle relaxant and mild laxative. Milk of Magnesia (magnesium hydroxide, $Mg(OH)_2$) is a popular remedy for indigestion, heartburn, and constipation.

Cases of magnesium deficiency do sometimes occur, for example, through alcoholism, and the symptoms of deficiency are similar to those of *delirium tremens*. Lack of magnesium manifests itself as lethargy, irritation, depression, and even personality changes. Dietary magnesium supplements are much less effective in treating magnesium deficiency because the metal is only absorbed slowly through the stomach and the gut, as shown by studies on concentration camp victims after World War II, who took a year to replenish the body's magnesium store.

Transfusions of magnesium salts are also reported to be beneficial in reducing deaths among people suffering heart attacks.

Dangerous element

There is no evidence that magnesium produces systemic poisoning although persistent dietary over indulgence of magnesium supplements and medicines can lead to muscle weakness, lethargy, and confusion. Using magnesium-based medicines too often and in too large doses can raise blood magnesium to dangerously high levels and this has even been known to cause death.

Element of war

In some air raids during World War II as many as half a million 2 kg magnesium incendiary bombs would be scattered over a city in the space of an hour. The result was massive conflagrations, and occasionally these would be so widespread that they caused a firestorm which sometimes engulfed most of the city.

Once magnesium starts to burn it is almost impossible to extinguish, and it will even burn in the absence of oxygen by combining with the nitrogen of the air (to form magnesium nitride, Mg_3N_2), and it will continue to burn when doused with water which reacts violently with it liberating hydrogen gas. The only way to control burning magnesium is to cover it with sand.

Magnesium metal is not easily ignited so this had to be done by a thermite reaction at the heart of the bomb. The thermite reaction, between aluminium powder and iron oxide, releases so much energy that it produces molten iron, and more than enough heat to cause the magnesium casing of the bomb to burn fiercely.

It was the need to produce such bombs that led to the development of a method of extracting magnesium from sea water, and processing plants to carry out this extraction were built in the USA and the UK during World War II.

Element of history

The first person to recognise that magnesium was an element was Joseph Black (1728–99) at Edinburgh, Scotland, in 1755. He distinguished magnesia (magnesium oxide) from lime (calcium oxide) although both were produced by heating their carbonate ores. Another magnesium mineral called meerschaum (magnesium silicate) was reported by Thomas Henry (1734–1816) in 1789, who said that it was much used in Turkey to make pipes for smoking tobacco. These became popular in Europe in the 1800s.

An impure form of metallic magnesium was first produced in 1792 by Anton Rupprecht (no dates available) who heated magnesia with charcoal, and he gave the element the name austrium after his native Austria. A tiny sample of the metal was first isolated in 1808 by Humphry Davy (1778–1829), by the electrolysis of magnesium oxide, and he proposed the name magnium, for the simple reason that he said the word magnesium was too like manganese and would lead to confusion. His suggestion was not taken up either, and it became known as magnesium.

The French scientist, Antoine-Alexandre-Brutus Bussy (1794–1882) made a sizeable amount of metallic magnesium in 1831 by reacting magnesium chloride with potassium, and he recorded its properties.

In 1906, Richard Willstätter (1872–1942) proved that magnesium was essential to green plants when he burned some purified chlorophyll and obtained about 1.7 percent of an ash residue that he showed was mainly magnesium oxide.

Even though it was known that seawater tasted slightly bitter on account of the magnesium it contained, it was not until 1941 that the Dow Chemical Company in Texas first devised a way of extracting it from this source and produced an ingot of the metal this way.

Economic element

More than 50 minerals are known which contain magnesium, such as dolomite (calcium magnesium carbonate ($CaMg(CO_3)_2$) named after the Dolomite Alps of Northern Italy where it was first found. The main ore is magnesite ($MgCO_3$) which is mined to the extent of 24 million tonnes per year, half of which is produced by China, with Russia, Turkey, and North Korea also producing in excess of 1 million tonnes a year. Magnesite is heated to convert it to magnesia (MgO), and this has several applications although most goes into making heat-resistant bricks for fireplaces and furnaces.[101] It is also added to fertilizers, and used as a supplement for cattle feed especially at the start of the grazing season. It is put into plastics as a bulking agent. Workable reserves of magnesite exceed 2 billion tonnes as ores. This pales in comparison with the more than one million billion tonnes of magnesium dissolved in the oceans, and from which a sizeable proportion of the world's magnesium metal is extracted.

Most magnesium metal is produced by two processes. The older one is the electrolysis of the molten chloride, but most is now made by the thermal reduction of magnesium oxide with silicon. This latter method can even extract magnesium from asbestos waste and other calcium magnesium silicate minerals. World production of the metal itself is around 850,000 tonnes per year. It is produced mainly in China (75 percent of the total), followed by Canada (5 percent), the USA (5 percent), Russia (4 percent), India (3 percent) and Kazakhstan (3 percent) plus a few other countries which produce it in small amounts.

Magnesium now takes third place behind iron and aluminium as useful metals, and like aluminium it combines strength with lightness. It is alloyed with up to 10 percent aluminium, plus traces of zinc and manganese, which improves its strength, corrosion resistance and welding qualities, and this alloy is used for car bodies and aircraft. More than half the magnesium produced ends up this way. It was first used in F1 racing cars in the 1950s and the Porsche car which won the 1971 Le Mans race had a magnesium frame.

Elektron alloys are the new high strength, lightweight alloys which are possible because of magnesium's low density ($1.8 g/cm^3$) and they have motor sport and aerospace applications. They consist of magnesium alloyed with 1–2 percent of zirconium, zinc, and lanthanoid elements such as neodymium and gadolinium. They can be cast, welded, and machined and they are very corrosion-resistant. Magnesium

---

[101] The temperature at which magnesium oxide is produced determines its eventual use. Higher temperatures are needed when it is to be made into refractory materials.

tubes and rods can be safely welded. Magnesium is increasingly used by car-makers because of its environmental benefits in making vehicles lighter and longer-lasting. Reducing the weight of a car not only means it uses less fuel but it causes less damage in accidents. At the end of its useful life, the magnesium in all these products can be recycled at very little cost.

The next major use of magnesium is in iron and steel production, where it is added to the molten metal to remove sulfur, and this accounts for about 20 percent of magnesium production. The rest ends up in a variety of products where its lighter weight is a bonus, such as car seats, laptop computers, cameras, luggage, lawnmowers, power tools, and disk drives. It was used to make German fighter aircraft in both world wars of the last century. In the 1980s an engineer Frank Kirk began to produce bicycles with a magnesium frame that was die cast as one piece, and consequently they were lightweight and strong. However, they had drawbacks and production ceased in 1992.

The chemical industry needs magnesium for various processes, and the dye industries sometime use magnesium sulfate as a mordant. Magnesium hydroxide is being seen as a safer replacement for caustic soda in neutralising waste acids, and it is being added as a fire-retardant to plastics.

Because it is an electropositive metal, magnesium can act as a 'sacrificial' electrode to protect iron and steel structures. It corrodes away preferentially when they are exposed to water which otherwise would cause rusting.

Environmental element

Magnesium in the environment	
EARTH'S CRUST	23,000 p.p.m. (2.3%)
	Magnesium is the 7th most abundant element.
SOILS	approx. 5 p.p.m.
SEAWATER	1200 p.p.m.
ATMOSPHERE	traces.

The table above refers only to the magnesium in the Earth's crust, but if the mantle is taken into account then magnesium becomes the third most abundant element because the mantle is largely composed of olivine and pyroxene, which are magnesium silicates. Magnesium is particularly abundant in the oceans and is the second metal behind sodium in this respect.

Magnesium appears to be environmentally benign in all contexts, although some soils, known as *serpentine* soils, produced by the weathering of igneous magnesium rocks can be toxic to plants on account of their high chromium and nickel content.

Water trapped inside salt crystals that were laid down at various periods in the Earth's geological history show that the composition of the oceans has varied significantly over geological times. For example in the Cretaceous period of around 140–65 million years ago, the ratio of magnesium to calcium was 2 to 1 or less, and much lower than today's ratio of 5 to 1. However, in the much earlier Late Precambrian period, older than 540 million years ago, the ratio was higher at 4 to 1 and much like that of today. The changing ratio of these elements in the seas determined which minerals were laid down as deposits.

Chemical element

Data file

CHEMICAL SYMBOL	Mg
ATOMIC NUMBER	12
ATOMIC WEIGHT	24.3050
MELTING POINT	649°C
BOILING POINT	1,090°C
DENSITY	1.74 g/cm³
OXIDE	MgO

Magnesium is a silvery white, lustrous, relatively soft metal that is a member of group 2 (row 3s) of the periodic table, also known as the alkaline earth group. Magnesium is protected by a layer of oxide on its surface but the metal is easy to ignite as powder or ribbon and it burns fiercely. Magnesium's preferred oxidation state is +2 and it is almost always to be found as the cation Mg^{2+}.

Magnesium has 22 known isotopes of magnesium of mass numbers 19 to 40, and no known isomers. There are three naturally occurring ones: magnesium-24 which makes up 79 percent; magnesium-26 (11 percent); and magnesium-25 (10 percent). None is radioactive. There are also traces of three radioactive magnesium isotopes, magnesium-28, magnesium-29, and magnesium-30, that are produced in minute amounts from the decay of naturally occurring uranium isotopes.[102]

The longest-lived radioactive isotope is magnesium-28 with a half-life of 21 hours followed by magnesium-27 with a half-life of 9.5 minutes. Magnesium-26 is geologically important; it comes from aluminium-26 which has a half-life of 717,000 years. Some meteorites contain this isotope in larger than expected amounts and this means that they originated in the cloud of gas and dust that surrounded the sun as it was forming. They are among the oldest objects in the solar system.

Element of surprises

1. Magnesium carbonate is the powder which weightlifters and gymnasts rub on to their hands in order to get a good grip of their apparatus.

2. The idea of using burning magnesium to produce brilliant light by which to take photographs was suggested by William Crookes (1832–1919) in *Photographic News* in October 1859. However, it was to be another five years before this was done

---

[102] The decay mode is known as cluster decay in which magnesium nuclei are emitted from uranium istopes 232 to 236.

successfully and that was a portrait taken by Alfred Brothers in 1864 in Manchester, England. For the next 100 years magnesium was an essential part of flash photography. To begin with this involved using a rubber bulb to blow magnesium powder into a candle flame. This gave way to flash bulbs which used a charge of electricity to ignite a magnesium filament.

3. At the Le Mans sports car race on 11 June 1955 one of the competitors lost control of his car which crashed into the stands killing the driver instantly and rupturing the car's fuel tank which burst into flames setting alight the car's magnesium body. The result was a terrible fire and 81 spectators died, some killed by flying debris, others by the fire which rescue workers tried to put out using water which of course only made things worse. The race continued on regardless and was won by Mike Hawthorn.

4. Magic candles on birthday cakes are those which surprisingly relight themselves once they have been blown out, and they do this by having small amounts of magnesium powder incorporated into the wick. Once this has been heated it easily reignites by reacting with oxygen from the atmosphere.

Manganese

Pronounced man-gan-eez, the name is derived either from the Latin *magnes*, meaning magnet, because its common mineral pyrolusite has slight magnetic properties, or from *magnesia nigri*, referring to black magnesia rock which is manganese dioxide, MnO_2. (Both manganese and magnesium derive their names from a region of Greece known as Magnesia where minerals of both metals were to be found in ancient times.)

French *manganèse*; German *mangan*; Italian *manganese*; Spanish *manganeso*; Portuguese *manganês*.

Essential element

All living things need this metal but too much can send you insane. Industrially, it is alloyed to make a very strong type of steel which is used for safes and prison bars.

Element of life

Manganese in the human body	
BLOOD	2–8 p.p.b.
BONE	varies between 0.2 and 100 p.p.m.
TISSUE	varies between 0.2 and 2 p.p.m.
TOTAL AMOUNT IN BODY	about 12 mg.

Manganese is an essential element for all species—indeed for some creatures, such as the red ant, it makes up 0.05 percent of their weight. Some organisms, such as diatoms, molluscs and sponges, accumulate manganese. Fish can have up to 5 p.p.m. and mammals up to 3 p.p.m. in their tissue, although normally they have around 1 p.p.m. In 1931, A.R. Kemmerer and co-workers proved manganese to be an essential requirement of mice and rats, and in 1936 it was shown that the bone disease perosis in chickens could be prevented by given them manganese. Manganese compounds are added to fertilizers and animal feedstuffs because it may be lacking in certain soils, and animals grazing on such land may suffer manganese deficiency.

Humans also need manganese, although this was only realised in the 1950s, perhaps because the requirement is so modest. It is needed for the working of various enzymes, and it has been demonstrated to be involved in glucose metabolism, the operation of vitamin B_1, and it is associated with RNA. Most manganese in the body is in the bones; other organs with above average amounts are the pituitary and mammary glands, the liver, and the pancreas.

Food element

The daily dietary intake of manganese averages 4 mg, with a range of 1–10 mg depending on the foods that are eaten. There is never any need for humans to take manganese supplements because we get more than enough of this element from food.

Milk is low in manganese, while liver can have as much as 10 p.p.m., but only a fraction of this, possibly as little as 5 percent, is absorbed into the body. The foods which provide most people with their supply of this element are cereal products and nuts. Beetroot has one of the highest levels of manganese (36–110 p.p.m. dry weight) while tree fruits, such as apples and oranges have the lowest (1–2 p.p.m.). Foods rich in manganese are sunflower seeds, coconuts, peanuts, almonds, Brazil nuts, blueberries, olives, avocados, corn, wheat, bran, rice, oats and tea. The French delicacy, snails, also contains a high level.

Dangerous element

The manganese ion, Mn^{2+}, is the normal form in which the element occurs, and as such is not poisonous, but the purple coloured permanganate ion, MnO_4^-, is toxic. While Mn^{2+} may be the most common form of manganese, the biologically active form is Mn^{3+}, and this is the least stable form of the element.

Exposure to dust or fumes from manganese is a health hazard and working conditions should not exceed 5 mg of the metal per cubic metre, even for short periods. Workers breathing in the fumes from hot manganese suffer fume fever, symptoms of which are fatigue, loss of weight, and impotence. Miners who were affected by this condition also displayed symptoms that came to be known as 'manganese madness' with involuntary laughing or crying, aggression, delusions, and hallucinations. These effects were first noted by James Couper in France in 1837 and their physical symptoms were seen as similar to those of Parkinson's disease, indicating damage to the central nervous system.

The dangers of manganese dust became particularly prevalent in the Philippines among those mining pyrolusite. The manganese passed from the lungs to the brain and after a year or so the person affected began to display symptoms such as hallucinations, emotional instability, shuffling gait, impaired speech, and aberrant behaviour. Those affected would recover after about three months but some of the symptoms might be permanent. Fortunately, manganese madness rarely now occurs.

In Victorian times, and for a long time thereafter, Condy's Disinfecting Liquid was a staple commodity of pharmacies, despite being highly dangerous if it were drunk. For this reason it was flavoured with lavender oil so it could not be mistaken for anything else. The active ingredient in Condy's solution was potassium permanganate. Despite the dangers, gargling with Condy's fluid was recommended for halitosis and sore throats, and to bathe wounds, because it was an effective antiseptic and deodorising solution. It is no longer available.

Historical element

Manganese was known long before it was isolated as an element. As the black ore manganese dioxide it was even used by the pre-historic cave painters of the Lascaux region of France which date back more than 30,000 years. In more recent times its minerals were mined in Germany, Italy and England, and were used commercially by glass makers for hundreds of years to remove the pale greenish tint of natural glass which is due to traces of iron ions (Fe^{2+}) in the sand from which it is made. The Roman author, Pliny the Elder, who perished when Pompeii was destroyed in 79 AD, wrote of a black powder that glass-makers used to make their product crystal clear. This was almost certainly pyrolusite (manganese dioxide); it was referred to as *sapo vitri*, glass soap.[103] It was also used as a black pigment by potters.

In 1740, the Berlin glass technologist Johann Heinrich Pott (1692–1777) investigated pyrolusite, showed that it contained no iron as has been assumed, and from it he produced potassium permanganate ($KMnO_4$), one of the strongest oxidising agents known to early chemists. Several chemists in the 1700s tried unsuccessfully to isolate the metal component in pyrolusite, and credit for first doing this successfully generally goes to the Swedish chemist and mineralogist Johan Gottlieb Gahn (1745–1818). The year was 1774. However, a student at Vienna, Ignatius Kaim, had already described how he had produced manganese metal, in his dissertation written in 1771.

Manganese dioxide was responsible for the discovery of the element chlorine, because when it is added to hydrochloric acid, chlorine gas is given off. This same effect was noted by the Swedish chemist Carl Wilhelm Scheele (1742–86) when he added hydrochloric acid to vegetable ashes and he deduced that these also must contain manganese dioxide. The same happened with the ashes of other plants and it became recognised that manganese is an essential element for all plants. In 1808 manganese was found in ox bones, in 1811 in human bones, and in 1830 it was detected in human blood.

Economic element

The most common manganese minerals are pyrolusite (manganese dioxide, MnO_2) and rhodochrosite (manganese carbonate ($MnCO_3$). Others are psilomelane (barium-containing MnO_2); cryptomelane (potassium-containing MnO_2) and manganite (basic manganese oxide, $MnO(OH)$). Around 10 million tonnes of the metal are mined every year, and reserves are estimated to exceed 3 billion tonnes. If these were ever to be exhausted then we would have to exploit the manganese nodules on the ocean floor—see *Element of surprises*.

The main mining areas for manganese ores are China (8 million tonnes), South Africa (6 million tonnes), Australia (5 million tonnes), Gabon (3 million tonnes),

[103] If too much was added then it would colour the glass purple, and indeed some early types of window glass produced in the American colonies, and which have been exposed to bright sunlight for centuries, now have a purple tinge due to the formation of permanganate; they are highly valued as such. The colour of true amethysts is similarly due to traces of this form of manganese in what would otherwise be colourless quartz crystals.

Kazakhstan (2.5 million tonnes), and India (2.5 million tonnes). The metal is obtained either by the thermal reduction of the oxide using other metals such as sodium, magnesium, or aluminium, or by the electrolysis of manganese sulfate.

Manganese metal is not used as such because it is too brittle, 95 percent of mined ore goes into alloys, mainly steel which contains about 1 percent to improve its strength, working properties and wear resistance. Manganese steel itself contains around 13 percent, and a patent for this was granted in 1883 to a 24-year-old metallurgist, Robert Hadfield (1858–1940), of Sheffield, England. This alloy is extremely strong and is used for railway tracks, earth-moving machinery, safes, helmets, rifle barrels, and prison bars. Other important alloys with a few percent of manganese are manganese bronze (mainly copper) and manganese nickel-silver (an alloy of manganese, copper, zinc and nickel). Adding 1.5 percent of manganese to aluminium improves its corrosion resistance and this is the kind of aluminium used for drinks cans.

Commercially, the two most important compounds of manganese are manganese dioxide, MnO_2, (around 180,000 tonnes per year) and manganese sulfate, $MnSO_4$, (about 120,000 tonnes per year). The former was added to the electrolyte paste in the older zinc-carbon batteries and it was there to prevent the build-up of hydrogen gas around the carbon electrode as the battery was discharged, thereby reducing the current. Today, MnO_2 is more likely to be found in rubber, or as a catalyst in industry. Manganese sulfate is needed for the electrochemical manufacture of manganese metal, and for manufacture of the fungicide Maneb (a manganese dithiocarbamate derivative).

Other compounds that find application are the lower oxide, MnO, manganese carbonate, $MnCO_3$, and potassium permanganate, $KMnO_4$. The first of these goes into fertilizers and ceramics, the second is the starting material for making other manganese compounds, and the third is used to remove organic impurities from waste gases and effluent water.

Environmental element

Manganese in the environment	
EARTH'S CRUST	1000 p.p.m. (0.1%).
	Manganese is the 12th most abundant element.
SOILS	av. 440 p.p.m., with a range from 7 to over 9,000 p.p.m.
SEAWATER	10 p.p.b.
ATMOSPHERE	0.01 microgram/m³, but can be 20 times this level in cities.

Manganese has caused no known environmental damage. It is one of the most abundant metals in soils where it occurs as oxides and hydroxides, and where it cycles through its various oxidation states, this activity being largely due to microbes. Soil may be deficient in manganese, so that animals grazing on such land may suffer, hence the need either to add manganese salts to fertilizers or manganese supplements to their feedstuffs. On the other hand, in some soils manganese can approach levels that are toxic to plants, especially in acidic or poorly aerated soils. Plants generally tolerate high levels in soils and this is reflected in the variable amounts they contain, for example, grasses can have between 17 and 300 p.p.m. (dry weight).

Chemical element

Data file	
CHEMICAL SYMBOL	Mn
ATOMIC NUMBER	25
ATOMIC WEIGHT	54.938049
MELTING POINT	1,244°C
BOILING POINT	1,962°C
DENSITY	7.4 g/cm³
OXIDES	MnO and MnO$_2$

Manganese is a hard, brittle, silvery metal and a member of group 7 (row 3d) of the periodic table. It is reactive when pure, and as a powder it will burn in oxygen. It reacts with water (it rusts rather like iron) and dissolves in dilute acids. Manganese can exist in several oxidation states, +2 to +7, of which the most common are +2 as in $MnCl_2$ and the ion Mn^{2+} in solution, and +4, as in the oxide MnO_2. The best known compound is potassium permanganate, $KMnO_4$, in which it exhibits oxidation state +7 and which is a powerful oxidising agent.

There are 26 known isotopes of manganese with masses 44 to 69, and seven known isomers. Natural manganese consists of a single isotope, manganese-55, which is not radioactive. The longest-lived radioactive isotope is manganese-53 with a half-life of 3.7 million years.

Element of surprises

1. The most surprising occurrence of manganese is on the ocean floor where there is an estimated trillion tonnes (10^{12}) of manganese-rich nodules scattered over large areas, with the North East Pacific being particularly rich in them. In 1876 the three-masted sailing ship, *Challenger*, was sent on a scientific expedition to explore the deep oceans. It returned with a number of curious, cone-shaped lumps that had been dredged up from the ocean floor at various sites around the world. These turned out to be mainly made of manganese with smaller amounts of copper, cobalt and nickel. They appear to have formed around sharks' teeth which are one of the few parts of living things that are capable of surviving the intense pressures at the bottom of the oceans.

During the Cold War the US built a deep-ocean exploration vessel ostensibly to investigate the possibility of mining these manganese-rich nodules. The event was much publicised—the ship was said to have been commissioned by the billionaire Howard Hughes and was called the *Hughes Glomar Explorer*—but it was later revealed that the vessel's real owner was the CIA and its purpose was to locate and recover a wrecked Soviet nuclear submarine, the K-129. This it eventually located, but it broke up as it was being raised to the surface and the section containing the sought-after secret information was said to have been lost.

2. The microbe *Deinococcus radiodurans* was discovered in the 1950s in meat that had been subjected to intense radioactivity. It appears to be able to survive such radiation by accumulating manganese in preference to iron, and it uses the manganese to destroy the damaging oxygen free radicals which radiation produces. In so doing the manganese supports the DNA repair mechanism of cells so that this in not overwhelmed by radiation damage. (Species of microbes that are most sensitive to radiation have low levels of manganese and high levels of iron.)

3. The human immune system fights invading pathogenic bacteria by denying them the iron they need for their enzymes. In 2010 it was discovered that these microbes then use manganese in place of iron and in particular for *ribonucleotide reductases* which are essential for making their RNA and repairing their DNA.

Meitnerium

Pronounced mite-neer-ee-um, the element is named after the Austrian physicist Lise Meitner (1878–1968), who in 1938 correctly explained the results of an experiment in which uranium atoms had undergone nuclear fission to produce other elements, thereby confirming earlier indications that this might happen.

French *meitnerium*; German *Meitnerium*; Italian *meitnerio*; Spanish *meitnerio*; Portuguese *meitnério*.

Element of creation and contention

Meitnerium was first made in 1982 at the German nuclear research facility, the Gesellschaft für Schwerionenforschung (GSI), in Darmstadt, by a group headed by Peter Armbruster and Gottfried Münzenberg. They bombarded a target of bismuth-209 with accelerated iron-58 ions. After a week, a single atom of element 109, isotope 266, was detected. This decayed after 5 milliseconds by α-emission to bohrium and then dubnium, and then, via electron capture, to rutherfordium-258.

Other isotopes of meitnerium have been observed as part of the decay sequence of even heavier elements.

Due to the controversies surrounding the discoveries of elements 104 to 107—see Transfermium Elements for a fuller account—no name was proposed for this element by its discoverers until 1992 when an IUPAC evaluation committee issued its report. The name meitnerium was approved in 1997 along with the names for elements 104–108.

Chemical element

Data file	
CHEMICAL SYMBOL	Mt
ATOMIC NUMBER	109
ATOMIC WEIGHT	278 (longest-lived known isotope)
MELTING POINT	not known, but should be higher than iridium's which is 2,410°C
BOILING POINT	not known
DENSITY	not known, but in excess of iridium's 22.5 g/cm³
OXIDE	none yet made, but likely to be Mt_2O_3

Meitnerium is a radioactive element and a member of group 9 (row 6d) of the periodic table, coming below iridium. It is most unlikely that enough meitnerium will ever be made to be visible, or for its physical properties to be investigated. However, predictions are that it would be a silvery metal susceptible to attack by air, steam, and acids. If it mirrors iridium then its preferred oxidation states will be +3 and +4 and possible +6 as MtF_6.

Meitnerium has seven known isotopes of mass numbers 266, 268, 270, 274–76, and 278 to 279, with one possible isomer. The longest-lived is meitnerium-278 with a half-life of eight seconds, although it is predicted that meitnerium-279 might have a half-life as long as six minutes.

Mendelevium

Pronounced men-del-eev-ee-um, the element is named after Dimitri Mendeleyev, who produced one of the first periodic tables.

French *mendelévium*; German *Mendelevium*; Italian *mendelevio*; Spanish *mendelevio*; Portuguese *mendelévio*.

Essential element

Essential only in extending the periodic table. Mendelevium has no useful applications, nor is it ever likely to have, and it is never encountered outside nuclear facilities or research laboratories.

Element of history

The first atoms of mendelevium were made in 1955 by Albert Ghiorso (1919–), Bernard G. Harvey, Gregory R. Chopin, Stanley G. Thompson (1894–1953), and Glenn T. Seaborg (1912–99). Only 17 atoms of mendelevium were produced during an all-night experiment on 18 February of that year. Using the 60-inch cyclotron at Berkeley, California, they bombarded einsteinium-253 with alpha-particles (helium ions) and mendelevium-256 was detected with a half-life of around 78 minutes. In later experiments, several thousand atoms were made, and now millions of atoms can be produced.

Economic element

Annual world production of mendelevium is estimated to be only a few micrograms per year. This metal is of no economic significance as yet.

Chemical element

Data file	
CHEMICAL SYMBOL	Md*
ATOMIC NUMBER	101
ATOMIC WEIGHT	258 (longest-lived known isotope)
MELTING POINT	830°C (*est.*)
Properties such as boiling point and density are never likely to be measured because sufficient amounts have never been made.	
OXIDE	likely to be Md_2O_3

* The original discoverers suggested Mv but this was overruled by IUPAC.

Mendelevium is a radioactive element and a member of the actinide group (row 5f) of the periodic table. It is most unlikely that enough mendelevium will ever be made to be visible or for its mechanical properties to be measured, or its chemistry investigated. However, predictions are that it would be a silvery metal susceptible to attack by air, steam, and acids. The fluoride, MdF_3, and hydroxide, $Md(OH)_3$, appear to be insoluble in that they co-precipitated with lanthanoid versions of these compounds which mendelevium's chemistry should resemble.

Enough atoms of mendelevium have been produced to show that its chemistry would be based on oxidation states +2 and +3, as Md^{2+} and Md^{3+} respectively. Its melting point has been estimated to be around 1,000K (ca 830°C).

Mendelevium has 17 known isotopes of mass numbers 245 to 261, and five known isomers. The longest-lived isotopes are mendelevium-258 with a half-life of 52 days, and the next longest-lived is mendelevium-260 with a half-life of 28 days. (No isotope has been theoretically calculated to have a longer half-life.)

Element of surprises

Research suggests that in addition to mendelevium in oxidation state +2 which will be relatively stable, this metal might even exhibit compounds in the +1 oxidation state.

Mercury

Pronounced merk-yoo-ree, it is named after the planet Mercury, and the chemical symbol, Hg, comes from the Latin *hydragyrum*, meaning liquid silver.[104]

French *mercure*; German *Quecksilber*; Italian *mercurio*; Spanish *mercurio*, Portuguese *mercúrio*.

Essential element

Now regarded as inherently damaging to human health, it was once widely prescribed in medications even for babies, and it continues to circulate naturally through all parts of the environment.

Element of life

Mercury in the human body	
BLOOD	8 p.p.b.
BONE	0.5 p.p.m.
TISSUE	0.2–0.7 p.p.m..
TOTAL AMOUNT IN BODY	6 mg.

Mercury has no biological role even though it is present in every living thing. It is widespread because it is present in the atmosphere due to its volatility, both as the metal and as the organo-mercury compounds which are formed by microorganisms.

The human intake of mercury is about 3 micrograms per day for adults and about 1 µg for babies and young children. Every mouthful of food we eat contains some.

Agricultural soils may hold as much as 0.2 p.p.m. of mercury and this finds its way into plants and food crops, in particular carrots, potatoes and mushrooms, these last can have as much as 1 p.p.m or more. Grass contains relatively little, around 4 p.p.b., so grazing animals are not contaminated, and meat and dairy products have low levels.

The nutritional guidelines regarding excess mercury levels in meat state that it should not exceed 0.05 p.p.m. However, ocean fish take up mercury, with tuna and sword fish concentrating it to levels 100,000 times those of the surrounding seawater, but this does not render them unsafe to eat. Some algae can concentrate mercury within them to a hundred times the concentration in the surrounding water.

---

[104] This was the description first given to it by Aristotle around 350 BC. Its common name of quicksilver derives from the old English word *cwic*, meaning living.

Creatures that live on the algae thus build up within their flesh significant amounts of the element so that fish may have up to 0.12 p.p.m. of mercury. The World Health Organisation proposed 0.05 p.p.m. as an acceptable upper limit for most foods. For fish, however, the upper safe limit was set 10 times larger at 0.5 p.p.m.

In the 1970s, there was a food scare in the USA when analysis of canned tuna revealed higher-than-expected levels of mercury, which were assumed to be caused by environmental pollution. Overnight, canned tuna disappeared from supermarket shelves, and millions of cans were destroyed. Then, analysis of a sample of tuna from the 1800s that had been preserved in a museum display case, showed the same level of mercury—and it also showed the presence of the protective element selenium which counteracts the mercury.

The levels of methyl mercury in fish have been monitored for many years and in 2002 it was reported that shark had most with 1.5 p.p.m., followed by swordfish with 1.4 p.p.m. Fresh tuna had much less at 0.4 p.p.m. and canned tuna had half this level of mercury at 0.2 p.p.m. The reason that canned tuna has only half the methyl mercury of fresh tuna is that during the canning process it loses much of its oil and with it goes a large part of the mercury. These levels of mercury should be compared to those of other popular ocean fish such as cod with 0.07 p.p.m., plaice with 0.06 p.p.m., and haddock with 0.04 p.p.m. Warnings have also been given about not eating king mackerel, tilefish, or marlin all of which have more than 1 p.p.m. mercury, although these are not widely available.

That mercury in fish might affect children appears not to have been borne out by a long-term study. This was carried out in the Seychelles and by 2006 it showed that children there, who on average eat 11 portions of fish a week and consequently have more mercury in their diet, were no less intelligent than children who ate very little fish, as in the USA.

Medical element

Despite its toxicity, mercury was much used by doctors down the ages. Calomel (mercury(I) chloride, aka mercurous chloride, Hg_2Cl_2) was prescribed as a laxative and diuretic, and corrosive sublimate (mercury(II) chloride, aka mercuric chloride, $HgCl_2$) was used as a disinfectant. The latter was known in India as early as the 1100s and was made by heating mercury, salt, brick dust and alum for three days in a closed earthenware pot, and then adding water to dissolve out the corrosive sublimate before crystallising it. Calomel was then made by grinding these crystals with fresh mercury metal. Only this less soluble form was deemed safe enough to be taken internally. When syphilis became a problem in Europe in the late 1400s the only known cure was calomel and it worked because the mercury killed the organism *Treponema pallidum* that caused the disease, but the 'cure' as it was known, was risky and almost as feared as the disease itself.

Mercury(I) chloride is so much less toxic than mercury(II) chloride that it was widely used and as calomel was on sale in over-the-counter medicaments, especially suitable for dogs (as a de-wormer), humans (as a laxative), and babies (as teething powder). Metallic mercury dispersed in fat was used as an ointment for skin complaints. This salve first appeared in the 1200s and its application was known to induce excess salivation, which we now realise meant that the mercury was being absorbed through the skin and into the blood stream. Later skin treatments used mercurochrome (also known as merbromin) which is a complex organic derivative that was developed in the 1800s as a household antibacterial and antiseptic agent.

Mercury is, or was, part of traditional Chinese medicine and two preparations, Antidotal pills and Cinnabar Sedative Pills contain around 4 percent of mercury compounds. The former is given at the rate of four pills three times a day for poisonous insect stings, delivering a dose of 0.2 g of mercury, the latter at five pills three times a day as a calmative, delivering as much as a gram.

Mercury alloy dental fillings were invented by a US dentist, Greene Vardiman Black (1836–1915), in 1895. He found that an alloy of 70 percent silver and 30 percent tin, when ground to a powder and mixed with mercury, formed a plastic mass that that could be pressed into tooth cavity and which would harden in about five to 10 minutes and which expanded as it solidified thereby exactly filling the hole. The modern versions of alloy powder consists either of silver, tin and copper, or an alloy of gallium, indium, and tin (aka Gallistan). Mercury amalgams are still being used in many countries and it has been estimated that even in mercury conscious USA there are still more than 900 tonnes of mercury amalgams in peoples' teeth. In fact there is no free mercury as such in dental amalgam if it is properly mixed.

A person's reaction to mercury is unpredictable. Some can tolerate it in large amounts without showing signs of poisoning, while others are so sensitive that when mercury-based drugs were injected into them they were dead within seconds of the injection.

In babies, mercury poisoning would manifest itself as 'pink disease', showing as a pink coloration of the fingers, toes, cheeks, nose and buttocks. Medically it was known as acrodynia or erthroedema. The cause was the use of teething powders containing calomel. In adults, the pink coloration of the skin was seen as a beauty aid and there used to be a Mexican beauty cream, Crema de Belleza Manning, which relied on calomel to achieve the desired effect. Even so it was poisonous.

Typical medicaments containing mercury were blue pills which were a mixture of finely ground sugar and mercury and grey powder (mercury and chalk) which were prescribed as laxatives, Golden eye ointment, which was mercury(II) oxide and used to treat conjunctivitis, mercury ointment which was mercury metal dispersed in beeswax and used to treat boils, and an ointment made of red mercury(II) iodide and lard which would cure ringworm. There were also mercury(II) chloride solutions that were used as antiseptics and for disinfecting a person's skin before surgery. An ethyl-mercury compound, thiomersal, was used for many years as a preservative for vaccines and blamed by some as the cause of autism.

It is tempting to assume that when famous people were treated with mercury the reason was syphilis and King Henry VIII of England and Ivan the Terrible of Russia were both dosed this way. Charles II of Great Britain almost certainly died of mercury poisoning but probably not as a result of medical treatment, although he too might well have needed it as a result of his notorious sex life. He was exposed to the metal when doing alchemical experiments which he carried out in a poorly ventilated room in the palace.

Compounds of the lower +1 oxidation state, mercury(I), are generally much less toxic that compounds of the higher +2 state, mercury(II), partly because the latter tend to be more soluble and they can pass through the gut wall within minutes of being swallowed. The upper limit for a medical dose of mercury(I) chloride, also known as calomel, was 30 times bigger than that of mercury(II) chloride, also known as mercury sublimate, and consequently calomel became the ingredient that doctors preferred to prescribe.

Dangerous element

Mercury is absorbed through the lungs, skin and digestive tract. It varies in toxicity depending on whether it is the metal, mercury(I), mercury(II) or an organo-mercury compound. Mercury(I) is much less toxic than mercury(II), because the former is much less soluble than the latter. Cinnabar (HgS) is the least soluble salt of all and a litre of water in contact with this dissolves only 10 nanogram (10^{-8} g).

Mercury poisoning was once relatively common but is now rare thanks to stringent health and safety regulations and the phasing out of many of its uses. All mercury compounds are toxic, methyl mercury extremely so. This can pass the blood/brain barrier and move across the placenta, with the result that mercury affects the central nervous system and can cause a foetus to be deformed. A fatal dose of methyl mercury is probably around 200 mg.

Mercury has a particular attraction to sulfur and it will attach itself to the sulfur atoms of certain amino acids. When these amino acids are part of an enzyme's protein it may result in that enzyme being rendered inactive. The enzyme which operates the so-called sodium pump (Na/K-ATPase) which is essential to the working of the central nervous system, is particularly sensitive to mercury and this is reflected in the most noticeable symptoms of mercury poisoning, the 'shakes' and the mental disturbances—see below. Most mercury in the body is found in the kidneys, followed by the liver, spleen and brain. The body rids itself of mercury not only through urine and faeces, but also via the lungs and sweat, as well as locking some away in nails and hair. There are various effective antidotes for treating mercury poisoning once it has been diagnosed.

The physical symptoms of acute mercury poisoning are a severe headache, nausea, vomiting, stomach pains, diarrhoea and a metallic taste in the mouth. After a few days there is excess salivation, swelling of the salivary glands and eventual loosening of the teeth.

Poisoning by smaller amounts of mercury over longer periods of time, as occurs with industrial exposure, has a different set of symptoms, most of which are due to effects on the brain, and the mental deterioration is known as erethism. It is characterised by symptoms such as timidity alternating with anger and aggression, lack of concentration, loss of memory, depression, insomnia, listlessness, irritability, and a paranoid belief that other people are persecuting you. There is a tremor of the hands, to the extent that their handwriting becomes spidery. One of the first symptoms is also increased salivation.

Liquid mercury is particularly insidious because the metal is slightly volatile and can be absorbed by the lungs, which is why the level in air should not exceed 0.1 mg per cubic metre. When the British sloop *Triumph* was transporting flasks of mercury from Spain to London in 1810 one of them broke open in a storm with the result that all 200 members of the crew were eventually affected, and three died, as did all the cattle and birds that were on board. Research by Michael Faraday (1791–1867) in the 1820s provided the explanation. It was he who first demonstrated the volatility of mercury, and we now know that the evaporation rate of mercury from the surface of the liquid is 800 mg per square metre per hour. (This can be prevented by covering the surface with water or oil.) It was breathing mercury vapour which had caused all the problems on board the *Triumph*.

'Mad as a hatter' is a phrase which owes its derivation to the use of mercury in the hat industry, and described the behaviour of those whose job it was to turn beaver and rabbit fur into felt, the raw material from which hats used to be made. In order to get the short hairs of this type of fur to mat together the pelts were dipped in a solution of mercury nitrate and then dried. Breathing the dust from subsequent processes involving the felt—and there were several steps in turning felt into hats—then caused health problems. The process used in matting the felt for hats was called carroting because of the colour of the mercury(II) nitrate. Workers in the industry often suffered from 'hatters shakes' and 'mercury madness'.

Mercury in the hair of the famous

Hair contains a lot of sulfur-containing amino acids and these attract mercury and provide a permanent record of the level of exposure to the element. It was often the custom in earlier times to preserve locks of hair from those who had died and when these have been analysed by modern methods curiously high levels of mercury have been found. This has led some to speculate that the person had been medically treated with calomel or that they had taken part in alchemical experiments and thereby breathed in mercury vapour.

The hair of Isaac Newton (1642–1727) has shown high levels of mercury, and these almost certainly came from his alchemical work because he was known to have been celibate all his life, while that in the hair of the Scottish poet Robert Burns (1759–96) suggests he had been treated for venereal disease.

Other workers most at risk from chronic mercury poisoning were gilders, dentists, those in the electrical industries—and detectives, see *Element of Surprises*. Most of these trades no longer use mercury, and in those that do it is strictly controlled so that the risks are now negligible.

Gold amalgam was used as a method of gilding other metals such as buttons. When these were heated with an amalgam of gold dissolved in mercury, the gold would deposit on the surface and the mercury would distil off. The advantage of the process was that a gram of gold would gild 500 buttons. Those engaged in the trade suffered from 'Gilders Palsy', and those living nearby were also exposed to mercury. In Birmingham in the early 1800s the metal collected in gutters in the streets surrounding such factories. Despite technical improvements for trapping out the mercury from the flues of gilding establishments, the process only ceased after 1840 when electroplating took its place.

In the 1920s, a seed dressing known as Ceresan was introduced and this was a 2 percent solution of ethyl mercury chloride. By the 1960s the procedure had become widespread with more than 150 proprietary products on the market. Sadly this form of crop protection led to several mass poisonings in developing countries when villagers made bread from the treated grain. In several incidents in northern Iraq, in the early 1970s, more than 5,000 peasants went down with mercury poisoning and 280 died.

Element of history

Cinnabar (aka vermilion, mercury sulfide, HgS), was used as a bright red pigment by the Paleolithic painters of 30,000 years ago to decorate caves in Spain and France with dramatic pictures of animals and humans. In that respect mercury has been part of human history for a very long time. Whether these artists of the Old Stone Age realised that it could be made to yield mercury metal by heating it we cannot know.

The first and greatest Chinese emperor, Qin, may have shortened his life by taking mercury based medications in the belief that these would give him eternal life. His tomb is supposed to have contained a model map of China with its rivers depicted with mercury. Mercury may yet be found in the great tomb that he had built and which is still being excavated. The Chinese alchemist Ko Hung (281–361 AD) wrote of the wonder of turning bright red cinnabar into silver mercury simply by heating.[105] The oldest known sample of mercury, contained in a coconut-shaped vessel, was found by the German archaeologist Heinrich Schliemann (1822–90) in an ancient Egyptian tomb at Kurna, which dated from around 1600 BC although there is reason to believe that it may have been placed there by Arabs at a much later date. Mercury was also known in India about this time. The Greek and Roman writers such as Aristotle, Pliny the Elder, and Vitruvius knew how mercury was obtained from cinnabar and that it could be purified by squeezing it through leather. They also knew it was poisonous.

The Almadén deposit in Spain has been exploited for 2,500 years and has yielded an estimated 200,000 tonnes of mercury, most of which has been lost down the centuries. Pliny reported that more than four tonnes of the metal were imported into Rome every year.

In the Americas, cinnabar was mined by the Incas and used as a cosmetic and war paint. The Spaniards discovered the large deposits of Huancavelica, in 1566, which they mined and smelted to get the mercury they needed to extract gold. In 1848 the miners of the Californian Gold Rush used mercury from the New Almaden Mines of California for the same purpose, to dissolve gold and thereby extract it.

Mercury was thought by the alchemists to be the key to transmuting other metals into gold, but despite centuries of investigation along these lines little that was useful was achieved. On the other hand, mercury was important to the Scientific Revolution which began in the 1500s. It was needed for barometers and thermometers. Mercury was the ideal fluid for these instruments because it did not adhere to glass and so moved freely as the pressure or temperature changed.

---

[105] The sulfur is oxidised by the air, forming sulfur dioxide gas, and mercury metal is left behind.

Mercury as its oxide, HgO, was also the key to the discovery of oxygen, one of the milestones in the development of chemistry. When Joseph Priestley (1733–1804) heated this in 1774 it decomposed into the elements and he collected the oxygen gas.

Economic element

Native mercury occurs naturally as tiny droplets in cinnabar deposits which are generally associated with volcanic rocks. Cinnabar is the chief ore and is mined mainly in China and Krygyzstan. Spain, which used to be the major producer no longer mines it. Another mercury ore is livingstonite (mercury antimony sulfide, $HgSb_4S_8$). World production of mercury is around 1,000 tonnes per annum and falling year by year as the dangers associated with this metal have become apparent. It is still traded in so-called 'flasks' a quantity that has been used since Roman times. A flask contains 34.5 kg or 76 lb imperial of mercury and is made of iron, this being one of the few metals not to form an amalgam with mercury. Vast quantities of mercury have been mined over the centuries and some countries have stockpiles of the metal; the US Government itself has 4,400 tonnes, most of which is stored at Somerville, New Jersey.

The use of mercury has declined significantly and is now mainly confined to the chemical industry which uses it in the form of catalysts, and this accounts for the major use, and for some electrical applications such as switches and rectifiers, which need to be reliable. Formerly a lot of mercury was used in batteries, thermometers, barometers, and for extracting gold. The rotating lenses which focussed the beam of light in a lighthouse floated on a pool of mercury which acted as a low friction bearing for the three tonnes of revolving glass.

By far the biggest use was as an electrode in the manufacture of chlorine and sodium hydroxide by means of the electrolysis of brine, although such plants are rapidly being phased out in favour of methods which do not require mercury. Mercury was also used to treat seed corn to make it resistant to fungus disease, but such seed dressings have now been discontinued. Older uses for mercury such as felt production, plating, tanning, dyeing, and as a de-worming powder have all been superseded by other substances.

Mercury dissolves some metals to give so-called amalgams. This happens with gold and silver, and in the past this has been the cause of much mercury poisoning. In the early 1800s the gilding of the dome of the cathedral of St Isaac at St Petersburg, Russia, required 100 kg of gold as amalgam to be applied to the copper sheets, with the result that 60 workmen lost their lives through mercury poisoning. Until the middle of the 1800s, mirrors were made using silver amalgam and those engaged in this trade suffered accordingly, until the German chemist Justus von Liebig (1803–73) showed that a silver mirror could be deposited on glass by chemical means that did not involve mercury.

Much less mercury is now used in consumer batteries and fluorescent lighting, but it has not been entirely eliminated. In the case of batteries, mercury is now restricted to button cells for hearing aids and other small electronic devices. The batteries consist of an outer zinc casing, which acts as the anode, filled with a paste of zinc hydroxide and mercury oxide surrounding a small steel cathode. The benefit of these batteries is that they provide a voltage of 1.35 and maintain this even as they age. In the case of fluorescent lights, a metre long fluorescent tube now contains only 10 mg of mercury as opposed to more than 35 mg in previous versions. These lamps can now be disposed of as non-hazardous waste. The Dutch lighting company Philips has now produced light bulb which require only 2 mg of mercury.

The compound used as a detonator is mercury fulminate which explodes when it is struck. This material is so sensitive because of its molecular structure which consists of a fragile straight line of seven atoms, i.e. ONCHgCNO. This was only revealed in 2007 when crystals were finally subjected to X-ray analysis. Mercury fulminate was first made by a British chemist Edward Howard in 1799, and he presented a paper on it, and demonstrated it, at the Royal Institution in London in 1800. It soon found use as a primer for percussion caps and explosives.

Exports of mercury are to be banned from Europe from 2011 and from the USA from 2013. In Europe the mercury waste from industrial processes such as that of chlorine and caustic soda plants will be stored in an underground salt mine in steel containers. These plants will all have been phased out by 2020.

Environmental element

Mercury in the environment	
EARTH'S CRUST	50 p.p.b.
	Mercury is the 68th most abundant element.
SOILS	0.01–0.5 p.p.b., but contaminated soils can have up to 0.2 p.p.m.
SEAWATER	40 p.p.t.
ATMOSPHERE	av. between 2–10 nanograms (ng) per cubic metre, although at locations over ore deposits it can be as high as 1,500 ng per cubic metre. Rainwater contains between 2–5 p.p.t.

Mercury cycles through the various compartments of the environment, primarily because there is an enormous natural input from volcanoes, soil erosion, and microbial release of organo-mercury compounds. In addition there is a large release from human activity, mainly from burning coal and oil, and from municipal incinerators where it comes mainly from discarded batteries and fluorescent tubes. Mercury is also emitted by crematoria and it has been estimated that such a place will emit around 5 kg of mercury per year from dental fillings, although in some countries such as Sweden the flue gases are filtered to remove it. It has been estimated that as much as 3 grams of mercury is released by the average corpse. However, these figures have been disputed because many who are cremated are elderly and have lost most or all of their teeth and so it is not a serious environmental problem. Further, it is possible to prevent the loss of mercury during cremation by adding a capsule of a selenium compound to the coffin and this reacts with the mercury vapour to form involatile mercury selenide.

One way to remove mercury from the flue gases of coal-burning power plants will be to inject activated charcoal into the gas stream which will absorb the metal and then be captured into the waste fly ash. An alternative method is to have the flue gases pass through honeycomb-like filters containing activated charcoal impregnated with sulfur.

Perhaps the worst example of environmental pollution by mercury was the Minamata Bay disaster in Japan in the 1950s. There, mercury had been discharged at the rate of 100 tonnes per year into the bay from a local chemical company for around 30 years and had built up in the sediment of the bay to such an extent that fish from the bay had high levels of organomercury compounds in their flesh; some fish contained 0.2 percent of mercury. These fish were eaten by many living in the area.

The result was that more than 10,000 people were crippled with 'Minamata disease' which was basically organomercury poisoning of the central nervous system. The symptoms of the victims were not the normal ones of mercury poisoning, and for a while other toxins were suspected. Eventually it was proved that organomercurials were responsible, and that these had been produced by microbes living in the silt of the bay. When a ban on fishing was imposed in 1956 there were no further cases. The factory then introduced a mercury-separating process to clean up its waste waters. However, it took 30 years before some of its victims were properly compensated.

Cathy Banic of the Meteorological Service of Canada has been tracking airborne mercury and reported in 2003 in the *Journal of Geophysical Research* that the Earth's atmosphere contains about 2,500 tonnes of the metal at any one time. Analysis of the sediment at the bottom of Lake Windermere in England has shown a rise in mercury from 0.15 p.p.m around 1000 AD, to 0.3 p.p.m. in the Middle Ages, then to 0.6 p.p.m. in Victorian times, and finally to 1 p.p.m today. Analysis of mercury levels in the peat of a Greenland fen showed the lowest levels in the Dark Ages (500–1000 AD) when they were less than 0.5 micrograms (µg) per cubic metre in the levels of those years. In the last century the levels peaked at 165 µg in 1953 although they have since declined to 14 µg in 1995 and they continue to fall.

Research carried out in the 1990s revealed that trees and soils emit *elemental* mercury. Steven Lindberg reported that the level of emission over a forest canopy was 100 nanograms (ng) per square metre per hour, and 7.5 ng per square metre per hour over soil—nor was this from soil near industrial sites, but from areas where there had been no possible contamination by human activity. Using specially grown saplings in laboratory chambers, it was shown that plants take up mercury from the soil in which they grow and release it through their leaves to the air. However, when the concentration of mercury in the air reaches a certain level, the process reverses and the plants re-absorb mercury.

In 1996 a team at the University of Georgia in Athens, USA, led by Richard Meagher, reported in the *Proceedings of the National Academy of Sciences* that they had genetically engineered plants that could detoxify land polluted with mercury compounds by absorbing them and reducing them to elemental mercury, which was then released from the plant's leaves. They had taken a gene from bacteria that can carry out this process and inserted it into a weed called arabidopsis, which was then shown to thrive on a growth medium that contained twice the level of mercury chloride that would kill normal plants.

The United Nations Environment Programme agreed at a meeting in Nairobi, Kenya, in 2003, that there should be a global effort to reduce mercury emissions. The report they issued identified the main sources as coal-fired power stations and waste incinerators, which together accounted for 70 percent of emissions, and that 60 percent of the mercury from these sources was being emitted by Asian countries. The global emission of mercury by human activity is estimated to be around 2,200 tonnes per year of which 1,500 come from power plants, 200 from metal production, 100 from cement production, 100 from waste disposal and 300 tonnes from small-scale gold mining. All of this contaminates the planet but it does little harm to us as individuals. The threats to our welfare come generally from more direct contact with this element.

Chemical element

Data file	
CHEMICAL SYMBOL	Hg
ATOMIC NUMBER	80
ATOMIC WEIGHT	200.59
MELTING POINT	−39°C
BOILING POINT	357°C
DENSITY	13.5 g/cm³
OXIDE	HgO

Mercury is a liquid, silvery metal and a member of group 12 (row 5d) of the periodic table. It is stable in air and with water, and is unreactive towards acids, except concentrated HNO_3, and alkalis. It exhibits two oxidation states: +1 as in insoluble Hg_2Cl_2, and as the more commonly +2 as in soluble $HgCl_2$ and Hg^{2+} in solution. Mercury in oxidation state +4 was reported in 2007 by chemists Lester Andrews and Xue-feng Wang of the University of Virginia. They made molecules of HgF_4 by reacting mercury and fluorine at very low temperatures and trapping the product in a matrix of solid argon and identifying it.

There are 40 known isotopes of mercury with mass numbers 171 to 210, and 16 known isomers. There are seven main naturally occurring isotopes: mercury-202 which accounts for 30 percent, mercury-200 (23 percent), mecury-199 (17 percent), mecury-201 (13 percent), mercury-198 (10 percent), mercury-204 (7 percent), and mercury-196 (0.1 percent). The last of these has been found to be radioactive with an incredibly long half-life of 2.5×10^{18} years (equivalent to 2.5 *billion* billion years). Indeed nuclear theory suggests that all isotopes of mercury should be radioactive but with half-lives that, as yet, are too long to be measured.

Of the known radioactive isotopes, mercury-194 has the next longest half-life of 440 years followed by mercury-203 with a half-life of 46.6 days.

There are minute traces of other radioactive isotopes in the Earth's crust, namely mercury-205 to -208 and these are continually produced in trace amounts from the naturally occurring isotopes of uranium and polonium.

Element of surprises

1. The first mercury thermometer was invented in 1657 in Florence, Italy. However, these did not become commonly used until the invention of narrow bore glass tubing in the 1690s which made it possible to register accurately the changes in the volume of mercury caused by small changes in temperature.

2. The alchemists believed that mercury was unique on account of its density and mobility. In their theory it was *the* element, a component of all metals, and so held the key to the transmutation of base metals into gold. It uniquely represented the quintessential property of *fluidity*. Therefore reports from Siberia, that mercury could freeze solid, and thereby lose its elemental fluidity, were discounted as little more than travellers' tales.

   So it came as a shock to the Russian scientist Mikhail Vasilyevich Lomonosov (1711–1765) and his colleague, A. Braun of St Petersburg when, on 26 December 1759, they experimented with snow to see how low a temperature they could achieve. A mixture of ice and salt can produce a temperature drop of many degrees below 0°C and they thought that mixing snow and acids might result in even lower temperatures, and so it did. Suddenly the mercury in their thermometer stopped moving, and it appeared to be solid. Curious to know what had happened, they broke away the glass and found that it had become a solid metal ball with the mercury from the tube protruding like a piece of wire, which they could bend, just like other metals. The belief that mercury was unique had been destroyed. Indeed nails made from solid mercury are strong enough to be hammered into wood.

3. Mercury metal can be diluted with solid materials, by carefully grinding them together in a pestle and mortar, and in this way pharmacists used to prepare grey powder and blue pills. The former consisted of finely divided mercury mixed with chalk, and it could be stirred into a drink like milk, while in the latter it was mixed with sugar and could be pressed into tablets. Both were seen as sure remedies for constipation, although they were introduced originally as part of the treatment for syphilis.

4. Detectives and forensic scientists whose job it was to find, highlight, and photograph fingerprints began to display the symptoms of chronic mercury poisoning, such as excessive salivation, stomach pains, insomnia, tremors, irritability, and depression, although for a long time these were not linked to their real cause, which was mercury. Indeed it was only in the 1940s that it was realised that they were suffering from chronic mercury poisoning. The dusting powder used to be made by grinding chalk and mercury together. Breathing the dust of the powder presented an occupational risk. The skill was to leave only the faintest film of powder in the surface, which meant brushing most of it in the air and there it could be breathed in. Mercury that entered the body in this way tended to stick in the lungs and be absorbed.

Molybdenum

Pronounced mol-ib-den-um, the name is derived from the Greek *molybdos*, meaning lead because its ores were originally believed to be of this metal.

French *molybdène*; German *Molybdän*; Italian *molibdeno*; Spanish *molibdeno*; Portuguese *molibdênio*.

Essential element

This is the metal of extremes; added to steel it produces an alloy suitable for tools like high speed drills, while its sulfide is a superb high temperature lubricant. And molybdenum turns out to be essential to early life in the seas and to the health of trees.

Element of life

Molybdenum in the human body	
BLOOD	1 p.p.b.
BONE	less than 0.7 p.p.m.
TISSUE	*av.* 20 p.p.b.
TOTAL AMOUNT IN BODY	5 mg.

Molybdenum may have been the limiting factor to life in the early oceans because it is not readily soluble except as the molybdate ion, MoO_4^{2-}, and this would not leach from minerals until there was free oxygen available. Eventually molybdenum was to become essential to all species because it is at the heart of the enzymes which convert nitrogen to ammonia and without this chemical, proteins cannot be produced.

There are around 50 molybdenum-containing enzymes, used by both plants and animals. The best-known is the nitrogen-fixing *nitrogenase*, found in the root nodules of legumes, such as beans. Algae also fix nitrogen, and they also have another molybdenum enzyme which helps them get rid of unwanted sulfur by converting it to dimethyl sulfide which is volatile. It is this gas which attracts sea birds to areas where the sea is rich in nutrients and likely to have lots of fish.

Other enzymes that require molybdenum are *aldehyde oxidase* which converts aldehydes into acids, and *sulfite oxidase* which detoxifies sulfite (SO_3^{2-}) by oxidising it to harmless sulfate (SO_4^{2-}). Both of these enzymes are to be found in the liver. *Aldehyde oxidase* is needed for the metabolism of alcohol, which is first converted by a zinc-containing enzyme to acetaldehyde, then to acetic acid by the molybdenum-containing enzyme. The acetic acid is then used by cells as a source of energy. In the human body molybdenum moves around as the molybdate ion. The parts of the body with most molybdenum are the bones, skin, liver and kidney.

Food element

The average intake is about 0.2 mg per day. The absolute minimum intake that is needed is not known, although it may be as low as 0.05 mg, but in any case the intake should not regularly exceed 0.4 mg because above this level it can provoke a toxic response.

Foods which have the most molybdenum are pork, lamb, and beef liver with 0.15 mg per 100 g (= 1.5 p.p.m.), while green beans have 0.7 p.p.m, and eggs 0.5 p.p.m. Other foods with above average amounts of molybdenum are sunflower seeds, wheat flour, soybeans, lentils, peas, and oats. Plant foods can vary between 0.002 and 1 p.p.m. (fresh weight) with fruits having the least and legume vegetables, such as beans, having the most. Cereal grains have 0.5 p.p.m. on average, sweet corn has 0.2, potatoes 0.05, tomatoes 0.03 and apples 0.002 p.p.m.

Dangerous element

Though essential in tiny amounts, molybdenum can be highly toxic in larger doses. Animal experiments have shown that too much molybdenum causes deformed foetuses. Exposure to too much molybdenum dust by metal workers can cause eye irritation, joint pains, and headaches.

Medical element

The enzyme, *xanthine oxidase*, contains molybdenum, and this produces uric acid, which is how the body excretes unwanted nitrogenous material. If this enzyme is too active it can lead to gout, the painful accumulation of sharp crystals of uric acid in the joints. Modern treatment for this condition targets this enzyme to depress its activity. See also lead.

Although the radioactive isotope molybdenum-99 (half-life 66 hours) is not used for medical diagnosis as such, it is produced as a necessary precursor to one which is to used and that is the isotope isomer technetium-99m.[106] This the most commonly used radioisotope in nuclear medicine diagnostics and treatment in hospitals. Supplies of molybdenum-99, however, had run low by 2009 and it was then that the US Department of Energy awarded contracts for its production from sources other than nuclear-weapon grade enriched uranium. Most molybdenum-99 has previously been extracted from nuclear reactors in the Netherlands and Canada.

Element of history

The soft black molybdenum mineral previously called molybdena, and now known as molybdenite (molybdenum sulfide, MoS_2), looks very like graphite. The two were often confused and both were used to make pencils.[107] Even when it was realised that

---

[106] Techntium-99m is given to patients because it collects in various organs of the body and highlights them for diagnosis—see technetium.

[107] The 'lead' of a pencil takes its name from the molybdenum mineral which was thought to be a lead-based ore.

they were different, the former was still mistaken for a lead ore until, in 1778, the Swedish chemist, Carl Scheele (1742–86), published an analysis of the mineral, and showed that it was neither lead nor graphite.

Others investigated it and speculated that it contained a new element but it proved difficult to reduce it to a metal, mainly because it was impossible to make an intimate mixture of the mineral and carbon by grinding, on account of molybdena's softness. These early chemists also reported that heating the mineral in air gave off sulfur dioxide fumes and left behind an oxide. Treating that with nitric acid formed white crystals of molybdic acid.

Scheele passed the problem over to his friend Peter Jacob Hjelm (1746–1813) who was based in Uppsala, Sweden. Hjelm decided to heat molybdic acid with carbon and ground the two together in linseed oil which formed a paste. This he heated to red heat in a closed crucible, whereupon the heat carbonised the oil and together with the charcoal it reduced the oxide to the metal. The new element was announced in the autumn of 1781.

About this time, Scheele discovered a simple and specific test for molybdenum. When this was oxidised to molybdate in which the metal is in oxidation state +6, it would form an intense blue colour when a reducing agent was added to the solution. Even minute amounts of the metal could be detected this way and the depth of colour was a measure it the concentration. For almost 200 years the test was used, despite the fact that chemists could not identify the agent responsible for the colour. In 1996 the puzzle was solved by a group of German chemists at the University of Bielefeld who showed it to consist of a cyclical cluster made up of 154 molybdenum atoms interlinked with oxygen atoms.

Element of war

The British Army were the first to use tanks in warfare in World War I. These were protected by 75 mm manganese steel plates, which unfortunately were not strong enough to protect the occupants against a direct hit by a heavy shell. The answer to this problem was to use armour plating made of molybdenum steel. This had been invented 20 years previously by the French company Schneider & Co. Not only did this provide much better protection, but it was much lighter and the protective sheeting could be reduced to a thickness of 25 mm, permitting more speed and manoeuvrability.

Economic element

Molybdenite (MoS_2) is the chief mineral ore, with wulfenite (lead molybdate, $PbMoO_4$) being less important. Some molybdenum is obtained as a by-product of tungsten and copper production. World production is around 200,000 tonnes per year, and reserves amount to 12 million tonnes of which 5 million tonnes are in the USA. One of the largest deposits of molybdenite, known as the Climax deposit, was found in Colorado, USA, in 1918. The main mining areas are the USA which produces around 60,000 tonnes, China (50,000 tonnes), Chile (40,000 tonnes), and Peru (20,000 tonnes).

Molybdenite is processed by roasting, which converts it to molybdenum oxide (MoO_3). This is then reduced to the metal in various ways, such as by converting the oxide into ammonium molybdate followed by treatment with hydrogen to reduce it to the metal. Some molybdenum roasters are equipped to recover rhenium metal as well and these are the only commercial source of that element.

Molybdenum is rarely made as the bulk metal because of its high melting point, and instead is produced and sold as a grey powder. Indeed it could not be cast as a molten metal until 1959 when a special crucible was developed for the purpose. Generally, molybdenum items are fabricated by compressing the powder at very high pressures.

Molybdenum is used in electrical and electronic devices, in engineering, and in glass manufacture, in which it is used for furnaces because it is not affected by molten glass. Most molybdenum (75 percent) goes into alloys, such as cast iron and 'moly steel'[108] where just a few percent of the element can confer desirable properties such as high strength up to 2,000°C, low expansion coefficient (it has the lowest coefficient of expansion of all the engineering metals), good electrical conductivity, high resistance to corrosion, and resistance to wear even though it is present only as a thin coating. Moly steel is used for automobile and aircraft engine parts, and even for rocket engines.

Other alloys are used for heating elements, radiation shields, glass furnace electrodes, and for the anodes in X-ray equipment. Molybdenum alloys make good tools, and the higher the stress they are under, the higher should be their molybdenum content, with high-speed tools containing as much as 7 percent. They are referred to as M-series high-speed alloys and they are used for drills, gear cutters, and saw blades.

Molybdenum disulfide is used as a lubricant and an anti-corrosion additive, because it forms strong, stable films on metal surfaces and will function under extremes of high and low temperatures and high pressures.

Molybdenum powders are used in circuit inks for circuit boards, and in microwave devices and heat sinks for solid state devices. Power rectifiers rely on molybdenum sheet plate, alloyed with nickel, copper or rhodium, to provide thermal expansion control and heat management, and these are used in the electric motors of trains and in industrial motor power supplies. An increasing amount of molybdenum is used in spray applications where it is blended with binders rich in chromium and nickel and then plasma-sprayed on to piston rings and other moving parts which are exposed to heavy wear.

In the chemical industry a molybdenum catalyst is required for the desulfurisation of fossil fuels and the formation of synthetic fibres and rubbers. Some is turned into sodium molybdate, ideal as a bright orange pigment for ceramics and plastics.

---

[108] The famous steel-and-glass *Pyramide du Louvre* in Paris is constructed of it.

Environmental element

Molybdenum in the environment	
EARTH'S CRUST	1.5 p.p.m.
	Molybdenum is the 54th most abundant element.
SOILS	av. 2 p.p.m., range 0.1–18 p.p.m.
SEAWATER	10 p.p.b.
ATMOSPHERE	traces.

Molybdenum differs from the other micronutrients in soils in that it is less soluble in acid soils and more soluble in alkaline soils, the result being that its availability to plants is sensitive to pH and drainage conditions. Liming soil is regarded as the better way to mobilise molybdenum, and better than applying it directly because too much molybdenum may then affect animals which eat the plants. Fodder with more than 10 p.p.m. of molybdenum would put most livestock at risk. However, some plants can have up to 500 p.p.m. of the metal when they grow on alkaline soils.

Chemical element

Data file	
CHEMICAL SYMBOL	Mo
ATOMIC NUMBER	42
ATOMIC WEIGHT	95.94
MELTING POINT	2,617°C
BOILING POINT	4,612°C
DENSITY	10.2 g/cm³
OXIDES	MoO_2, Mo_2O_3, MoO_3

Molybdenum is a lustrous, silvery metal and a member of group 6 (row 4d) of the periodic table. The metal is fairly soft when pure, and it is attacked slowly by acids. It can exhibit several oxidation states, namely –2, and +1 to +6. The most stable ones are +4, as in $MoCl_4$, and +6 as in MoF_6 and the soluble molybdate ion MoO_4^{2-}.

Molybdenum has 35 known isotopes of mass numbers 83 to 117, and 8 known isomers. There are seven naturally occurring isotopes: molybdenum-98 is the most abundant at 24.1 percent, molybdenum-96 (16.7 percent), molybdenum-95 (15.9 percent), molybdenum-92 (14.8 percent), molybdenum-97 (9.6 percent), molybdenum-100 (9.6 percent), and molybdenum-94 (9.3 percent). Molybdenum-92 and molybdenum-100 are radioactive but only weakly so, the former has a half-life of 1.9 × 10²¹ years and the latter of 8.5 × 10¹⁸ years, which is still more than a billion times longer than the age of the Earth.[109] The next longest-lived radioactive molybdenum isotope is molybdenum-93 with a half-life of 4,000 years.

[109] There are also minute traces of radioactive molybdenum isotopes in uranium minerals that have been produced as fragments of nuclear fission.

Element of surprises

1. The blades of some Japanese swords of the 1300s have been found to contain enough molybdenum for them to be considered as a type of moly steel, with all the strength and corrosion resistance this implies. Yet molybdenum was not recognised as a metal until the 1700s, not tested as an alloy for steel until the late 1800s, and not widely used until the last century. Whichever Japanese blacksmith chanced upon the benefits of adding molybdenum to iron, he kept the secret to himself and it appears to have died with him.

2. Molybdenum is the limiting factor for forest growth, not phosphorus as was previously assumed. This was realised only in 2009. Nitrogen uptake by trees depends on soil bacteria converting atmospheric nitrogen to ammonia and these bacteria need molybdenum for enzymes with which to carry out this essential process. Simply adding phosphate to the soil does not improve growth unless it contains traces of molybdenum, which normal fertilizer phosphate does in fact contain, and it was this which confused earlier researchers who were looking into the factors which affect tropical rainforests.

3. In 2010 a molybdenum catalyst was reported which can generate hydrogen gas directly from seawater using the energy of sunlight. Other methods of generating this gas this way—see titanium—require pure water to be used.

Neodymium

Pronounced nee-oh-dim-ee-um, the name is derived from the Greek *neos didymos*, meaning new twin.

French *néodyme*; German *Neodym*; Italian *neodimio*; Spanish *neodimio*; Portuguese *neodímio*.

Neodymium is one of 15 chemically similar elements referred to as the lanthanoid or rare earth elements, which extend from element atomic number 57 (lanthanum) to element atomic number 71 (lutetium). The term rare earth is a misnomer because some are not rare. The minerals from which they are extracted, and the properties and uses they have in common, are discussed under Lanthanoids.

Essential element

Once it was of interest only to chemists, but this century it has become a key component of devices ranging from mobile phones to wind turbines.

Cosmic element

The amounts of neodymium isotopes in rocks and lava been used to study how the geochemical evolution of the Earth's interior has evolved. Comparing the ratio of neodymium isotopes in surface rocks with the ratio in meteorites shows differences that indicate that the Earth's mantle divided into two layers and that neodymium moved to the lower layer. This happened within 30 million years of the planet forming.

Element of life

It is difficult to separate out the various amounts of the different lanthanoids in the human body, but they are present; the levels are highest in bone, with smaller amounts being present in the liver and kidneys. It is impossible to estimate the amount of neodymium in an average adult, and it is difficult to judge how much we take in every day in our diet, but it is probably only a few milligrams a year.

Neodymium has no biological role but it can have profound effects on parts of the body. For example, neodymium dust and salts are very irritating to the eyes. However, ingested neodymium salts are regarded as only slightly toxic if they are soluble, and non-toxic if they are insoluble.

Neodymium is not readily taken up by plant roots so not much gets into the human food chain. However, there tends to be more of this element in plants than of the other lanthanoids, and vegetables have on average 10 p.p.b. (dry weight) and some plants can have as much as 3,000 p.p.b. The level of neodymium in sewage is less that most of the other lanthanoids suggesting that what is taken into the body is preferentially retained.

Medical element

Tests on rats showed that an injection of 60 milligrams per kilogram of body weight of neodymium nitrate causes blood clotting to cease altogether, probably by interfering with the calcium ions that are needed for this to occur. It was thought that this anticoagulant property could be useful in treating human conditions, and much work was done trying to put this discovery to use, ultimately to no avail.

Element of history

Neodymium was discovered in Vienna, Austria in 1885 by Karl Auer (1858–1929) but it came at the end of a bewildering sequence of discoveries.

Didymium, the element which the Swede, Carl Gustav Mosander (1797–1858), claimed to have separated from cerium in 1839, turned out to be a mixture of lanthanoid elements. In 1879 samarium was isolated from didymium, followed a year later by gadolinium. In 1885 Auer obtained neodymium (that is, new didymium) and praseodymium (that is, green didymium) from it, confirming what had already been suspected from its atomic spectrum. This had been studied by the Czech chemist Bohuslav Brauner (1855–1935) at Prague in 1882 and was shown to vary according to the mineral from which didymium was extracted. At the time he made his discovery, Auer was a research student of the great German chemist, Robert Bunsen (of bunsen burner fame). Bunsen had become recognised as the world expert on the element didymium, and he accepted Auer's discovery immediately, whereas other chemists were to remain sceptical for some time.

A sample of the pure metal was first produced in 1925.

Economic element

Neodymium is found in minerals that include all the lanthanoid elements. The most important ore for neodymium is monazite but some bastnäsite ores are rich in it as well. China is currently the dominant producer, but there are large unexploited deposits in the USA, Brazil, India, Australia, Greenland, and Tanzania. Global reserves of neodymium are estimated to be in excess of 8 million tonnes. World production of neodymium oxide is about 20,000 tonnes a year, and the pure metal is obtained by reacting neodymium fluoride with calcium.

Adding a little neodymium metal to magnesium alloys greatly strengthens them, but this use has been overshadowed by a more important alloy, that of neodymium, iron and boron (often referred to as NIB). This was first produced in 1983 and was found to make excellent permanent magnets. This discovery made possible the miniaturisation of all kind of electronic devices and their manufacture now accounts for most neodymium that is produced with this element being found in mobile phones, microphones, loudspeakers, electronic musical instruments, and computer hard disks. It is also an essential part of starter motors, car window mechanisms, door locks, windscreen wipers, and fuel pumps. Wind turbines require powerful magnets and these are NIB magnets, and for every megawatt of electricity they generate, they require between 0.7 and 1 tonne of NIB.

Neodymium glass is made by adding the oxide Nd_2O_3 to molten glass and the result is a pleasing pale to deep lavender colour depending on the amount of neodymium added. Such glass was first made in the 1920s and described with names such as Alexandrite glass (after the colour of the gem stone), Wisteria glass, and Twilight glass. Early examples are highly prized. Neodymium produces this characteristic colour by absorbing light of wavelengths corresponding to yellow and green so we see the complementary colour which is purple. Neodymium is also included in the glass of tanning booths because it keeps out the heat of infrared rays while letting through those UV rays which cause the skin to darken.

Neodymium glass today is used for powerful lasers. Those known as Nd-YAG have crystals of yttrium aluminium garnet doped with about 1 percent of neodymium and have a wide variety of applications in eye surgery, cosmetic surgery and the removal of skin cancers. They are also used in range finders, and for cutting metals. In addition there are the HELEN lasers (high energy lasers embodying neodymium) which are used for research and can create plasmas with temperatures of around a million degrees.

In the chemicals industry neodymium oxide and nitrate are used as catalysts in the polymerising of so-called dienes which are used for manufacturing rubber.

Environmental element

Neodymium in the environment	
EARTH'S CRUST	38 p.p.m.
	Neodymium is the 26th most abundant element.
SOILS	average 20 p.p.m. (range 4–120 p.p.m.)
SEAWATER	4 p.p.t.
ATMOSPHERE	virtually nil.

Neodymium is the second most abundant of the rare earth elements (after cerium) and is almost as abundant as copper. It poses no environmental threat to plants or animals.

Neodymium plays an important role in geological analysis on both an historical timeline and for predicting a future event. Both involve analysing the ratio of neodymium isotopes to those of other elements. The future predictor refers to the analysis of recent lava from a volcano which can reveal information about the likelihood of a possible large scale eruption.

The historical dating record can be read from the ratio of neodymium to samarium, the former coming from radioactive atoms of the latter. This can date not only rocks but seabed sediments and in this last case has been able to show that changes in ocean currents—referred to as the oceans' conveyor belt system—do not trigger global warming as formerly suggested but that these currents only respond to changes in the climate after hundreds of years. The reason that neodymium provides such reliable information is that it is never complicated by being part of biological processes.

Chemical element

Data file

CHEMICAL SYMBOL	Nd
ATOMIC NUMBER	60
ATOMIC WEIGHT	144.24
MELTING POINT	1,021°C
BOILING POINT	3,070°C
DENSITY	7.0 g/cm³
OXIDE	NdO and Nd_2O_3

Neodymium is a bright silvery-white metal, and a member of the lanthanoid group (row 4f) of the periodic table. It quickly tarnishes in air, and so has to be protected either by storing it under oil or casing it in plastic. It reacts slowly with cold water, and rapidly with hot. Neodymium differs from most of the other lanthanoid elements in that it has three oxidation states: +2 (for example, $NdCl_2$), +3 (the principal one, for example, NdF_3), and +4 (for example, Na_3NdF_7).

There are 38 known isotopes neodymium with mass numbers 124 to 161, and 13 known isomers. There are seven naturally occurring isotopes of neodymium in ores: neodymium-142 accounting for 27 percent, neodymnium-144 (24 percent), neodymium-146 (17 percent), neodymium-143 (12 percent), neodymium-145 (8 percent), neodymium-148 (6 percent), and neodymium –150 (6 percent). Three of these are radioactive: neodymium-144 which has a half-life of 2.29×10^{15} years (2.29 million billion years), neodymium-148, half-life 3×10^{18} years (3 billion billion years), and neodymium-150, half-life 6.7×10^{18} years (6.7 billion billion years), Neodymium-144 is radioactive, albeit weakly so, with a half-life of 2,000 thousand billion years (2×10^{15}). The next longest-lived radioactive isotope is neodymium-147 with a half-life of 11 days.[110]

Element of surprises

1. Neodymium-iron-boron (NIB) magnets are so powerful that those handling them must wear protective glasses because they fly together with such force that they can shatter and send splinters flying in all directions. At times young people have used these industrial magnets to attach ornaments to their cheeks by putting one of the small magnets on the inside of the mouth. However, they have then proved impossible to pull apart, sometimes necessitating a visit to a hospital for surgical removal.

2. In the USA, NIB magnets have been used to check for counterfeit currency because they are powerful enough to detect the magnetic particles in the ink used to print real currency.

---

[110] There are also minute traces of other radioactive neodymium isotopes in uranium ores, where they have been formed by nuclear fission.

Neon

Pronounced nee-on, the name is derived from the Greek *neos,* meaning new.
French *néon*; German *Neon*; Italian *neo*; Spanish *neón*; Portuguese *neônio*.

Essential element

Neon in signs may brighten our lives but the rarity of this gas limits its use in all other respects.

Cosmic element

Neon is the fifth most common element in the Universe and it forms from the fusing of helium and oxygen. It is rare in the Earth's atmosphere, and in that of Jupiter.

Stars in our part of the galaxy that are like the Sun puzzled astronomers in that they appeared to have more neon than the Sun. That anomaly was solved in 2005 using spectral data from the Chandra X-ray Observatory. (This circles our planet and is located 139,000 km above the Earth's surface.) It showed that the abundance of neon on the Sun was the same as that of the stars.

Element of life

Neon is a harmless gas which can have no biological role on account of its inability to react with anything.

Element of history

In 1898, at University College London, William Ramsay (1852–1916) and Morris Travers (1872–1961) isolated krypton gas by evaporating liquid argon. They had been expecting to find a lighter gas which would fit a niche above argon in the periodic table of the elements. In June of that year they modified and repeated their experiment and allowed solid argon, surrounded by liquid air, to evaporate slowly under reduced pressure and collected the gas which came off first. This time they were successful and when they put a sample of the new gas into their atomic spectrometer the effect startled them by its brilliance. As Travers wrote:

The blaze of crimson light from the tube told its own story, and it was a sight to dwell upon and never forget. It was worth the struggle of the previous two years...

Ramsay's son, Willie, first suggested a name for the new gas, novum, from the Latin word *novus*, meaning new, and while his father liked the idea, he preferred to base it on the Greek word *neos,* and so it became neon.

Economic element

Neon is extracted from liquid air by fractional distillation and it comes off as a gas mixed with helium, traces of which are removed from it by being absorbed on to activated charcoal. Commercial needs for neon are met by only a few tonnes of the gas per year, although more could be produced because there are 65 billion tonnes of neon in the Earth's atmosphere.

Neon is used mainly in so-called neon signs, although only red ones are pure neon. These in particular are renowned for their intensity of colour which is why they are used as beacon lights, especially where there is likely to be fog. The first neon sign was made by Georges Claude in 1910, and demonstrated in Paris. They were ideal for advertising and by the 1930s they were to be seen in many towns and cities. They are quite robust and will operate without attention for up to 20 years or more.

There are other uses for neon, such as in diving equipment, lasers, high-voltage switching gear, and as a refrigerant for very low temperatures. While liquid neon cannot achieve the temperature of liquid helium (−269°C), it can maintain a temperature of −246°C and it has many times the refrigerant capacity of helium, but it is much more expensive with the result that vey little is used.

The earliest lasers were helium-neon ones although these have been replaced by laser diodes for most applications.

Environmental element

Neon in the environment	
EARTH'S CRUST	70 p.p.t.
	Neon is the 82nd most abundant element; in other words it is extremely rare.
SOILS	nil.
SEAWATER	0.2 p.p.m.
ATMOSPHERE	18 p.p.m. by volume.

Neon poses no threat to the environment, and indeed can have no impact at all because it is chemically unreactive and forms no compounds. It is not only to be found in the Earth's atmosphere. In 1909, Armand Gautier collected gas bubbling up from fumaroles near Vesuvius, and from hot springs near Naples, and showed that these contained neon.

Chemical element

Data file	
CHEMICAL SYMBOL	Ne
ATOMIC NUMBER	10
ATOMIC WEIGHT	20.1797
MELTING POINT	−249°C
BOILING POINT	−246°C

DENSITY	0.9 g per litre
OXIDE	none

Neon is a colourless, odourless gas and is one of the group of elements known as the noble gases which comprise group 18 (row 2p) of the periodic table. (The others are helium, argon, krypton, xenon, radon, and ununoctium.) Neon is unreactive towards all known chemicals, including fluorine. However, in the rarified atmosphere of a gas discharge tube, species such as NeH+ and other ions of transient existence have been observed.

There are 19 known isotopes of neon with mass numbers 16 to 34, and no known isomers. There are three main naturally occurring isotopes: neon-20, the most abundant at 90.5 percent, neon-22 (9.2 percent), and neon-21 (0.3 percent). In neon trapped in minerals of the Earth's crust the ratio of the various isotopes varies somewhat. For example, the neon trapped in granite contains relatively more of the heavier isotopes neon-21 and neon-22, whereas that trapped in diamonds, and that emitted by volcanoes is mainly neon-20.

There are also three radioactive isotopes which occur in minute amounts: neon-24, half-life 3.4 minutes and which is the longest lived radioactive isotope, neon-25 (half-life 602 milliseconds) and neon-26 (half-life 197 milliseconds). They are produced by the decay mode known as cluster decay and which was discovered in the 1980s. The element isotopes producing these are thorium-230, protactinium-231, and uranium-232 to 236, which occur in uranium bearing deposits.

Neon was important in proving the existence of isotopes. In 1913, J.J. Thompson (1856–1940) observed that a stream of neon atoms appeared to give two beams when exposed to a magnetic field and concluded that some neon atoms were heavier than others. He realised that these were the proof that isotopes existed, in other words atoms of the same element could have different atomic weights. The concept of isotopes had recently been suggested by Frederick Soddy as a way to explain the products of radioactive decay.

Element of surprises

1. J. Norman Collie (1859–1942) also laid claim to the discovery of neon, and is reputed to have said: 'If anyone happens to write an obituary for me I want two things said: I first discovered neon and I took the first X-ray photograph.' His assertion to have first produced neon is rather suspect, but Collie was a skilled research chemist based at University College London (UCL) where he did collaborate with Ramsay. Indeed he may have been among the first to *observe* neon's red glow.

    However, a book about his life, *The Snows of Yesteryear: J. Norman Collie*, written by William Taylor in 1973, says that Collie stumbled across neon independently while investigating the atomic spectrum of hydrogen at low pressures. If it were so, then it was unsubstantiated, and it is difficult to reconcile this with the well documented evidence that neon was first isolated by Ramsay and Travers. Yet Collie was a highly respected academic, and eventually became Professor of Organic Chemistry at UCL.

Such was Collie's fame as an investigator, and pioneer mountaineer, that he is reputed to have been the character on whom Conan Doyle based his famous fictional detective Sherlock Holmes, who was also a competent chemist.

2. Plasma globes contain neon. The dramatic effect of these was discovered accidentally by Bill Parker at the Massachusetts Institute of Technology in 1971 when neon gas escaped into a test chamber and came under the influence of a high voltage generated by an alternating current, and it had to be an alternating current. Suddenly coloured streams of light filled the chamber as the neon's electrons are excited to high energy orbitals when the current flows, and as the current reverses they fall back they release the light. Despite the high voltages, the glass spheres are safe to touch and then the streamers of plasma flow towards any finger which touches the globe.

Neptunium

Pronounced nep-tyoon-ee-um. It was named after the planet Neptune because the element preceding it, uranium, had been named after Uranus. Just as the planet Neptune comes after Uranus in the Solar system, so neptunium comes after uranium in the periodic table of the elements. Neptune was discovered in 1846 and named after the Roman god of the sea.

French *neptunium*; German *Neptunium*; Italian *nettunio*; Spanish *neptunio*; Portuguese *neptúnio*.

Essential element

Neptunium-237 is converted to plutonium-238 which is used as an energy source for spacecraft.

Element of life

Neptunium has no role to play in living things, and it is never encountered outside nuclear facilities or research laboratories. Tests on animals showed that it was not absorbed through the digestive tract, and when injected into the body it tended to accumulate in the bones, from where it was only slowly released.

Element of history

In 1934, Odolen Koblic (1897–1959) of Czechoslovakia thought he had extracted a small amount of a new element from the wash water from the roasted uranium ore pitchblende. He said it was element 93, and he named it bohemium (symbol Bo) in honour of his native land, Bohemia. He sent samples of his material to Ida Tacke-Noddack (1896–1978), an expert in chemical analysis, who carried out chemical and spectroscopic investigations and informed him that it was a mixture of tungsten and vanadium. Although we now know that this mineral contains minute traces of neptunium, it is most unlikely that Koblic had separated any of it from the ore.

In early 1934, Enrico Fermi (1901–54) in Italy, tried to produce elements 93 and 94 by bombarding uranium with neutrons, and claimed to have been successful. He proposed the name ausonium (symbol Ao) for element 93, this being the Classical Greek name for his homeland Italy. In September of that year Ida Tacke-Noddack questioned Fermi's discovery of elements 93 and 94, pointing out that not only had he not discovered them, but that he hadn't done a complete analysis of what he had actually achieved, which was the fission of uranium into lighter elements. Fermi had in fact discovered nuclear fission but had not realised it.

In 1938, Romanian physicist Horia Hulubei (1896–1972) and French chemist Yvette Cauchois (1908–99) claimed to have discovered element 93, basing this on the spectra of elements in mineral samples. The proposed the name sequanium (symbol Sq) derived from Sequana, the Latin name for the river Seine. This claim was criticised on the grounds that element 93 did not occur naturally so there was no reason to believe them. However, as we now know that neptunium is naturally occurring, it is not impossible for these two researchers to have found proof of its existence.

Neptunium was discovered by Edwin M. McMillan (1907–91), with the help of Philip H. Abelson (1913–2004), at Berkeley, California.[111] McMillan had bombarded a uranium target with slow neutrons in the cyclotron and noted that some unusual beta-rays were then emitted from it, showing a new isotope was present. It was left to Abelson, who spent a week there in May 1940, to prove they came from a new element. He was the right person to do this, having worked in the Radiation Laboratory at Berkeley for his PhD, which he had been awarded in 1939. Within three days of being given a sample of material by McMillan, he had proved that it was a new element and within the week they had sent off a paper to the journal *Physical Review* announcing their discovery, and this was published later that year.

McMillan had made neptunium from uranium-238 by converting this to uranium-239 with neutron addition after which it underwent loss of a negative electron from the nucleus (beta-emission) thereby increasing the nuclear charge by one so that element 92 became element 93. McMillan named it neptunium. This isotope, neptunium-239, had a half-life of 2.4 days. A cloak of secrecy then descended on the work because of national security, although during World War II work continued on neptunium. In 1944, the first pure compound, the oxide NpO_2, was made using a few micrograms of the element that had been extracted from a nuclear reactor.

McMillan shared the 1951 Nobel Prize for chemistry with Glenn T. Seaborg (1912–99) for their discoveries of transuranium elements.

Economic element

Today, neptunium-237 is extracted in kilogram quantities from the spent uranium fuel rods of nuclear reactors or as a by-product of plutonium production. This isotope is long-lived, gives off only weak radioactivity (in the form of alpha-radiation) and can be handled in an ordinary laboratory given the necessary safety measures. Neptunium-237 has been used in neutron detectors. Some neptunium is irradiated with neutrons to form plutonium-238 which is used as an energy source for spacecraft.

Chemical element

Data file	
CHEMICAL SYMBOL	Np
ATOMIC NUMBER	93
ATOMIC WEIGHT	237.0 (isotope Np-237)

[111] Abelson became best known as the editor of *Science,* a post he held from 1961 to 1984.

MELTING POINT	640°C
BOILING POINT	3902°C
DENSITY	20.3 g/cm³
OXIDE	NpO, NpO_2, Np_2O_5, NpO_3

Neptunium is a radioactive member of the actinide group (row 5f) of the periodic table of the elements. The metal itself, which is silver in colour, can be made from neptunium fluoride (NpF_3) by reaction with molten lithium or barium at 1,200°C. Chemically it is very reactive and is attacked by oxygen, steam, and acids, but not alkalis. It can exist in several oxidation states: +3, +4, +5, and +6 all of which give different coloured solutions, namely purple Np^{3+}, yellow Np^{4+}, green NpO_2^+, and pink NpO_2^{2+}. Typical compounds are NpF_3, $NpCl_4$, NpF_5, etc., and in exhibiting these various oxidation states it is chemically similar to uranium.

There are four neptunium isotopes that occur naturally and these are the isotopes neptunium-237 to 240. They occur in uranium-bearing deposits where they are produced by a series on neutron-capture and beta-decay reactions. Neptunium-239 and 240 are the main ones.[112]

There are 20 known isotopes of neptunium with mass numbers 225 to 244, and four known isomers. The longest-lived isotopes are neptunium-237 with a half-life of 2.14 million years, neptunium-236 with a half-life of 155,000 years, and neptunium-235 with a half-life of 396 days. All the other isotopes have half-lives of less than a week.

Neptunium-237 is fissile and several countries considered the possibility of using this isotope for nuclear weapons. However, no test explosion using it has ever been carried out.

Element of surprises

1. Although neptunium-237 has a half-life of 2.14 million years, this does not mean that any of the neptunium that was around when the Earth was formed, 4.57 billion years ago, could still be around to day. Even if there had been a *trillion* tonnes of neptunium (10^{12} tonnes) it would all have undergone radioactive decay within 240 million years. (A trillion tonnes of neptunium contains 2.5×10^{33} atoms. After 2.14 million years this would be reduced by a half to 1.25×10^{33}, after 4.28 million years it would halved again to 6.25×10^{32}, and so on, until after 111 half-lives, which amount to 237 million years, it would be down to a single atom.)

2. Every home with a smoke detector will possess some neptunium, because this is the element to which the americium in such detectors is transforming as it emits the alpha particles that detect the presence of smoke. About 0.2 percent per year of the americium converts to neptunium.

---

[112] The various processes which produce neptunium are as follows. Uranium-238 captures a neutron to become uranium-239 which then emits a beta particle to become neptunium-239 and this isotope can then capture a neutron to become neptunium-240. Naturally occurring primordial plutonium-244 alpha decays to uranium-240 which then beta decays to neptunium-240. Uranium-237, which occurs in small amounts, beta decays to neptunium-237. Natrually occurring americium-241 alpha decays to neptunium-237. Neptunium-238 is occasionally produced by neutron capture by neptunium-237.

Nickel

Pronounced nik-el, the name is a shortened form of the German *kupfernickel*, meaning either the Devil's copper or St Nicholas's copper.

French *nickel*; German *Nickel*; Italian *nichel*; Spanish *niquel*; Portuguese *niquel*.

Essential element

Nickel alloys are remarkable metals and can resist corrosion, work at very high temperatures, do not expand on heating, and remember previous shapes. And nickel is the battery material of the future.

Cosmic element

A lot of the nickel that is mined on Earth arrived here in the form of giant meteorites. One of these hit the Sudbury region of Ontario, Canada, hundreds of millions of years ago and is estimated to have contained 200 million tonnes of the metal. This deposit was discovered in 1856 by construction workers who were building the Pacific Railway and it was originally regarded as a potential source of copper. However, by 1905 it had become the world's largest source of nickel, once a method of extracting this metal and refining it had been developed. To access this deposit 17,000 holes were drilled into the rock and each charged with dynamite—the preparations took more than a year—and then exploded together releasing 4 million tonnes of rock for mining.

In 2005 the Mars rover called Opportunity found a nickel-iron meteorite, the first such object ever identified on another planet.

Element of life

Nickel in the human body	
BLOOD	levels vary between 10–50 p.p.b.
BONE	not more than 0.7 p.p.m.
TISSUE	1–2 p.p.m.
TOTAL AMOUNT IN BODY	15 mg.

Nickel has been shown to be essential to some species, and it is related to growth. It may even be an essential element for humans, but if so, its requirement could be as little as 5 micrograms (μg) a day, although the daily intake is estimated to be around 150 μg. In some species it becomes incorporated into enzymes and in the soil bacterium *Streptomyces* it is part of *superoxide dismutases*, enzymes whose role is to convert dangerous superoxide radicals to safer molecules such as hydrogen peroxide and water.

Nickel occurs in some beans, such as those used to make canned baked beans, where it is a component of enzymes. One such enzyme is the *Jack bean urease,* each molecule of which contains 12 nickel atoms. Another relatively rich source of nickel is tea which has 7.6 p.p.m. in the dried leaves. Other plants generally have less than half this amount.

Once it enters the body, nickel binds itself to albumin and thereby moves around. Most tends to accumulate in the lungs, liver, and kidneys. Unwanted nickel is excreted via the urine.

Dangerous element

Nickel presents some health hazards to humans, and these are of three kinds: close contact with nickel-containing metal can cause dermatitis; breathing in nickel dust can cause lung and nasal cancer; inhaling nickel carbonyl vapour, even in tiny amounts, can kill. This last chemical immediately produces a painful sensation in the throat and chest followed by more serious symptoms depending on the degree of exposure, but rapid treatment with the antidote DCC (short for diethyldithiocarbamate) will usually save a person's life.

The second and third of these threats are mining and industrial hazards respectively, but the first hazard is one that affects ordinary people, some of whom are very sensitive to nickel. Even the nickel in stainless steel is able to cause dermatitis, which begins as an irritating itch known as 'nickel itch'. It can be caused when stainless steel wrist watches, spectacle frames, garment fasteners, and jewellery come in contact with the skin. It is exacerbated by sweat, the acids of which dissolve a little of the nickel.

Nickel is thought to cause cancer by substituting for zinc and magnesium atoms in *DNA polymerase.* The nickel ion is slightly different and so affects the behaviour of this enzyme perhaps causing it to bind the wrong nucleotide, resulting in the formation of a rogue sequence of DNA and a cancerous cell.

Poisoning by ingesting nickel compounds is almost unknown, but a two-year-old child who ate 15 grams of nickel sulfate crystals died within a few hours.

Element of history

Meteorites contain both iron and nickel, and earlier ages used them as a superior form of iron, working them into tools and swords which were not unlike the stainless steel of today. Because the metal did not rust, it was regarded by the natives of Peru as a kind of silver. A zinc-nickel alloy called *pai-t'ung* (white copper) was in use in China as long ago as 200 BC and exported to the Middle East, from where some even reached Europe. The ore from which it was made came from Yunnan in southern China where there are nickel sulfide deposits.

A reddish-brown ore was known to German copper miners and they gave it the nickname *kupfernickel* which most probably meant devil's copper because it was good for nothing except colouring glass green. Early mineralogists tried in vain to extract copper from it. Then, in 1751, Alex Fredrik Cronstedt (1722–65), working in Stockholm, investigated a new mineral, now called nickeline (NiAs), from a cobalt mine at Los, Hälsingland, Sweden. He too thought it might contain copper but he could not extract any of this metal from it. What he did eventually get from this (and from kupfernickel) was a new metal which he christened nickel in 1754. For many years many chemists would not accept his claim that it was a new metal element and believed it was an alloy of cobalt, arsenic, iron and copper. This seemed likely as these elements were present in the new metal as trace contaminants. It was not until 1775 that absolutely pure nickel was produced by Torbern Bergman (1735–84) and this settled any doubts about its elemental nature.

There was little demand for nickel until 1844 when silver plating was developed, and then nickel was found to be the most desirable base metal on which to deposit the more expensive metal. Thereafter electroplated nickel silver cutlery (EPNS) became popular.

Economic element

Nickel has a particular affinity for sulfur, and sulfide ores are an important source of the metal. Millerite (NiS) is one example but this ore is rare. Most ores from which nickel is extracted are iron/nickel sulfides, such as pentlanite, $(Fe,Ni)_9S_8$. Nickel was first extracted from an ore called nickeline (nickel arsenide, NiAs) but this is also now rare.

Nickel minerals are mined in more than 20 countries and especially Russia (300,000 tonnes), Canada (250,000 tonnes), Indonesia (190,000 tonnes), Australia (190,000 tonnes), New Caledonia[113] (125,000 tonnes), and Colombia (100,000 tonnes). The Ontario deposit mentioned in the *Cosmic element* section above still supplies about 15 percent of the world demand for the metal. Global annual production is around 1,500,000 tonnes and easily worked reserves amount to around 140 million tonnes.

The extraction and manufacture of pure nickel from its ores was greatly improved in 1888 when the industrialist Ludwig Mond (1839–1909) and his assistant Carl Langer (1859–1935) investigated the problem of leaky valves made of nickel through which carbon monoxide gas was passing. They discovered that the gas was reacting with the nickel to form nickel carbonyl, $Ni(CO)_4$, which is volatile and boils at 43°C. From this they developed a process, the Mond Nickel Process, in which impure nickel is treated with CO gas to form the nickel carbonyl. This is then passed into a chamber where pellets of nickel are kept at a temperature high enough to release the nickel which deposits on their surface, so they grow in size. (The CO is reused.) The process is potentially very hazardous but it was brought into commercial production, despite several cases of poisoning, including some deaths, among operatives.

About half the nickel produced each year ends up alloyed with steel to give stainless steel, whose composition can vary but is typically iron with around 18 percent

---

[113] This country supplied almost all the world's nickel for 40 years from 1875 to 1915.

chromium and 8 percent nickel. Famous buildings may be partly clad in it such as the iconic Chrysler Building in New York, and today it can be seen around the world on roofs of galleries and airport terminals, and especially for lifts (elevators). Although stainless steel is generally a shiny grey metal, its surface can be almost any colour, depending on how it is treated.

Nickel is easy to work and can be drawn into wire. It resists corrosion even at high temperatures and for this reason it is used in gas turbines and rocket engines. Other important uses of nickel are in metal plating, industrial catalysts, and for coinage. Nickel coins were introduced in Belgium in 1860, and today many countries still use coins containing nickel. The five-cent US coin, the so-called 'nickel', is copper alloyed with 25 percent nickel, and was first produced in 1865.[114]

Most nickel ends up in alloys, some of which have remarkable properties. Invar (short for invariable) is 64 percent iron and 36 percent nickel; it got its name because it has the curious ability of not expanding when heated. Invar was discovered by Charles Guillaume of the Bureau of Weights and Measures in Paris who was seeking a metal with which to make copies of the standard metre. It was also ideal for use in tape measures, and as parts of clock mechanisms and chronometers.

Nichrome is a nickel/chromium alloy with between 11 and 22 percent chromium, plus small amounts of silicon, manganese and iron. Nichrome does not oxidise even at red heat and is used in appliances such as toasters and electric ovens. Monel is an alloy of nickel and copper (for example, 70 percent nickel, 30 percent copper, with traces of iron, manganese and silicon) which is not only hard but will resist corrosion by seawater, so that it is ideal for propeller shafts in boats and in desalination plants. It is also unaffected by acids, so that it is used in the chemical and food processing industries.

INCO-276 is made of 57 percent nickel with 16 percent each of chromium and molybdenum, plus a few percent of other metals. This alloy resists corrosion by hydrogen sulfide gas, which normally attacks stainless steel, and it is used for boring wells that go deep into the Earth's crust where this gas is often encountered.

The alloy MM002 is made of 60 percent nickel with around 10 percent each of cobalt, tungsten and chromium plus small amounts of several other metals. It is used for the new generation of turbine blades which are grown as single crystals in special moulds.

The alloy N09706 is a so-called super-alloy and is composed of nickel (41 percent), iron (39 percent) and chromium (15 percent) with small amounts of other metals like niobium, manganese, and titanium. It is being produced at more than 12,000 tonnes a year for heavy-duty land-based gas turbines which can operate at 1,300°C.

Finely powdered nickel is used as a catalyst in the hydrogenation of oils to fats. In the 1890s the French chemist Paul Sabatier (1854–1941) showed that it was possible to convert edible oils into fats for making margarine.[115] Nickel powder is also added to paints that coat the housing of sensitive electronic instruments because it shields

---

[114] Today the coin's metal content is worth twice its face value, but melting these coins down is a serious criminal offence in the USA.

[115] Margarine had first been produced by Hippolyte Mège-Mouriés (1817–80) and it name is derived from *margaron*, the Greek word for a pearl.

them from stray radiation. Rechargeable nickel-cadmium batteries are made from sintered nickel powder.

NiMH (nickel metal hydride) batteries are used in hybrid vehicles and they are compact in size, reliable, and have a long life-expectancy. They use nickel oxyhydroxide doped with various other elements as the positive electrode; the electrolyte is aqueous potassium hydroxide; and the negative electrode is a nickel hydride alloy. Hydrogen is absorbed at the negative electrode to form the nickel metal hydride during charging. During discharge, the hydrogen is released and oxidised to water.

Environmental element

Nickel in the environment	
EARTH'S CRUST	av. 80 p.p.m.
	Nickel is the 23rd most abundant element in the Earth's crust.
SOILS	av. 20 p.p.m.
SEAWATER	0.5 p.p.b.
ATMOSPHERE	insignificant.

Most nickel on Earth is inaccessible because it is locked away in the planet's iron-nickel molten core which is 10 percent nickel. Indeed, a lot of the nickel that is found naturally in soils and in the seas has been deposited on the Earth from the 'rain' of space dust and meteorites that the planet scoops up each year in its journey round the sun. The total amount of nickel dissolved in the sea has been calculated to be around 8 billion tonnes (8×10^{12} kg). Organic matter has an ability to absorb the metal which is why coal and oil contain measurable amounts.

The nickel in soil can be as low as 0.2 p.p.m. and as high as 450 p.p.m. in some clay and loamy soils. The average is around 20 p.p.m. Heavily fertilized soil may have its nickel content raised since phosphate fertilizers contain traces of the metal and farmland near oil and coal burning industries can have raised levels. Sewage sludge from industrial areas may increase the level, and in some soils fertilized with it, nickel can exceed 800 p.p.m. In parts of Canada near metal-processing industry it has been recorded as high as 26,000 p.p.m. (2.6 percent).

Plants can tolerate a certain amount of nickel in the soil in which they grow but on heavily contaminated soil some crops will fail. Beans and maize are very sensitive to nickel and even watering them with a 40 p.p.m. solution of nickel will kill them. Clover will not grow on soil that has 80 p.p.m. of nickel, but oats will cope with such levels quite happily and absorb some of the nickel while doing so. Liming such soil, which raises the pH, will reduce nickel absorption dramatically. Some plants will tolerate extremely high levels of nickel and the flower *Hybanthus floribundus* will accumulate as much as 6,000 p.p.m., which is 0.6 percent (dry weight) in its leaves and stems, suggesting that this might some day in the future be used to decontaminate polluted soil while acting as a source of reclaimable nickel.

Chemical element

Data file	
CHEMICAL SYMBOL	Ni
ATOMIC NUMBER	28
ATOMIC WEIGHT	58.6934
MELTING POINT	1,453°C
BOILING POINT	2,732°C
DENSITY	8.9 g/cm³
OXIDES	NiO, Ni_2O_3

Nickel is a silvery, lustrous, malleable, ductile metal and a member of group 10 (row 3d) of the periodic table. It resists oxidation, but is soluble in acids (except concentrated HNO_3), yet unaffected by alkalis. The bulk metal is ferromagnetic. Nickel's preferred oxidation state is +2, as in $NiCl_2$, $NiSO_4$, and Ni^{2+} in solution. Its compounds can also exhibit other oxidation state such as 0 in $Ni(CO)_4$, and +3 in Ni_2O_3 which is used in some rechargeable batteries.

There are 32 known isotopes of nickel of mass numbers 48 to 79, and eight known isomers. There are five naturally occurring isotopes, four of which are stable and one is radioactive, albeit with an incredibly long half-life. These isotopes are nickel-58 which accounts for 68 percent, nickel-60 (26 percent), nickel-62 (4 percent), nickel-61 (1 percent), and nickel-64 (1 percent). The radioactive isotope is the most abundant one, and nickel-58 has a half-life of 7×10^{20} years, which is almost a trillion times longer than the age of the Earth. The next longest-lived radioactive isotope is nickel-59 with a half-life of 75,000 years. The isotope nickel-48, first discovered in 1999, is the most proton-rich isotope known of any element, having 28 protons to 20 neutrons.[116] Both these numbers are so-called 'magic numbers' for protons and neutrons. The half-life of nickel-48 is around a microsecond.

Element of surprises

1. Perhaps the most remarkable alloy of nickel is the one it forms with aluminium, called nickel aluminide, which has the exact chemical composition Ni_3Al and as such has an ordered array of atoms in its structure. It was first developed at the US Government's Oak Ridge National Laboratory at Tennessee, but it was found to be too brittle and research on it was abandoned until it was discovered that adding a fraction of a percent of boron remedied this failing and created a material of superior strength. What makes nickel aluminide so special is its unique property at high temperatures. It is six times stronger than stainless steel, and it actually gets stronger the hotter it gets. At 800°C it is twice as strong as it is at room temperature. The hotter an engine runs the more efficient it becomes, and this alloy

---

[116] Stable isotopes generally have the same number of neutrons as protons or even more, and the heavier the element the greater the excess of neutrons, e.g. lead has 82 protons and 126 neutrons.

will allow engines to run at red heat temperatures, above 1,000°C. It may one day even become part of the engine of the family motor car. Before then it will be incorporated into rockets, high performance jet engines, and heat-exchangers.

2. The alloy of nickel, manganese, and gallium (Ni_2MnGa) is known as memory shape alloy because when distorted it will snap back to its original shape if heated or exposed to a strong magnetic field. Another curious alloy of nickel which can also remember a previous shape is nitinol (55 percent nickel, 45 percent titanium which corresponds to the chemical formula NiTi) which was developed in the USA in the 1960s and has been commercially successful. The name is an acronym of Nickel Titanium Naval Ordnance Laboratory. Spectacle frames made from nitinol can be bent and twisted and when released they will jump back to their original shape. In 2006, General Motors of the USA took out a patent (US 7,063,377) for nitinol car hoods (bonnets) which would help minimise accident crash damage by absorbing a lot of the energy of impact.

3. A nickel disc has been developed which will store up to 100,000 pages of data and should remain stable for 1,000 years—and it will also resist temperatures of 800°C. The disc, called HD-Rosetta, stores information as actual words of text, which means that it will always be possible to extract its information, albeit with the aid of a powerful microscope.

Niobium

Pronounced ny-o-bee-um, and called after *Niobe* of Greek mythology.[117] In fact the discoverer of niobium called it columbium after the mineral from which it was first obtained and this name continued to be used for well over 150 years until, in 1950, the International Union of Pure and Applied Chemistry (IUPAC) decided that niobium, the alternative name suggested in 1844, should be the official one. Columbium is still used in some areas, such as engineering.

French *niobium*; German *Niob*; Italian *niobio*; Spanish *niobio*; Portuguese *nióbio*.

Essential element

This element is used in superconducting magnets and is also alloyed to other metals to make them heat resistant, hence its use for the nozzles of space rockets.

Element of life

Niobium in the human body	
BLOOD	5 p.p.b.
BONE	70 p.p.b.
TISSUE	140 p.p.b.
HAIR	2 p.p.m.
TOTAL AMOUNT IN BODY	1.5 mg.

Niobium has no known biological role although there is enough in humans for it to be measurable. Apart from measuring its concentration, no research on such niobium has been undertaken, but some animals were given radioactive niobium-95 to see how this element was digested. These tests showed that less than 5 percent of the niobium was absorbed into the body, and in most animals only 1 percent was absorbed. That which did pass through the gut wall tended to accumulate mainly in the bone, and also in the liver and kidneys. Niobium was excreted via the urine.

Dangerous element

Niobium and its compounds are regarded as somewhat toxic but there are no reports of humans being poisoned by it, although in Russia there is a limit of 10 p.p.b. for niobium in drinking water. Niobium dust causes eye and skin irritation.

---

[117] She was the daughter of king *Tantalus* after whom tantalum, the metal below niobium is named.

Medical element

The alloys of niobium with tin or titanium are superconducting, albeit only below −250°C, but when they were discovered they revolutionised the technique of magnetic resonance which requires very strong magnetic fields. This is used by research chemists to analyse compounds and by hospital doctors for full-body diagnostic MRI scans. These alloys are noted for their superconductivity even when carrying high current density, whereas the metal oxide superconducting materials, which can operate at much higher temperatures, tend to lose this necessary property when exposed to high electrical current.

Niobium alloys are used in surgical implants, such as pacemakers, because these metals are tolerated by human tissue. When treated with sodium hydroxide, a porous layer forms on their surface which is compatible with bone and will bond to it.

Element of history

Niobium was discovered in 1801 by the English chemist Charles Hatchett (1765–1847) in London. When examining some minerals in the British Museum collection he was intrigued by a small, dark, heavy specimen labelled columbite, which had been part of the collection of Hans Sloane (1660–1753), who had bequeathed it to the museum many years before and which had been sent to him by John Winthrop of Connecticut early in the 1700s.

Hatchett heated the mineral with potassium carbonate and then dissolved the product in water from which an oxide precipitated when the solution was neutralised with acid. The filtered solution contained iron and manganese, which he also confirmed were present, but it was the precipitate on which he focused his interest and which he deduced must contain a new metal although he failed to extract this in a pure form.

He announced his discovery to the Royal Society of London in November 1801, naming the element columbium, the poetic name for America. For many years it was disputed whether columbium was a true element or just another form of tantalum, which was discovered the following year. The metals are chemically similar, occur together in nature, and are difficult to separate. In 1844 the German chemist Heinrich Rose (1795–1864) proved that the mineral columbite contained both elements and he renamed columbium niobium in honour of Niobe, the Greek goddess of grief.[118] Nevertheless, confusion continued and also involved an element then known as ilmenium, purported to have been discovered in 1846. This later turned out to be a mixture of niobium and tantalum, but ilmenium only disappeared from the chemical literature in 1871.

---

[118] Perhaps understandably so because all her 12 children were murdered.

Hatchett died before a sample of the pure metal was produced. That was achieved only in 1864 by the Swedish chemist Christian Wilhlem Blomstrand (1826–97) who reduced niobium chloride to niobium metal by heating it with hydrogen gas.

Economic element

Niobium and tantalum are mined together. The chief mineral is columbite, also known as niobite, and it composition is approximately $(Fe,Mn)(Nb,Ta)_2O_6$ which can vary according to the proportions of the four metals that are present. Another mined mineral is pyrochlore, $(Na,Ca)_2Nb_2O_6(OH,F)$, and some niobium is obtained as a by-product of tin-mining. Niobium and tantalum are separated by chemical means, such as solvent extraction. Total world production of these ores is 230,000 tonnes of which the niobium content is 60,000 tonnes. (The tantalum component amounts to 1,000 tonnes.)

The main mining areas are Brazil, which produces 210,000 tonnes of ore, Canada (10,000 tonnes), and Australia (1,000 tonnes). The known reserves are reported to be 3–4 million tonnes, mainly as pyrochlore.

Niobium is added in small amounts to improve stainless steel, especially if it is to be welded, and welding rods also contain niobium. Small amounts of niobium impart greater strength to other metals, especially those that are exposed to low temperatures. The alloy with zirconium is particularly resistant to chemical attack, especially to corrosive alkalis, and even to molten lithium and molten sodium. Because niobium has low neutron-capture it was the metal of choice for tubing carrying these molten metals which were used in fast breeder nuclear reactors.

Niobium coated with platinum group metals are used as industrial anodes, while niobium-coated copper anodes are used to give cathodic protection to structures which are in contact with harsh chloride-containing solutions and these anodes have been found to last up to 20 years. Rocket nozzles are made of niobium alloys because they resist corrosion even when they become red hot.

Environmental element

Niobium in the environment	
EARTH'S CRUST	20 p.p.m.
	Niobium is the 33rd most abundant element.
SOILS	av. 24 p.p.m., maximum 300 p.p.m.
SEAWATER	0.9 p.p.t.
ATMOSPHERE	virtually nil.

Plants generally show only traces of niobium and many have none at all, although some mosses and lichens can have 0.45 p.p.m. However, plants growing near niobium deposits can accumulate the metal to levels above 1 p.p.m.

Chemical element

Data file	
CHEMICAL SYMBOL	Nb
ATOMIC NUMBER	41
ATOMIC WEIGHT	92.90638
MELTING POINT	2,468°C
BOILING POINT	4,742°C
DENSITY	8.6 g/cm³
OXIDES	Nb_2O_5 (also NbO and NbO_2)

Niobium is a shiny, steel-grey metal, which is soft when pure. It is a member of group 5 (row 4d) of the periodic table. Niobium resists corrosion due to an oxide film on the surface. The metal is inert to acids, even to aqua regia, at room temperature. However, it is attacked by hot, concentrated acids, and especially by alkalis and oxidising agents. Its preferred oxidation state is +5 as in $NbCl_5$ although oxidation states +3 and +4 are also known such as in NbF_3 and NbF_4.

There are 35 known isotopes of niobium with mass numbers 81 to 115, and 25 known isomers. There is only one main naturally occurring isotope, niobium-93, which is not radioactive. There are minute traces of other isotopes in uranium ores where they are produced as a result of nuclear fission. Among them is niobium-91 with a half-life of 680 years, and niobium-94 with a half-life of 29,300 years, as well as the longest-lived radioactive isotope, niobium-92, which has a half-life of 34.7 million years. There are also traces of niobium-95 and niobium-97.

Element of surprises

Niobium is used in jewellery and is popular with some people because, unlike some other metals, it is hypoallergenic. Its surface can be given a variety of colours by anodising so that it shimmers with iridescent colours. The metal is protected by a thin film of oxide on its surface and which is normally transparent. However, if it is thickened it will reflect light with an interference pattern that make it appear multi-coloured. The interference comes because the waves of light reflected from the oxide surface are not in phase with those reflected from the underlying metal surface. The degree of interference depends on the thickness of the oxide layer and the angle at which it is viewed. The oxide layer is artificially thickened by the process of anodising in which the item of jewellery acts as the anode in an electrolytic cell (with ammonium sulfate as the electrolyte solution). Depending on the length of time it is exposed to the current almost any colour can be produced, but the niobium blues and purples are particularly attractive. By masking parts of the object, it is possible to produce items with more sophisticated colours and patterns.

Nitrogen

Pronounced ny-troh-jen, the name is derived from the Greek words *nitron genes*, meaning nitre forming. Nitre was the old name for potassium nitrate, KNO_3, which was also known as saltpetre.

French *azote*; German *Stickstoff*; Italian *azoto*; Spanish *nitrógeno*; Portuguese *nitrogênio*. (The German word derives from *ersticken* and *stoff* meaning 'suffocating substance.' Nitrogen was also known as 'choke-damp' in England until a Dr Chaptal suggested the name nitrogen.)

Other words associated with nitrogen are azo, azide, ammonia, amino, and amine. The first two come from the French word *azote* meaning 'without life' and was chosen by Lavoisier, and it is from this root that words like azo-dye ultimately derive. Ammonia is the name of the gas NH_3 and the word is derived from Ammon, a god of ancient Egypt, because it was from there that ammonium chloride was to be obtained; this was called *sal ammoniac*, meaning salt of Ammon. The words amino acid and amine refer to derivatives of ammonia.

Essential element

This abundant but unreactive gas is needed by every living thing as well as being a key component of many highly reactive molecules. Making nitrogen react with hydrogen was a major breakthrough for the chemical and agrochemical industries to the extent that a third of all the food crops grown now rely on the ammonia that the reaction produces.

Cosmic element

Nitrogen as N_2 molecules has been detected in interstellar space, as have some simple nitrogen compounds. Like the atmosphere of planet Earth, nitrogen gas is also the chief component of the atmosphere of Saturn's moon Titan, but the atmosphere of Mars contains only 2.6 percent.

Element of life

Nitrogen in the human body	
BLOOD	34,000 p.p.m. (3.4%)
BONE	43,000 p.p.m. (4.3%)
TISSUE	72 p.p.m. (7.2%)
TOTAL AMOUNT IN BODY	1.8 kg.

Although nitrogen is essential to life, the dilemma is that its most stable form is unreactive nitrogen gas (N_2) which accounts for 78 percent of the atmosphere. Life depends on this being converted to ammonia, a process known as nitrogen fixation, and it is from this molecule that almost all other nitrogen-containing molecules are derived.

Nitrogen is a constituent of DNA and, as such, is part of the genetic code. Nitrogen is also a key part of the many amino acids that form enzymes and other proteins. Amino acids are so called because they contain both an amino group (NH_2) and an acid group (CO_2H). They can polymerise to long chains of protein by reacting the amino group of one molecule with the acid group of another.[119] Nitrogen is a component of many other biologically important molecules such as the haem molecule that makes up haemoglobin, and acetylcholine, which is the neurotransmitter that passes messages from one nerve ending to another.

Food element

Plants need either ammonia or nitrate, and there is a natural 'reservoir' of these in the soil in the form plant debris (compost), animal excreta (manure), soil fauna, and microbes which live in the soil. A little nitrate even arrives with rainwater, which dissolves the nitrogen oxides produced in thunderstorms and from combustion. Some plants can draw on atmospheric nitrogen directly, such as legumes like peas and beans, thanks to rhizomes growing as nodules on their roots. These take carbohydrate from the plant in return for providing ammonia. The key to boosting crop yields for human consumption is to add ammonium nitrate, NH_4NO_3, directly to the soil as fertilizer, and 80 percent of its production is used for this purpose to the extent that around 2 billion humans are now indirectly reliant on this source of dietary nitrogen.

Nitrate may be part of the natural system, but for many years it was suspected of causing stomach cancer because it can be reduced to nitrite (NO_2^-) and thence to nitrosamines which are known carcinogens. This led to the passing of laws limiting the amount of nitrate in drinking water although the dangers were much exaggerated. Nitrate occurs naturally in many vegetables, such as lettuce and spinach, and it is even produced by microbes in the human gut, with the result that only a small part of the nitrate in the body comes from drinking water.

Dangerous element

Nitrogen is non-toxic unless it excludes oxygen from the lungs. However, it can cause problems when inhaled at high pressures by deep-sea divers. The gas is slightly soluble in blood and when the pressure is then released it forms bubbles of the gas which can block blood vessels, a condition is known as the 'bends'. Some nitrogen compounds are highly toxic, such as hydrogen cyanide (aka cyanic acid, HCN) which can kill within minutes by blocking the enzyme that provides the heart muscle with energy.

---

[119] These chains are then attracted to one another by hydrogen-bonds—see hydrogen—and it is these that enable the protein to adopt and hold a particular shape.

Nitrogen's most dangerous substances are those involved in explosives. Ammonium nitrate can be explosive and cause widespread damage and loss of life. The first such explosion on a large scale occurred at Oppau, Germany, where, on 21 September 1921, it destroyed the Haber Bosch plant killing 430 people. An even larger disaster occurred on 15 April 1947, when a ship in Texas City harbour caught fire and its load of 5,000 tonnes of ammonium nitrate exploded, killing 552 people and injuring 3,000. On 21 September 2001 a fertilizer warehouse, in which more than 200 tonnes of ammonium nitrate was being stored, exploded at Toulouse, France, killing 31 people, injuring 2,442, and causing damage in excess of 2 billion euros ($2.5 billion). The blast was so great it broke windows two miles away. The most recent disaster was in North Korea on 22 April 2004 when a train load of ammonium nitrate exploded killing 162 and injuring more than 3000.

Medical element

Liquid nitrogen is needed to freeze blood or to preserve genetic material, such as eggs and sperm, or for performing so-called cryogenic surgery when unwanted tissue can be excised simply by freezing to very low temperatures.

From a metabolic point of view, nitric oxide (NO) is much more important. It can dramatically affect bodily functions and is an important part of the drugs that treat angina, which is the painful condition in which the flow of blood to the heart is restricted. The earliest treatment for this condition has been found in ancient Chinese manuscripts which described how crystals of nitre (aka potassium nitrate, KNO_3) should be sucked to relieve the condition. We now know why this might work because bacteria in the mouth can reduce nitrate (NO_3^-) to nitrite (NO_2^-) and this then releases the NO which relaxes the blood vessels.

In 1987, Salvador Moncada discovered that NO was a vital body messenger for relaxing muscles, and today we know that it is involved in the cardiovascular system, the immune system, the central nervous system, and the peripheral nervous system. The nitric oxide producing enzyme, *NO-synthase*, is abundant in the brain. NO is now known to activate more than 100 kinds of sites within the body such as enzymes, respiratory proteins, membrane receptors, and ion channels. NO controls many physiological processes and it does this by means of its effect on the messenger molecule known as cGMP (short for cyclic-guanosine monophosphate) and in some cases it actually forms a compound with this molecule such as 8-nitro-cGMP. Other molecules and ions related to NO are produce in the body such as the oxides N_2O and N_2O_3, $ONOO^-$ (aka peroxynitrite) which causes inflammation, and the much more active entity HNO (aka nitroxyl).

Although NO is relatively short-lived, it can diffuse through membranes to carry out its functions. In 1991, a team headed by K.-E. Andersson of Lund University Hospital, Sweden, showed that NO activates an erection by relaxing the muscle that normally resticts the flow of blood into the penis. The drug Viagra works by releasing NO to produce the same effect.

The indirect effect of substances which release NO had been noted for more than a century, although not that they acted by producing this molecule. The benefits of breathing amyl nitrite vapour on blood pressure were first noted as long ago as

1867, and sniffing this was used to relieve the symptoms of angina. In World War I, doctors noticed that ammunition workers packing shells with nitro-glycerine had very low blood pressures and this also led to the use of this compound as a vasodilator.[120]

Sometimes, so much NO may be formed in the body that it threatens life. A leading cause of death among patients in intensive care is septic shock. As the body generates NO to fight the infection it may also lower the blood pressure to dangerous levels. Inhibitors which block the NO-forming enzymes are needed to prevent this happening.

The radioactive isotope nitrogen-13 is used in the modern technique known as positron emission tomography, or PET scanning. This isotope has a half-life of only 10 minutes and when it decays it emits a positron, which is the positive equivalent of the negative electron (its anti-particle). When nitrogen-13 is administered to a patient the released positrons are annihilated by electrons; this process emits radiation resembling X-rays. These are monitored to produce a 3D image of the body; PET is particularly useful for examining the brain.

Element of war

Conventional explosives rely on nitrogen compounds. Gunpowder is a mixture of charcoal (carbon), sulfur, and potassium nitrate, and was first made in China. Potassium nitrate crystals were scraped from the walls of cellars and latrines where they form as a result of bacterial action on nitrogen-rich waste; this was the only source of saltpetre for hundreds of years.

The next useful explosive to be discovered, in 1846, was nitroglycerine, which was made by reacting glycerine (today known as glycerol) with nitric and sulfuric acids. It was ideal for blasting in quarries and for tunnelling, but was notoriously temperamental. Then, in 1867, the Swedish chemist Alfred Nobel (1833–96) found that by absorbing it on a type of clay it became safe to handle and would only explode when initiated by a detonator. Thus dynamite was invented and Nobel became very rich, eventually using his wealth to set up and fund the Nobel Prizes.

The third nitrogen compound to be used as an explosive was TNT, short for trinitrotoluene, discovered in 1863.[121] Today there are much more sophisticated nitrogen-based explosives, in that they are safer to handle, such as HMX (chemical name tetranitrotetraazacyclooctane) and RDX (chemical name trinitrohexhydrotrazine) and these are widely used for munitions.

What makes all such compounds able suddenly to release a lot of energy is the eagerness of the nitrogens they contain to revert violently to nitrogen gas with the evolution of large amounts of heat which causes the gas being generated to expand violently.

---

[120] The vasodilators amyl nitrite and butyl nitrite are also used by gay men to enhance sexual pleasure.

[121] The largest explosion of this occurred accidentally on 6 December 1917 at Halifax, Nova Scotia, Canada, when a ship carrying 2,750 tonnes collided with another vessel and thereby destroyed most of the town, killing 1,600 and injuring 9,000.

Element of history

Sal ammoniac (aka ammonium chloride, NH_4Cl) intrigued the alchemists who were fascinated by its volatility. Robert Boyle rightly said in 1661 that it was composed of ammonia, which he called volatile alkali, and hydrochloric acid, which he called muratic acid. It was manufactured in Egypt by heating a mixture of camel dung and cow dung with salt and urine. Later it was made by heating the antlers of deer ('hartshorn') and other nitrogen-rich natural protein materials such as wool and silk. Sal ammonia was popularly used as smelling salts.

In the 1760s, Henry Cavendish (1731–1810) and Joseph Priestley (1733–1804) each carried out experiments to remove the oxygen from air, noting that this decreased the volume by about a fifth and left behind an 'air' in which a lighted candle was immediately extinguished and in which a mouse would soon die. Both men speculated in letters to each other on what this 'air' might be, but neither concluded it was a substance in its own right, nor an element. Stephen Hales (1677–1761), a clergyman and scientist, had previously carried out similar research but again did not come to the obvious conclusion. So it was left to a young student Daniel Rutherford (1749–1819) to propose that air consisted mainly of nitrogen gas.[122] He proposed this in his doctorate thesis on 12 September 1772 at Edinburgh, Scotland, only a few months before Priestley published a paper on it. His thesis, written in Latin, was entitled *De aere fixo dicto aut mephiticio* (On the air called fixed of mephite).

Economic element

Nitrate minerals are known, such as nitratine (aka Chile saltpetre, sodium nitrate, $NaNO_3$), nitrobarite (barium nitrate, $Ba(NO_3)_2$), and nitrocalcite (calcium nitrate hydrate, $(Ca(NO_3)_2.4H_2O)$, but they are uncommon.

There are more than 3 million billion tonnes (4×10^{15}) of nitrogen in the Earth's atmosphere and this is the source of nitrogen used in industry, with around 45 million tonnes a year worldwide being extracted. After air has been filtered through molecular sieves,[123] to remove carbon dioxide and water vapour, it is cooled to liquefaction temperature and separated into its component parts by fractional distillation, a process in which the nitrogen boils off first while the oxygen remains behind. Highly purified nitrogen obtained this way can have less than 10 p.p.m. of other gases and ultrapure nitrogen, used in the electronics industries, can have as little as 10 p.p.b.

Other methods for producing nitrogen do not require liquefaction, and are employed to generate the gas on site, such as for oil wells. Pressure swing absorption uses molecular sieves to absorb oxygen, water vapour and carbon dioxide, and can yield 99 percent pure nitrogen, while another process uses a semi-permeable membrane that allows the other gases in the air to pass through but prevents nitrogen from escaping.

---

[122] Daniel Rutherford was the uncle of Walter Scott, author of *Ivanhoe* and other historical novels.
[123] These are zeolite materials (types of silicates) which have tiny pores that can trap specific kinds of molecules.

Nitrogen is primarily used in three different ways: as an inert atmosphere; as liquid nitrogen coolant; and as a raw material for the chemicals industry.

As an inert atmosphere its job is simply to exclude oxygen from materials and systems which are sensitive to it. For example, apples will keep in cold storage for up to two years if the container is filled with nitrogen gas. Nitrogen gas is pumped into old oil wells to force out more oil, whereas ordinary air would oxidise some of the oil, producing unwanted by-products. Nitrogen is used to inflate the tyres of aircraft and of some motor cars.

As a low temperature coolant, liquid nitrogen's boiling point of $-196°C$ means most things cooled by it become solid and this prevents any kind of chemical reaction from occurring, thereby preserving all manner of sensitive biological material.

As a feedstock for the chemical industry nitrogen is invaluable; there are hundreds of chemical compounds that contain this element. The sequence of converting the nitrogen of the air to the products we need is first to convert it to ammonia, then to nitric acid and later on to more sophisticated materials such as plastics like poly-urethane, fibres like nylon, and the brilliantly coloured azo-dyes.

Other nitrogen compounds produced industrially are hydrazine (N_2H_4) which is used as an antioxidant, and hydroxylamine (NH_2OH), a mild reducing agent which is used in dyeing.

Early attempts by chemists to react nitrogen and hydrogen failed to produce any ammonia (NH_3), no matter how strongly they were heated together and no matter how high the pressure. What the reaction needed was a catalyst and that's what the German chemist Fritz Haber (1868–1934) discovered in 1905,[124] and chemical engineer Carl Bosch turned it into a commercial process. Although the yield of ammonia was only a few percent, it was possible to circulate the unreacted gases through the reaction chamber again and again. On 3 July 1909, the first successful Haber-Bosch chemical plant started up at Oppau, Germany and operated at 450°C and 100 atmospheres pressure. (As mentioned above, this plant came to a disastrous end 12 years later).

Today, there are Haber-Bosch plants around the world producing more than 150 million tonnes of ammonia a year, the vast majority of which goes into making fertilizers. (The nitrate is made from ammonia by reacting the gas with oxygen and water—see later). Today, thanks to improvements in the catalysts, yields of 20 percent of NH_3 can be produced at pressures of 150 atmospheres.

For this achievement, Haber was awarded the Nobel Prize for chemistry in 1918. The Haber process also increased the supply of nitric acid because ammonia was the feedstock of another process discovered by the German chemist Wilhelm Ostwald (1853–1932),[125] who had found that ammonia could be oxidised to nitric acid (HNO_3) by heating it with oxygen in the presence of a platinum catalyst.

A newer process makes ammonia from natural gas and air and operates at 380°C and 80 atmospheres—it requires only half the energy to make the same amount of ammonia as the Haber process.[126]

---

[124] The catalyst is iron activated by potassium hydroxide. Haber's first catalysts were either osmium or uranium which worked spectacularly well but were too expensive.

[125] Ostwald received the Nobel Prize for chemistry in 1909.

[126] The overall reaction for the process is $7CH_4 + 10H_2O + 8N_2 + 2O_2 \rightarrow 7CO_2 + 16NH_3$ and it is carried out in several stages.

An alternative method of 'fixing' nitrogen is to pass an electric spark through air and simulate the reaction that occurs when lighting passes through air: the reaction of oxygen and nitrogen to form nitric oxide and then nitrogen dioxide. This is known as the Birkeland-Eyde process, named after the two Norwegians who developed it, and the nitrogen oxides it produces are converted to nitric acid. This process is only economic when there is a plentiful supply of cheap electricity.

Environmental element

Nitrogen in the environment	
EARTH'S CRUST	25 p.p.m.
	Nitrogen is the 30th most abundant element.
SOILS	av. 5 p.p.m.
SEAWATER	only 0.1 p.p.b at the surface, but 0.5 p.p.m. in the deepest regions.
ATMOSPHERE	78% in dry air.

The nitrogen of the atmosphere came as a result of the out-gassing of the molten Earth and even today some escapes this way when volcanoes erupt. Most of the planet's nitrogen is present as this gas although there are some deposits of nitrates that were formed by the evaporation of natural waters in which they built up.

The Nitrogen Cycle is the term used to describe the movement of this element through the various phases of the environment as an essential part of the living planet. When nitrogen is 'fixed' it gets incorporated into living things from which it may emerge as excreta from animals or from micro-organisms which decompose dead matter. As a result of all this activity there is a significant amount of nitrogen in the soil in the form of amino acids, humus, ammonium salts, nitrates and nitrites; these last two can be converted back to nitrogen gas or nitrous oxide and returned to the atmosphere, of which this latter gas makes up 0.3 p.p.m. Ten million tonnes of this are released as a by-product of microbial action each year, as well as from human activity. Other nitrogen gases are the nitric oxides NO and NO_2 often referred to as NO_x which make up 0.05 p.p.b. Although 100 million tonnes of these enter the atmosphere each year as a result of combustion, they are quickly washed out of the atmosphere in rain. NO_x can be particularly unpleasant in sunny climates, where these molecules react with hydrocarbons in the atmosphere, such as traces of unburnt fuel, to produce photochemical smog over cities. Catalytic convertors on cars have more or less solved this problem.

Bacteria in the soil and algae in the oceans are the primary 'fixers' of nitrogen. The cyanobacteria in soil form nodules on certain plant roots, particularly clover, legumes such as beans, soya, and alfalfa. Nitrogen fixation is carried out by the enzyme system *nitrogenase* which consists of two large proteins; the smaller one contains four iron atoms, while the larger has twelve irons and two molybdenums. Both proteins are needed to fix nitrogen. Attempts to replicate this process in the laboratory, with a view to finding a low-energy means of fixing nitrogen has so far met with little sucess. In 2003, a molybdenum compound was discovered which was capable of turning

N_2 to NH_3 under similarly mild conditions as found in Nature. In 2007, a tantalum catalyst which will also do this was reported.

A microbe has been discovered in hydrothermal vents at the bottom of the Pacific Ocean which can convert N_2 to NH_3 at 92°C and it may have been essential to fixing nitrogen at the earliest stages of life on Earth. These, and other deep-sea organisms such as archaea and bacteria, are also capable of turning N_2 into nitrate and cyanide ions. Much yet remains to be discovered about the Nitrogen Cycle.

Nitrogen trifluoride (NF_3) is a gas used as a cleaning agent for chambers in which chemical vapour deposition is used in the manufacture of microchips. It converts silicon traces to the gas silicon tetrafluoride (SiF_4) so that these can be removed. However, NF_3 is a powerful and persistent greenhouse gas with infrared absorption bands in regions of the spectrum that are not currently absorbing and those who use this gas, of which more than 3,000 tonnes are manufactured every year, have a duty to see that as little as possible escapes into the atmosphere. Some does escape because the measured level has risen from 0.02 p.p.t. in 1978 to 0.45 p.p.t. in 2008.

Chemical element

Data file	
CHEMICAL SYMBOL	N
ATOMIC NUMBER	7
ATOMIC WEIGHT	14.00674
MELTING POINT	−210°C
BOILING POINT	−196°C
DENSITY	1.25 g per litre
OXIDES	N_2O, NO, NO_2/N_2O_4

Nitrogen is a colourless, odourless gas and the first member of group 15 (row 2p) of the periodic table. Nitrogen is generally unreactive at normal temperatures although it will react with lithium metal to form lithium nitride, Li_3N, and similarly with magnesium metal to form Mg_3N_2, and this latter metal will actually burn in an atmosphere of nitrogen. Nitrogen has an extensive chemistry, and its compounds exist in a wide range of oxidation states from −3 of ammonia to the +5 of nitrate (NO_3^-) and nitric acid, HNO_3.

Nitrogen has 17 known isotopes of masses 10 to 25, and one known isomer. It occurs naturally as nitrogen-14 which accounts for 99.6% and nitrogen-15 which represents only 0.4 percent. Neither is radioactive. The longest-lived radioactive isotope is nitrogen-13 which has a half-life of 10 minutes. The radioactive isotope nitrogen-16 is produced inside nuclear reactors from the oxygen of water. Its half-life is seven seconds and its detection outside the reactor is a way of finding leaks.

Element of surprises

1. Not all explosives are used to destroy and kill: sodium azide (NaN_3) is used to save lives. This is the explosive in airbags that protect passengers in cars when they crash and there is around 200 g in each bag. If the impact velocity is above 10 mph, this will register in a sensor in the electronic control unit which inflates the airbag within 25 thousandths of a second.[127] The explosion is ignited by an electrical impulse which generates a temperature of 300°C.

2. In 1999, the Edwards Airforce Base Laboratory in California revealed that it has made the ion N_5^+ which consists of five nitrogen atoms arranged in a V-shape. What surprised the chemical community was that the ion appeared to be quite stable. The hope that it could form a salt with the azide ion as $N_5^+N_3^-$ (formula N_8) has proved elusive. However, the compound HN_5 has been made although it is stable only at −40°C.

3. Atomic nitrogen, in other words single atoms of the element, has recently been produced at the Carnegie Institution of Washington, but to produce it required pressures in excess of 150 gigapascals, which is 1.5 million times atmospheric pressure. Atomic nitrogen appears to be semiconducting.

4. Nitrous oxide (N_2O) induces mild hysteria and was formerly known as laughing gas. It is still used as an anaesthetic. It is also the propellant gas for cans of whipped cream on account of its solubility in fats.

5. The horseshoe vetch (*Hippocrepis comosa*) not only has microbes that turn N_2 to NH_3 but they go even further and convert it to 3-nitropropionic acid (O_2NCH_2. CH_2CO_2H) which protects the plants against being eaten because it is unpleasant tasting. When the plant eventually dies it releases nitrate to the soil.

6. Nitrogen gas is the preferred gas for pumping alcoholic drinks from storage vessels to taps in bars. In the case of some beers it will even produce a drink with a smoother head.

7. Nitrogen oxide N_2O_4 and its reaction with hydrazine N_2H_4 has been used as a fuel for rocket engines in the Titan rockets and some other space exploration vehicles.

---

[127] Five times faster than the blink of an eye.

Nobelium

Pronounced no-beel-ee-um, it is named after Alfred Nobel, the Swedish chemist, industrialist, and founder of the Nobel Prizes.

French *nobélium*; German *Nobelium*; Italian *nobelio*; Spanish *nobelio*; Portuguese *nobélio*.

Essential element

The most newsworthy aspect of this element were the dramas and controversies surrounding its discovery.

Element of creation and contention

Three research groups claimed discovery of element 102. The issue of whom should be recognised as its discoverer, and so be able to give it a name, was not resolved until 1997, almost 40 years after it was first made.

The conflict has its origins in the earliest days of heavy-element research, when physicists started to hunt for elements beyond fermium (element 100) using particle accelerators. Previously, newly-discovered elements had been extracted from minerals or produced within nuclear reactors, provided they had half-lives long enough to allow them to be separated and identified by chemical means.

Most of the transfermium-element isotopes first discovered had half-lives too short (minutes or seconds) and identification had to rely on measured decay energies and the detection of decay products that were produced by the newly-claimed element. Isotopes of other elements could be produced in the many reactions that occurred in particle accelerators but the bombarded target materials could be confusing and led to misinterpretations. Much of this was not well understood in the early days of transfermium-element research, and led to wrongly identified isotopes because data was not properly interpreted.

Four laboratories sought element 102. The prestige of discovering a new element became part of the Cold War rivalries between the USA and the USSR (Russia). In 1956, a team led by Georgy Flerov at the Institute of Atomic Energy (IAE), Moscow, synthesised element 102 by bombarding a target of plutonium-241 with oxygen-16 and got atoms of isotope-252 (originally thought to be isotope-253). They also tried plutonium-239 with oxygen-16 in an unsuccessful attempt. They did not file a claim because of less than accurate measurements, but they had thought of a name for it and that was joliotium (symbol Jo) in honour of Irene Joliot-Curie (1897–1956) who had recently succumbed to radiation poisoning (see polonium). Essentially no attention was paid to this effort at the time.

In 1957, a team of international researchers at the Nobel Institute of Physics in Stockholm bombarded a mixed target of curium-244 to 247 with carbon, claiming to get isotope-253, or possibly isotope-251. They announced that the isotope's alpha-decay energy was 8.5 MeV with a half-life of about 10 minutes. They proposed the name nobelium (symbol No) in honour of Alfred Nobel. This was accepted somewhat prematurely by the International Union of Pure and Applied Chemistry (IUPAC), only to be retracted later. The chemical symbol 'No' was a symbolic warning of what was to come.

In 1958, a team led by Albert Ghiorso (1915–) at the Lawrence Berkeley Laboratory (LBL)[128] bombarded a mixed-isotope target of curium-244/245/246 with carbon-12, claiming to get isotope-254 with a half-life of 55 seconds. This claim was also to attract criticism. They also tried to reproduce the claimed results of the Nobel Institute using essentially the same target composition of isotopes and the same projectiles. They were unable to produce any 102 or indeed detect anything that was claimed. In 1959, the Nobel Institute published a paper in response to LBL's announcement that they failed to reproduce their claimed results. They criticised LBL's effort, saying that it was not run at a high-enough energy. They did, however, suggest that they had probably produced isotope-255, not isotope-253, as originally claimed.

For several years, both the Nobel Institute and the LBL adjusted their data and changed their claimed isotope in response to criticisms of their interpretations and evidence that they had misinterpreted data. It appeared they had even engaged in guesswork in an effort to identify their isotope and thereby justify discovery of element 102. The LBL continued to do this until 1967, when they learned of experiments on element 102 coming from the USSR.

The new Russian Joint Instituted of Nuclear Research (JINR), based at Dubna, began element research in the early 1960s and all work on element discovery shifted to the JINR from the IAE. From 1962 to 1966 its physicists did research on element 102, synthesising isotopes 252 to 256. They demonstrated that the LBL did not synthesise the element 102 isotope as claimed—that work had been flawed with inaccuracies. Nor could the claim of the Nobel Institute be reproduced.

In 1962, the JINR first produced element 102, bombarding uranium-238 with neon-22 to get isotope-256, but they felt that the results were not fully convincing. The first unambiguous production took place in 1963 when they bombarded americium-243 with nitrogen-15 to get isotope-254. The JINR's work during this time proved that element 102 had definitely been discovered in 1956 at the IAE, but that work had been marred by less than accurate measurements. The JINR physicists published their results in October 1966 and later submitted a discovery claim, again proposing the name joliotium (symbol Jo) for the new element.

In December 1966, LBL re-ran all of the experiments mentioned in the newly published JINR publication; they got the same results and confirmed the JINR's work. In the light of this new data, LBL re-evaluated and re-interpreted the results of their original work from 1958 to 1961. The information from the JINR made it possible for them finally to define the isotopes that they had formerly synthesised but were unable to correctly identify at the time.

In 1967, Ghiorso published a paper that used the re-interpreted data to claim that from 1958–61 the LBL had actually discovered element 102 before the JINR and

[128] This was later renamed as the Lawrence Berkeley National Laboratory (LBNL).

should be recognised as the discoverers. A claim was submitted to IUPAC, and they proposed that the Swedish name nobelium be retained because it was already commonly used. This was approved by IUPAC in 1971, somewhat surprisingly because of the lack of credibility in the original work regarding its discovery.

Other controversies now emerged between the JINR and LBL over naming elements 104, 105, and 106. In the mid-1970s, IUPAC tried to set up a committee to study and adjudicate these protracted controversies. Along with the International Union of Pure and Applied Physics (IUPAP) they finally succeeded in 1985 and set up the Transfermium Working Group (TWG)—see Transfermium Elements for more details. The TWG recommended that the JINR be awarded discovery for element 102 and in 1997 IUPAC officially awarded discovery priority to the JINR for their 1963 work which was the first to unambiguously synthesise elements 102. However, IUPAC retained the name nobelium for element 102 because of its long-term use and familiarity.

Chemical element

Data file

CHEMICAL SYMBOL	No
ATOMIC NUMBER	102
ATOMIC WEIGHT	259.1—see *Element of surprises*

Properties such as melting point, boiling point, and density, are likely never to be measured because there will never be more than a few atoms produced at any one time.

OXIDE	none yet made, but likely to be No_2O_3

Nobelium is a radioactive element and a member of the actinide group (row 4f) of the periodic table. So far not enough nobelium has been made at one time to be visible by the naked eye, but predictions are that it would be a silvery metal susceptible to attack by air, steam, and acids. Research has shown, somewhat unexpectedly, that the chemistry of nobelium in solution would be based on a stable +2 oxidation state, No^{2+}, and a less stable +3 state, No^{3+}, rather than the other way round.

Nobelium has 12 known isotopes with mass numbers ranging from 250 to 260 and 262, and four known isomers. The longest-lived known isotope is nobelium-259 whose half-life is 58 minutes. (Nobelium-261 is theoretically calculated to have a half-life of 2.78 hours)

Element of surprises

1. In 1998, physicists at the Argonne National Laboratory, Illinois, were able to make an atom of nobelium isotope-254, which decayed after 55 seconds, and they generated this by bombarding lead with calcium ions. They then trapped it in a silicon detector and studied its gamma spectrum. From this they were able to deduce that the nucleus of the atom was a squashed sphere with a high spin.

2. In 2010, an international team at the GSI Helmholtz Centre for Heavy Ion Research at Darmstadt in Germany actually weighed individual nobelium atoms even though these had half-lives of less than a second. The atoms were created in a fusion reactor and emerged at a rate of around one atom per second but moving too fast to analyse, so they were slowed down by forcing them to go through helium before moving to a magnetic chamber where their mass could be measured.

Osmium

Pronounced oz-mee-um, the name is derived from the Greek word *osme*, meaning smell, because the metal surface gives off volatile osmium tetroxide (OsO_4) which has an irritating, pungent odour.

French *osmium*; German *Osmium*; Italian *osmio*; Spanish *osmio*; Portuguese *osmio*.

Essential element

This is the heaviest of all the elements and yet it produces a volatile oxide. Once used for exotic fountain pen nibs, today it finds few commercial outlets.

Element of life

Osmium has no biological role, and it has been detected in humans but at levels that are insignificant.

Dangerous element

The metal itself is not toxic, but its volatile oxide is and this melts at 40°C and boils at 130°C. Even breathing air containing only 0.11 micrograms per cubic metre of osmium tetroxide is enough to irritate the lungs, eyes, and skin of some people, and workers should not be exposed to concentrations of greater than 2 micrograms per cubic metre. The vapour is also known to cause intense headaches and impaired vision.

Element of history

In 1803, the English chemist Smithson Tennant (1761–1815) dissolved crude platinum in dilute aqua regia and observed, as others had done before him, that not all the metal went into solution. Earlier experimenters had assumed that the powder which remained behind was just graphite, but his tests showed it was not, and he began to investigate it further. By a combination of acid and alkali treatments he eventually separated it into two new metal elements, which he named iridium and osmium, christening the latter on account of the strong odour it gave off. Although it was recognised as a new metal, little use was made of it because it was rare and difficult to work with.

Economic element

Osmium can exist as the uncombined metal, and as alloys with iridium. These are known either as iridosmine or osmiridium. Osmium ores are rare but some deposits from which it can be extracted are to be found in Russia, Canada, South Africa, and the USA. Most osmium is obtained as a by-product of nickel refining, extracted from the Canadian nickel ores of Ontario. Less than 100 kg of osmium are produced each year. There is little demand for the metal, which is difficult to fabricate, but osmium powder can be worked by sintering at 2000°C and employing high pressures. The powder is produced by extracting the osmium from the anode mud that settles at the bottom of the electrochemical cells used to purify nickel and copper. The osmium in the mud is oxidised to osmium tetroxide, which is distilled off, and then reduced to the metal with hydrogen.

Osmium is used in a few alloys and in industry as a catalyst. At one time it was to be encountered in the nibs of high quality fountain pens, compass needles, long-life gramophone needles, and clock bearings, where its extreme hardness and corrosion resistance were appreciated. It was used as the filament for a successful electric light bulb devised by Karl Auer in 1897, but the filament was very fragile and would break if the bulbs were not handled carefully. Osmium filaments were replaced by tantalum in 1905 and those were eventually replaced by the tungsten ones which were more successful. However, these kinds of light bulbs are now being phased out because they generate far more heat than light.

Osmium tetroxide is used for staining tissue that is to be examined by electron microcopy. It works by cross-linking the fatty molecules within membranes and thereby fixing them in place, and the osmium itself helps by highlighting contrasting features within the sample.

Just as there are anti-cancer drugs based on platinum, so there are some which have been produced from osmium and these too appear to be effective against ovarian and colon cancers which do not respond to platinum drugs like cisplatin. The osmium-based drugs were announced in 2009, but they have yet to be screened for human treatment.

Environmental element

Osmium in the environment	
EARTH'S CRUST	*av.* 1 p.p.b.
	Osmium is the 81st most abundant element, i.e. it is exceedingly rare.
SOILS	not recorded.
SEAWATER	not known, but very low.
ATMOSPHERE	nil.

Studies show that there is little or no environmental threat from osmium, even near processing plants. The analysis of osmium in layers of the peat bogs in the Xistral Mountains of northwest Spain in 2009 revealed an increase in osmium dating to around 2700 BC, which corresponded to the start of early mining in the area, followed by a larger increase around the time when the Romans occupied the Iberian Peninsula in the first century BC and began more extensive mining. A third increase in osmium can be dated to around 1750 AD which was the start of the industrial age.

Chemical element

Data file	
CHEMICAL SYMBOL	Os
ATOMIC NUMBER	76
ATOMIC WEIGHT	190.23
MELTING POINT	3,054°C
BOILING POINT	5,027°C
DENSITY	22.6 g/cm³
OXIDE	OsO_4, also OsO_2

Osmium is a lustrous, silvery metal, and one of the so-called platinum group which consists of the metals of groups 8, 9, and 10, namely ruthenium osmium, rhodium, iridium, palladium, and platinum. It can occur as a native metal. Osmium is a member of group 8 (row 5d) of the periodic table. It is the densest metal known, although only by the narrowest of margins: osmium has a density at 20°C of 22.588 g/cm³, and for iridium it is 22.562 g/cm³. Calculated densities based on the crystal structures of these two metal confirm that this is the correct order and they give values of 22.66 for osmium and 22.65 for iridium. Osmium is unaffected by water, and acids, but dissolves in molten alkalis. Osmium powder reacts slowly with the oxygen of the air and gives off detectable amounts of osmium tetroxide vapour. Osmium's oxidation states range from −1 as in $Na_2Os(CO)_4$, to +8 as in OsO_4. Its more common oxidation states are +2 as in $OsCl_2$, +3 as in $OsCl_3$, and +4 as in $OsCl_4$.

There are 36 known isotopes of osmium with masses 162 to 197, and nine known isomers. There are seven naturally occurring isotopes: osmium-192 is the most abundant at 41 percent, osmium-190 (26.5 percent), osmium-189 (16 percent), osmium-188 (13.5 percent), osmium-187 (1.5 percent), osmium-186 (also at 1.5 percent), and osmium-184 (0.02 percent). Three of these are radioactive including the most abundant, osmium-192, which has a half-life of 9.8×10^{12} years (9.8 trillion years). Also radioactive are osmium-186 with a half-life of 2×10^{15} (2,000 trillion years) and osmium-184 with a half-life of 5.6×10^{13} years (56 trillion years). Nuclear theory suggests that all isotopes of osmium should be radioactive and that those we currently regard as stable have in fact incredibly long half-lives.

Osmium-187 is the product of the radioactive decay of rhenium-187 which has a half life of 4.56×10^{10} years and by measuring the ratio of these two metals in a terrestrial rock sample, or in a meteorite, it is possible to date it.

Apart from the above radioactive isotopes, the longest-lived one is osmium-185 which has a half-life of 94 days.

Element of surprises

At one time osmium tetroxide was employed by forensic chemists to highlight fingerprints because the vapour reacts to form a black deposit of the lower oxide, OsO_2, when it comes in contact with the minute traces of oils left by a finger touching a surface.

Oxygen

Pronounced ok-si-jen, from the Greek, *oxy genes* meaning acid forming.
French *oxygène*; German *Sauerstoff*; Italian *ossigeno*; Spanish *oxígeno*; Portuguese *oxigénio*.

Essential element

This element displays some bizarre properties and yet it is the gas we need a constant supply of but even then it must not be too little or too much. And the oxygen that is part of our teeth can actually reveal which part of the world we come from.

Cosmic element

Oxygen is the third most abundant element in the universe, after hydrogen and helium. In stars like the Sun its formation is part of the fuel cycle that provides their energy. When a star has burned up its hydrogen to form helium, and its helium to form carbon, then carbon-burning begins converting this element to oxygen. The amount of oxygen produced depends on the mass of the star. The more massive a star is, the higher its internal temperature, and the more oxygen is formed.

Oxygen-burning will also occur provided the temperature is a billion degrees. The Sun is not massive enough for this to happen, but in a sufficiently massive star further elements are formed from the fusing of two oxygens to produce the nuclei of the elements silicon, phosphorus, and sulfur.

In the solar system, the Earth is unusual in having a high level of oxygen gas (O_2) in its atmosphere, a sure sign of advanced forms of life. The only other planets in the solar system with any oxygen are Mars, whose atmosphere contains 0.15 percent, and Venus which has even less. Their oxygen is produced by UV radiation acting on other molecules. Oxygen levels in the atmosphere built up slowly starting about 2 billion years ago and then increased more rapidly, even reaching around 30 percent, before falling to the levels we have today. The higher levels are seen as the explanation of the giant flying insects which were around in the Carboniferous era 300 million years ago.

In 2004, astronomers were able to show that a large Jupiter-sized planet circling a star 150 light years from Earth had oxygen in its atmosphere. However, as the planet is close to its parent star it seems that the chances of there being life on it are slim.

Element of life

Oxygen in the human body	
BLOOD	constituent of water
BONE	28%
TISSUE	16%
TOTAL AMOUNT IN BODY	ca.43 kg and it is present mainly as water.

Oxygen is essential for all forms of life since it is a constituent of DNA, and almost all biologically important compounds contain it. It is even more dramatically essential, in that animals must have a minute-by-minute supply of the gas in order to survive. Oxygen in the lungs is picked up by the iron atom at the centre of haemoglobin in the blood, and thereby transported to where it is needed to release energy by chemical reactions. Because of haemoglobin, a litre of blood will dissolve 200 cc of oxygen, much more than will dissolve in the same volume of water.[129] The haemoglobin passes its oxygen to an enzyme, *monooxygenase*, which also has an iron atom at its active centre. This enzyme is the catalyst for various oxidation processes in the body including synthesising new molecules and detoxifying others.

The percentage of oxygen in the air is mid-way between two extremes that would make life on Earth impossible for humans: below 17 percent we would die through lack of it, and above 35 percent all organic material would be highly flammable. The three astronauts who were destined for the first manned Apollo flight were burned alive in minutes in their spacecraft on 27 January 1967, when fire started in the oxygen-enriched air of the cabin.

Oxygen gas makes up a fifth of the atmosphere, amounting to more than a *million* billion tons. It is fairly soluble in water which makes life in rivers, lakes, and oceans possible. When we breath in oxygen it reacts with energy stores in our bodies to provide the warmth and muscular action which keeps us alive. We return it to the atmosphere as carbon dioxide, where it is captured by plants and, with the help of sunlight, is turned into carbohydrates. This process, known as photosynthesis, releases oxygen back into the atmosphere.

Oxygen has the capacity to *oxidise*. With organic material this can be rapid and intense as in a fire, or slow and gentle as in a living being, but in both cases it is oxidising carbon compounds to carbon dioxide. What has allowed oxygen to build up in the atmosphere is its relative lack of reactivity as a molecule, although once a reaction has been initiated it will proceed rapidly and indeed oxygen will react with almost all other elements and organic substances.

Medical element, dangerous element

The brain must have oxygen to function and, if denied oxygen, it will begin to die within minutes, but it is not generally realised that too much oxygen will poison it. The threat is not appreciated by many sports divers, who need to be warned against

---

[129] Not all species use haemoglobin as the oxygen carrier—spiders and lobsters use a copper-containing version, which is why their blood is blue; sea worms use an entirely different protein called haemerythrin.

breathing pure oxygen below 10 metres since this can lead to convulsions and, because of this, several divers have drowned. It is better to use 'nitrox' which is air with increased oxygen content.

Premature babies used to be kept in an oxygen-rich incubators until it was realised that some became blind because of it. Nevertheless, oxygen masks are a key part of medical treatment and are needed to supply oxygen-rich air—about 30 percent oxygen—to patients whose lungs or heart are impaired in some way.

The process of using oxygen goes through many stages, intermediate among them is its transformation to two undesirable compounds: the highly dangerous superoxide radical, which is an oxygen molecule that has picked up a single electron, that is, O_2^-, and the slightly less dangerous hydrogen peroxide, H_2O_2. Both can damage living cells and so there are special enzymes to neutralise them. *Superoxide dismutase* converts the superoxide to hydrogen peroxide, then *glutathionine peroxidase*, and similar enzymes, reduce this to water. Superoxide may be dangerous, but some cells in the body purposely make it in order to destroy invading organisms.

In the early years of this century there was a craze for oxygen therapy which could be bought in pressurised cans so that deep breaths of this could be taken and the effect would be an instant lift by instantly boosting the body's energy. In theory they could re-supply the blood with oxygen after strenuous energy has depleted it, but the effect is minimal. Other products were marketed that claimed all kind of benefits could be obtained in the form of oxygen moisturiser for the skin, oxygen capsules that would kill bad bacteria in the gut, and oxygen suits which allowed the skin to breath it in. None of these products could have produced the benefits being claimed.

An emergency supply of oxygen automatically becomes available for the passengers in an aircraft when it suddenly experiences a pressure drop. This is not stored as oxygen gas but as the chemical sodium chlorate. The oxygen masks which drop down above each seat in the stricken aircraft are activated by a small detonation which mixes the chlorate with iron filings and the two chemicals react to release the gas.

Element of history

In 1624, the first ever submarine was demonstrated to King James I of England. It was manned by 12 rowers whose oars protruded through sealed ports and it travelled under water for two hours. On board was its Dutch inventor, Cornelius Drebbel (1572–1633), plus a few passengers, one of whom later reported that when the air in the submarine had been consumed it was refreshed from a container. It seems obvious that it must have contained oxygen and in 1608 Drebbel had shown that gently heating saltpetre (aka potassium nitrate, KNO_3) released a gas, which we now know was oxygen although it was not identified as such at the time.

What made it difficult for early investigators to understand the nature of air was the belief that it was an element, so the idea of its being a mixture of gases was quite alien to the established way of thinking. Yet, in the Middle Ages, Leonardo da Vinci (1452–1519) had noted that air must contain something essential to life and that when this had been used up then neither a candle flame nor a living animal could survive.

In the 1600s, scientists like the Englishman, Robert Hooke (1635–1703), the Russian, Mikhail Lomonosov (1711–65), the Dane, Ole Borch (1626–90), and the Frenchman, Pierre Bayen (1725–98), all produced oxygen, but none of them identified it as such. The credit for discovering oxygen is now shared by three other chemists: the Englishman Joseph Priestley (1733–1804), the Swede Carl Wilhelm Scheele (1742–86) and the great French chemist, Antoine Lavoisier (1743–94).

Priestley was the first to publish an account of oxygen, having made it in 1774 by focussing sunlight on to mercuric oxide (HgO) confined in a glass tube, and collecting over water the gas which was evolved. He noted that a candle burned more brightly in the new 'air' and that a mouse lived longer and appeared more lively, when breathing it. He also tried some on himself, and wrote:

The feeling of it to my lungs was not sensibly different from that of common air, but I fancied that my breast felt peculiarly light and easy for some time afterwards.

However, unknown to Priestley, Scheele had produced oxygen three years earlier in June 1771 also by heating mercuric oxide, as well as heating various nitrates. He had written an account of his discovery in a manuscript which he sent to his publisher in 1775, but which did not appear until 1777. It is clear from his preserved note books that not only had he identified the gas as a new substance, but he had even sent a letter to Lavoisier, posted on 30 September 1774, describing what he had done; a copy of the letter was found among Scheele's papers after his death.

Because Lavoisier never acknowledged this letter, it has always been assumed it went astray, and that autumn Lavoisier too claimed to have made oxygen and that he had done this independently of Priestley. However, we know that Priestley had visited him in October and also told him how he had made oxygen, even though he had not yet published details of his discovery.

What Lavoisier did was far more important than claiming priority for the discovery. He realised the importance of oxygen as the key to a different way of looking at chemistry, and it sounded the death knell for phlogiston that had been the basis of chemical theory for a hundred years. In that theory, when things burned they lost phlogiston to the air and if they burned in a confined space the air could become saturated with phlogiston and so the flame went out. Not so, said Lavoisier, in fact just the opposite: burning took oxygen *from* the air.[130]

Subsequently Lavoisier proposed that the new gas be called oxygène, because he fancied it was an essential component of all acids. The English were opposed to the name, but it entered the language anyway, thanks to a poem called 'Oxygen' that was an ode in praise of the new gas. It had been written by Erasmus Darwin[131] and published in 1791 in his popular book *The Botanic Garden*.

---

[130] Priestley had even named the new gas dephlogisticated air, on the assumption that it was capable of absorbing a maximum amount of phlogiston which is why things burned longer in it.
[131] The grandfather of Charles Darwin.

The story of the three claimants to oxygen's discovery is told in a play by Carl Djerassi and Roald Hoffman, called 'Oxygen', in which a panel of experts has the job of deciding to award a posthumous Nobel Prize to one of the three protagonists. This was first performed at the Tricycle Theatre, Kilburn, London, in February 2000.

Economic element

All oxygen used industrially comes from the air with 100 million tonnes being extracted every year. Oxygen is produced in one of two ways. The preferred method for large scale production is from liquefied air, from which nitrogen distils off preferentially, leaving liquid oxygen behind. Oxygen is shipped as a liquid in specially insulated tankers (one litre of liquid oxygen is equivalent to 840 litres of the gas). Oxygen is also stored and transported in cylinders of the compressed gas, for use in metal cutting and welding, or for medical purposes.

Another method of producing oxygen is to pass a stream of clean, dry air through a bed of zeolite molecular sieves which absorb the nitrogen giving a stream of gas that is 90–93 percent oxygen which can be pumped down a pipeline to where it is needed. When the zeolite is saturated with nitrogen, it is regenerated by reducing the pressure which allows the nitrogen to escape. While this is going on, the flow of air is directed through a second bed of refreshed zeolite so that the process will deliver a continuous supply of oxygen.

Newer methods that give almost pure oxygen use a ceramic membrane based on zirconium oxide through which oxygen gas will preferentially diffuse, forced through it either by high pressure or by an electric current.

Commercially produced oxygen gas is destined mainly for steel-making (55 percent ends up this way) and the chemical industry (25 percent), but significant amounts are used in hospitals, water treatment, rocket launches, and metal cutting.

Injecting oxygen through a special high pressure lance into molten iron removes the chief impurities of sulfur and carbon, converting the iron to steel, and as it does this it raises the temperature of the metal to around 1,700°C. Cutting through even the toughest steel can be done quickly with an oxygen-acetylene torch whose heat from a burning mixture of the two gases is sufficient to raise the temperature to 3,100°C, high enough to melt the steel.

In the chemical industry, oxygen is reacted with ethylene to form ethylene oxide which is then turned into materials ranging from antifreeze to polyester bottles and fabrics.

The water in rivers and lakes needs to have a regular supply of oxygen. When this gets depleted the water will no longer support fish and other aquatic species. Polluted water is assessed by its biochemical oxygen demand (BOD), the amount needed to restore its ecological balance. When neither the plants that live in water, nor the slow absorption of oxygen from the atmosphere can provide enough oxygen, the BOD can be met by bubbling the gas through the water, and this is done to rejuvenate polluted rivers and to treat sewage.

Environmental element

Oxygen in the environment	
EARTH'S CRUST	47%
	Oxygen is the most abundant element.
SOILS	present as minerals and soil water.
SEAWATER	constituent element of water.
ATMOSPHERE	21% in dry air.

The crust of the Earth is composed mainly of silicon-oxygen minerals, and many other elements are there as their oxides. Nearly half the known elements are to be found as one form of oxide or another. Some are direct oxide minerals such as quartz (aka silica, silicon oxide, SiO_2), rutile (aka titania, titanium oxide, TiO_2), and haematite (iron oxide, Fe_2O_3). In others the oxygen is part of the non-metal component such as carbonate (CO_3^{2-}), silicate (SiO_4^{4-}), or phosphate (PO_4^{3-}) so there are minerals like calcite (calcium carbonate), fluoroapatite (calcium fluoride phosphate), and kaolinite (an aluminium silicate). The composition of the continental crust is mainly made up of silicon oxide and silicates which account for 62 percent.

The oxygen in the Earth's atmosphere comes from the photosynthesis of plants, and has built up over eons of time as they utilised the abundant supply of carbon dioxide in the early atmosphere and released oxygen. Photosynthesis began about 3 billion years ago, but the oxygen then produced was mainly used up in oxidising iron(II) to iron(III). This continued for about a billion years and then the level in the atmosphere began slowly to rise until it reached around 1 percent about 2 billion years ago, when blue-green algae developed. About 500 million years ago it rose relatively rapidly to around 20 percent when the first land plants started to appear. Although for a time it exceeded this amount, even reaching 30 percent, it fell back to around 20 percent and has remained at this level.

Although animals carry out the reverse process of converting oxygen to carbon dioxide, the average oxygen atom now only takes part in this cycle once every 3,000 years. Even the burning of 7 billion tonnes of fossil fuels a year makes almost no impression on the amount of oxygen in the atmosphere. Human activity, however, can have profound effects on another gaseous form of oxygen and that is ozone—see box.

Ozone (aka trioxygen, O_3)

This gas consists of three oxygen atoms joined together in a V-shaped molecule. It was first identified in 1840 at the University of Basel in Switzerland by the German chemist, Christian Friedrich Schönbein (1799–1868) who based its name on the Greek word for smell.[132] At first he thought he had discovered a new element similar to chlorine, but eventually he proved that it was simply another form of oxygen. Ozone is not a stable

---

[132] This Greek word also gave its name to the element osmium, and for the same reason.

form of oxygen and reverts back to ordinary oxygen. There is a natural low level of ozone in the air we breathe, about 0.02 ppm.

Ozone plays two roles in the Earth's atmosphere: in the troposphere, the lower part of the atmosphere, it is a pollutant; in the stratosphere, the region between 20 and 40 km above the surface, it acts as a shield, protecting the planet from the damaging ultraviolet rays from the sun.

In the stratosphere, the ozone layer is a vital barrier which absorbs ultraviolet light, both in its formation from O_2, and in its decomposition back to O_2. Ultraviolet light of wavelengths less than 240 nm reacts with oxygen and cleaves this into two atoms, which then combine with other oxygen molecules to form O_3. This in its turn absorbs other ultraviolet rays of wavelengths 230–290 nm, which decompose it back to $O_2 + O$. Without the ozone layer, dangerous radiation capable of harming living cells would penetrate to the Earth's surface.

There are other components of the stratosphere that destroy ozone, such as chlorine atoms, and it was the use of chlorofluorocarbon gases in aerosols, plastic foams, and refrigerators that posed the greatest threat because these were long-lived in the atmosphere and were a rich source of chlorine atoms. These gases have now been outlawed.

Ozone is also an industrial resource, used for purifying drinking water (and especially bottled water), and for public swimming pools. Ozone is a more potent disinfectant than commonly used chlorine gas, but the protection it provides is not as long-lasting as that of chlorine which is why this is better for purifying mains water supplies. If it is necessary to remove ozone from air or water then this process can be speeded up by passing the air or water over charcoal which catalyses the decomposition.

There are two ways of making ozone industrially. The usual method is to pass air through concentric glass tubes with metallised surfaces across which is applied a discharge of 15 kilovolts and 50 Hz. The gas emerging from such treatment contains 2 percent ozone. Another method, used when only low concentrations of ozone are needed for sterilising or disinfecting, is to expose air to ultraviolet light.

Ozone can be condensed into a blue liquid or frozen into a violet-black solid; both are dangerously explosive.

Chemical element

Data file

CHEMICAL SYMBOL	O
ATOMIC NUMBER	8
ATOMIC WEIGHT	15.9994
MELTING POINT	−218°C
BOILING POINT	−183°C
DENSITY	1.43 g per litre

Oxygen is a non-metallic element and a member of group 16 (row 2p) of the periodic table. It is a colourless, odourless, two-atom gas, aka dioxygen O_2. Oxygen is reactive and will form oxides with all other elements except helium, neon, argon and kyrpton. It is moderately soluble in water (30 ccs per litre). Oxygen's preferred oxidation state is −2 (as in H_2O), but it can be −1 (H_2O_2), and as high as +2 (OF_2). There are numerous organic oxygen compounds such as alcohols (ROH), ethers (R_2O), ketones (R_2CO), and acids (RCO_2H), as well as more complex biological molecules like carbohydrates, for example, glucose ($C_6H_{12}O_6$). Many powerful oxidising agents contain a lot of oxygen such as the perchlorates ($MClO_4$), dichromates ($M_2Cr_2O_7$), permanganates ($MMnO_4$), and nitrates (MNO_3) where M is a univalent metal ion such as potassium K^+.

There are 17 known isotopes of oxygen of masses 12 to 28, with no known isomers. There are three stable naturally occurring isotopes of oxygen: oxygen-16 accounts for 99.76 percent, oxygen-17 (0.04 percent), and oxygen-18 (0.2 percent). There is also radioactive isotope-20 which is produced in minute amounts as the emitted nuclei from the decay of naturally occurring thorium-228.[133] The longest-lived radioactive isotope is oxygen-15 with a half-life of 122 seconds. Rather unexpectedly, the Sun has a higher proportion of the oxygen-16 isotope than oxygen here on Earth.

Because oxygen-18 is 12 percent heavier than oxygen-16, it can influence the behaviour of water and this has implications for the global environment. The ratio of oxygen-18 to oxygen-16 in the world's oceans has varied slightly over geological time and this has left an imprint on parts of the environment, providing evidence of past climates. When the world is in a cooler period, water molecules with oxygen-16 evaporate more easily than their heavier oxygen-18 counterparts and so the snow that falls is very slightly richer in the former, and the water that is left behind is very slightly richer in the latter. Marine creatures therefore lay down shells that have more oxygen-18 than expected and these are preserved in sediments. Analysing the oxygen-18/oxygen-16 ratio in such deposits reveals the cycle of global cooling and warming that has characterised the past half million years with its five ice ages.

The oxygen gas of the atmosphere is also marginally enriched in oxygen-18, although plants release oxygen with the same ratio as occurs naturally. The reason is that oxygen-breathing life forms consume the oxygen-16 at a slightly faster rate.

The ratio of the two isotopes in the remains of more recent skeletons has been used to show the regions from which they originally came. The nearer a person is to the Equator, the more oxygen-18 there is in their enamel of their teeth. For example, this has enabled the remains of people buried in England during the Roman period to be identified as having come from southern parts of the Empire.

Element of surprises

1. It was the lack of oxygen which brought the Biosphere project in Arizona to a premature end in the mid-1990s. Eight people had been sealed into the glass-walled eco-system to see if it was possible for humans to sustain life on the Moon

---

[133] The decay mode is known technically as cluster decay.

or Mars. Within a month they were gasping for breath as the oxygen in the air fell to 17 percent. Somehow 30 tonnes of the gas had disappeared and was thought to have reacted with iron in the soil.

2. Oxygen condenses to a pale blue liquid with magnetic properties. Oxygen was first liquefied in 1877 by the Frenchman Louis Cailletet (1832–1913), although he produced only a few drops, and not enough to observe this unusual property. It was left to the Scot James Dewar (1842–1923) in 1891 to make liquid oxygen in large enough quantities to study its properties. The reason for this strange behaviour is the unique chemical bonding between the two oxygen atoms, in which two of the bonding electrons of O_2 refuse to pair up as in normal chemical bonds, and these provide the magnetic effect.

3. When solid oxygen is put under increasing pressure it goes through six different crystal phases one of which, called the ε-phase, is dark red in colour and no longer magnetic. This phase exists at pressures above 80,000 atmospheres and in 2006 it was reported that the oxygen is present as O_8 but this is not the ring molecule corresponding to its sister element's S_8, Instead it consists of four O_2 molecules stacked sideways forming a boxlike structure. (In 2001, molecules of O_4 had been reported, but again these were really only a pair of loosely-bound O_2 molecules.)

   When oxygen is subjected to a sudden pressure of 1.2 million atmospheres it becomes a fluid that will conduct electricity and on this basis it can be classed as metal under these conditions.

4. Oxygen is very soluble in liquid perfluorocarbons, and bottles and sprays of such solutions may become available as emergence treatments capable of delivering oxygen to various organs of the body including the lungs. They also appear to be able to rejuvenate skin and remove fine lines and wrinkles.

Palladium

Pronounced pal-ay-dee-um, it is named after the asteroid Pallas, which was itself named after *Pallas,* the Greek goddess of wisdom.

French *palladium*; German *Palladium*; Italian *palladio*; Spanish *paladio*; Portuguese *paládio*.

Essential element

Palladium has a key role in the catalytic convertors of car exhausts and in electronic devices such as computers and mobile phones.

Element of life

The amount of palladium in the human body is not recorded but is very low. The element has no known biological role although it has been detected in mammalian tissue, albeit at levels of only 2 p.p.b.

Medical element, dangerous element

Palladium is regarded as being of low toxicity, being poorly absorbed by the body when ingested. Palladium chloride ($PdCl_2$) was formerly prescribed as a treatment for tuberculosis and was given in doses of 65 mg per day and without apparent ill effects—or benefits. Palladium at high doses would be poisonous, and tests on rodents have shown it to be carcinogenic, but there is no evidence that it has this effect on humans.

Element of history

Palladium is part of a wider story that resulted in the discovery of four platinum group metals—palladium, rhodium, osmium, iridium—by two London chemists, William Hyde Wollaston (1766–1828), and his friend and collaborator, Smithson Tennant (1761–1815).

As early as 1700, miners in Brazil were aware of *ouro podre,* 'worthless gold,' which was in fact a native alloy of palladium and gold. However, it was not from this source that palladium was first extracted, but from platinum, and this was carried out in 1803 by William Hyde Wollaston, in rather mysterious circumstances.

Wollaston, one of 14 children of a Norfolkshire clergyman, was educated at Cambridge University, where he got a medical degree, and he practised as a doctor for a few years. He also had an interest in chemistry. When he became partially sighted in 1800, he decided to devote more of his time to chemistry. He formed a working partnership with Tennant in 1800 to engage in various chemical enterprises, one of which was to investigate the production of pure platinum. They kept their partnership, and the work they were doing, secret. To begin with, their business struggled yet it was in these early years that the four new metals were discovered. Eventually their company was profitable, but it appears to have ceased trading around 1820.

Like others before them, who had dissolved platinum in aqua regia, they knew that not all of it went into solution and that it left behind a black residue. Tennant decided to investigate this residue and from it he eventually isolated osmium and iridium. Meanwhile Wollaston concentrated on the solution of platinum from which he extracted palladium and rhodium. First Wollaston reclaimed the platinum from solution by neutralising the acid with sodium hydroxide and then adding ammonium chloride. This precipitated the platinum as ammonium hexachloroplatinate, free of all other metal impurities. Heating this recovered the platinum.

Then by adding mercury cyanide, $Hg(CN)_2$, drop-by-drop to the platinum-free solution he obtained another, yellow, precipitate, which turned out to be palladium cyanide. He filtered this off and heated it strongly whereupon it decomposed to the metal.[134] His note book for 1802 shows that he thought first of naming it ceresium, after the asteroid Ceres which had been first observed in 1801,[135] but decided instead to honour the asteroid Pallas which had been discovered in March 1802 around the time that palladium was first made, and so it became palladium.

Wollaston estimated that there was about 0.5 percent palladium in the original platinum, but having separated this impurity and proved to his own satisfaction that it was a new metal, he was faced with a quandary. If he announced the discovery, the venture would become public knowledge, and the two partners did not want this to happen. Yet the chemist in him could not ignore what he had found, so he took most of his palladium to a couple who specialised in minerals, a Mr and Mrs Forster, whose shop was at 26 Gerrard Street, Soho, London, and there in April 1803 they advertised it for sale. Handbills referred to it as the 'new silver' and listed its properties and its price. It was on sale at a shilling a grain (65 mg) which was six times the price of gold. Wollaston had left more than 1,300 grains (about 80 g) with them. The editor of the *Journal of Natural Philosophy, Chemistry and the Arts,* one William Nicholson, also published details of the new metal, again without revealing its discoverer.

A noted chemist, Richard Chenevix (1774–1803), was sufficiently curious to purchase a sample of the 'new silver' and was so intrigued by it that he eventually returned to the shop and purchased more. He investigated its properties and declared at a meeting of the Royal Society that it was nothing less than an alloy of mercury and platinum. Yet, when other chemists tried to make it from these ingredients they invariably failed. Wollaston then issued a challenge, via Nicholson's journal, offering £20 (equivalent to £5,000 or $7,500 at today's values) to anyone who could make palladium from other metals.

Naturally the prize remained unclaimed, and in February 1805 Wollaston eventually revealed himself as the discoverer of palladium and in July of that year he read a paper to the Royal Society giving a full account of the metal and its properties. He admitted his secrecy, saying that he hoped to 'take advantage of it, as I have a right to

---

[134] The remaining solution still contained rhodium, which he obtained as rose red crystals of rhodium chloride.
[135] That asteroid gave its name to cerium.

do.' In fact there was little money to be made—even the Forsters had sold only a quarter of what he had given them—and eventually he reclaimed most of his palladium from them and deposited it with the Royal Society.

The only use to which palladium was put in its early years was for navigational instruments because it appeared resistant to corrosion. Later in the 1800s an alloy with 20 percent silver was introduced into dentistry.

Brazilian gold, which was imported into London in the early 1800s for making coins, sometimes gave coins that were much paler than normal and were suspected of being counterfeit. They were, in fact, the *ouro podre* mentioned above which consists of 86 percent gold, 10 percent palladium and 4 percent silver. There was little use for the palladium that had to be extracted from this gold, despite attempts to popularise it as untarnishable silver and showing that it was ideal for medals, some of which were cast to commemorate important events and presented to royalty.

Economic element

Specimens of native palladium have been found in Brazil, and there are some minerals rich in palladium, such as stibiopalladinite (palladium antimonide, Pd_5Sb_2) and braggite (a mixed palladium, platinum, nickel sulfide) but most palladium is extracted as a by-product from nickel refining in Canada, and copper and zinc refining in South Africa, North America, Russia, and Zimbabwe. World production is around 200 tonnes[136] and reserves are estimated to be 24,000 tonnes. A lot of palladium is recovered by recycling, amounting to around 45 tonnes per year.

Most palladium (125 tonnes) goes into catalytic convertors for cars, while the remainder is used for jewellery (25 tonnes), industrial catalysts (10 tonnes), electronic devices (40 tonnes), dental fittings (20 tonnes), and coins (2 tonnes) such as the Canadian Maple Leaf coin of weight 1 troy ounce (31 grams). A tiny amount of palladium is used to illustrate manuscripts in preference to silver leaf which tends to tarnish and need cleaning. Palladium objects can now be hallmarked thereby guaranteeing their quality.

In 1990, most catalytic convertors relied on platinum to reduce emissions from car exhausts, and while this metal is still important, palladium is now the main ingredient because this is even more efficient at removing unburnt and partially burnt hydrocarbons from the fuel. Moreover, it copes better with higher temperatures and it was instrumental in solving the problem of the high levels of unburnt fuel emitted when an engine starts from cold.

China is the largest market for palladium jewellery, but there is a growing demand for men's wedding rings to be made of palladium especially in the USA. In the chemical industry, palladium catalysts are used in the production of hydrogen peroxide, nitric acid, terephthalic acid, and vinyl acetate monomer. Terephthalic acid is used to make polyethylene terephthalate from which polyester fibre and plastic bottles are made, and vinyl acetate is turned into polyvinyl acetate, which is the basis of most paints and surface coverings.

---

[136] The trade measures output in troy ounces, which in this case is around 8 million ounces. (1 troy ounce is equivalent to 31 g.)

In electronics, palladium is part of the multi-layer ceramic capacitors that are needed for laptop computers and mobile phones. Hundreds of billions of these capacitors are manufactured each year; and the average digital appliance has 150 or more of them. They consist of layers of palladium sandwiched between insulating layers of ceramic, sometimes with as many as 50 layers interleaved.

Palladium metal has the unusual ability to absorb hydrogen gas, and it will do this to the extent of more than 900 times its own volume of the gas at room temperature. When a hydrogen molecule encounters the surface of the metal it breaks into its component atoms which then rapidly penetrate inwards. Hydrogen gas will even diffuse through the metal, to emerge and recombine as H_2, making this a useful means of separating and purifying hydrogen. The amount of palladium which goes into hydrogen purification is around 2 tonnes a year, but this should grow as the importance of hydrogen in the economy grows.

Palladium is quite malleable but cold working increases its strength and hardness. It resists corrosion except in sulfur-contaminated environments.

Palladium is used in dental alloys (with silver) mainly in Japan, although there is now a trend back to using gold, as the price of palladium has increased following demand from the automobile makers. Palladium is also used in other ways: to coat electrical contacts in switches; for non-magnetic springs in clock and watches; for special mirrors; and in jewellery where its alloy with gold is known as white gold.

Environmental element

Palladium in the environment	
EARTH'S CRUST	av. 0.6 p.p.b.
	Palladium is the 76th most abundant element, i.e. it is very rare.
SOILS	0.5 to 30 p.p.b.
SEAWATER	0.04 p.p.t.
ATMOSPHERE	nil.

Palladium has little environmental impact. It is present at low levels in some soils and the leaves of trees have been found to contain 0.4 p.p.m. (fresh weight). Some plants, such as the water hyacinth, are killed by low levels of palladium salts but most plants tolerate it, although tests indicate that at levels above 3 p.p.m. growth is affected.

Chemical element

Data file	
CHEMICAL SYMBOL	Pd
ATOMIC NUMBER	46
ATOMIC WEIGHT	106.42
MELTING POINT	1,552°C
BOILING POINT	3,140°C

DENSITY	12.0 g/cm³
OXIDES	PdO and PdO$_2$

Palladium is a lustrous, silvery-white, malleable and ductile metal and a member of group 10 (row 4d) of the periodic table. It is one of the so-called platinum group metals, namely ruthenium osmium, rhodium, iridium, palladium, and platinum, but is the least dense of these metals and has the lowest melting point. It resists corrosion, but dissolves in oxidising acids, namely sulfuric and nitric acids, and in molten alkalis. Like gold, palladium can be beaten into leaves as thin as 1 μm (micron) or less. Its preferred oxidation state is +2 (as in PdCl$_2$) and there are some compounds of oxidation state +4, such as PdF$_4$ and K$_2$PdF$_6$.

There are 38 known isotopes of palladium with masses of 91 to 128, and 11 known isomers. There are six naturally occurring isotopes: palladium-106 is the most abundant at 27 percent, palladium-108 (26 percent), palladium-105 (22 percent), palladium-110 (12 percent), palladium-104 (11 percent), and palladium-102 (1 percent). Palladium-110 is radioactive with a half-life of 6×10^{17} years. There are also traces of palladium-107 which is radioactive with a half-life of 6.5 million years, as well as traces of other radioactive isotopes, mainly palladium-109, produced as fission products in uranium-bearing deposits.

Element of surprises—and delusion

1. The Nobel Prize for chemistry in 2010 was awarded to Richard Heck, Ei-ichi Negishi, and Akira Suzuki for developing a palladium-catalysed reaction which forms carbon-to-carbon bonds, and which led to innumerable new compounds that were important not only for chemistry, but also for industry, medicine, agriculture, and electronics.

2. In March 1989, Martin Fleischmann and Stanley Pons held a press conference to announce that they had achieved 'cold fusion' by which they meant they had found a way of generating energy from a nuclear reaction under ordinary laboratory conditions. What they had done was to pass a current through an electrochemical cell containing lithium deuteroxide (LiOD). This liberated deuterium, the heavier isotope of hydrogen gas, at the palladium cathode and, like hydrogen, it can diffuse into the metal. They believed that there it underwent nuclear fusion to form helium with an accompanying release of energy in the form of heat.

   While the announcement created a flurry of interest at the time, carefully controlled experiments by others could not verify their claims which were eventually discredited. Clearly another source of heat was being tapped, but what this was remains unclear. Nevertheless, there are a number of other researchers who continue to explore this phenomenon. After many years and many experiments, they claim that they continue to get positive results. These reactions are referred to as LENR/CANR, which stand for Low Energy Nuclear Reactions/ Chemically Activated Nuclear Reactions.

Phosphorus

Pronounced fos-for-us, the name is derived from the Greek *phosphoros*, meaning bringer of light, and it was also the name given to the Morning Star (Venus).

French *phosphore*; German *Phosphor*; Italian *fosforo*; Spanish *fósforo*; Portuguese *fósforo*.

Essential element

When surrounded by four oxygen atoms this element as phosphate (PO_4^{3-}) is safe and a key part of both the global and biological components of life on Earth as well as providing some useful chemicals, but when the oxygens are stripped away all kinds of devils are let loose.

Cosmic element

Phosphorus nuclei are formed in large stars—much larger than our Sun—when two oxygens fused together followed by loss of a proton ($8 + 8 - 1 = 15$).

A lot of the phosphorus on Earth has come from meteorites the influx of which amounts to around 100 tonnes per day. These have an average content of 0.1 percent of phosphorus (mainly as the mineral schreiberite) so that altogether this represents an input of 35 tonnes of phosphorus per year. Over the lifetime of the Earth, which is 4.6 billion years, the amount of phosphorus added to the environment has been around 100 billion tonnes, helping to replenish that which living things use but which is continually being lost to the depths of the oceans—see *Environmental element*.

Element of life

Phosphorus in the human body	
BLOOD	345 p.p.m.
BONE	av. 70 000 p.p.m. (7%)
TISSUE	3,000–8,000 p.p.m. (0.3–0.8%)
TOTAL AMOUNT IN BODY	780 g.

In the natural world phosphorus is never encountered as the element itself, only as phosphate. This can exist as the negatively charged phosphate ion (PO_4^{3-}), which is how it occurs in minerals, or as organophosphates in which there are organic molecules attached to one or two of the oxygens, which is how it occurs in living cells. The phosphate in blood is 92 percent organophosphate and only 8 percent is present as the simple phosphate ion. Blood levels of the latter are around 30 p.p.m. in adults, but double this in children, showing how important this element is for their growth and development. The body organ relatively rich in phosphate is brain tissue, although most phosphate is present in the skeleton because bone is largely calcium phosphate.

Phosphate is essential to life as a component of DNA, albeit only a small part. Other organophosphates that are essential are the energy molecule ATP (adenosine triphosphate), the messenger molecule GMP (guanosine metaphosphate), and the phospholipids which make up cell membranes. ATP is used by the body on a truly remarkable scale, more than 1 kg per hour of ATP is produced, used, and recycled, drawing its energy from the breakdown of glucose. During strenuous exercise, the requirement for ATP can rise to 1 kg every two minutes.

DNA, ATP, GMP, and most organophosphates are negatively charged and as such they must be balanced by a positive charge, the preferred one being the magnesium ion Mg^{2+}.

Food element

Phosphate is a dietary requirement, the recommended intake being 800 mg per day, but it is not something that a person need worry about. A normal diet provides between 1,000–2,000 mg per day and depends on the extent to which phosphate-rich foods are consumed, such as tuna, salmon, sardines, liver, turkey, chicken, eggs and cheese, all of which have more than 200 mg per 100 g.

The amount of phosphorus that is naturally present in food varies considerably but it can be as high as 370 mg per 100 g in liver, or it can be very low, as it is in vegetable oils. Lean meat generally contains 180 mg per 100 g and eggs are also a good source of phosphorus with 220 mg per 100 g. Certain types of processed cheese can have as much as 500 mg per 100 g, but this is high because of the 2 percent of disodium phosphate that is added to emulsify the fat in the cheese and prevent it from separating. Other foods processed with phosphates are sausages and cooked meats.

Not all phosphate-rich foods are a source of this nutrient, however. Plants store phosphate in their seeds as inositol hexaphosphate (IHP) so that they can germinate and put down roots without relying on phosphate from the soil, although it is ultimately from the soil that the plant gets its phosphate for growth and to provide its own seeds with IHP. Foods like nuts are rich in this kind of phosphate, but IHP is not a form of phosphate that is digestible. Wholemeal flour has 340 mg phosphate per 100 g but a lot of this is IHP. (White flour has much less, 130 mg per 100 g, but this is mostly digestible.)

Since phosphate is an essential part of the diet, it is perhaps not surprising to find that phosphates are considered safe when used as food additives. The permitted phosphate additives used most frequently in foods are various sodium, calcium, and potassium salts of phosphoric acid (H_3PO_4), and even the acid itself is added to colas. Disodium phosphate is added to evaporated milk to emulsify the fat and it also gives the necessary calcium-to-phosphate balance that prevents this type of milk from becoming semi-solid when it is stored. Phosphate food additives were first introduced in the nineteenth century as leavening agents to give cakes, pastries, and biscuits a lighter texture. They consisted of calcium dihydrogen phosphate and sodium bicarbonate, and they are still the key ingredient in self-raising flour.

Phosphate food additives that have been approved for use are identified by their E-code numbers, thus sodium phosphates are coded E339, potassium phosphates are E340, calcium phosphates are E341, and magnesium phosphates are E343. Ammonium phosphates are E342 but these are not permitted in some countries. Phosphoric acid (H_3PO_4) is coded E338.

The skeleton represents a massive reserve of phosphate that the body can draw on, and it continues to add and subtract from this store throughout life. Bone may appear to be inert but it is constantly being deposited and re-dissolved at millions of sites on its surface—see below. Too little phosphate in the body is a condition known as hypophosphatemia and can be caused by too little phosphate in the diet.

Dangerous element[137]

The form in which the element itself is best known is white phosphorus. This is dangerously flammable and a deadly poison—indeed as little as 100 mg may be a fatal dose for a human. As a poison, phosphorus quickly attacks the liver and death will result within a week.

Breathing in phosphorus vapour over a long period of time led to the industrial disease known as phossy jaw, which slowly ate away the victim's jaw bone. This condition afflicted those who worked making phosphorus matches in the 1800s, but cases of the disease continued to crop up in the last century among those who worked with it.

White phosphorus is transported as the liquid and trains carrying this dangerous chemical have been involved in serious accidents. At Brownston, Nebraska, on 1 April 1978, a rail tanker containing 50,000 litres of molten phosphorus broke open during a crash and burned for three days, destroying 30 acres of crops. A similar tanker was derailed at Miamisburg, Ohio, on 8 July 1988, causing the evacuation of all 15,000 residents of that town while the tanker burned itself out. On 17 July 2007, a tanker carrying phosphorus was part of a freight train en route from Kazakhstan to Poland and it became derailed and caught fire at Lviv on the Ukraine/Poland border. Thankfully there were no casualties. (There is a safer form of phosphorus: red phosphorus which is not flammable. It is made by heating white phosphorus in a sealed vessel for several days.)

There are some highly toxic organophosphates capable of blocking the enzyme, *acetylcholinesterase*, which is a key part of the way the body transmits signals from one nerve ending to another. When the enzyme is blocked there is malfunctioning of major body organs, such as the heart, resulting in death. It is these kinds of organophosphates which are used as chemical warfare agents, the most infamous of which are tabun and sarin.

Medical element

Despite the dangers, phosphorus was a widely used pharmaceutical, and books like *Free Phosphorus in Medicine*, published in 1874, extolled its supposed benefits. It was given only in doses of a 1/20th of a grain (3 mg) but was particularly recommended

---

[137] For a fuller account of the disasters associated with this element down the centuries, see my book *The Shocking History of Phosphorus* (Macmillan), also published as *The 13th Element* (USA: John Wiley & Sons).

for various mental conditions, such as nervous breakdown, depression, migraine, epilepsy, psychiatric disorders, and even erectile dysfunction. It was also given to treat physical conditions such as stroke, pneumonia, alcoholism, tuberculosis, cholera, and cataracts. Phosphorus tonics and toothache treatments were sold as over-the-counter medicines well into the last century, and it was still prescribed for certain bone conditions in the 1920s, although by 1930 it had been eliminated from medical pharmacopoeias.

Much safer than prescribing phosphorus was to give a partially oxidised form known as hypophosphorous acid (H_3PO_2) or its salts, known as hypophosphites. These came into vogue after 1857 and were popular as a treatment for various conditions such as neurasthenia, a condition characterised by a general lack of energy, and even prescribed for tuberculosis. They too were useless, but were sold as over-the-counter medication for decades in the form of Scott's Emulsion which was a blend of cod liver oil and calcium hypophosphite.

Osteoblasts create bone; osteoclasts digest it to release calcium so that its components can be recycled to other sites in the body. However, with age comes loss of bone which becomes much weaker leading to osteoporosis and its associated risk of fracture during a fall. There are phosphorus-containing drugs to treat this condition. They are the bis-phosphonates, whose molecules consist of two phosphates linked through a carbon, that is, $[O_3P\text{-}C\text{-}PO_3]^{4-}$, the carbon having two other groups attached to it (not shown). Bisphosphonates are used to treat other conditions that result in bone wastage such as Paget's disease and bone cancers. The drugs work by binding strongly to the calcium in bone and this interferes with the action of the osteoclasts. However, there are side effects associated with the bisphosphonates which some claim can cause joint pain. On the other hand, bisphosphonate has been found to boost the immune system and destroy cancer cells.

Poly-phosphates are chains made up of many phosphate groups joint together and these have a role in blood clotting at wounds. The platelets which rush to control bleeding carry these inorganic polymers.

Element of history

Phosphorus was first produced by Hennig Brandt (c. 1630–1710) at Hamburg in 1669 when he evaporated urine and heated the residue until it was red hot, whereupon phosphorus vapour distilled from it and this he collected by condensing it in water. Brandt tried to keep his discovery secret, thinking he might have discovered the Philosopher's Stone that could turn base metals into gold, but to no avail. Eventually he was reduced to selling it for a living and he supplied phosphorus to Daniel Kraft (1624–97) who exhibited it around Europe. He offered to sell the secret to Johann Kunckel (1630–1703) of the University of Wittenberg, but the latter was unwilling to pay the price and went off and discovered how to make it himself. Brandt also earned money by showing Gottfried Leibnitz (1646–1716) how to make it. By these various acts and his own reticence, he even forfeited the claim to be its discoverer, because these three men were often given the credit for it.

The most successful phosphorus producer of the 1700s was Ambrose Godfrey (1660–1741) laboratory assistant of the London-based scientist Robert Boyle

(1626–91). Boyle was fascinated by phosphorus and he too discovered how to extract it from urine, but he carried out chemical investigations into it. Godfrey saw it as a means of becoming rich and he went into the business of making and selling phosphorus and supplied most of Europe with it from his premises near Covent Garden in London. Phosphorus was sufficiently expensive for its use to be mainly medicinal.

It was not until it was realised that bone was calcium phosphate, and that this too could be used to make phosphorus, that it became more widely available. In the early 1800s the phosphorus match was invented, and this spurred demand for the element. Most was manufactured by heating phosphoric acid (produced by dissolving bone in sulfuric acid) with charcoal. By the end of that century, phosphorus was being extracted from mineral phosphates by heating these with coke in an electric furnace.

Element of war

White phosphorus was used in the wars of the last century in tracer bullets, incendiary bombs, and smoke grenades. Phosphorus shells, ostensibly fired to act as markers for aiming mortars, proved to be a weapon of terror because of the terrible wounds that burning phosphorus produced. Such weapons are not forbidden by international law when used to create smoke screens or directed against military targets.

The scattering of phosphorus fire bombs over cities in World War II was to cause widespread destruction. The week of air raids on Hamburg in July 1943 is now notorious for use of phosphorus bombs, when as many as 25,000 of them were dropped at a time, each weighing 14 kg (30 lbs). Almost 2,000 tonnes of burning phosphorus fell on that hapless city. More recently, phosphorus shells were used by Israeli armed forces as in the Lebanon War of 2006 and the Gaza conflict of 2009.

Red phosphorus might have prospects in munitions of the future and it is being promoted as a primer for small arms cartridges in place of lead styphnate, which like all lead compounds is now seen as a threat to the environment. The new primer is manufactured by ATK in the USA and sold under the trade name P$_4$rimer.

Another group of warfare agents based on phosphorus are the nerve gases such as tabun, sarin, and the VX agents which have been stockpiled in the past, and sometimes used by terrorists and dictators.

Economic element

There are many phosphate minerals, the most abundant being forms of apatite $Ca_5(PO_4)_3X$ where X can be fluoride, hydroxy (OH), or chloride.[138] Fluoroapatite provides the most extensively mined deposits. The chief mining areas are China (45 million tonnes), USA (30 million tonnes), Morocco (28 million tonnes), Russia (11 million tonnes), and Tunisia (8 million tonnes), but phosphate rock is mined in 30 other countries including the Philippines where is comes in the form of weathered deposits of accumulated bird droppings. World production is 160 million

---

[138] The gem stone turquoise is a copper aluminium iron phosphate mineral of composition $Cu(Al/Fe^{3+})_6(PO_4)_4(OH)_8$.

tonnes per year and reserves amount to around 6 billion tonnes of easily accessible ore, although there are vast deposits of poorer quality ore.

Phosphate rock is processed in one of two ways: either it is dissolved in sulfuric acid, which converts the phosphate to phosphoric acid (H_3PO_4), or it is heated with coke and sand in an electric furnace which releases white phosphorus. This method has now been mainly phased out in the West, leaving China as the main producer of the elemental phosphorus needed for making raw materials such as phosphorus trichloride (PCl_3) and phosphorus pentasulfide (P_2S_5) which are used to make many other products. For example PCl_3 is the starting material for the manufacture of the popular herbicide glyphosate.

Most phosphate rock ends up as phosphoric acid, destined to be used to make fertilizers, and around 50 million tonnes of the acid are made each year. Some phosphoric acid that is produced directly from apatite can now be purified sufficiently to enable it to be used in foods, but previous food-grade material was made from phosphoric acid made from white phosphorus. This was converted to phosphorus pentoxide by burning, and then to the acid by dissolving the oxide in water. Purified acid is used to make food additives, animal feed supplements, surface cleaners, detergents, and dishwasher powders. Lower grade phosphoric acid is used industrially to give metals a protective layer and as a rust-remover and rust-preventer.

Early in the 1800s, it was found that treating bone or mineral phosphates with sulfuric acid produced a soluble form of calcium phosphate, called 'superphosphate', and that applying this to farmland boosted crop production. All fertilizers need to contain phosphate, either in the form of calcium hydrogen phosphate, which is soluble and therefore available to plants, or ammonium phosphate, which is also soluble and, in addition, provides the essential element nitrogen.

There was increased demand for phosphorus in the two World Wars, but the biggest boost to production came after 1945 with the discovery of the benefits of adding sodium tripolyphosphate ($Na_5P_3O_{10}$) to laundry products. Eventually phosphorus production reached a million tonnes a year by the 1970s, although it has since steadily declined from this high level. The phosphate in detergents has been replaced by zeolites (see silicon) but some sodium tripolyphosphate is still used in dishwasher powders. This softens water by neutralising calcium.

Some of the elemental phosphorus which is converted to phosphorus trichloride (PCl_3) goes to making phosphorous acid (H_3PO_3) needed for flame retardants, insecticides, and weed-killers. A little is turned into phosphorus sulfides which go into oil additives, as corrosion inhibitors and to reduce engine wear, and into strike-anywhere matches. Other compounds made from elemental phosphorus are the gas phosphine (PH_3) for making flame-proofing agents and biocides, zinc phosphide (Zn_3P_2) which is used as a rat poison, and magnesium phosphide (Mg_3P_2) which is the basis of warning flares used at sea. In the flares the magnesium phosphide is combined with calcium carbide and when the mixture gets wet, the phosphide forms the spontaneously flammable gas diphosphine (P_2H_4). This ignites the acetylene gas (C_2H_2) given off by the calcium carbide as it too reacts with water.

Disodium phosphate (Na_2HPO_4) is used for making the special glass needed for sodium lamps, for ceramics such as bone china, in leather tanning, for water softeners, and for dye manufacture. Calcium hydrogen phosphate is used in toothpaste and sodium monofluorophosphate (Na_2PO_3F) is added to this to provide fluoride.

Some synthetic fibres, such as polyester, can be made flame retardant by incorporating a phosphorus-containing compound into the polymer chain itself.

High grade phosphorus is used to make metal phosphides, such as those of gallium and indium which are used in light emitting diodes. The phosphorus for this has to be 99.9999 percent pure, in other words have only one impurity atom per million atoms of phosphorus, and around 7 tonnes of such high grade phosphorus are produced each year.

Environmental element

Phosphorus in the environment	
EARTH'S CRUST	1,000 p.p.m.
	Phosphorus is the 11th most abundant element.
SOILS	0.65 p.p.m.
SEAWATER	surface waters have 1.5 p.p.b., deep ocean waters have 60 p.p.b.
ATMOSPHERE	trace amounts.

In 2010, controversial research was published which indicated that the temperature of the seas on the early Earth around 3.5 billion years ago, were not 85°C as previously assumed, but were nearer 40°C and therefore more conducive to life. This was deduced by examining the oxygen isotopes in phosphate deposits which were laid down at the time.

Phosphorus as phosphate is of crucial importance to the environment because this element is one of the limiting factors controlling the population of species in the sea and on land. In the oceans, the concentration of phosphate is very low, particularly at the surface. The reason lies partly with the insolubility of calcium, aluminium, and iron phosphates. In the oceans phosphate is quickly used up and falls to the deep as organic debris. Nevertheless, there can be excess phosphate in rivers and lakes, resulting in excess slimy green algae which can choke out other living things. For this reason there was once a campaign against the use of phosphates and laws were passed in many countries reducing the amount permitted in detergents and fertilizers—but see below.

Phosphorus cycles through the environment but this particular cycle is unlike the other natural cycles involving key elements, such as carbon and nitrogen, because phosphate cannot circulate via the atmosphere and its major environmental movement is from soil to rivers, to oceans, to bottom sediment, where it accumulates until it is moved by geological uplift to becoming dry land again, a process that can take hundreds of millions of years. During the course of this downward drift the phosphate is cycled and recycled through the many plants, microbes, and animals of the various eco-systems.

The accessible phosphate on land amounts to around 200 million tonnes, of which around 3 million tonnes (1.5 percent) are in living things, and from where 20,000 tonnes a year of phosphate are washed to the seas. There it joins the 80 million tonnes already present, of which only 100,000 tonnes (0.12 percent) is in living things. Slowly the phosphate sinks to the bottom of the oceans to become part of the trillion tonnes of phosphate rock of the Earth's crust. While all these numbers seem large in absolute terms, on closer analysis their relative amounts show how sensitive the planet is to this element. The oceans of the world could support a vast population were all its phosphate to be accessible to living organisms, but most of this is at a depth to which sunlight cannot penetrate so there is no energy input to enable marine plants to flourish and thereby sustain a food chain. For the seas, all the sunlight is at the top and all the phosphate is at the bottom. The result is that the oceans of the world are really marine 'deserts' in that they support relatively little life except where there are up-welling ocean currents which bring phosphate-rich water to the surface, enabling marine organisms like algae and fish to flourish. (Iron is also needed by marine algae—see page 260.)

Phosphorus as phosphate was for many years regarded as the worst polluter of lakes and inland seas, especially where these were fed by rivers whose waters had been enriched by industrial effluent, domestic sewage, and run-off from over-fertilized land, all of which carried high levels of this nutrient. The result was eutrophication, which produced vast mats of blue-green algae and made life impossible for other aquatic creatures. Phosphate was assumed to be the cause as this is the nutrient that limits growth under normal conditions.

Rivers like the Rhine, and the Great Lakes of North America, suffered most. This led to campaigns against the use of phosphates in detergents, these being seen as the primary cause because the onset of the problem seemed to coincide with the introduction of these products which contained up to a third of their weight as phosphate. In fact, research in the 1990s showed that phosphate was not as environmentally damaging to natural waters as once believed—it did encourage algae to flourish, but these were not held in check because the species that normally fed on them, the zooplankton, had been killed by other pollutants such as heavy metals and pesticides.

The largest environmental disaster involving elemental phosphorus was that at Placentia Bay, Newfoundland in 1968. The cause was waste waters from a newly built phosphorus facility which resulted in the deaths of millions of fish over a period of several months. What puzzled investigators was that lobsters in pots that were sited near the waste water outlets of the plant came were completely unharmed. What was not realised at the time was that though the level of phosphorus in the waste water was tiny, and at very low levels when this had been diluted by seawater, it could nevertheless accumulate in the livers of fish until it reached levels that were toxic. The problem was solved by releasing no waste water to the bay but recycling it instead. Even so, that phosphorus plant has now closed down.

Chemical element

Data file	
CHEMICAL SYMBOL	P
ATOMIC NUMBER	15

ATOMIC WEIGHT	30.973761
MELTING POINT	44°C (white phosphorus)
BOILING POINT	280°C (white phosphorus)
DENSITY	1.8 g/cm³ (white), 2.2 g/cm³ (red), 2.7 g/cm³ (black)
OXIDES	P_4O_6 and P_4O_{10}

Phosphorus is a non-metal and a member of group 15 (row 3p) of the periodic table. There are several forms of phosphorus, called white, red, and black phosphorus although their colours are more likely to be slightly different from those implied by these descriptions. White phosphorus is the kind manufactured industrially and as we have seen this is the form which glows in the dark, is spontaneously flammable when exposed to the air, and is a deadly poison. It consists of P_4 molecules. The glow of white phosphorus comes from a slow chemical reaction in its surface in which two transient light-emitting molecules are formed, namely HPO and P_2O_2. The lower oxide of phosphorus, P_4O_6, also glows but very faintly.

Red phosphorus is made by heating white phosphorus for several days in a closed vessel; it does not glow, is stable in air, and is not poisonous. It consists of randomly interlinked phosphorus atoms. Red phosphorus can vary in colour from orange to purple, due to slight variations in its chemical structure. The third form, black phosphorus, is made under high pressure, looks like graphite and, like graphite, has the ability to conduct electricity. Little is made because no use has yet been found for it.

Phosphorus can exhibit several oxidation states ranging from −3 (as in PH_3) to +5 (as in PCl_5). The most common states are +3 as in PCl_3, and HPO_3^{2-} in solution, and +5 as PCl_5, H_3PO_4, PO_4^{3-} in minerals, and HPO_4^{2-} in solution.

There are 24 known isotopes of phosphorus of masses 24 to 46, and no known isomers. Natural phosphorus consists almost entirely of the isotope phosphorus-31 which is not radioactive, with traces of radioactive phosphorus-32 which is produced by cosmic ray spallation of argon in the upper atmosphere. Phosphorus-32 is also produced for research purposes, and this has a half-life of 14 days. The longest-lived radioactive isotope is phosphorus-33 with a half-life of 25 days.

Element of surprises

1. In the Sherlock Holmes story, *The Hound of the Baskervilles*, the large dog which appears so frightening has been painted with white phosphorus to make it glow like a ghostly apparition. There are reports of people in Victorian times applying a solution of white phosphorus to the skin to produce the same theatrical effect.

2. There is now a legal requirement to remove phosphate from sewage in many parts of the world and this may one day result in almost all the phosphate needed for farming and industry being reclaimed from this source. New sewage technology allows phosphate to be recovered effectively and in some respects this is a cleaner source of material because it is free of contaminants such as cadmium and uranium that have to be removed from phosphate rock before it can be used. One day, the phosphates used in dishwashers and even in processed

foods, might well derive from the waste we dispose of down kitchen sinks and bathroom toilets.

3. In 2004, Japanese researchers discovered that molten red phosphorus at 1,000°C (under pressure) behaved really oddly: it could exist as two distinct liquids at the same time. Such a situation had always been assumed to be impossible because liquids, like gases, are homogeneous states of matter with their molecules endlessly in motion and intermingling. However, in liquid red phosphorus there were round drops of high density liquid phosphorus co-existing within lower density liquid phosphorus.

4. For 40 years from 1862 onwards, hundreds of people made their living in Bedfordshire and neighbouring counties in England by digging up fossilized dinosaur bones and selling them for fertilizer manufacture. These lay about a metre below the surface of the soil and consisted of a bed of rock two metres deep. This stretched for more than 100 miles along what had once been coastal strip where all these creatures had drowned following a meteor impact around 95 million years ago.

5. In 2010 the rare and highly reactive molecule P_2 was produced.

Platinum

Pronounced pla-tin-um the name is derived from the Spanish *platina*, meaning little silver. French *platine*; German *Platin*; Italian *platino*; Spanish *platino*; Portuguese *platina*.

Essential element

Valuable in its own right and used as bullion and for expensive jewellery, this metal is more important as catalysts both for car exhausts and for the chemical industry.

Cosmic element

Platinum levels are higher in rocks brought back from the Moon, and found in meteorites, than in the Earth's crust. Meteor impact sites on Earth, such as that of the Sudbury Basin in Canada, also have associated deposits containing platinum which can be mined economically, but most of the platinum-bearing rocks welled up from the Earth's mantle during tectonic eruptions billions of years ago.

Element of life

The amount of platinum in the human body is not known, but is low. It has no known biological role. Plants growing on the soil over platinum-bearing rocks can absorb up to 6 p.p.m., but it stays mainly within their roots. In the muscles of mammals the level of platinum is around 0.2 p.p.b.

Dangerous element

Platinum metal is non-toxic and platinum implants are generally well tolerated by the body, although some platinum compounds are poisonous. About half of those who come into regular contact with platinum chemicals eventually suffer an allergic reaction, known as platinosis, and experience symptoms akin to those of asthma and the common cold. Those afflicted are few in number because few people handle such compounds.

Medical element

In 1962, Barnett Rosenberg was investigating the effects of electromagnetic fields on cell division and noticed the formation of giant filaments growing from *E. coli* bacteria. Curiously, the cells were not dividing. It took him three years to find out why, but at the end of that time he had a chemotherapy drug that went to clinical trials in the early 1970s, and in 1978 it was approved for use. It has since saved the lives of thousands of cancer patients with tumours of the testicles, breast, head, and neck. It is especially effective in treating children, and its cure rate exceeds 90 percent.

Rosenberg had been using platinum because he assumed it would be completely inert. In fact a small amount of it had reacted with the chloride and ammonium ions in the culture medium to form the compound cis-dichlorodiamminoplatinum(II), a chemical which had first been made in 1845 and was known as Peyrone's chloride, $PtCl_2(NH_3)_2$. This drug now has a generic name, cisplatin, and a trade name, Platinol.

Cisplatin, and similar platinum anti-cancer drugs, work by latching on to DNA at any point where it has two guanine bases next to each other. This distorts the DNA, thereby making replication impossible, nor can the cisplatin molecules be dislodged by DNA repair enzymes. Cisplatin affects not only cancer cells but other organs of the body, in particular the kidneys and bone marrow. Many of those treated experience side effects such as vomiting and numbness in the hands and feet but less potent variants of the drug, such as carboplatin and oxaliplatin, cause fewer side effects and are less damaging to the kidneys. Carboplatin is the preferred treatment for ovarian cancer.

Element of history

Probably the oldest worked specimen of platinum is that from an ancient Egyptian casket of the seventh century BC, which was unearthed at Thebes and was dedicated to the Queen Shapenapit. However, it appears to be the only example from Egypt and there is no evidence that the Greeks and Romans (nor the Chinese) knew of this metal, but across the Atlantic it was known and used. Native people of South America, living on the Pacific coast of what is now the Columbia/Ecuador border, were able to work platinum, as shown by burial goods dating back 2,000 years.

The Spaniards conquered this area in the 1500s but they were more interested in gold and it was not until the 1700s that they began to take an interest in the curious metal that could be polished like silver, but unlike silver it never tarnished. Some alluvial deposits in South America were rich in both metals, especially those of the Chocó region and around the river Pinto, in Columbia, and platinum was often called *platina del Pinto*. Nuggets from these locales were 85 percent platinum alloyed with around 8 percent iron, plus a few percent of the other platinum metals such as palladium, rhodium and iridium.

It is impossible to know who discovered platinum, and the question really is: who first recognised it as a new metal in its own right? In 1557 an Italian scholar, Julius Scaliger, wrote of a metal from Spanish Central America that could not be made to melt and this must have been platinum. A manuscript dated 1726 and written by José Sánchez de la Torre y Armas described the separation of gold and platinum. However, the man generally credited with bringing platinum to people's attention is Antonio Ulloa (1717–95). In 1735 he set sail for Panama where he first encountered the new metal and he described it in detail in his diary, later published. The French ship that Don Antonio embarked on to return to Spain, was captured by the Royal Navy and he ended up in England. There he was looked after by members of the Royal Society, to which he was elected a fellow.

Platinum also came to the attention of Charles Wood (1702–74) who came across it in Jamaica in 1741 and sent a sample to the Royal Society, but not until 1750. That year the society's journal *Philosophical Transactions* contains several papers on the new metal from William Watson, reporting work done by William Brownrigg (1711–1800) and Wood. Others who noted the new metal were the Swede Henrik Scheffer in 1752, the Englishman William Lewis (1714–81) in 1754, the German Andreas Marggraf (1709–82) in 1757, and two Frenchmen Pierre Macquer (1718–84) and Antoine Baumé (1728–1804) in 1758.

In the 1700s, the Spanish authorities tried to ban platinum because of the adulteration of gold with it, and for a time they closed the mines where it was to be found and decreed that existing stocks of the metal were to be disposed of into deep water. Only later did they realise that *platina* was a useful metal and in the years 1788–1805 they shipped three tonnes across the Atlantic to Spain. Platinum was given to several European scientists, who were keen to investigate the new metal, in the hope that they might discover uses for it. However, platinum could not be cast as a metal because its melting point is too high and in the 1700s and 1800s it was worked by hammering hot platinum sponge (made by heating ammonium hexachloroplatinate). This produced objects that looked as if they had been cast from the molten metal.

Economic element

It has been said that 20 percent of all consumer goods either contain some platinum or are produced using it. Indeed this metal is involved in oil refining, car exhausts, fibre optic cables, computer hard disks, fertilizers, paints, jewellery, anti-cancer drugs, laboratory equipment, and pacemakers. It is expensive and the price reflects the fact that generally demand exceeds supply. Platinum is traded in troy ounces and world production in 2009 was around 6 million ounces. (A troy ounce is 31 grams which means production worldwide amounts to 186 tonnes.) That was not a typical year because of the global financial crisis. Of the 239 tonnes of platinum traded in 2006, the last year before the crisis began, 130 tonnes were for automobile catalytic convertors, 49 for jewellery, 13 for electronics, 11 for the chemical industry, and a similar amount for medical and biomedical treatments, as well as several smaller markets.

Around 75 percent of the world's platinum comes from South Africa, where it occurs as cooperite (platinum sulfide, PtS), while Russia is the second largest producer accounting for 10 percent, followed by Canada, where sperrylite (platinum arsenide, $PtAs_2$) occurs in the nickel-bearing deposits of Ontario.[139] Zimbabwe and the USA also produce platinum. This is a difficult metal to extract; even the richest deposits contain only 5 g of the metal per tonne of rock processed. Some platinum is extracted as a by-product of copper and nickel refining. World platinum reserves are estimated at 27,000 tonnes, three quarters of which are in the 150-mile long Merensky Reef, a volcanic region in South Africa. This was discovered in 1924, but it could not be mined until a successful method of extraction was developed in the 1970s.

[139] Other platinum ores are braggite (platinum palladium sulfide, PtPdS) and moncheite (platinum telluride, $PtTe_2$).

Nuggets of platinum occur naturally, as does an alloy of platinum and iridium known as platiniridium. One nugget of platinum found at Kondyor in eastern Russia weighed 3.5 kg, and was worth around $100,000. Often platinum is alloyed with a small amount of iridium which increases its durability and hardness.

Platinum rings, earrings, and necklaces are particularly popular in Japan which accounts for half the world demand for platinum jewellery. The metal was also fashionable in Europe during 'La Belle Epoque', the period before World War I. Platinum jewellery began to be hallmarked around 50 years ago as a guarantee of its quality. Ring quality platinum is an alloy of 95 percent platinum and 5 percent of other metals including copper, titanium, and palladium. Some upmarket watches have casings made of platinum. When King George VI was crowned king emperor in 1937, his wife Queen Elizabeth was crowned with a new crown made entirely of platinum and decorated with the Koh-i-Noor diamond. It is now on display in the Tower of London.

Such was the respect for platinum's stability and resistance to corrosion, that it was chosen for the standard metre and the standard kilogram, both housed at the International Bureau of Weights and Measures, Paris. A duplicate kilogram weight is kept in Russia, and was originally standardised against the Paris one. Strangely it was found to have decreased by 17 micrograms (0.0017 percent) when it was checked after a century of resting in a sealed underground vault. The loss was probably due to traces of osmium which can be oxidised by the air to form volatile osmium tetroxide.

A lot of the platinum which goes into catalytic convertors is recovered, with the USA being particularly geared for such recycling. A typical catalyst contains less than 2 g but this covers a surface area inside the convertor that is greater than the area of a football playing field. The catalytic convertor removes unburnt hydrocarbons, carbon monoxide, and nitrogen oxide emissions by converting them to carbon dioxide, water, and nitrogen gas.

The industries which need platinum are the chemical, electrical, glass, and aircraft industries, each accounting for about 10 tonnes of the metal per year. In the chemical industry a fine platinum gauze catalyst is required for the manufacture of nitric acid from ammonia. This has to be replaced constantly because the metal slowly becomes used up; about 0.3 g of platinum is lost per tonne of nitric acid produced. Silicone production, and that of benzene and xylenes, also relies on platinum catalysts. The fine holes (spinnerets) through which molten glass and rayon are forced in order to convert them to fibres are made of platinum.

Electrical industries use platinum coatings for computer hard disks, thermocouples, and fuel cells. Computers have a hard disk made either from aluminium or high-quality glass coated with several layers, one of which is a cobalt-platinum alloy with magnetic properties in which 'bits' are stored in a series of circular tracks. Demand for these disks is now consuming around 15 tonnes of platinum a year.

Proton exchange membrane (PEM) fuel cells are likely to become important for electric vehicles later this century, because they operate at low temperatures, start-up from cold and are compact. Each vehicle will need about 10 g of platinum. The cells consist of platinum-coated carbon electrodes separated by a polymer membrane. At the anode, the fuel (hydrogen gas) releases its electrons and the stripped hydrogen nuclei (the protons) then migrate through the polymer to the cathode where they react with oxygen from the air to form water and pick up electrons. The net result is a flow of current.

There are many other uses for platinum. As well as for making optical fibres and liquid crystal display glass, especially for laptop and hand-held computers, platinum is used to coat turbine blades in jet engines, and spark-plugs in automobiles. For balloon catheters, pacemakers, and dental alloys, a 90/10 platinum/osmium alloy is used. Like gold, platinum can be hammered into sheets of less than a micron in thickness and such foil has been used as a coating for the cones of space missiles. Platinum has the same thermal expansion as glass so it is used to make sealed electrodes in glass systems. An alloy of 77 percent platinum and 23 percent cobalt is an extremely powerful magnet. Platinum wires are used for the heating element in high-temperature electric furnaces. Even Post-it notes rely on platinum which is used as a catalyst in the production of their silicon-based adhesive.

Some platinum, around 6 tonnes a year, is sold as bullion to investors and collectors in the form of small bars and coins. The American Eagle, the USA's first platinum coin, with a face value of 25 dollars, was introduced in 1997 and is the most traded.

Environmental element

Platinum in the environment	
EARTH'S CRUST	*approx.* 1 p.p.b.
	Platinum is the 75th most abundant element, in other words it is rare.
SOILS	varies from 0.5 to 75 p.p.b.
SEAWATER	*approx.* 20 p.p.t.
ATMOSPHERE	nil.

Platinum is at such low levels in the environment that it poses no threat to plants or animals. The analysis of snow in Greenland has shown that the level of platinum increased six-fold once mining began in the 1800s compared to snow which fell 7,000 years ago. Today it is 45 times higher due to the widespread use of catalytic convertors for cars.

Chemical element

Data file	
CHEMICAL SYMBOL	Pt
ATOMIC NUMBER	78
ATOMIC WEIGHT	195.078
MELTING POINT	1,772°C
BOILING POINT	*av.* 3800°C
DENSITY	21.5 g/cm^3
OXIDE	most stable is PtO_2 but also PtO and PtO_3

Platinum is a lustrous, silvery-white, malleable metal and a member of group 10 (row 5d) of the periodic table. It has the third highest density, behind osmium and iridium. Platinum is unaffected by air and water, but will dissolve in hot aqua-regia, hot concentrated phosphoric and sulfuric acids, and in molten alkali. It is as resistant as gold to corrosion and tarnishing. Indeed platinum will not oxidise in air no matter how strongly it is heated. Its preferred oxidation states in chemical compounds are +2 and +4.

There are six naturally occurring isotopes: platinum-194, which accounts for 33 percent, platinum-195 (34 percent), platinum-196 (25 percent), platinum-198 (7 percent), platinum-192 (1 percent), and platinum 190 (0.01 percent) of which platinum-198 is radioactive with a half life of 3.20×10^{14} (320 trillion years), and platinum-190 is radioactive with a half life of 7×10^{11} (700 billion years), in other words they are near stable. The other four are considered stable, although nuclear theory indicates that they are also radioactive but with incredibly long half-lives.

There are 37 known isotopes of platinum with masses 166 to 202, and 13 known isomers. The other long-lived radioactive isotopes are platinum-193 which has a half-life of 50 years, and platinum-188 with a half-life of 10.2 days.

Element of surprises

1. Platinum has the capability, as a catalyst, to initiate some chemical reactions at room temperature. For example, if a platinum wire is held in methanol (methyl alcohol) vapour it soon glows red hot as it catalyses the oxidation of this chemical to formaldehyde. Platinum will cause a mixture of oxygen and hydrogen to explode just as if it had been sparked. In the last century, a cigarette lighter was manufactured which did away with flints and instead had a rim of platinum around the fuel outlet which would induce the fuel to react instantly with the oxygen of the air.

2. At one time it looked as though platinum might be as abundant as silver and gold, especially when deposits were found in the Urals in 1822. In 1828 the Russian Government began issuing rouble coins of the metal and around 15 tonnes of platinum went into making 1.5 million coins. These ceased production in 1846, when mining began in Colombia, because this caused the price of the metal to fall to levels that made the coins attractive to counterfeiters. In fact we now know that the counterfeited coins contained more platinum than those of legal tender.

3. A black and white photograph of the Moon rising over a lake was printed on platinum by Edward Steichen in 1904. It was sold at auction in New York in 2006 for $3 million.

Plutonium

Pronounced ploo-toh-nee-um, it was named in the same way as the two elements preceding it in the periodic table: uranium (atomic number 92) had been called after the planet Uranus, and neptunium (93) after Neptune, so this element (94) was named after Pluto which had been discovered in 1930. In fact Pluto was downgraded from a planet to a dwarf planet in 2006 by the International Astronomical Union.[140]

But should the name be plutium or plutonium? The latter sounded better so it was chosen. Its chemical formula should have been Pl, but the discoverers preferred Pu because they said it sounded rather rude.

French *plutonium*; German *Plutonium*; Italian *plutonio*; Spanish *plutonio*; Portuguese *plutónio*.

Essential element

So much of this element has been man-made that it poses a threat to future generations unless it is stored safely, but it can be turned into fuel for nuclear reactors.

Element of life

Plutonium has no role to play in living things, and it is never encountered in bulk outside nuclear facilities, nuclear arsenals, research laboratories, and underground storage vaults. Nevertheless, because of nuclear power accidents and nuclear weapons testing there has been enough environmental pollution by plutonium to ensure that we all have a few of its atoms in our body. It is highly dangerous because of the alpha particles it emits. Normally these pose no threat because they are unable to penetrate even the thinnest of material, so it is safe to handle a container of plutonium, and even to hold the bare metal in gloved hands—see *Element of surprises*.

Dangerous element

As a chemical poison, plutonium is less deadly than other better known poisons such as arsenic. However, it's not its chemistry which makes plutonium deadly but its radiation. If plutonium oxide is ingested, it is not likely to be retained permanently

---

[140] The name plutonium was first given by the mineralogist Edward Daniel Clarke (1769–1822) of the University of Cambridge to what he thought was a new metal element. This he had obtained from barite ore ($BaSO_4$) by heating it with the intense flame of a hydrogen/oxygen torch. He named it honour of plutonism, a theory of the 1790s which said that the origin of many minerals was from the heat of volcanic activity, a theory which Clarke supported. He reported his new element in the *New Monthly Magazine*, of 1 November 1816. What he had obtained we shall never know, but it was certainly not a new metal element, nor was it even barium metal itself.

because it is so insoluble—only about 0.04 percent is absorbed by the body—but the little that is absorbed tends to end up in the bone marrow and there it poses a serious threat. The plutonium which is absorbed by the cell walls of the gut is continually sloughed off into the mass of material passing through the intestines because these cells are replaced every five days in any case.

Animal studies in the 1940s showed that plutonium was lethal when injected in milligram doses. Tests were even carried out on people who were said to be terminally ill. It was as a result of these experiments that we now know that most plutonium ends up in bone, from which it leaches out only at a very slow rate. In fact those who took part in these tests survived for up to 38 years, showing that even as a radioactive toxin the dangers of plutonium are not so bad as generally feared. In fact plutonium is 20 times less dangerous than radium which has been widely used in medicine.

Medical element

Plutonium-238 is used as the energy source in heart pacemakers.

Element of history

In early 1934, Enrico Fermi (1901–54) bombarded uranium with neutrons in an attempt to produce elements 93 and 94. He claimed to have discovered these, proposing the name hesperium for element 94 and thereby seeking to honour his homeland. (He suggested the chemical symbol Es from its Latin name *Esperia*.) In September that year, Ida Tacke-Noddack (1896–1978) published a paper that analysed Fermi's claims and pointed out that he had not done a complete analysis of his findings and that what he'd actually achieved was nuclear fission of the uranium but that he had not recognized the fact.

Plutonium was first made on 14 December 1940 at Berkeley, California, by Glenn T. Seaborg (1912–99), Arthur C. Wahl (1917–2006), Joseph W. Kennedy (1916–57), and Edwin McMillan (1907–91; Nobel Prize winner 1951) who produced it by bombarding uranium-238 with deuteron ions (heavy hydrogen ions). This first produced an isotope of the element neptunium which had been discovered earlier that year. This new isotope, neptunium-238, had a half-life of two days and decayed with emission of beta particles to form element 94 (plutonium). Within a couple of months element 94 had been conclusively identified and its basic chemistry shown to be like that of uranium.

A paper was dispatched to the journal *Physical Review* announcing the discovery in March 1941, but it was immediately withdrawn when it was realised that another isotope of element 94, isotope-239, was capable of undergoing nuclear fission in a way that could fuel an explosive chain reaction and might be used to make an atomic bomb.[141]

To begin with, the amounts of plutonium produced were tiny, and invisible to the eye, but by August 1942 there was enough to see, and a few weeks later there was enough to weigh, although it amounted to only 3 micrograms. However, by 1945

---

[141] The paper was eventually published in 1946.

the Americans had several kilograms, and enough plutonium to make three atomic bombs.

Element of war

Plutonium is one of three fissile elements, along with thorium and uranium, and as such has the potential to deliver a massive release of energy. When a nucleus of plutonium-239 is hit by a neutron it not only undergoes fission, releasing a lot of energy, but it also releases several more neutrons, each of which can strike a neighbouring atom causing it to fission and release yet more energy and yet more neutrons, and so on. In a fraction of a second, the chain reaction runs out of control and a colossal explosion ensues, large enough to destroy a city. To produce such an explosion requires a certain 'critical mass' of the metal.

In World War II, the US Government set up a Plutonium Project as part of their atomic weapons research programme. As a result, a great deal of chemistry was uncovered about this new element, but that information was largely incidental to the project and the research focussed mainly on ways in which it might best be separated from other elements. By the end of 1943 a pilot plant at Oak Ridge was in operation, making plutonium from uranium by the series of nuclear reactions which are summarised thus:

$$^{238}U + \text{neutron} \rightarrow {}^{239}U \ (t_{\frac{1}{2}} = 24 \text{ minutes})$$
$$^{239}U - \beta^- \rightarrow {}^{239}Np \ (t_{\frac{1}{2}} = 2.3 \text{ days})$$
$$^{239}Np - \beta^- \rightarrow {}^{239}Pu \ (t_{\frac{1}{2}} = 24,100 \text{ years})$$

To begin with, the production was about a gram a day but this soon increased. After six months the uranium fuel rods were removed and the plutonium extracted by dissolving them in nitric acid which formed plutonium nitrate, a convenient way to handle and transport the element. The new element, which was known to be element 94 was given the secret code 49.

Making a uranium bomb was relatively simple. It consisted of firing one lump of uranium at another and they would explode—it didn't need to be tested; it would work. This was the bomb that was used to destroy Hiroshima on 6 August 1945.

To produce a nuclear explosion using plutonium required another technique. This was the implosion method. A sub-critical sphere of plutonium metal was surrounded by precisely placed conventional explosive charges and when these were detonated they produced a shock wave that compressed the sub-critical plutonium sphere to increase its density and thereby create a critical mass which resulted in a runaway chain reaction and nuclear explosion. The plutonium sphere was 9.2 cm in diameter with a small 2.5 cm hole at its centre. It was manufactured as two hemispheres so that the centre hole could be made. This hole contained a pellet of initiator material composed of polonium-210 and beryllium-9 to provide a flux of neutrons to help initiate and boost the chain reactions. The two hemispheres were then fitted together to form a sphere. This method needed the components to be precisely calibrated, with the proper amount of plutonium and just the right amount of conventional explosive charges all of which had to be set off simultaneously. A test of this system was needed to see if it would work. On 16 July 1945, a device—code name Gadget—which used 6 kg of plutonium—was successfully tested in the desert of Alamagordo, New Mexico.

The second plutonium explosion was in the form of a bomb, code-named Fat Man, which was dropped on the Japanese city of Nagasaki at 11.02 am on 9 August 1945. The explosive capacity was equivalent to 21,000 tonnes of TNT, killing about 70,000 citizens and wounding 100,000 others. A second plutonium bomb was to be dropped on Kumagaya two days later, but the bomb was not quite ready, so 6,000 tonnes of conventional bombs were used instead to destroy that city. The war ended on 14 August, and the city of Kokura, scheduled for nuclear obliteration on 17 August, was spared.[142]

The Imperial Japanese Army had its own atomic weapons programme, according to Tatsuasaburo Suzuki, one of the last remaining members of the research team who was still alive in the 1990s. It was thwarted by a shortage of materials although they had deduced that such a bomb could be made using uranium-235. They even had a target for their first bomb: Saipan, the Japanese island captured by the Americans in 1944, and which was being used as a bomber base.

A hydrogen bomb is triggered by a plutonium bomb whose explosion generates temperatures high enough to cause hydrogen atoms, in the form of the heavier isotopes, deuterium or tritium, to fuse together, as hydrogen does in the Sun. The result releases vastly more energy than an atomic bomb and give explosions equivalent to millions of tonnes of TNT.

Economic element

Plutonium metal is obtained by heating together plutonium tetrafluoride (PuF_4) with a metal like calcium at 1,200°C. More plutonium has been produced than any other synthesised element, primarily because of its nuclear weapon capabilities. It is also being produced as a by-product of nuclear power plants. In theory it may one day be used to fuel such power plants. Annual world production of plutonium is probably in excess of 20 tonnes and there may be around 1,000 tonnes of the metal in storage, either as bombs or as metal rods.

A nuclear reactor relies on the fission of fuel rods enriched with uranium-235, which has to be partly separated from the more abundant uranium-238 isotope before it can be used. The nuclear fuel produces heat in abundance, but after about three years the fuel rods have to be reprocessed to remove the 3 percent of waste products that have accumulated. When this has been done, the uranium and the plutonium, which has also formed, can be used as fuel again.

In so-called 'breeder' reactors, which are used to make plutonium for bombs, uranium-238 is bombarded with neutrons to form plutonium-239. Many of the bombs stockpiled during the Cold War—and there were about 80,000 of them, amounting to more than 200 tonnes of the metal—are now being dismantled, but the prob-

---

[142] Kokura had been the original target for the first plutonium bomb. Because it was obscured by clouds, the bomber went on to its back-up target, Nagasaki, where there were the Mitsubishi shipyards and torpedo works. Sadly the bomb fell a few miles to the north of the intended aiming point and exploded over the city's main Christian church wiping out the largest Christian community in Japan.

lem is what to do with the plutonium they contain. This has to be stored safely, and prevented from falling into the wrong hands. It could be used as fuel for nuclear reactors, which would slowly use it up. For a variety of reasons—political, economic and environmental—this is unlikely to happen for some time.

Plutonium-239 can also be used as a fuel in the new generation of fast nuclear reactors which burn a mixed oxide (MOX) fuel consisting of uranium and plutonium. MOX fuel contains around 60 kg of plutonium per tonne (1,000 kg) of fuel, and after its four-year life span such a fuel element would have burned up three quarters of its plutonium. One gram of plutonium used in a conventional nuclear reactor has the potential to release as much energy as a tonne of oil.

The plutonium in a spent nuclear fuel rod is also mainly plutonium-239 (around 58 percent), but there is also plutonium-240 (24 percent), plutonium-241 (11 percent), plutonium-242 (5 percent), and plutonium-238 (2 percent). The plutonium and uranium are separated by chemical means based on the differential solubility of their salts in water compared to kerosene (paraffin) containing tributyl phosphate.

Plutonium-238 can be used as a power source, but in a different way. Although it could be separated from spent fuel rods, it is easier to obtain it by neutron bombardment of neptunium-237. It too is an alpha emitter but without emitting γ-rays. Therefore, it is entirely safe so long as it is enclosed in a steel casing and the heat it produces is used in deep-sea diving suits and spacecraft.

Plutonium-238 was used on the Apollo 14 lunar flight of 1971 to power seismic devices and other equipment left on the Moon, and it was also the power supply used by the two Voyager spacecraft launched in 1977. These were designed to take advantage of a rare celestial conjunction of planets that occurs only once every 175 years and which enabled scientists to study the four outer planets, Jupiter, Saturn, Neptune, and Uranus. These were visited over a period of 10 years, during which masses of data and pictures were transmitted back to Earth.

Plutonium-238 mixed with beryllium generates neutrons and is used for research purposes.

Environmental element

There are seven plutonium isotopes that occur naturally on Earth and these are plutonium-238 to 244. Plutonium-244 is the primordial isotope with a half-life of 80.8 million years (8.08×10^7 years). Its decay chain produces plutonium-240, while the other isotopes are generated from uranium and curium.[143]

It is not this natural plutonium which poses a threat to the environment, but that which has been released from nuclear bombs tested above ground, a practice that was prohibited by international agreement among the major powers at the time (the UK, USA, and Soviet Russia) in 1963 by the Limited Test Ban Treaty. Not all nuclear-

---

[143] Plutonium-244 alpha-decays to uranium-240 which beta-decays to neptunium-240, which beta-decays to plutonium-240. Uranium-238 has been found to decay by the rare decay mode of double beta decay to produce plutonium-238. Naturally occurring curium-244 alpha-decays to produce plutonium-238. The other isotopes are produced in trace amounts in uranium-bearing deposits by neutron capture and beta-decay reactions. Uranium-238 captures neutrons to become uranium-239 which beta decays to plutonium-239; then by successive neutron captures, plutonium-240, 241, 242, and 243 are produced.

weapons testing nations signed the treaty, and nuclear testing in the atmosphere continued throughout the 1970s. France did not cease testing until 1974. China and Israel tested in the 1970s with China testing throughout that decade with its last atmospheric test taking place in 1980. From 1945 to 1980 there were about 525 atmospheric nuclear explosions.

In the USA, the plutonium salvaged from unwanted atomic bombs is melted into glass logs as plutonium oxide, and these were being buried in Nevada's Yucca Mountains 100 miles north-east of Las Vegas but this was halted by President Obama in 2010. These two tonne logs are made from borosilicate glass containing cadmium and gadolinium which, like boron, are powerful neutron absorbers. The use of gadolinium zirconium oxide glass ($Gd_2Zr_2O_7$) has also been proposed; this can safely immobilise plutonium for up to 30 million years. The plan is to encase the logs in stainless steel and store them deep underground, possibly four kilometres deep with their bore-holes plugged with concrete. Some people fear that in the distant future slow corrosion and water seepage might leach enough plutonium and concentrate it in a rock cavity to produce a critical mass. This seems an unlikely event because plutonium oxide is the least soluble of oxides—a million litres of water will only dissolve one atom of it.

Chemical element

Data file	
CHEMICAL SYMBOL	Pu
ATOMIC NUMBER	94
ATOMIC WEIGHT	244.0642 (plutonium-244)
MELTING POINT	640°C
BOILING POINT	3,330°C
DENSITY	19.8 g/cm³ and when the metal melts it becomes even more dense, so that solid plutonium floats on it.
OXIDE	the main one is PuO_2 but there are others.

Plutonium is a silvery, radioactive metal, a member of the actinide group (row 5f) of the periodic table. The metal will ignite in air at 135°C and explodes in contact with the solvent carbon tetrachloride. A piece of plutonium is warm to touch because of the energy given off by the alpha decay, and a large lump of plutonium will produce enough heat to boil water. In a 1 gram sample of plutonium-239, 12 billion atoms decay every second. Plutonium is attacked by oxygen, steam and acids, but not alkalis. Plutonium can exhibit 4 oxidation states, namely +3 as in PuF_3 and $PuCl_3$ which are lavender coloured, +4 as in PuF_4 which is yellow, +5 as PuO_2^+ ions in solution, and +6 as PuO_2^{2+} ions in solution.

There are 20 known isotopes of plutonium with mass numbers ranging from 228 to 247, and seven known isomers. The most abundant isotope is plutonium-239 with a half-life of 24,100 years. There are longer-lived isotopes such as plutonium-244 (80 million years) and plutonium-242 (376,000 years), but most isotopes have much shorter half-lives, such as the useful plutonium-238 (88 years). There are still a few primordial atoms of plutonium-244 around, and some plutonium-239 is produced in uranium ores. In 1971, Darleane Hoffman discovered naturally occurring plutonium-244 when she extracted 8×10^{-15} grams (= 8 million billionths) of plutonium from this ore in which there is only 10^{-15} grams per tonne.

Element of surprises

1. Plutonium oxide occupies 40 percent more volume that plutonium metal, and this is always a threat when storing the metal. In the USA at the Los Alamos National Laboratory in New Mexico, a canister which held 2.5 kg of plutonium metal had not been properly welded air-tight in 1983, and the oxygen that seeped in reacted with the metal to form plutonium oxide. Over a period of 10 years this escaped as dust into the polythene bag in which the canister was contained. There the radioactivity caused the polythene to become brittle and give off some of its hydrogen atoms which then diffused into the plutonium and formed plutonium hydride which acted as a catalyst for the oxidation. When the polythene bag finally burst it let in more air and further oxidation occurred even more rapidly. Had the process gone on much longer then the expanding plutonium oxide would have ruptured the outer canister and contaminated the whole storage facility.

2. The first sample of plutonium ever made was put in an old cigar box belonging to Professor G.N. Lewis (1875–1946), the chemist who was famed for his theory of chemical bonding and molecular structure, and for 25 years it languished forgotten in a cupboard. Then in 1966 it was discovered during a clear-out at the University of California, Berkeley, and presented to the Smithsonian Institution, where it is now kept.

3. When the young Queen Elizabeth II visited the UK's Atomic Energy Authority at Harwell in the 1950s she was handed a lump of plutonium metal in a plastic bag and she commented on how warm it felt. The heat was being generated by the radioactive decay of the metal.

4. Everyone thinks that plutonium-239, the weapons-grade isotope, is the only plutonium isotope that can be used to make nuclear weapons. However, in 1962 the USA successfully conducted a nuclear test explosion using reactor-grade plutonium-240. This was detonated underground in Nevada.

Polonium

Pronounced pol-oh-nee-um, the name is derived from Poland, the native country of Marie Curie who first isolated the element.

French *polonium*; German *Polonium*; Italian *polonio*; Spanish *polonio*; Portuguese *polónio*.

Essential element

Used positively in antistatic safety devices and negatively as an agent of assassination.

Cosmic element

Polonium-210 was included in the Lunokhod rovers which were sent to explore the Moon by the Soviet Russia in 1970–4. The heat which it generates was designed to keep the instruments on board warm and working during the cold lunar nights.

Element of life

Polonium has no biological role but billions of its atoms are present in our bodies and they come from natural sources such as the food we eat (all soil contains traces of uranium and this produces polonium) and the air we breathe (this contains traces of radon gas which decomposes to polonium). Polonium has even been detected in tobacco smoke. Polonium targets no particular organ of the body but, because it is an alpha-emitter, wherever it ends up it has the potential to damage cells.

Dangerous element

Polonium is highly dangerous because of its intense radioactivity and this overrides other toxicity considerations which might follow from its chemical similarity to tellurium, the noxious element immediately above it in the periodic table. It is regarded as one of the deadliest substances known and the maximum safe body burden is only 7 picograms (7×10^{-12} g) making it more than a billion times more toxic, weight for weight, than hydrogen cyanide.

The amount of polonium excreted by the average person per day is 1×10^{-16} gram (10 million billionths of a gram), which is about 300 billion atoms. The biological half-life of the polonium which has been absorbed into the human body is about 50 days. Polonium was extensively researched in the 1940s as part of the Manhattan Project for making atomic bombs in World War II, and then more thoroughly in the 1950s and 1960s as part of nuclear energy projects in the USA. Tests on animals revealed how dangerous polonium was once it got inside the body because of its alpha-emissions. This research was carried out by J. Newell Stannard and George W. Casarett at the University of Rochester and reported in the official publication *Radiation Research. Supplement 5. Alpha Particle Emitter, Polonium-210.* This reveals that ingesting a mere microgram of polonium-210 can be fatal and this amount is as small as a particle of dust and barely visible.

Polonium-210 became infamous as the poison used by members of the Russian Secret Police, the FSB, to murder a former member of their group, 43-year-old Alexander Litvinenko. He had defected to the UK, been granted asylum, and been given a new identity. Nevertheless, on 1 November 2006 at a London hotel he met a former Russian colleague and drank a cup of green tea which contained a few micrograms of this isotope. The radioactivity to which his body was then subjected made him ill within days and although he was admitted to hospital he slowly deteriorated and died on 23 November. Only towards the end of his life was it clear he was dying of radiation sickness and that was then identified as due to polonium-210.

From an assassin's point of view, polonium-210 appears to be a perfect poison. Although it is a radioactive isotope it does not emit gamma rays so escapes detection at airports. Only a tiny dose of the poison is needed to kill and the symptoms the victim displays will be mistaken for other diseases. What this poison reveals, however, once it has been detected, is the identity of the murderer because polonium-210 always leaves a trail of a few atoms which are easily confirmed and in the Litvinenko assassination these were to be found on aircraft seats, in hotel bedrooms, and at tables where the assassin had taken his meals. Fifty locations around London were eventually found to have been contaminated with polonium-210.

No medical treatment could have saved Litvinenko because there is no antidote for alpha particles. Had it been known within a few hours that he had been poisoned with polonium-210 then it would have been possible to remove the unabsorbed polonium-210 still in his stomach and intestines and thereby mitigated some of the damage to his system. However, once it has passed through the gut wall and entered the blood stream, polonium will move around the body and pass into the cells of various organs. Cells which have a high rate of division, such as the hair follicles, will be quickly affected, but wherever it ends up it will wreak havoc. The immune system will break down, and there will be a dramatic drop in white blood cells. The polonium-210 which goes into the liver damages the one organ that is seeking to remove this alien chemical from the body.

Element of history

Uranium ores contain about 100 micrograms of polonium per tonne (0.00000001 percent), and in 1898, at Paris, Marie Skłodowska Curie (1867–1934), working with her husband Pierre Curie (1859–1906), obtained the first sample of the element from this source after months of painstaking work.[144] They were motivated in wanting to discover what was in the uranium mineral pitchblende that caused it to emit four times more radioactivity than could be explained by the uranium itself. They were able to purchase several tonnes of ore residues from a uranium mine at Joachimsthal,

---

[144] Pierre Curie died in a street accident. He fell while crossing the road and was run over by a cart.

Bohemia, courtesy of the Austrian Government to whom they had appealed for help because they could not afford to buy pitchblende.

The existence of this element had been forecast by the great Mendeleyev in 1891 who could see from his periodic table that there might well be the element that followed bismuth and come below tellurium; he even predicted it would have an atomic weight of 212. He was nearly right. In fact the Curies had extracted the isotope polonium-209 which has a half-life of 103 years.

Before the advent of nuclear reactors the only source of polonium was its tedious extraction from uranium ore but that did not prevent its being used. In the early part of the last century it was to be found in textile mills and in factories making photographic plates. It was designed to reduce static electricity, which in mills caused unpleasant electric shocks to operatives and in the case of photographic plates caused dust to cling to them. Such anti-static devices, which are still very much in use, present no threat to humans because alpha-rays cannot travel more than a few centimetres before colliding with molecules in the air, nor can they penetrate skin. A sheet of paper will stop them. Alpha-particles are only dangerous when they are generated inside the human body.

Economic element

No one now extracts polonium from uranium ores. Instead, it is made in gram quantities—probably about 100 g per year—by bombarding bismuth with neutrons in a nuclear reactor. A capsule of polonium containing only one gram will reach a temperature of 500°C because of the intense alpha radiation it emits, and for this reason polonium is used as a lightweight heat supply for space satellites. Such a capsule will generate 520 kilojoules of energy per hour.

All the world's polonium for commercial use is produced in Russia at the Ozersk nuclear reactor near the city of Chelyabinsk, east of the Ural Mountains. From there it goes to the Sarov nuclear facility near the city of Samara in southeast Russia where the polonium is separated from the bismuth and sealed into capsules. Almost all of these are taken to St Petersburg to be exported by air to the USA for distribution to users, most of whom needed it for antistatic devices. The polonium is electroplated on to the surface of metal foil and these devices remove a charge of static electricity, which is a build-up of electrons, by virtue of the alpha particles. These are the nuclei of helium atoms and as such are positively charged and each is in need of two electrons to become helium gas so they mop up any electrons they come across.

Polonium is used as a source of alpha radiation for research and, alloyed with beryllium, it can even act as a portable source of neutrons, which normally only access to a nuclear reactor can provide.

Environmental element

Polonium is present in minute traces in some minerals and it is among the 10 least abundant metals on Earth. It poses no real threat to the environment because there is so little of it and its half-life is short.

Chemical element

Data file	
CHEMICAL SYMBOL	Po
ATOMIC NUMBER	84
ATOMIC WEIGHT	209
MELTING POINT	254°C
BOILING POINT	962°C
DENSITY	9.3 g/cm³
OXIDE	PoO_2, and PoO_3

Polonium is a silver-grey radioactive semi-metal and a member of group 16 (row 6p) of the periodic table. It dissolves in dilute acids, and solution so formed is pink, the colour of the Po^{2+} ion, but rapidly turns yellow as the alpha radiation forms oxidising species from water molecules and these convert it to Po^{4+}. Polonium's oxidation states range from −2 as in PoH_2, to +6 as in PoF_6, with intermediate oxidation states +2 as in $PoCl_2$, and +4 as in PoF_4 which is more stable.

There are 33 known isotopes of polonium with masses of 188 to 220, and 28 known isomers. There are eight naturally occurring isotopes, polonium-210 to 216 and 218, which are produced in trace amounts by the radioactive decay of uranium, thorium, and plutonium. The longest-lived isotopes are polonium-209 with a half-life of 103 years, polonium-208, half-life of 2.9 years, and polonium-210, half-life of 138 days. Polonium-209 and polonium-208 are made by bombarding lead or bismuth with protons, deuterons, or alpha particles in a cyclotron, but they are more costly to produce.

Element of surprises

1. Marie Curie died of leukaemia in 1934, caused by her exposure to radium over many years. Her daughter, Irène Joliot-Curie, who also won a Nobel Prize, likewise died of leukaemia, in 1956, but that was attributed to her exposure to polonium as a result of a capsule of the element exploding on her laboratory bench 15 years earlier.

2. Although its boiling point is 962°C, a sample of polonium will evaporate if left in the open and within three days more than half of it will have disappeared. This is believed to happen because the alpha particles being emitted are able to knock off the surface layers of the element. Even when polonium is kept in a container it will continue to lose a few atoms which always escape, and in this way it reveals where it has been, as we saw in the Litvinenko case.

Potassium

Pronounced poh-tass-ee-um, name is derived from the English word potash. The chemical symbol K comes from *kalium* the Medieval Latin for potash, which may have derived from the Arabic word *qali*, meaning alkali.

French *potassium*; German *Kalium*; Italian *potassio*; Spanish *potasio*; Portuguese *potássio*.

Essential element

Potassium is the key to passing messages along nerves, and is an essential component of fertilizers.

Element of life

Potassium in the human body	
BLOOD	plasma 400 p.p.m., red blood cells 4,000 p.p.m. (0.4%)
BONE	2,100 p.p.m. (0.21%)
TISSUE	16,000 p.p.m. (1.6%)
TOTAL AMOUNT IN BODY	varies within range 110–140 g depending on the weight of muscle.

Potassium is an essential element for almost all living things, except possibly a few bacteria. Because one of its commonly occurring isotopes is radioactive, it has been suggested that this element is the key to natural genetic modification in plants and animals, and indeed around 4,000 potassium atoms undergo radioactive decay in the average human every *second*. Red blood cells have most potassium, but muscle and brain tissue also has a lot.

In the body, potassium has many functions, the more important of which are regulating intracellular fluids, soulblising proteins, operating nerve impulses, and contracting muscles. Potassium as the ion K⁺ concentrates inside cells, and 95 percent of the body's potassium is so located, unlike sodium and calcium, which are more abundant outside cells.[145] The ratio of potassium between cell and plasma in the human body is 27-to-1. That which is in the plasma acts as an electrolyte, helping maintain plasma viscosity and osmotic pressure.

The movement of sodium and potassium across nerve cell membranes is responsible for the transmission of nerve impulses, because this lateral motion of charge

---

[145] The concentration of potassium inside cells is high relative to sodium and thought to reflect the ratio of these elements in the seas of the pre-Cambrian period in which life evolved. Today there is much more sodium than potassium in seawater.

passes like a wave along the direction of the fibre, as if it were an electric current. Cell membranes have channels through which sodium and potassium ions flow selectively and against a concentration gradient which would normally induce them to flow from locations of higher to those of lower concentration. These so-called sodium-potassium pumps are at a density of more than 100 per square micrometre of membrane surface,[146] and each can transfer 200 sodium and potassium ions per second in and out of the cell.

Some channels only permit potassium to pass through. Potassium ions enter the cell via a chain of seven special sites through the channel which are occupied by these ions and they jump from one site to the other when required to move. (Only potassium ions can occupy these sites, and this is what keeps out other ions.) The movement of potassium backwards and forwards along a channel requires very little energy.

Food element

It is not generally appreciated that the need for potassium salts in the diet is much greater than for sodium salts, and indeed it has been suggested that increasing the former relative to the latter would be beneficial. Some studies claimed that taking extra potassium can reduce blood pressure by increasing sodium excretion, but other research has failed to confirm this. The recommended daily intake of potassium is 3.5 g for both men and women, more than the daily intake recommended for sodium. Vegetarians take in a lot more potassium than non-vegetarians because potassium is abundant in all plants.

Around 90 percent of dietary potassium is absorbed by the body's lower intestine and we must have a regular supply of dietary potassium because we have no mechanism for storing it in the body. The average adult loses about a gram a day in their urine. Some foods are particularly rich in potassium, such as raisins and almonds, which have 860 mg per 100g, dates and currants (750), peanuts (680), rhubarb (430), and bananas (350). Other common foods with above average amounts are potatoes, bacon, Cantaloupe melon, bran, mushrooms, chocolate, and fruit juices. In some processed foods potassium comprises more than 1 percent of their weight, such as All-bran (1.1 percent), butter beans (1.7 percent), dried apricots (1.9 percent), yeast extracts[147] (2.6 percent), and instant coffee (4.0 percent). The only foods with virtually no potassium are sugar, vegetable oils, butter, and margarine.

There are several potassium-based food additives such as potassium sodium tartrate ($KNaC_4H_4O_6$, aka E337) which is a component of baking powder, potassium bromate ($KBrO_3$, aka E924) which has been used to improve the dough used for bread making, potassium hydrogen sulfite ($KHSO_3$, aka E228) which is added to wines to check rogue yeasts, and potassium benzoate ($C_6H_5CO_2K$, aka E212) which is a preservative used to protect against microbes.

---

[146] Equivalent to 100 trillion per square cm.
[147] Know by various trade names such as Marmite and Vegemite.

Dangerous element

The toxin of the black mamba kills its victim by specifically blocking the potassium channels. The same paralysis can be achieved by injecting a concentrated solution of potassium chloride into the blood stream which prevents the movement of potassium out of the cell because there is already too much potassium on the outside pushing to get in. All body functions are affected, but none more dramatically than the heart muscle which stops beating.

Murders have been committed with potassium chloride, and doctors and nurses have been known to end the lives of terminally ill patients by giving them injections of potassium chloride solution. Such injections have been used in the USA to execute criminals if a condemned man agrees to donate his organs for transplants. Potassium chloride leaves the body's organs undamaged, unlike execution by poison gas or the electric chair.

Potassium chloride was the chosen murder weapon of one notorious serial killer whose victims were babies and young children. She was Beverley Allitt who was employed as a children's nurse at the Grantham and Kesteven General Hospital in Lincolnshire, England. Over a period of 10 weeks she injected 10 children with potassium chloride and murdered four of them before she was found out.

Potassium chloride is used as a dietary supplement, and is sold in the form of LoSalt[148] to those who wish to reduce the amount of sodium chloride in their diet. There are rare cases of death by excess ingestion of this salt and someone who consumed 15 g died as a result, although the normal amount to cause a serious toxic response would have to be more than twice this amount.

Medical element

The human body needs a constant supply of potassium to make lean tissue and to keep the kidneys working. If there is a deficiency, then the person experiences muscular weakness, which can affect the heart muscle, causing irregular beats and even cardiac arrest. There is even a condition known as hyperkalenia which is a genetic defect that results in attacks of partial paralysis due to the malfunctioning of potassium metabolism, but the condition is extremely rare and those afflicted can lead normal lives.

There are some conditions which lead to potassium deficiency, such as starvation, kidney malfunction, and the long-term taking of diuretics to stimulate urine loss. Diuretics often have extra potassium included in their formulation to counter this loss. Severe attacks of diarrhoea lead to a temporary lack of potassium and consequently a feeling of weakness. Chronic potassium deficiency leads to depression and confusion. However, it is rare for potassium deficiency to lead to ill health because it is abundant in food, particularly in vegetables and fruits.

At the other extreme, an excess of potassium in the body depresses the central nervous system, and large doses of several grams of potassium chloride will paralyse it causing convulsions, diarrhoea, kidney failure, and even heart attack.

---

[148] This is 60% potassium chloride and 40% sodium chloride.

Element of history

People knew of potassium salts long before potassium was known as an element. Native Americans used the ash of the saltwort to flavour food, and they preserved meats with wood ash, both of which have a high level of potassium carbonate (aka potash, K_2CO_3). Similar practices went on in Asia. Potassium salts in the form of potassium nitrate (aka saltpetre, KNO_3) and potassium aluminium sulfate (aka alum, $KAl(SO_4)_2$) have been known for centuries. The diarist, John Aubrey (1626–97), reported that spreading plant ash was a way of improving soil, a practice that had first been advocated by Edward Broughton of Kington in Herefordshire in the early 1600s.

When Lavoisier drew up the first list of chemical elements, he included among them various 'earths', of which potassium oxide was one, and which could not at the time be broken down into simpler components and so they qualified as elements according to Lavoisier's definition. We now know all his earths as metal oxides.

Humphry Davy (1778–1829) turned his attention to these earths with a view to seeing whether they could be further decomposed. He attempted to do this using electricity, and at first he dissolved them in water but found that passing an electric current through the solutions gave off only oxygen and hydrogen. On 6 October 1807 he decided to do away with the water and simply took some fresh potash that had become moist by exposure to air and placed this on a platinum disc connected to the negative pole of his battery. He then put a platinum wire, which was connected to the positive pole, in contact with the potash and within minutes observed the formation of metallic globules on the platinum disc. These were a new element, potassium, and the first time a metal had been isolated using electrolysis. (This technique had become available following the invention in 1799 of the electric battery, then known as the voltaic pile.) To prove that these globules were not coming from the platinum, he repeated the experiment using other metals as the electrodes and the results were the same. The new element was immediately christened potassium by Davy after the source from which it came.

Davy noted that when the metallic globules were thrown into water they skimmed about on the surface, burning with a lavender-coloured flame. This colour is typical of potassium and can be seen by putting a tiny sample of a potassium salt in a Bunsen burner flame.

The French chemists Louis-Josef Gay-Lussac (1778–1850) and Louis-Jacques Thénard (1777–1857) were also able in 1808 to prepare potassium by heating a mixture of potassium hydroxide and iron filings to a high temperature, and this method was the one subsequently used by Davy to obtain more of the metal.

Economic element

At the start of the Industrial Era in Britain in the mid 1700s, there was a need for more potash for making soap for washing wool than local trees could provide. This had to be imported, first from Russia and then from the American colonies, where forest clearances yielded enough potash to supply the world's need for a hundred years.

Minerals mined for their potassium are sylvite (potassium chloride, KCl), carnallite (a mixed potassium magnesium chloride, $KCl.MgCl_2$), and alunite (potassium aluminium sulfate, $KAl_3(SO_4)_2(OH)_6$). The main mining area used to be Germany, which before World War I had a monopoly on potassium, mined as carnallite at Stassfurt from 1861.[149] Orthoclase (potassium aluminium silicate, $KAlSi_3O_8$) is also mined extensively, but not for its potassium, or aluminium, but because if of its use in porcelain, ceramics, and glass. Adding potassium to glass makes is stronger and more scratch-resistant.

World production of potassium ores is about 35 million tonnes, and reserves are vast and in excess of 10 billion tonnes. Today most potassium minerals come from Canada (11.5 million tonnes), Russia (6.5 million tonnes), Belarus (5 million tonnes), Germany (3.5 million tonnes), and the USA (1 million tonnes). The Dead Sea brines of Israel and Jordan also yield 3 million tonnes.

The process of extracting potassium from sylvite involves crushing the ore and then separating the potassium chloride from the other minerals present by using a concentrated solution of sodium chloride of exactly the right composition to ensure only the impurities in the ore, and not the potassium chloride, dissolves.

Most potassium (around 95 percent) goes into fertilizers, and the rest goes mainly into making potassium hydroxide (KOH), by the electrolysis of potassium chloride solution. Some of this is converted to potassium carbonate (K_2CO_3). Potassium carbonate goes into glass manufacture, while potassium hydroxide is used to make liquid soaps and detergents. A little potassium chloride goes into pharmaceuticals, medical drips, and saline injections.

Other potassium salts are used in baking, tanning leather, and to make iodised salt, but in these cases it is the negative anion, not the potassium, that is the key to their use, and that is also the case with potassium chlorate, produced to make the heads of matches and for fireworks, and with potassium nitrate, which is needed for gunpowder.

Environmental element

Potassium in the environment	
EARTH'S CRUST	21,000 p.p.m. (2.1%)
	Potassium is the 8th most abundant element.
SOILS	av. 14 p.p.m.
SEA WATER	380 p.p.m.
ATMOSPHERE	traces.

Most potassium occurs in the Earth's crust as silicate minerals, such as feldspars and clays. Potassium is leached from these by weathering, which explains why there is quite a lot of this element in the sea, although not as much as sodium because potassium tends to end up in ocean sediment. The concentration of sodium in the sea, which in terms of sodium chloride, is 27 g per litre, is more than 35 times the concentration of potassium chloride, which is only 0.75 g per litre.

[149] Production ceased in 1972.

Potassium is a key plant element, and although it is soluble in water, little is lost from undisturbed soil because as it is released from dead plants or animal excrement it quickly becomes strongly bound to clay particles, and is retained ready to be reabsorbed by the roots of another plant. Ploughing upsets this system of nutrient recycling and potassium is lost from the land, which is why it has to be replaced by fertilizers. Wood ash used to be the main source of fertilizer potassium, but today it is potassium chloride of mineral origin (aka MoP, muriate of potash)[150] which accounts for almost all potassium in fertilizers.

Chemical element

Data file	
CHEMICAL SYMBOL	K
ATOMIC NUMBER	19
ATOMIC WEIGHT	39.0983
MELTING POINT	64°C
BOILING POINT	774°C
DENSITY	0.86 g/cm³
OXIDE	K_2O, also KO_2, and K_2O_2.

Potassium is a soft, silvery-white metal and a member of group 1 (row 4s), also known as the alkali metal group, of the periodic table. The metal itself is obtained from the reaction of sodium metal with potassium chloride—see below. Potassium is silvery when first cut but it oxidises rapidly in air and tarnishes within minutes; it is generally stored under oil or grease. It is light enough to float on water with which it reacts instantly to release hydrogen, which burns with a lilac flame. It will even react with ice at −100°C. However, the metal will dissolve unchanged in liquid ammonia to form blue solutions that are used as strong reducing agents.

Potassium is low melting as the data file shows, but a potassium-sodium alloy, known as NaK, and consisting of 78 percent potassium and 22 percent sodium is liquid down to *minus* 10°C, and a three-component alloy of potassium-caesium-sodium will remain liquid down to *minus* 78°C. Potassium's only oxidation state is +1 as in K⁺, indeed the chemistry of potassium is almost entirely that of the potassium ion, K⁺.

There are 25 known isotopes of potassium with masses 32 to 56, and four known isomers. There are three naturally occurring isotopes: potassium-39, which comprises 93 percent of the total, potassium-41 (7 percent), and potassium-40 (0.012 percent). The last of these is radioactive and is a pure beta-emitter, with a half-life of 1.3 billion years and it is responsible for almost all of the argon gas which accounts for 1 percent of the atmosphere.[151] Potassium-40 can be used to date rocks by comparing

---

[150] Muriate is the traditional name for chloride, the term then in use when fertilizers with potassium were first introduced.

[151] Potassium-40 decays by three routes, two of which (beta-plus decay and electron capture) produce argon-40; the other route converts potassium-40 to calcium-40 by beta-minus decay.

the ratio of potassium to argon within them, the so-called potassium/argon dating method. The energy released by the radioactive decay of potassium-40 contributes to the warming of the Earth and may play a part in the convection currents in the mantle.

The second longest-lived radioactive isotope is potassium-42 which has a half-life of 12.4 hours.

Element of surprises

1. World production of potassium metal itself is 200 tonnes per year, and it is obtained by passing sodium vapour up a column at 870°C down which molten potassium chloride flows. They react and release potassium vapour which is condensed at the top of the column. Potassium metal is produced in order to manufacture potassium superoxide (KO_2) which it forms when it burns in oxygen gas. This material is kept in mines, submarines, and space vehicles in order to regenerate the oxygen in the air when this has become depleted. This oxide reacts with carbon dioxide to form potassium carbonate (K_2CO_3) and in so doing it releases oxygen gas.

2. When a scorpion stings its prey it injects a venom consisting of potassium salts and potassium channel blocking proteins, and this has the effect of paralysing and killing. The sting may not kill larger creatures but it is sufficiently painful to deter predators.

3. A rather unusual use for potassium chloride has been suggested, and that is to increase rainfall over regions prone to drought. Normally, clouds release only about a third of their moisture as rain, but this can be doubled if they are seeded with fine particles of potassium chloride. These are released by flares mounted on the wings of aircraft flying beneath the clouds. The potassium chloride drifts up into the cloud and down comes heavy rain.

4. The amount of potassium in the average human is 140 grams and while the amount of this which is the radioactive isotope potassium-40 is small, around 17 milligrams, nevertheless 4,400 of its atoms are undergoing radioactive decay every *second*.

Praseodymium

This is pronounced prah-zee-oh-dim-ee-um, and the name is derived from the Greek *prasios didymos*, meaning green twin because it was one of two elements that made up the supposed element didymium—see below—and its oxide was a striking green colour.

French *praséodyme*; German *Praseodym*; Italian *praseodimio*; Spanish *praseodimio*; Portuguese *praseodímio*.

Praseodymium is one of 15 chemically similar elements referred to as the lanthanoid or rare earth elements, which extend from element atomic number 57 (lanthanum) to element atomic number 71 (lutetium). The term rare earth is a misnomer because some are not rare. The minerals from which they are extracted, and the properties and uses they have in common, are discussed under Lanthanoids.

Essential element

Best known for the clear yellow colour it imparts to glass and pottery glazes, praseodymium has also helped scientists almost to reach the absolute zero of temperature.

Element of life

It is difficult to distinguish the various amounts of the individual lanthanoids in the human body, but they are present, and the levels are highest in bone, with smaller amounts being present in the liver and kidneys. There is no estimate of the amount of praseodymium in an average adult, and no one has monitored diet for praseodymium content, so it is difficult to judge how much the average person takes in, but it is probably only a milligram or so per year. Praseodymium is not taken up by plant roots to any great extent, and the amount in vegetables is on average only 1–2 p.p.b. (dry weight), so only a little gets into the human food chain. The metal has no biological role, but it has been noted that praseodymium salts stimulate metabolism.

Dangerous element

Soluble praseodymium salts are mildly toxic by ingestion, but insoluble salts are non-toxic. They are skin and eye irritants, and this is especially true of praseodymium metal powder and dust.

Element of history

The new element didymium was announced in 1841 by Carl Mosander (1797–1858). It was jokingly said that he chose the name because he was the father of two sets of twins. He had discovered that cerium harboured two other elements and he had extracted these, one of which was lanthanum and the other he called didymium. Whereas the first of these was a true element, the second was not, although it was accepted as such for more than 40 years.

Later chemists wondered whether didymium too might consist of more than one element, and their suspicions were confirmed when Bohuslav Brauner (1855–1935) of Prague showed in 1882 that its atomic spectrum was not that of a pure metal. The Austrian chemist, Carl Auer von Welsbach (1858–1929) took up the challenge and in June 1885 he announced to the Vienna Academy of Sciences that he had succeeded in splitting didymium into its two components: neodymium and praseodymium in the form of their oxides.

A pure sample of metallic praseodymium was first produced in 1931.

Economic element

The most important ores in which praseodymium occurs are monazite and bast-näsite, and these are discussed in more detail together under the lanthanoids. China, USA, Brazil, India, Australia, and Greenland have large reserves of praseodymium that are estimated to be in excess of 2 million tonnes. World production of praseo-dymium oxide is about 6,000 tonnes per year, and the metal itself is produced by the reaction of praseodymium fluoride and calcium.

Praseodymium oxide is best known for its ability to give glass a 'pure' yellow colour, and this is still called didymium glass. It is used for the goggles which protect the eyes of welders and glass-blowers because it filters out infrared (heat) radiation. Praseo-dymium is used in ceramics and glazes, where it produces brilliant pastel colours, especially shades of green and yellow.

Praseodymium metal makes up the core material in carbon-arc electrodes for film studio lights, searchlights, and flood lighting, and it is used in alloys for permanent magnets.

Another use of praseodymium is to alloy with magnesium and thereby increase its strength sufficient for it to be used in aircraft engines.

Environmental element

Praseodymium in the environment	
EARTH'S CRUST	9.5 p.p.m.
	Praseodymium is the 39th most abundant element.
SOILS	approx. 8 p.p.m. (dry weight) with range 1–15 p.p.m.
SEAWATER	1 p.p.t.
ATMOSPHERE	virtually nil.

Praseodymium is one of the more abundant of rare earth elements and it is four times more abundant than tin. It poses no environmental threat to plants or animals.

Chemical element

Data file	
CHEMICAL SYMBOL	Pr
ATOMIC NUMBER	59
ATOMIC WEIGHT	140.90765
MELTING POINT	931°C
BOILING POINT	3,510°C
DENSITY	6.8 g/cm³
OXIDE	Pr_2O_3

Praseodymium is a soft, malleable, silvery metal and a member of the lanthanoid group (row 4f) of the periodic table. It reacts slowly with oxygen forming a flaky green oxide layer, and while it is less readily oxidised than other rare earth elements, it still needs to be stored under oil or coated with plastic. It reacts rapidly with water. It preferred oxidation state is +3, and this is the only one which can exist in aqueous solutions which are green in colour by virtue of the Pr^{3+} ion. There are compounds with oxidation states +2 and +4 and fluorides are known for all three states: PrF_2, PrF_3, and PrF_4.

There are 39 known isotopes of praseodymium of atomic mass 121 to 159, and 15 known isomers. Naturally occurring praseodymium consists almost entirely of one stable isotope and that is praseodymium-141. There are minute traces of other naturally occurring radioactive praseodymium isotopes produced in uranium ores as fission fragments. The longest-lived radioactive isotope is praseodymium-143 with a half-life of 13.6 days.

Element of surprises

1. Praseodymium alloyed with nickel, of composition $PrNi_5$, responds to being magnetised by lowering its temperature. This unusual property has made it possible for scientists to approach to within 1/1000th of a degree of absolute zero (−273.15°C).

2. Praseodymium has more syllables (six) in its name than any other element.

Promethium

Pronounced proh-mee-thee-um, the name was suggested by Grace Mary Coryell, whose husband Charles was one of its discoverers—see below. She proposed that it be called prometheum after Prometheus, the Greek who stole fire from the Gods and gave it to humans, and this was accepted although later its spelling was changed by IUPAC to promethium in accord with most other metals.

French *prométhium*; German *Promethium*; Italian *prometo*; Spanish *prometio*; Portuguese *promécio*.

Promethium is one of 15 chemically similar elements referred to as the lanthanoid or rare earth elements, which extend from element atomic number 57 (lanthanum) to element atomic number 71 (lutetium). The term rare earth is a misnomer because some are not rare, although promethium is almost non-existent on Earth because all its isotopes are radioactive with short half-lives.

Essential element

This radioactive element is used in button-sized atomic batteries that can power equipment such as guided missiles and pacemakers.

Cosmic element

See *Element of surprises.*

Element of life

Promethium has no role to play in living things.

Medical element

When promethium was injected into animals it became localised on the surface of bones from which it was only slowly removed, behaving in this respect like other rare earth elements. Consequently it was suggested that promethium-147 could be used to investigate bone conditions by monitoring the X-rays it generates. What has prevented this was the difficulty in selectively separating this isotope. Then, in 2004, researchers led by Weisheng Liu at Lanzhou University, China, reported a large ligand molecule which was shown to attach itself to promethium-147 in preference to other metals and so could be used to extract it from solution.

Historical element

Element 61 was first predicted in 1902 by the Czech scientist Bohuslav Branner (1855–1935); he speculated that there should be an element between neodymium and samarium. This was confirmed by Henry G. J. Moseley (1887–1915) in 1914 when he proved that there must be an element with this atomic number, and that it was still to be found. Attempts were made to discover it from 1912 onwards, with several claims being made in the 1920s and 1930s. Although none of the early work was officially recognised, some of those working on it might possibly have detected this element.

In 1924, Italian chemists Luigi Rolla (1882–1960) and Lorenzo Fernandes (1902–77) at the Royal University in Florence, analysed samples of the Brazilian rare earth mineral monazite. They separated a concentrated nitrate solution of the many lanthanoid elements it contains and by fractional crystallisation obtained a solution containing mostly samarium. It gave X-ray spectra mainly of samarium but with extra lines which they attributed to element 61. They published their results in 1926 and proposed the name florentium (symbol Fr) for the new element. Rolla even claimed priority in its discovery saying that they had found element 61 in 1924 and locked their results away in a sealed vault in the Academi of Lincei.

In 1926, US chemists B. Smith Hopkins (1873–1952), J. Allen Harris and Leonard Francis Yntema (1892–1976) of the University of Illinois, thought they had found element 61 in material extracted from neodymium (element 60) and samarium (element 62). Again their evidence was based on lines in the atomic spectra. They proposed the name illium (symbol Il) after the state of Illinois. However, other scientists, and in particular Tacke-Noddack and Walter Noddack, were unable to reproduce their results and the spectral lines were thought to be due to traces of other known elements.

Also in 1926, US chemist Charles James (1880–1928), who was one of three independent discoverers of lutetium, at the University of New Hampshire published an article claiming the discovery of element 61, although he did not propose a name for it. The previous year he had obtained an X-ray analysis supposedly showing seven L-series lines between the corresponding lines of elements 60 and 62. James's interest in element 61 dated back to 1912 when he corresponded with William Crookes (discoverer of thallium) and Henry Moseley (see the chapter on the periodic table) in an effort to gather as much information as he could from these leading researchers on filling the missing gap in the periodic table.

James spent the next 14 years gathering information, collecting samples, refining techniques, and analysing minerals. He submitted his X-ray spectroscopic analysis to the two leading researchers in the world on this technique, both of whom agreed with his findings. James had been a referee for the B. Smith Hopkins *et al* paper mentioned above in early 1926 which claimed discovery of element 61 and he had approved that paper for publication. James was a modest individual and he never engaged in disputes nor denigrated the work of others. He decided to withdraw the paper he had submitted to the *Journal of the American Chemical Society*—and in which he had previously published 60 papers—and sent it instead to an obscure journal which explains why his claim went unnoticed.

We now know that none of these claims could have been justified, for the simple reason that all isotopes of promethium are radioactive with half-lives too short for them to have survived from the time the Earth was formed. There was no way that this element could have been successfully extracted from terrestrial sources. Tiny amounts of promethium do occur in uranium ores as a result of fission, so in theory it could be extracted from this source, but the task would have been technically impossible on account of the tiny amount that is present, calculated to be around a picogram (10^{-12} g) per tonne of ore.

A more realistic claim to have obtained element 61 was made in 1938 by H.B. Law, M.L. Pool, J.D. Kurbatov and L.L. Quill at Ohio State University who had bombarded praseodymium and neodymium with neutrons, deuterons and alpha particles in a cyclotron and detected element 61 in the debris. They proposed the name cyclonium (symbol Cy), but this was not accepted as a discovery as there was no *chemical* proof that the missing element had been made. In fact in 1935 a method of extracting promethium became technically possible thanks to the work of Olavi Erämetsä of the Institute of Technology at Helsinki, Finland, who developed a chromatographic method of separation.[152]

In 1939, the 60-inch cyclotron was completed at the University of California. Some of the first to use it were Italian physicist Emilio Segrè (1905–89) and Chinese physicist Chien-Shiung Wu (1912–97) who bombarded rare earth elements to get element 61. Segrè claimed that they had almost certainly produced isotopes of element 61, but they could not prove it to their satisfaction. In fact many element discoveries took place soon thereafter stimulated by the wartime development of the atomic bomb, and the Segrè and Wu work is now rarely referred to.

Finally proof of element 61 was forthcoming in 1945 from the work of Jacob A. Marinsky (1918–2005), Lawrence E. Glendenin (1918–2008), and Charles D. Coryell (1912–71) at Oak Ridge, Tennessee, USA, because they had the new technique of ion-exchange chromatography at their disposal and with it they were able to separate out samples of isotope-147 from the fission products of uranium fuel from a nuclear reactor. Their work was reported in 1947. They were going to call the element clintonium after the Clinton Laboratories in which the work was done, until Coryell's wife suggested prometheum. This name was changed to promethium by IUPAC when they approved the name two years later.

Economic element

Promethium is obtained in milligram quantities from the fission products of nuclear reactors. Most promethium is used only for research purposes, but some finds useful applications. Promethium-147 is commercially available, and is employed as a low-energy beta-emitter in luminous paint, where it glows with a pale blue or greenish glow. It is also used in so-called 'atomic batteries' which are specialised miniature batteries about the size of a drawing pin; these are used for guided missiles, watches, pacemakers, and radios. They have a useful life of around five years. The beta par-

---

[152] Olavi Erämetsä contacted me after reading the first edition of *Nature's Building Blocks* and kindly sent me an account of his early work in this area.

ticles emitted by promethium-147 are made use of in meters that can measure the thickness of sheet steel—and even of paper.

Promethium isotopes will generate X-rays when their beta-rays impinge on heavy metals and it may one day be possible to produce portable X-ray units making use of this feature. There are also expectations that it could be used as a power source for satellites and space probes.

Environmental element

All the promethium which might once have existed on Earth when it first formed, has long since radioactively decayed. Even if there had been as much promethium as cerium, which is the most abundant of the rare-earth elements, making up 68 p.p.m of the crust, we can calculate that it would all have vanished within 10 thousand years.[153]

Chemical element

Data file	
CHEMICAL SYMBOL	Pm
ATOMIC NUMBER	61
ATOMIC WEIGHT	145
MELTING POINT	1,168°C
BOILING POINT	*approx.* 2,700°C
DENSITY	7.2 g/cm³
OXIDE	Pm_2O_3

Promethium is a radioactive metal, and a member of the lanthanoid group (row 4f) of the periodic table. A microgram sample of the metal itself was obtained in 1963 by F. Weigel of Munich who reacted promethium fluoride (PmF_3) with lithium at 800°C. Its preferred oxidation state is +3 and in solution it exists as the Pm^{3+} ion. Several compounds have been made such as PmF_3 and $PmCl_3$. Promethium salts generally have a pink or red colour, all of which cause the surrounding air to glow with a pale blue-green light.

There are 38 known isotopes of promethium with mass numbers ranging from 126 to 163, and 19 known isomers. There are minute traces of some radioactive promethium isotopes in uranium ores, where they have been formed by nuclear fission. The longest-lived radioactive isotope is promethium-145, with a half-life of 17.7 years and this undergoes radioactive decay by the nucleus capturing one of the innermost electrons (the decay mode known as electron capture) and emitting gamma radiation.

---

[153] Assuming it was promethium-145, which has a half-life of 17.7 years, this would mean that 1 million tonnes of promethium would become 500,000 tonnes after 17.7 years, 250,000 after 35.4 years, 125,000 after 53.1 years, and so on down to 1 tonne after around 350 years. Even a *million* million tonnes would be reduced to a single tonne after only 700 years, and down to a mere gram in a little over 1,000 years.

The more common promethium-147 has a half-life of 2.62 years and is a beta-emitter with only weak gamma-radiation.

Element of surprises

Promethium may be found only as the occasional few atoms in our solar system, coming as it does from the decay of uranium, but it has been detected in the spectrum of the star HR465 in Andromeda. Considering that no isotope longer-lived than promethium-145 can exist, this means that the star is manufacturing the element in vast quantities and on its surface. How this is happening is as yet unexplained.

91
Pa

Protactinium

Pronounced pro-tak-tin-ee-um, the name is derived from the Greek *protos*, meaning first, as a prefix to the name actinium, because it is a precursor to this element, forming it as it undergoes radioactive decay. The element was also called brevium and uranium-X by those who first observed a short-lived isotope, but the name protoactinium was chosen for a longer-lived isotope and, by convention, took precedence, although it was shortened to protactinium in 1949 by the International Union of Pure and Applied Chemistry (IUPAC).

French *protactinium*; German *Protactinium*; Italian *protoattinio*; Spanish *protactinio*; Portuguese *protacnídio*.

Essential element

Natural, radioactive element which is too short-lived to be of any use.

Dangerous element

Protactinium is highly dangerous because of its intense radioactivity, and this overrides other toxicity considerations, but conventional toxicity is thought to be low. Special precautions have to be taken when dealing with it, because it is an alpha-emitter.

Element of history

In 1871, Mendeleyev suggested that there might be an element between thorium and uranium. However, this missing element evaded detection until radioactivity had been discovered in the 1890s, and that first resulted in the new elements radium and polonium being separated by the Curies. The missing element between thorium and uranium remained elusive.

In 1900, the English chemist William Crookes (1832–1919) separated an intensely radioactive material from uranium, but did not identify it as a new element. He simply named it uranium-X. In 1913, at Karlsruhe, Germany, the Polish chemists Kasimir Fajans (1887–1975) and Otto Göhring showed that this decayed by beta-emission and they named it brevium because it existed only fleetingly. We now know it is one of the members of the sequence of elements through which uranium-238 decays on its way to lead; brevium was the isotope protactinium-234.

A longer-lived isotope of the same element was physically separated from uranium ore in 1918 by Lise Meitner (1878–1968) at the Kaiser-Wilhelm Institute for Chemistry, Berlin, where she worked with Otto Hahn (1879–1968), although he was serving in the army at the time. In 1917, Meitner had only 21 g of pitchblende to work with and was unable to import more because of wartime restrictions. She ground up the sample and extracted it with hot concentrated nitric acid. This treatment dis-solved all but 2 g, which was in the form of an insoluble silicate. This she dissolved in hydrofluoric acid and was eventually to show that the precursor to actinium was present, in other words there was a radioactive element present that was transmuting into actinium. This was the longer-lived isotope protactinium-231, which is part of the decay series of uranium-235 as it transmutes into lead.

Meitner knew she had found a new element but needed more material if she was to isolate a sample of it. A colleague contacted Friedrich Giesel, who was associated with a radium-producing firm in Braunschweig, and as a result she obtained a further 100 g of pitchblende residues (from which the uranium and radium had already been extracted). By early 1918, she had enough evidence for the new element to suggest to Hahn that they submit a paper to the journal *Physikalishce Zeitschrift*, which they did in March that year. Titled 'The mother substance of actinium, a new radioactive element of long half-life' it proposed the name protoactinium.

When Lise Meitner and Otto Hahn investigated the chemistry of the new element they found it to be like tantalum, the metal that came above an empty slot in group 5 of Mendeleyev's periodic table. He had given it the temporary name of eka-tantalum for this very reason. In some respects the chemistry of protactinium did resemble tantalum, for example, its more stable oxide has the same formula as that of tantalum oxide, Ta_2O_5, but the chemical similarity was purely fortuitous and, in a modern periodic table, protactinium is put in its rightful place as a member of the actinide series of elements.

In June 1918, protactinium was also reported by Kasimir Fajan (1887–1975), and also by the English chemist Frederick Soddy (1877–1956) in Glasgow, Scotland. In 1927, Aristid von Grosse (1905–85) managed to extract 2 mg of protactinium as the oxide and in 1934 he reduced this to metallic protactinium by converting it to the iodide, PaI_5, and decomposing this in a vacuum by means of a heated filament.

Economic element

Protactinium-231 occurs naturally in uranium ores such as pitchblende, to the extent of 3 p.p.m. In 1961, the UK Atomic Energy Authority extracted 125 grams of 99.9 per-cent pure protactinium from 60 tonnes of spent uranium fuel elements; this became the major global stock of this element. It has been supplied to laboratories around the world enabling the chemistry of the element to be studied. However, no commercial use has yet been found for it.

Environmental element

Protactinium in the environment	
EARTH'S CRUST	traces
	Protactinium is among the 10 least abundant elements.
SOILS	traces
SEAWATER	0.00002 p.p.t., i.e. barely detectable.
ATMOSPHERE	nil.

Protactinium-231 and thorium-230 can be used to date marine sediments. They derive from different isotopes of uranium but they have very similar properties and appear to be precipitated at the same rate. While each by itself would be a guide to the age of a sediment, the ratio of the two is a better measure of geological time. Their respective half-lives are 32,500 and 80,000 years which means they can be reliably used to date sediments as old as 175,000 years.

Chemical element

Data file	
CHEMICAL SYMBOL	Pa
ATOMIC NUMBER	91
ATOMIC WEIGHT	231.03588
MELTING POINT	1,840°C
BOILING POINT	about 4,000°C
DENSITY	15 g/cm^3 (est.)
OXIDE	PaO, Pa$_2$O$_5$ and PaO$_2$

Protactinium is a silvery, radioactive metal that is part of the actinide group (row 5f) of the periodic table. It is attacked by oxygen, steam, and acids, but not alkalis. The element becomes superconducting below 1.4 K. It is of research interest only. The element displays the oxidation states +2 as in PaO, +3 as in PaI$_3$, +4 as in PaF$_4$, and +5 as in PaCl$_5$.

There are 29 known isotopes of protactinium with masses 212 to 240, and three known isomers. The radioactive decay series of uranium-235 and uranium-238 produce protactinium as a secondary decay product, as do the decay series of neptunium and plutonium. As a result there are five naturally occurring protactinium isotopes and these are protactinium-231, -233, -234, -235, and -236. Protactinium-234 has a half-life of 6 hours, 42 minutes, which decays by beta-emission, and which comes from the more abundant uranium-238 isotope. Protactinium-231, with a half-life of 32,500 years, decays by alpha-emission and it derives from uranium-235. Other longish-lived radioactive isotopes are protactinium-233 with a half-life of 27 days, and protactinium-230 with a half-life of 17.4 days.

Element of surprises

1. Research into protactinium was undertaken by Alfred 'Alfie' Maddock (1917–2009) as part of the UK's World War II project with the aim of making atomic weapons. It was thought at the time, but wrongly as it turned out, that protactinium-231 might be used to make an atomic bomb. Work on it began in Cambridge, then transferred to Montreal, and continued after the war back at Cambridge in a disused aircraft hangar. There the element was separated from sludge provided by the Atomic Energy Authority and its chemistry was investigated. When Princess Margaret came to open the new chemistry laboratories at Cambridge in 1958 she

was shown a sample of pure protactinium bromide (PaBr$_3$) which they had prepared. Research on this element continued at Cambridge until 1984.

2. Anyone who has a smoke detector in their home also has a minute amount of protactinium. The detector contains americium-241, which is the source of the alpha particles on which the smoke detection relies. This decays to neptunium-237 and this in its turn emits an alpha particle and thereby converts to protactinium-233.

Radium

Pronounced ray-dee-um, the name is derived from the Latin *radius*, meaning ray.
French *radium*; German *Radium*; Italian *radio*; Spanish *radio*; Portuguese *rádio*.

Essential element

A hundred years ago this element was seen as almost miraculous, finding outlets such as luminous dials and health cures, today it is regarded as too dangerous to be used.

Element of life

Radium in the human body	
BLOOD	0.007 p.p.t.
BONE	0.004 p.p.t.
TISSUE	0.0002 p.p.t.
TOTAL AMOUNT IN BODY	30 picograms (30×10^{-12} g).

Radium has no biological role. Nevertheless it is part of the sequence of elements through which thorium and uranium decay and so it is part of the environment—and the human diet. Because radium is present in soils, it gets into the food chain via vegetation, and levels in plants have been measured at 0.03 to 1.6 p.p.t. We have an average daily intake of 2 picograms although most passes through us unabsorbed.

Dangerous element

Radium is highly dangerous because of its intense radioactivity as an alpha-emitter, and this greatly overrides other toxicity considerations, such as its similarity to barium. The maximum permissible body burden for radium-226 is not measured in weight but in the dose of radiation it delivers, and this is set at 7,400 becquerel (Bq); 1 Bq is one decay per second.[154] The natural amount of radium in the average person delivers around 1 Bq.

The dangers of radium were apparent from the start. The first case of so-called 'radium dermatitis' was reported in 1900, only two years after the element's discovery. Henri Becquerel (1852–1908) carried a small ampoule of radium around in his

---

[154] The curie (Ci) is another, older, unit of radioactivity, defined as that amount of radioactivity with the same decay rate as 1 g of radium-226, which is 3.7×10^{10} decays per second.

waistcoat pocket for six hours and reported that his skin became ulcerated. Marie Curie experimented with a tiny sample that she kept in contact with her skin for 10 hours and noted how an ulcer appeared, although not for several days.

More serious illness came through the large-scale use of radium in the manufacture of luminous alarm clocks and watches, the dials of which were painted by hand. Those involved in their production were very much at risk. The work was done mainly by young girls who had a habit of licking the fine brushes they used in order to form a sharp point. The luminous paint contained only 70 micrograms per gram of radium but this was enough to cause severe problems when it got into the mouth and body, generally causing cancerous growths.

The dangers came to public notice in the well-publicised case of the 'Radium Girls' who sued their former employer, US Radium. The company tried devious methods to prevent the case coming to trial and it became a *cause célèbre* in the press. The plaintiffs were clearly dying, and all of them were quite young. In the end, the company settled out of court awarding each girl $10,000, but they were all dead within a few years. Some of the girls from the US Radium factory were so contaminated with radium that their hair, faces, hands and arms glowed luminously in the dark. In the New Jersey factory, of the 800 who worked there in the years 1917–24, 48 succumbed to radiation sickness of whom 18 died.

Luminous dials were needed for aircraft, compasses, gun sights, and electrical instruments in World War II, and the girls who painted them worked in a very different environment protected by screens and good ventilation, with all surfaces in the workshops being scrubbed down daily.

Medical element

For many years, and especially in the first half of the last century, radium was an essential part of the medical treatment for cancer. Surgeons implanted 'radium needles' (usually containing radium chloride or radium bromide) in tumours and then allowed the intense radiation they emitted to destroy the cancerous cells. This form of treatment is now rarely used, partly because it put at risk the lives of those who made the needles. Workers at the London Radium Institute, where they were prepared, invariably had low white-cell counts and several of them eventually died of radium exposure. Today other radioisotopes are preferred such as cobalt-60.

Element of history

Radium was discovered in Paris in 1898 by Marie Skłodowska Curie (1867–1934), working with her husband Pierre Curie (1859–1906). They managed to extract 1 mg of radium from 10 tonnes of the uranium ore pitchblende, a considerable feat bearing in mind the chemically primitive methods of separation available to them. They knew that they had discovered a new element because its atomic spectrum revealed new lines in the red region. They named the new element after the rays (*radius*) of faint blue light with which it glowed in the dark. The glow was caused by its radioactivity exciting the surrounding air. In the words of Marie Curie:

One of our joys was to go into our workroom at night when we perceived the feebly luminous silhouettes of the bottles and capsules containing our products. It was really a lovely sight and always new to us. The glowing tubes looked like faint fairy lights.

The metal itself was only isolated by her and André Debierne (1874–1949) in 1911, by the electrolysis of $RaCl_2$. At Debierne's suggestion they used a mercury cathode, with which the radium that was liberated formed an amalgam. This was then heated to remove the mercury leaving the radium behind.

Economic element

Early in the last century, radium was regularly extracted from uranium ores for commercial use. The amount of radium in such ores varies between 150 and 350 mg per tonne, depending on its source, with those from Zaire and Canada having the most. Other uranium ores which were exploited for their radium were from Colorado and the Congo, and indeed the wastes from any uranium processing facility could be used. Today radium is extracted from spent nuclear fuels which are now the source of any radium that is required, although the use of this element has declined greatly.

Luminous paint consisted of a mixture of radium bromide and zinc sulfide and the latter glowed by being activated by alpha rays from the decay of the radium. The owner of a luminous watch or clock is in no danger because the alpha rays cannot penetrate the glass face nor the casing.

Radium and beryllium were once used as a portable source of neutrons. The beryllium-8 nucleus will capture an alpha particle and then undergo spontaneous decomposition to carbon-12, releasing a neutron as it does. Today other alpha emitters are preferred such as polonium.

Environmental element

Radium in the environment	
EARTH'S CRUST	0.6 p.p.t.
	Radium is the 86th most abundant element, i.e. it is very rare.
SOILS	0.8 p.p.t. (although some have as much as 0.3 p.p.b.)
SEAWATER	less than 0.001 p.p.t. (but detectable).
ATMOSPHERE	nil.

Radium contributes to the background level of radiation that makes the Earth a naturally radioactive planet. It has been estimated that each square kilometre of the surface (to a depth of 40 cm) contains 1 g of radium. The amount of radium in the environment has increased as a result of human activity, notably through the use of phosphate fertilizers because phosphate rock contains uranium and therefore radium. Other sources of radium contamination are cement production and coal burning.

In parts of North America there are places where well water has measurable amounts of radium to the extent that they exceed the Environmental Protection

Agency's limit for safe drinking water which is 5 picocuries (pCi) [155] per litre (approx. 0.2 Bq per litre).

Chemical element

Data file	
CHEMICAL SYMBOL	Ra
ATOMIC NUMBER	88
ATOMIC WEIGHT	226
MELTING POINT	700°C
BOILING POINT	1,140°C
DENSITY	*ca.* 5 g/cm³
OXIDE	RaO

Radium is silvery, lustrous, soft, radioactive metal and a member of group 2 (row 7s), also known as the alkaline-earth group, of the periodic table. Although it is the heaviest member of that group it is the most volatile. It is bright when freshly prepared, but darkens on exposure to air. Radium reacts with oxygen and water. Its preferred oxidation state is +2 as in compounds like $RaCl_2$ and the oxide RaO. When put in a flame, radium compounds produce a crimson colour. One gram of radium will release 0.001 ml of radon gas per day and this has to be taken into account when storing the metal. A gram of radium also releases 4,000 kJ of energy per year.

There are 33 known isotopes of radium of masses 202 to 234, and 12 known isomers. There are nine naturally occurring isotopes: radium-220 to 228, and they are produced by the radioactive decay of elements of higher atomic number than radium (88) and they come mainly from uranium, thorium, and plutonium. There are three main isotopes: radium-223, which is part of the decay series of uranium-235 and which has a half-life of 11.5 days; radium-224, which is a member of the thorium-232 decay series, half-life 3.7 days; and radium-226 which is not only the longest-lived isotope with a half-life of 1,600 years, but comes from the decay of the most abundant uranium isotope, uranium-238. Another relatively long-lived radioactive isotope is radium-228 with a half-life of 5.75 years.

Element of surprises

Soon after its discovery, radium was regarded almost as if it were a new wonder drug, as radiation was actually seen as beneficial to the body. All kinds of quack cures were promised by those peddling radium treatments, and indeed the USA saw a 'radium craze' which began in 1903 and went on for almost 30 years. Typical of such radiation cures were the Cosmos Bag which was to be applied to relieve arthritic joints. Most popular of all was Raithor, a weak solution of radium salts that was claimed to be a general preventative of disease, and even to cure stomach cancer and mental illness.

---

[155] A picocurie (pCi) is a trillionth part of a curie, i.e. 1×10^{-12} Ci, and it is still sometimes used.

Those who found Raithor too expensive could buy a Revigorator, a flask lined with radium in which you stored water overnight to be drunk every morning.

The most famous case of death from Raithor was that of Eben Beyers, a steel magnate of Pittsburgh, who drank a bottle a day for about four years, at the end of which he was suffering from severe radiation sickness and cancer of the jaw. This necessitated extensive surgery in 1931, not that it could save him. This widely publicised case marked the end of radium water cures, and other over-the-counter treatments.

Radon

Pronounced ray-don, the name is derived from radium. This gaseous element was first called radium emanation, because it was emitted by radium. An alternative name was niton from the Latin *nitens* meaning 'shining'.[156] A similar emanation given off by thorium was called thoron, and that from actinium was called actinon. In fact, radon was to have several names because the various gases being emitted by uranium, thorium, actinium, and radium were thought to be different elements. The confusion arose because they were all short lived, so little could be gleaned about their chemistry. In 1923, the situation was rationalised and the name radon approved by the International Union of Pure and Applied Chemistry (IUPAC), since it was clear that they were isotopes of a noble gas and should be named accordingly, that is, the name should end in '-on'.[157]

French *radon*; German *Radon*; Italian *radon* (*emanio*); Spanish *radón*; Portuguese *rádon*.

Essential element

There are spas where breathing in this radioactive gas is actively encouraged as beneficial, but for the rest of us this is an element to be avoided as much as possible.

Element of life

Radon may have had an indirect biological role on the evolution of life on this planet because it has been estimated that it is the major contributor to natural background radiation which may be responsible for genetic modifications and evolutionary changes.

Dangerous element

Radon is encountered in everyday life as part of the atmosphere and it diffuses out of rocks, where it is formed during the natural decay sequences of uranium and thorium. It is a human health hazard for many underground miners, and not only those working in uranium mines. It had been noted in the Middle Ages that miners in some areas, such as Bohemia, had short lives and died of lung disease (cancer). However, see *Element of surprises*.

The danger of radon lies with the non-volatile radioactive isotopes that it transmutes to. These can adhere to dust particles and so enter the lungs. One of the decay products is lead-210 which has a half-life of 22 years, and this will tend to linger long in the body, like other lead atoms.

---

[156] The name and symbol Nt for niton was actually approved by the International Commission of Atomic Weights in 1912.

[157] The only noble gas which does not follow this naming convention is helium, and although it has been suggested that this should be called 'helion' the name has not been taken up.

In the 1980s and 1990s, it became apparent that in certain localities the amount of radon escaping from the ground, or from buildings constructed of granite, was much higher than average and that this could accumulate indoors, putting the occupants at risk. Radon could build up in basement rooms or even in ground floor rooms with cracked concrete floors, unless there was adequate ventilation to ensure its dispersal. The result of breathing air contaminated with radon was that it exposed occupants to a higher risk of lung cancer.

Government agencies issued warnings of the dangers and booklets were distributed to those areas most at risk explaining how the problem could be prevented, such as sealing cracks and gaps in floors and walls and around service pipes, and ensuring good ventilation in rooms. In severe cases of pollution, the answer was to create a space below a concrete floor where the radon could collect and then to extract it with a fan. In the majority of homes the exposure to radon radiation amounted to around 20 becquerel per cubic metre (Bq/m^3) which is very low, but some homes have levels as high as the $400 Bq/m^3$ which is considered the limit for industrial exposure. However, in some uranium mines levels of 10 million Bq/m^3 have been recorded.[158]

The US Environmental Protection Agency released a comprehensive document produced by the US National Academy of Sciences in February 1998. It was Part VI of *the Biological Effects of Ionizing Radiation (BEIR)* series of reports and was entitled 'The health effects of exposure to indoor radon.' It confirmed that radon could be the cause of lung cancer, a conclusion borne out by epidemiological studies on humans and experimental studies on animals. An analysis of the health data of 68,000 underground miners, and not only uranium miners, showed 2,700 had died of the disease. Radon is now thought to account for about 10 percent of cases of lung cancer, and this would explain why those who neither smoke, nor are exposed to cancer-causing fumes, are also prone to the disease.

Radon is not easily monitored. One analytical method employs a special polymer film that is exposed for up to a year and then analysed microscopically for the visible tracks left in the plastic by the intense alpha particles emitted by the gas as it radioactively decays. A quicker method is to measure the radioactivity of the various isotopes that radon-222 transmutes to,[159] and this is done by exposing charcoal to the air. This is left to absorb the radon for a set time, such as 48 hours, and then the γ-rays emitted by the 'daughter' elements into which it has transmuted can be detected. Home testing kits are available which are sent away for analysis after being exposed to the suspect environment.

Element of history

Radon was first observed by Ernest Rutherford (1871–1937) and R.B. Owens (1870–1937) in 1899. They had detected a radioactive gas coming from their samples of thorium. Even though radon atoms have short half-lives, the gas was

---

[158] 1 Becquerel is one radioactive decay per second.
[159] The sequence of elements is as follows: radon-222 (half-life 4 days) → polonium-228 (3 minutes) → lead-214 (27 minutes) → bismith-214 (2 minutes) → lead-210 (22 years) → bismuth-210 (5 days) → polonium-210 (20 weeks) → lead-206 (stable).

accumulating faster than it was decaying. Also in 1899, Pierre and Marie Curie detected a radioactive gas coming from their radium experiments. They noted that it took some time for the radioactivity to lessen, indicating that what they had observed was the isotope radon-222. They correctly concluded that it was a new radioactive element. Recent historians have suggested that this was really the discovery of radon.

The following year Friedrich Ernst Dorn (1848–1916) in Halle, Germany, noted that a gas was accumulating inside ampoules of radium, but he took no steps to discover what it was. Dorn merely repeated the experiments of the Curies with more varied radium compounds. It was in fact the longer-lived isotope radon-222 with a half-life of 3.8 days, whereas the radon that Rutherford had observed was radon-220 with a half-life of 56 seconds. In 1900, while he was at McGill University, in Canada, Rutherford devoted himself to investigating the gas more thoroughly, and along with Frederick Soddy (1877–1956) they showed that it was possible to condense it to a liquid, using liquid air.

In 1908, William Ramsay (1852–1916) and Robert Whytlaw-Gray (1877–1958) at University College, London, collected enough radon to determine several of its properties, such as its density, noting that it was the heaviest gas known. They gave it the name niton in 1910, from the Latin *nitens* meaning 'shining' which is what it appeared to do in the dark.

Medical element

Radon collects over samples of radium-226 at the rate of around 0.001 ml of radon per day per gram of radium. It was sometimes used in hospitals to treat cancer and was produced as needed and delivered in sealed gold needles.

Environmental element

Radon in the environment	
EARTH'S CRUST	traces
	Radon is among the 10 least abundant elements.
SOILS	traces
SEAWATER	10^{-8} p.p.t.
ATMOSPHERE	10^{-9} p.p.t., i.e. 1 part in 10^{21} parts of air.

There is a detectable amount of radon in the atmosphere although the amount swirling around the planet at any one time totals less than 100 grams. Nevertheless, the gas can accumulate in certain localities. Natural radon is almost totally composed of the two isotopes radon-220 and radon-222. These, and their radioactive decay products, contribute to atmospheric electricity to the extent of accounting for half of the ionization of the air near the surface of the Earth.

Chemical element

Data file	
CHEMICAL SYMBOL	Rn
ATOMIC NUMBER	86
ATOMIC WEIGHT	222
MELTING POINT	−71°C
BOILING POINT	−62°C
DENSITY	9.7 g per litre
OXIDE	none identified

Radon is a colourless gas and one of the group 18 elements (row 6p), also known as the noble gases, of the periodic table. It should be like xenon and chemically unreactive, forming only a few compounds with bonds to fluorine and oxygen. It forms radon bifluoride (RnF_2) and the cation RnF^+ although these have only a fleeting existence. When radon is cooled below its freezing point, it phosphoresces brightly and turns a deep orange colour. It is little studied, partly because it is a noble gas and is therefore reluctant to form molecules, and partly because its intense radiation is likely to destroy any compound that it might form.

There are 35 known isotopes of radon of masses 195 to 229, and 12 known isomers. Radon is part of the radioactive decay series of naturally occurring uranium and thorium, and of actinium and plutonium. Of the natural isotopes, uranium-235 produces radon-219 which has a half-life of only four seconds, and thorium-232 produces radon-220 which is only slightly longer-lived with a half-life of 56 seconds. It is the more abundant uranium-238 which produces the longest lived radon isotope, radon-222, which has a half-life of 3.8 days. Another relatively long-lived radioactive isotope is radon-211 with a half-life of 14.6 hours.

Element of surprises

1. Not all epidemiological research supports there being a link between radon and lung cancer in the general population. While such a link is not disputed for underground miners, who are exposed to high levels of radon, it has been suggested that the very low levels to which the general public is exposed in homes and offices pose nothing like the same level of risk. Indeed, some studies in the USA, China and Scandinavia seem to show that there is an inverse relationship, with there being fewer cases of lung cancer in regions with the higher levels of radon.

   One village in the Austrian Tyrol, Umhausen, has been built on a fall of rock that took place around 8,700 years ago. The rocks were granite, and the fall fractured them so that radon from the uranium they contain, easily escapes once it forms. The average level of radon in homes built there is around 2,000 Bq/m³ and in one it was more than 250,000. The incidence of lung cancer in the village was said to be statistically higher than expected but numbers involved were so small that this has been disputed.

2. Some people still believe that breathing radon may be beneficial. In the little spa town of Le-Mont-Dore in the Auvergne, France, it is possible to take radon-rich air in the form of 'nasal irrigation' with a tube inserted up one nostril so that visitors can breath in a gas that is drawn from a nearby natural hot spring. While most of this gas is carbon dioxide, it also contains a level of radon well above average, and it is this which is supposed to activate the blood, combat allergies, improve digestion, and stimulate the immune system.

3. Another resort which promotes the benefits of radon is Badgastein in Austria, whose waters contain radium and radon and which people can bathe in or drink. They can even breathe in the radon by spending time in an old gold mine where the temperature is around 40°C and where the air contains radon generating a radiation level of around 160,000 Bq/m³. The treatment is said to benefit many conditions and especially rheumatism, arthritis, and stress-related disorders.[160]

[160] Information about Badgastein and its radon treatment was kindly provided by Frank Harper-Jones.

Rhenium

*Pronounced ree-nee-um, it is named after *Rhenus*, the Latin name for the Rhine.*
*French *rhénium*; German *Rhenium*; Italian *renio*; Spanish *renio*; Portuguese *rénio*.*

Essential element

Alloyed with nickel, this element is a vital component of aero engines and turbine blades.

Element of life

The amount of rhenium in the human body has not been measured, but is likely to be very low. It has no biological role and little is known about its toxicity although this is believed to be very low. Indeed, there are no reported cases of humans being affected by rhenium. Radioactive isotopes rhenium-186 (half-life 89 hours) and rhenium-188 (half-life 17 hours) are sometimes used in the treatment of liver cancer.

Element of history

At the start of the last century, the periodic table of the chemical elements had almost been completed, but two members of one particular group, group 7, were still oddly missing, except for manganese, the element at the head the group. The second member of the group, technetium (element 43), was to remain undiscovered for the simple reason that there are no technetium minerals occurring naturally on Earth because all its isotopes are radioactive and have long since disintegrated. (Technetium atoms are to be found among the fission fragments of uranium but these were not detectable by methods available at the time.) In fact technetium was discovered in 1925.

The third member of the group, however, did exist but remained undiscovered even though its properties could be forecast from its position in the periodic table. For example, it would be a metal of high density, and would have a range of oxides.

Despite several attempts to detect the missing element in manganese ores, element 75 remained elusive, although there is evidence that it had been actually detected in 1905. Indeed it was the last stable, non-radioactive, naturally-occurring element to be discovered. There is some evidence that the Japanese chemist Masataka Ogawa discovered it in 1905. He had been sent to London in 1904 by the Japanese Government to work with William Ramsay (the discoverer of several noble gas elements) at University College. In 1905, Ogawa discovered a new element in the mineral thorianite which had been discovered in Ceylon (now called Sri Lanka) the previous year. He did this by analysing the spectrographic lines in its atomic spectrum and realised that a hitherto undiscovered element was present. He claimed it was the missing element 43, and at Ramsay's suggestion he named it nipponium (symbol Np) after his homeland Japan. In 1906, Ogawa returned to Japan and in 1908 he announced his discovery. Sadly he had misidentified it and his claim was soon shown to be incorrect. Nevertheless, research in the 1990s in which Ogawa's original photographic spectra were reanalysed showed that he had actually discovered element 75.

The isolation of rhenium was finally achieved in May 1925 by Walter Noddack (1893–1960) and Ida Tacke (1896–1978),[161] working at the Physico-Technical Testing Laboratory in Berlin.[162] There they concentrated it from the ore gadolinite in which it was an impurity, but only in trace amounts at 10 p.p.m. That they had obtained the missing element was confirmed by Otto Berg (1873–1939) of the Siemens & Halske Company in Berlin who examined the atomic spectrum of the concentrate they obtained and found several new lines that could only be due to a new element. The discovery was finally announced later that year by Tacke at a chemical meeting in Nuremberg.

Following the discovery, Noddack and Tacke continued researching the new element and even obtained a gram of the metal from 660 kg of a Scandinavian ore, molybdenite (molybdenum sulfide). This was the most abundant source they could find. In the 1930s a molybdenum mine in northern Wisconsin was found to offer a better source of rhenium, although even the ore from this mine contained less than 2 percent.

Economic element

Rhenium does not occur as the free uncombined metal, and no mineable mineral ore has been found. It comes mainly from copper deposits where that metal is mined in Chile and in the USA, and these locations are along the Pacific fault which runs through North and South America. It is also to be found where copper is mined in Mongolia, Iran, Armenia, Kazakhstan, and Finland. The ores gadolinite (see beryllium) and molybdenite (see molybdenum), may contain a little and it is from the latter that it is extracted via the flue dusts of molybdenum smelters. Although there was some production of rhenium in the years following its discovery, it was only in the 1950s that this became commercially worthwhile, when its use in alloys and as a catalyst created a demand. World annual production is now around 55 tonnes and the estimated reserves of rhenium are 3,500 tonnes.

Rhenium metal itself is produced by reducing ammonium perrhenate, NH_4ReO_4, and is usually obtained as a grey powder. By pressure stamping this under vacuum and heating in the presence of hydrogen, it is possible to fabricate pure rhenium objects although there is little demand for such items.

Rhenium is added to tungsten and molybdenum to form alloys that are used as filaments for ovens and X-ray machines (alloyed with tungsten). It is also employed in thermocouples that can measure temperatures above 2,000°C, and for electrical contacts which stand up well to electric arcs. It has occasionally been used for plating

---

[161] They were subsequently married.
[162] They also probably detected technetium in 1925, see page 526.

jewellery. Electroplating with rhenium was first achieved in 1934 and it was shown to give a bright, hard deposit. However, the metal is susceptible to oxidation and its surface needs to be protected by a coating of iridium.

Rhenium is also used as a catalyst in the chemical industry, especially in the production of high-octane gasoline, and in other hydrogen-addition processes. It is particularly valued because it is not easily 'poisoned' (that is, deactivated) by traces of sulfur and phosphorus.

More than 40 tonnes a year of rhenium goes into the making of single-crystal blades for aero engines and turbines. These can operate at temperatures as high as 1,600°C. These so-called super-alloys are mainly made of nickel with 6 percent rhenium. This use of the metal will ensure that rhenium will continue to be needed.

Environmental element

Rhenium in the environment	
EARTH'S CRUST	about 0.5 p.p.b.
	Rhenium is the 77th most abundant element, i.e. it is very rare.
SOILS	no data
SEAWATER	4 p.p.t.
ATMOSPHERE	nil.

There is so little rhenium in the environment that virtually nothing is known of how it behaves in soil, plants, crops or animals. There are no instances of pollution by rhenium salts from mining or industry.

Chemical element

Data file	
CHEMICAL SYMBOL	Re
ATOMIC NUMBER	75
ATOMIC WEIGHT	186.207
MELTING POINT	3,180°C
BOILING POINT	5,625°C
DENSITY	21.0 g/cm³
OXIDE	most stable oxide is Re_2O_3, but ReO_2, Re_2O_5, ReO_3 and Re_2O_7 are also known.

Rhenium is a silvery metal with the third highest melting point after carbon and tungsten. It is a member of group 7 (row 5d) of the periodic table. Rhenium has the fourth highest density of the elements, behind iridium, osmium, and platinum. Rhenium resists corrosion but slowly tarnishes in moist air. It dissolves in concentrated nitric acid but not in aqua regia. The metal's easy oxidation is demonstrated by its ready dissolution in bromine water. Rhenium's preferred oxidation state is +7, as in

ReF_7 and the perrhenate ion ReO_4^- in solution, but it can also exhibit other oxidation states from +3 to +6, such as $ReCl_3$, $ReCl_4$, and $ReCl_6$. Rhenium will also form an unusual hydride ion ReH_9^{2-} which can be separated as the potassium compound K_2ReH_9.

There are 35 known isotopes of rhenium of masses 160 to 194, and 21 known isomers. There are two naturally occurring isotopes: rhenium-187 which comprises 63 percent, and is weakly radioactive, and rhenium-185 (37 percent) which is not radioactive and is rhenium's only stable isotope. Rhenium-187 is a beta-emitter with a half-life of 41 billion years which means that 96 percent of the element that was present when the Earth formed 4.57 billion years ago is still around. A gram of rhenium undergoes only 4.5 disintegrations a second. The next longest-lived, radioactive isotope is rhenium-183 with a half-life of 70 days.

Element of surprises

1. Until 1994, there was no evidence for any natural rhenium mineral, and it was assumed that it only occurred as trace amounts in other ores, such as those of molybdenum. Then in that year, four Russian mineralogists who were studying the Kudriavy volcano on Iturup, a Russian island off the north east coast of Japan, came upon a soft whitish-grey material that had a metallic lustre and was flaky, rather like graphite. Yet, unlike graphite, it had a density of 7.5g/cm3 which is remarkably high for a mineral. It turned out to be almost pure rhenium sulfide.

   Its formation could not be explained. Volcanic gases contain traces of rhenium, and the superheated steam that emerged from Kudriavy had a few parts per billion, but why this was depositing as rhenium sulfide at a rate of several grams per day, could not be deduced.

2. Rhenium diboride (ReB_2) is metallic in nature, non-compressible, and extremely hard. In fact, it is so hard that it will scratch diamond.

3. Nuclear theory indicated that there should be no stable isotopes of rhenium, and that rhenium's only isotope listed as stable, that is, rhenium-185, is actually radioactive but with an incredibly long half-life.

Rhodium

Pronounced roh-dee-um, the name is derived from the Greek *rhodon*, meaning rose. French *rhodium*; German *Rhodium*; Italian *rodio*; Spanish *rodio*; Portuguese *ródio*.

Essential element

This element is a vital part of catalytic convertors for car exhausts and optical fibres.

Element of life

The amount of rhodium in the human body is not known but is very low. No biological role for rhodium has so far been discovered, nor is there ever likely to be one because this element is so rare. Most rhodium compounds are only slightly toxic by ingestion—rats required doses of 200 mg per kg body weight to be poisoned by it. This is equivalent to 14 grams for an average 70 kg adult human. Mice raised on drinking water containing 5 p.p.m. of dissolved rhodium, developed leukemia. There are almost no reported cases of humans being affected by this element in any way.

Element of history

Rhodium was discovered in 1803 by William Hyde Wollaston (1766–1828); the story is recounted in more detail under palladium. Wollaston dissolved in acid a sample of platinum, which came from South America. From the solution he recovered platinum, and palladium by precipitating them. He was then left with a beautiful red solution from which he obtained rose red crystals. These turned out to be sodium rhodium chloride, Na_3RhCl_6, and these he reduced to the metal itself by heating with hydrogen gas. He realised it was a new element and eventually reported it as such—see palladium for more details of Wollaston's work.

Economic element

Rhodium occurs as rare deposits of the uncombined metal, and it is found like this in Montana, USA, and as rare minerals, such as the rhodium-lead sulfide, rhodplumsite ($Rh_3Pb_2S_2$). Rhodium is available commercially and it comes as a by-product of the refining of certain copper and nickel ores which can contain up to 0.1 percent rhodium. Most rhodium comes from South Africa, the Urals in Russia, and a little is extracted from the copper-nickel ores mined in Ontario, Canada. World production of rhodium is around 30 tonnes per year and estimated reserves are 3,000 tonnes.

Most rhodium (80 percent) goes into automobile catalytic convertors because rhodium is excellent at reducing NO_x emissions. Reclaimed rhodium from exhausted convertors now amounts to almost 6 tonnes per year.

The next biggest demand for rhodium comes from the chemical industry which accounts for around 10 percent, and there it is used in the catalysts required for the making of nitric acid, acetic acid, and oxo-alcohols, and in reactions which add hydrogen such as the conversion of benzene (C_6H_6) to cyclohexane (C_6H_{12}). One rhodium compound used for this last kind of process was discovered by the Nobel Prizewinner, Geoffrey Wilkinson (1921–96), and is actually called Wilkinson's catalyst.[163] A new type of rhodium-quinone compound has been discovered by Dwight Sweigart at Brown University, Rhode Island, USA, and this promises to be an equally useful catalyst.

The remaining 10 percent of rhodium is needed for a variety of products such as a coating for optical fibres, and optical mirrors, as well as for thermocouples, laboratory crucibles, and the reflectors of headlights. The reason for this use is the metal's excellent reflectivity as a thin surface film. A little rhodium is used to coat electrical contacts such as in spark plugs for aircraft. Rhodium is also used in the detectors of nuclear reactors which measure neutron flux.

Environmental element

Rhodium in the environment	
EARTH'S CRUST	*av.* 0.2 p.p.b.
	Rhodium is the 79th most abundant element, i.e. it is extremely rare.
SOILS	not detected.
SEAWATER	barely detectable.
ATMOSPHERE	nil.

Rhodium is too rare to be assessed in either soils or natural waters, and so its effect on the environment can be assumed to be nil. Tests on plants show it is the least toxic member of the platinum group of metals.

Chemical element

Data file	
CHEMICAL SYMBOL	Rh
ATOMIC NUMBER	45
ATOMIC WEIGHT	102.90550
MELTING POINT	1,966°C
BOILING POINT	3,725°C
DENSITY	12.4 g/cm³
OXIDES	Rh_2O_3 (most stable), RhO, and RhO_2

Rhodium is a rare, lustrous, silvery, hard metal and a member of group 9 (row 4d) of the periodic table. It is one of the so-called platinum group of elements (which are

---

[163] Its chemical composition is $Rh(P(C_6H_5)_3)_3Cl$, aka chlorotris(triphenylphosphine)rhodium(I).

ruthenium, rhodium, palladium, osmium, iridium and platinum). It is unaffected by air and water up to 600°C, and unaffected by acids including aqua regia below 100°C. It is attacked by molten alkalis. The most common oxidation state of rhodium is +3, as in $RhCl_3$, but there are also compounds of oxidation states +1 to +6 and examples of the higher states are the fluorides RhF_4, RhF_5, and RhF_6. Wilkinson's catalyst—see *Economic element* above—has rhodium in oxidation state +1.

There are 35 known isotopes of rhodium of masses 88 to 122, and 24 known isomers. There is one main naturally occurring isotope of rhodium and that is rhodium-103, which is not radioactive.[164] The longest-lived, radioactive isotopes are rhodium-101 which has a half-life of 3.3 years, and rhodium-102m (an isomer) which has a half-life of 3.7 years.

Element of surprises

1. When rhodium is heated up to its melting point it absorbs oxygen from the atmosphere but it does not become transformed into one of its oxides, because as it solidifies it releases the oxygen again.

2. When Paul McCartney was honoured in 1979 as the best-selling song writer and recording artist at that time he was presented not with a gold or platinum-plated disc, which was the usual way of recognising sales of a million or of many millions. He was deemed worthy of a rhodium-plated disc.

---

[164] There are also tiny amounts of radioactive rhodium isotopes produced as fission products in uranium minerals.

Roentgenium

Pronounced ro-ent-gee-nee-um it is named after the German physicist Wilhelm Conrad Röntgen (1845–1923) who discovered X-rays in 1895. He was awarded the first Nobel Prize in physics in 1901. The name of the element is based on the English spelling of his name which is Roentgen.

French *roentgenium*; German *Roentgenium*; Italian *roentgenio*; Spanish *roentgenio*; Portuguese *roentgênio*

Essential element

Important only in pushing forward the upper limit of the periodic table.

Element of history

In 1986, physicists at the Russian Joint Institute for Nuclear Research (JINR), bombarded bismuth-209 with nickel-64 in an unsuccessful attempt to make element 111. In 1994, a team led by Peter Armbruster and Gottfred Munzenberg at the German Geselleschaft für Schwerionenforschung (GSI), were successful when they bombarded bismth-209 with nickel-64 and obtained few atoms of isotope 272 which has a half-life of 1.5 milliseconds. It decayed through a chain of alpha-emissions to meitnerium, bohrium, dubnium, and lawrencium.

The GSI was awarded priority of discovery in 2003 by the Joint Working Party committee of IUPAC/IUPAP which had been set up in 1998 to evaluate such claims. The name roentgenium was proposed in early 2004 and approved on 2 October that year at a IUPAC Bureau meeting, in order to fast-track the process.[165]

Environmental element

It is generally accepted that roentgenium does not exist naturally on Earth. However, there has been a claim that it might occur naturally in trace amounts as a primordial element. The Israeli physicist Amnon Marinov claimed detection in 2006 of two isotopes, roentgenium-261 and roentgenium-265 in natural gold samples. These were identified by the presence of unusual decay products. These were isotopes from spontaneous fission that could have come only from roentgenium because it was said that their presence could not be explained in any other way. Gold was chosen as the likely place to search for this element because roentgenium comes below gold in group 11 of the periodic table. These elements would have similar properties and would be likely to occur together. If they did exist, then the two isotopes would be neutron-

---

[165] The approval of new names was usually done at the Council meeting in odd-numbered years.

deficient deformed isomers. Moreover, they would need to have half-lives of 10^8 years or more if they really have been around since the time of the Earth's formation which was 4.57×10^9 years ago.

Chemical element

Data file	
Chemical symbol	Rg
Atomic number	111
Atomic weight	281 (longest-lived isotope known)
Properties such as melting point, boiling point, and density, are likely never to be measured because there will never be more than a few atoms produced at any one time.	
Oxide	likely to be non existent

Roentgenium is a radioactive element and a member of group 11 (row 6d) of the periodic table. This element should have the physical properties of a noble metal and, were it long-enough lived it should be possible to make compounds of it although, like gold, it might be reluctant to form them. However, were it to form compounds then its oxidation state would most likely be +3.

There are seven known isotopes of roentgenium with masses 272, 274, and 278 to 282, and no known isomers. There are no stable isotopes; all are radioactive. The half-life of the longest-lived known isotope, roentgenium-281 is 22.8 seconds. (Roentgeium-283 is theoretically calculated to have a half-life of 10 minutes.)

Rubidium

Pronounced roo-bid-ee-um, the name is derived from the Latin *rubidius*, meaning deepest red (ruby).

French *rubidium*; German *Rubidium*; Italian *rubidio*; Spanish *rubidio*; Portuguese *rubídio*.

Essential element

A unique role for rubidium has yet to be found and it applications are few, and even those are mainly for scientific research.

Element of life

Rubidium in the human body	
BLOOD	2.5 p.p.m.
BONE	0.1–5 p.p.m.
TISSUE	20–70 p.p.m.
TOTAL AMOUNT IN BODY	680 mg.

Rubidium has no known biological role but it has a slight stimulatory effect on metabolism, probably because rubidium can mimic potassium. The two elements are found together in minerals and soils, although potassium is much more abundant than rubidium. Plants will absorb rubidium quite readily and some plants, such as sugar beet, if stressed by a deficiency of potassium, will respond to the addition of rubidium.

Plants such as soya beans contain 220 p.p.m. rubidium, grass has 130 p.p.m., apples have 50 p.p.m., but sweetcorn has only 3 p.p.m. and onions even less at 1 p.p.m. Tea and coffee also contain measurable amounts of rubidium. In these various ways it enters the food chain and so contributes to a daily intake of between 1 and 5 mg, which is tiny compared to the recommended daily intake of potassium (3,500 mg).

Rubidium is absorbed easily from the gut and becomes distributed around the body, but there is no site where it prefers to accumulate, although very little gets into the bones or teeth. It is excreted in the urine.

Rubidium salts are not considered to be toxic.

Medical element

Rubidium is also like potassium in that it is slightly radioactive, and in this way its progress in the body can be monitored and used for medical research. For some reason, rubidium tends to be attracted to cancer cell membranes and surgeons have been able to use it to locate brain tumours at an early stage, and to treat the condition using other rubidium radio-isotopes.

Element of history

Rubidium was announced to the Berlin Academy at a meeting on 23 February 1861 by Robert Wilhelm Bunsen (1811–99) and Gustav Robert Kirchhoff (1824–87) of the University of Heidelberg, Germany. They had discovered it in lepidolite, a mineral that was first reported by a Jesuit priest and collector, Nicolaus Poda (1723–98), a century earlier. Poda's new mineral was known for its curious properties, for example when it is thrown onto glowing coals it froths to begin with, and then becomes glass-like, indicating its main component is silica.

Bunsen and Kirchhoff knew that lepidolite contained lithium, but suspected it could harbour another alkali metal as well. They extracted the metal components into solution and then added platinum chloride, which precipitated the potassium that was present. It precipitated the rubidium as well. This was only revealed by washing the precipitate with boiling water many times which slowly dissolved the potassium chloroplatinate and left a tiny amount of a residue whose atomic spectrum showed two intense ruby red lines that had never been seen before.[166] These indicated a previously undiscovered element and they named it rubidium after the colour of its spectrum.

A sample of pure rubidium metal was not prepared until 1928, when a chemist called Hackspill produced it. Samples of less than pure rubidium metal had been made in 1888 by heating rubidium hydroxide (RbOH) with aluminium powder, and a little had been made even earlier by the electrolysis of molten RbOH.

Economic element

No rubidium minerals are known, but rubidium is present in significant amounts in lepidolite, at about 1.5 percent (see lithium), and in pollucite (see caesium) and carnal-lite (see potassium). It is also present in traces of other minerals such as zinnwaldite, (a potassium lithium iron aluminium silicate) and leucite (a potassium aluminium silicate). Some natural brines contain as much as 6 p.p.m. rubidium. Rubidium is found in the extensive deposits of pollucite at Bernic Lake, Manitoba.

The amount of rubidium produced each year is small, and what demand there is can be met from a stock of a mixed carbonate by-product that is collected during the extraction of lithium from lepidolite. While this mixture is mainly potassium carbonate it also contains around 23 percent rubidium carbonate and 3 percent caesium carbonate. When it is heated with metallic sodium it reacts to form sodium carbonate and liberates the other metals, which are then separated by distillation. This yields a product that is 99.5 percent pure rubidium.

The little rubidium that is produced is used for research purposes only. There is no incentive to seek commercial outlets for the metal, because what it can do can almost certainly be done equally well by sodium, which is more than 5,000 times

---

[166] For the same reason, rubidium has been used in fireworks to produce a purple colour.

cheaper, or by potassium, which is 300 times cheaper. Rubidium metal itself can cost as much as \$20,000 per kg, but demand is small.

There have been occasional uses of rubidium as a 'getter' in vacuum tubes, a device that is put there to mop up minute traces of oxygen; and of rubidium carbonate to make special types of glass. It has also been demonstrated that rubidium could be used to turn heat energy into electricity via a so-called thermoelectric motor. In this, the rubidium is heated and loses an electron to form the positive ion Rb^+ which then passes through a magnetic field and in so doing generates an electric current in surrounding coils.

Environmental element

Rubidium in the environment	
EARTH'S CRUST	at least 90 p.p.m. and possibly more.
	Rubidium is the 16th most abundant element—see below.
SOILS	varies with a general range of 30–250 p.p.m., although some soils have very little, but a few exceed 400 p.p.m.
SEAWATER	0.1 p.p.m.
ATMOSPHERE	insignificant.

The relative abundance of rubidium has been reassessed in recent years and it is now suspected of being more plentiful than previously calculated. It is more abundant than copper, zinc, or lead. There are no environments where rubidium is seen as a threat.

Rocks that contain rubidium can be dated by the so-called rubidium-strontium dating method, which involves analysing the extent to which its radioactive isotope, rubidium-87, had transmuted into strontium-87.

Chemical element

Data file	
CHEMICAL SYMBOL	Rb
ATOMIC NUMBER	37
ATOMIC WEIGHT	85.4678
MELTING POINT	39°C
BOILING POINT	688°C
DENSITY	1.5 g/cm³
OXIDES	Rb_2O is the most stable, but there are other oxides such as the peroxide (Rb_2O_2), superoxide (RbO_2), and others—see below.

Rubidium is a soft, white metal and a member of group 1 (row 5s) of the periodic table, and they are collectively referred to as the alkali metals. Its low melting point means that if it is melted, it will often remain in this state when cooled to room temperature.

Rubidium is silvery when first cut but it soon ignites, so has to be stored under oil or grease. Unlike the other alkali metals, when rubidium is partially oxidised at low temperatures it forms two other oxides: Rb_6O and Rb_9O_2. It reacts violently with water and even with ice at $-100°C$. Rubidium metal is obtained from rubidium chloride by the reaction of this with calcium or potassium metal. The dominant, indeed almost exclusive, oxidation state of rubidium is +1 as in RbCl and the Rb^+ ion in solution. Rubidium can also accept an electron and become Rb^- under certain conditions.

There are 33 known isotopes of rubidium with mass numbers 71 to 103, and 13 known isomers. There are two naturally occurring isotopes: rubidium-85 which accounts for 72 percent and is not radioactive; and rubidium-87 which accounts for 28 percent and is radioactive, with a half-life of 49 billion years. It is a beta-emitter but not a gamma-emitter. While this suggests it is only weakly radioactive, it does contribute to the Earth's background radiation, and a sample of rubidium will emit enough radiation to blacken a photographic plate within two months. There are a number of other radioactive isotopes produce as fission products in uranium-bearing deposits. Otherwise the longest-lived radioactive isotope is rubidium-84 which has a half-life of 33 days.

Element of surprises

1. A rather curious property of rubidium silver iodide ($RbAg_4I_5$) is that, unlike most salts, this solid is highly conducting. Indeed it has the capacity to conduct electricity like some electrolyte solutions. It might one day find applications in thin film batteries.

2. In 1995, rubidium was used in the discovery of a so-called Bose-Einstein condensate (BEC). When Bose-Einstein conditions apply, then the wave functions of all the atoms overlap and all behave as a single entity. Two thousand atoms of rubidium were cooled to a temperature of a few billionths of a degree above absolute zero and this produced the BEC. Rubidium was used because it is easily vapourised and its atoms manipulated by lasers.

   In 2002 a group at the Max-Planck Institute in Germany went even further and confined a rubidium BEC within six lasers. As the intensity of these increased they produced what has been called a new state of matter and one in which the wave properties of the atoms stopped interacting with one another.

Ruthenium

Pronounced roo-thee-nee-um, the name is derived from *Ruthenia,* the Latin name for the region of Eastern Europe that now encompasses Ukraine, Belarus, and Russia.

French *ruthénium*; German *Ruthenium*; Italian *rutenio*; Spanish *rutenio*; Portuguese *ruténio*.

Essential element

Ruthenium is a key part of various electrical components and alloys because of its remarkable ability to resist corrosion, and it promises to have a future in dye-sensitised solar cells.

Element of life

The amount of ruthenium in the human body is not known, but is very low. It has no known biological role. Most ruthenium compounds are non-poisonous, and any ruthenium that is ingested tends to be retained in the bones and be there for a long time. The volatile oxide, RuO_4, is highly toxic by inhalation, but is never encountered outside research laboratories.

Medical element

The radio-isotope ruthenium-106 is used in the treatment of eye tumours. Ruthenium-based chemotherapy drugs are being tested and might one day be used in the treatment of colon cancer against which they appear to be effective.

Element of history

In 1807, the Polish chemist Jedrzej Andrei Sniadecki (1768–1838), who was based at the University of Vilno, began investigating some crude platinum ores from South America. He was hoping to find yet another new metal in them, just as William Hyde Wollaston (1766–1828) and Smithson Tennant (1761–1813) had found when they isolated the elements rhodium, palladium, osmium and iridium from this source. In May 1808 his efforts were rewarded by the discovery of a new metal and this he called vestium, after the asteroid Vesta, which had been first observed the previous year.[167] Sniadecki published his finding in a Russian journal.

When leading French chemists tried to repeat his work they were unable to find any vestium in the platinum ore they had available, and when Sniadecki learned of their

---

[167] The naming of elements after heavenly bodies was in vogue in the early 1800s, witness cerium named after the asteroid Ceres, and palladium after the asteroid Pallas.

work he dropped his claim to having discovered a new element, believing that he had been mistaken in his findings. Then, in 1825, the great chemist Jöns Jacob Berzelius (1779–1848), based at Stockholm, and Gottfried Wilhelm Osann (1797–1866), of the University of Dorpat (now Tartu) on the Baltic, investigated some platinum from the Ural mountains in Russia, where a large deposit of ore had been discovered. And though Berzelius could only detect the metals already known, Osann reported finding *three* new ones which he named pluranium, polonium, and ruthenium.

While the first two of these were never to be verified, the third was proved to be a new metal in 1840 by a former colleague of Osann's, Karl Karlovich Klaus (1796–1864) working at the University of Kazan. He showed that Osann's sample of ruthenium was very impure, and set about extracting and purifying his own sample and investigating its properties. The result is that today Klaus is cited as the discoverer, but he kept Osann's name of ruthenium. As Klaus made each new compound of ruthenium, he sent a sample to Stockholm to Berzelius who was still sceptical of its existence, until the weight of evidence forced him to acknowledge that it was indeed a new element.

Economic element

Ruthenium metal has been found in the free state, and there are a few ruthenium minerals, such as laurite, which is ruthenium sulfide, and ruthenarsenite, which is ruthenium nickel arsenide. All are rare and none acts as a commercial source of the metal. Ruthenium is obtained as a by-product of nickel refining in Ontario, while in South Africa it comprises around 10 percent of the platinum group metals that are mined there. World production is 12 tonnes per year and reserves are estimated to be around 5,000 tonnes.

Demand for ruthenium is rising as new uses are found for it. For example, ruthenium compounds are used in dye-sensitised solar cells which can harvest light energy from indirect sunlight and turn it into electricity. Mostly it finds use in the electronic industry (50 percent ends up this way) and the chemical industry (40 percent), with smaller amounts being used in alloying platinum and titanium.

In electronics some is used for electrical contacts but most goes into chip resistors. In the chemical industry it is used in the anodes for chlorine production in electrochemical cells. These anodes, known as dimensionally stable anodes, or DSAs, are made from titanium coated with ruthenium oxide which gives them a long service life under corrosive conditions, as well as an improved release of chlorine gas. Ruthenium is also part of the catalysts needed for the production of ammonia from natural gas, and the production of acetic acid from methanol.

Ruthenium is added to platinum for jewellery purposes, to make it harder, and is added to the titanium for deep-water pipes to make them corrosion resistant.

Environmental element

Ruthenium in the environment	
EARTH'S CRUST	about.1 p.p.b.
	Ruthenium is the 74th most abundant element, i.e. it is very rare.
SOILS	0.5 to 30 p.p.b.
SEAWATER	less than 0.005 p.p.t. (barely detectable).
ATMOSPHERE	nil.

Ruthenium is one of the rarest of metals on Earth. Very little data are available on its environmental impact on plants and estimates of its uptake have deduced levels of 5 p.p.b. or less, although algae appear to concentrate it—see *Element of surprises*.

Chemical element

Data file	
Chemical symbol	Ru
Atomic number	44
Atomic weight	101.07
Melting point	2,310°C
Boiling point	3,900°C
Density	12.4 g/cm³
Oxide	most stable is Ru_2O_3, but RuO_2, RuO_3 and RuO_4* are also known.

* This oxide can react explosively and is very toxic.

Ruthenium is a lustrous, silvery metal of the so-called platinum group[168], and a member of group 8, row 4d, of the periodic table. It is unaffected by air, water, and acids, but dissolves in molten alkalis. It exhibits various oxidation states of which the most common are +2, as in $RuCl_2$, +3, as in $RuCl_3$, and +4, as in RuO_2. Other oxidation states are +6, as in K_2RuO_4, +7, as in $KRuO_4$, and +8, as in the oxide RuO_4.

There are 38 known isotopes of ruthenium with masses 87 to 124, and seven known isomers. There are seven naturally occurring isotopes of ruthenium of which six are stable: ruthenium-102 is the most abundant at 31.5 percent, ruthenium-104 (18.6 percent), ruthenium-101 (17 percent), ruthenium-99 (12.8 percent), ruthenium-100 (12.6 percent), and ruthenium-98 (2 percent). Also naturally occurring is ruthenium-96 (5.5 percent), which is radioactive with a half-life of 67×10^{15} years (67 million billion years).[169] The next longest-lived, radioactive isotope is ruthenium-106 which has a half-life of 374 days.

---

[168] These are the six rare metals: ruthenium, rhodium, palladium, osmium, iridium, and platinum.
[169] There are also minute traces of radioactive ruthenium isotopes formed as fission products in uranium ores.

Element of surprises

1. The radioactive isotope ruthenium-106, which has a half-life of 374 days, is produced by nuclear reactors and was released from above-ground nuclear explosions and in theory some of this could enter the human food chain, although the only food compromised by this has been the edible seaweed *Porphyra*. This grows in the Irish Sea and is harvested off the coast of South Wales to be eaten as laverbread. At one time it became contaminated with radioactive ruthenium from discharges from the nuclear reprocessing plant at Sellafield in the UK. Why this algae absorbs ruthenium is not understood.

2. The famous Parker 51 fountain pen has a nib made of 14 carat gold tipped with ruthenium which in theory will never wear out.

3. Researchers Mehmet Zahmakiran, Yalçin Tonbul, and Saim Özkar at the Middle East Technical University in Ankara, Turkey, have produced a ruthenium catalyst consisting of nano-clusters in a nano-zeolite framework which is capable of adding hydrogen gas to benzene and similar compounds at room temperature, converting it to cyclohexane, and it offers an alternative process to that normally used by industry.

Rutherfordium

Pronounced ruth-er-ford-ee-um, the element is named after the New Zealand chemist Ernest Rutherford (1871–1937) who was one of the first to explain the structure of atoms and who won the Nobel Prize for chemistry in 1908.

French *rutherfordium*; German *Rutherfordium*; Italian *rutherfordio*; Spanish *rutherfordio*; Portuguese *ruterfórdio*.

Essential element

Notable only for its name and how this was acquired.

Element of life

Rutherfordium can have no role in living things, and it only exists fleetingly in nuclear research facilities.

Element of creation and contention

In 1964, a team led by Georgy Flerov (1913–90) at the Russian Joint Institute for Nuclear Research (JINR) in Dubna, bombarded plutonium-242 with neon-22 and produced isotope 259. They were cautious of announcing their success because of less than complete data. They also tried hitting plutonium-242 with neon-20 in an unsuccessful attempt. In 1966, they repeated the original experiment successfully, this time with completely credible data which confirmed the 1964 results. They submitted a discovery claim proposing the name kurchatovium (symbol Ku) for Igor Kurchatov (1903–60), former head of the Soviet nuclear research programme.

In 1969, a team led by Albert Ghiorso (1915–) at the Californian Lawrence Berkeley Laboratory (LBL)[170] made three successful attempts to produce element 104. They bombarded curium-248 with oxygen-16, and got isotope 260, californium-249 with carbon-12 and got isotope 257, and californium-249 with carbon-13 and got isotope 258. They filed a claim proposing the name rutherfordium (symbol Rf). This name had been originally proposed by the JINR in 1967 for element 103.

Element 104 now had two names, but which should it be? A dispute over priority of discovery followed, one that came to include other synthesised elements. (See Transfermium elements for a fuller account of the naming controversy.) The International Union of Pure and Applied Chemistry (IUPAC) proposed a temporary compromise for names of all new elements, and these would be based on the atomic number and element 104 became unnilquadium (un-nil-quad-ium).

---

[170] Now called the Lawrence Berkeley National Laboratory (LBNL).

In 1992, the working group set up by IUPAC to resolve the names controversy concluded that both the Russian and American researchers had some justification for making their claims, but decided that element 104 would be called rutherfordium.

Chemical element

Data file	
CHEMICAL SYMBOL	Rf
ATOMIC NUMBER	104
ATOMIC WEIGHT	267

Properties such as melting point, boiling point, and density, are likely never to be measured because there will never be more than a few atoms produced at any one time.

OXIDE	likely to be RfO_2

Rutherfordium is a radioactive element and a member of group 4 (row 6d) of the periodic table, coming below hafnium. So far not enough of this element has been made at one time to be visible to the naked eye, but predictions are that it would be a silvery metal susceptible to attack by air, steam, and acids. Its only oxidation state appears to be +4 and in solution it forms Rf^{4+} and RfO_2^{2+} ions. There is also evidence for its forming the species $RfCl_4$, $RfOCl_2$, $RfCl_6^{2-}$ and $RfBr_4$.

Rutherfordium has 14 known isotopes of mass number 253 to 263 and 266 to 268, and four known isomers. The longest-lived is rutherfordium-267 with a half-life of 1.3 hours. (No isotope has yet been theoretically calculated to have a longer half-life.)

Element of surprises

Rutherfordium is not like the higher members of group 4 in its behaviour with fluoride ions. Like hafnium and zirconium, it forms a complex ion with six fluorides: RfF_6^{2-} but it does not form RfF_7^{3-} whereas the other two metals do. The reason appears to be that relativistic effects are coming into play and these affect the behaviour of its bonding electrons.

Samarium

Pronounced sam-ayr-ee-um, it is named after samarskite, the mineral from which it was first extracted, and this was named after Vasili Samarskij-Byhove (1803–70) who was head of the Russian Corps of Mining Engineers from many years.

French *samarium*; German *Samarium*; Italian *samario*; Spanish *samario*; Portuguese *samário*.

Samarium is one of 15 chemically similar elements referred to as the lanthanoid or rare earth elements, which extend from element atomic number 57 (lanthanum) to element atomic number 71 (lutetium). The term rare earth is a misnomer because some are not rare. The minerals from which they are extracted, and the properties and uses they have in common, are discussed under Lanthanoids.

Essential element

Samarium-cobalt magnets retain their magnetism at high temperature and have microwave applications. This metal is also important for powerful lasers and powerful lights.

Element of life

Samarium in the human body	
BLOOD	approx. 8 micrograms per litre
BONE	n/a
TISSUE	n/a
TOTAL AMOUNT IN BODY	at least 50 micrograms

The samarium in humans is tiny and the metal has no biological role, but it has been noted that samarium salts stimulate metabolism. It is difficult to separate out the various amounts of the different lanthanoids in the human body, but they are present, and the levels are generally highest in bone, with small amounts being present in the liver and kidneys.

No one has monitored diet for samarium content, so it is difficult to judge how much we take in every day but it is probably less than 1 mg a year. Samarium is not taken up by plant roots to any great extent so does not get into the human food chain, although some vegetables can register 100 p.p.b. (dry weight) and some plants as much as eight times this figure.

Samarium salts are mildly toxic by ingestion, but insoluble salts are non-toxic. Exposure to samarium salts causes skin and eye irritation.

Medical element

The isotope samarium-153 is a beta-emitter with a half-life of two days. It was approved for use as a radiopharmaceutical in the USA in 1997 and is known as the drug Quadramet. It is used to treat the severe pain that is a feature of bone cancers.

Element of history

Samarium was discovered by Paul-Émile Lecoq de Boisbaudran (1838–1912) in Paris in 1879. Cerium, which had been discovered in 1803, was suspected of harbouring other metals, and these were identified in 1839 by Carl Gustav Mosander (1797–1858) as lanthanum and didymium. The latter in its turn was suspected of being a mixture of elements when its atomic spectrum was observed in 1879 and was noted to vary slightly according to the mineral from which it was obtained. In that year Boisbaudran extracted didymium from the mineral samarskite. He then made a solution of didymium nitrate, added ammonium hydroxide to it, and noticed that the precipitate which formed appeared to come down in two stages. He devised a way of separating the first precipitate and measured its spectrum. This proved to be the hydroxide of a new element which he named samarium. Samarium was eventually to yield another rare earth, europium, extracted from it by Eugène-Anatole Demarçay (1852–1904) in 1901.

Economic element

Samarium is found in minerals that include all the lanthanoid elements, and these are discussed under Lanthanoids, page 274. Samarium-containing ores are mainly mined in China, but there are large deposits in the USA, Brazil, India, Australia, Greenland, and Tanzania. The most important is monazite which contains up to 3 percent by weight of samarium. Samarium is also present in bastnäsite, and in the so-called ionic clays found in China. Worldwide reserves of samarium are estimated to be around two million tonnes.

World production of samarium oxide (Sm_2O_3) is about 1,300 tonnes per year, and the metal is produced by heating this with barium or lanthanum to temperatures high enough to drive off the samarium as a vapour. It can also be produced by the electrolysis of a molten mixture of samarium chloride ($SmCl_3$) and calcium chloride ($CaCl_2$).

One outlet for samarium is in permanent magnets, where its alloy with cobalt, either as $SmCo_5$ or Sm_2Co_{17}, produces magnets that are not only 10 thousand times more powerful than iron magnets, but have the highest resistance to demagnetisation of any known material. These magnets have allowed the miniaturisation of devices like motors and headphones. Without samarium, personal stereos would not have been possible. In recent years, samarium-cobalt magnets have been overshadowed by NIB magnets (see under neodymium) although they are still used in microwave applications because they retain their magnetism to temperatures above 700°C.

Samarium oxide finds specialised use in ceramics and for making glass which absorbs infrared rays. Calcium chloride crystals doped with samarium are used in lasers and masers (the microwave equivalent of lasers) and it is the latter which are capable of cutting through steel and have been bounced off the surface of the moon.

In some countries the chemical industry converts surplus ethanol to the gas ethene, which is the feedstock for many plastics, and this conversion is aided by samarium catalysts. Other uses for samarium are in carbon-arc electrodes for film studio arc lights, in infrared sensitive phosphors, and as an excellent neutron absorber for the control rods which regulate the reactor core in nuclear power plants.

Environmental element

Samarium in the environment	
EARTH'S CRUST	8 p.p.m.
	Samarium is the 40th most abundant element.
SOILS	5 p.p.m. (dry weight), range 2–23 p.p.m.
SEAWATER	0.8 p.p.t.
ATMOSPHERE	virtually nil.

Samarium is the second most abundant of the rare earth elements and it is almost four times as common as tin. It poses no environmental threat to plants or animals.

Chemical element

Data file	
CHEMICAL SYMBOL	Sm
ATOMIC NUMBER	62
ATOMIC WEIGHT	150.36
MELTING POINT	1,077°C
BOILING POINT	1,790°C
DENSITY	7.52 g/cm³
OXIDE	Sm_2O_3

Samarium is a silvery-white metal and one of the lanthanoid group (row 4f) of the periodic table. It is relatively stable in dry air, but in moist air an oxide coating forms. Heated to 150°C, the metal ignites spontaneously. Its preferred oxidation state is +3 as shown by compounds like SmF_3 and $SmCl_3$, although like its fellow lanthanoid element, europium, samarium has a lower oxidation state +2, albeit less stable. One of the lower oxidation compounds is the intense blue samarium iodide (SmI_2) which is used by organic research chemists to make synthetic versions of natural products because of the way it can produce cyclic molecules with the right spatial arrangement of atoms. It is also good at de-chlorinating environmental pollutants such as PCBs at temperatures as low as 60°C and so renders them harmless. In solution samarium exists as the Sm^{3+} ion.

There are 38 known isotopes of samarium with mass numbers ranging from 128 to 165, and 12 known isomers. Naturally occurring samarium in minerals has seven isotopes of which three are stable: samarium-152 which accounts for 27 percent; samarium-150 (7 percent); and samarium-144 (3 percent). The other four isotopes are weakly radioactive and have very long half-lives: samarium-154 which accounts for 23 percent and has a half-life of 2.3×10^{18} years (that is, 2.3 billion billion years) samarium-147 (15 percent) which has a half-life of 1.06×10^{11} years (that is, 106 billion years), samarium-148 (11 percent) which has a half-life of 7×10^{15} million years (= 7 million billion years), and samarium-149 (14 percent) which has a half-life of 2×10^{15} years (= 2 million billion years). There are also minute traces of some radioactive samarium isotopes in uranium ores, where they have been formed by nuclear fission. One of these fission products is samarium-146, with a half-life of 103 million years which is present deep in the Earth's mantle, of which some traces are detectable at the surface. This isotope is also produced by the alpha decay of naturally occurring gadolinium-164.

Element of surprise

The oddest samarium compound is samarium(II) sulfide (SmS) which exists as black crystals with semiconductor properties. When these are scratched they immediately transform into golden crystals that conduct like a metal, and the same thing happens when they are put under high pressure.

Scandium

Pronounced skan-dee-um, the name derives from *Scandia* the Latin name for Scandinavia.

French *scandium*; German *Scandium*; Italian *scandio*; Spanish *escandio*; Portuguese *escândio*.

Essential element

A relatively abundant metal with few uses except as an alloy with aluminium, but one use of this has been deemed unsporting.

Cosmic element

Scandium is more abundant in the heavens that it is down here on Earth. This observation was made as long ago as 1908 when Sir William Crookes (1832–1919) and G. Eberhard examined the visible spectra of the Sun and certain stars, and identified strong bands arising from this element.

Element of life

Scandium in the human body	
BLOOD	approx. 8 p.p.b.
BONE	approx. 1 p.p.b.
TISSUE	approx. 1 p.p.b.
TOTAL AMOUNT IN BODY	0.2 mg.

There is no known biological role for scandium. This might not always have been so because ash residues from coal contain up to 0.1 percent suggesting that some environmental enrichment may have occurred in plants of the Carboniferous period, 300–350 million years ago.

Only trace amounts of scandium enter the food chain so that the average person's daily intake is less than 0.1 microgram. Only about 3 percent of the plants that have been analysed for scandium showed its presence, and even so amounts were tiny, with vegetables having only 5 p.p.b. although grass has 70 p.p.b. Tea leaves showed more than this, with an average of 140 p.p.b., which is perhaps understandable since this plant requires aluminium, and scandium is chemically very similar.

Scandium is not toxic, although there have been suggestions that some of its compounds might be carcinogenic.

Element of history

When Dimitri Mendeleyev devised his periodic table of the elements in 1869 he noticed that there was a gap in atomic weights between calcium (40) and titanium (48) and he predicted that there should be another element of intermediate atomic weight (44) which he referred to as eka-boron, since he thought it would come below boron in the same group III of his table. Consequently he forecast that its oxide would have the formula X_2O_3.

Scandium was discovered 10 years later, in 1879, by Lars Frederik Nilson (1840–99) who was the Professor of Analytical Chemistry at the University of Uppsala, Sweden. He extracted it from euxenite, a complex mineral containing as many as eight metals. He thought that this was only to be found in Scandinavia, so he called the new metal scandium. Nilson had extracted erbium oxide from euxenite, and from this oxide he obtained both ytterbium oxide and another oxide of a lighter element that he could not identify. Its atomic spectrum showed lines not previously reported for any known metal, thus proving that he had stumbled across a new element. He studied its chemistry and determined its atomic weight to be 44, and showed that its oxide had the formula Sc_2O_3. His colleague at the university, Professor Per Theodor Cleve, pointed out that this was the missing eka-boron predicted by Mendeleyev.

Nilson never saw a sample of the metal itself. This was only made in 1937 by the electrolysis of molten scandium chloride dissolved in a melt of other metal chlorides at 800°C.

Economic element

There are a few pure scandium mineral specimens and these are very rare and highly sought after by collectors. Crystals of the greenish-black thortveitite, $Sc_2Si_2O_7$, are particularly expensive. (The mineral was named after Olav Thor Tveit who discovered it.) The mineral also contains yttrium, and it is found mainly in Norway.

Scandium comes chiefly from three mines which are located in the Ukraine, Russia, and China, while some comes from uranium mill tailings. Russia has a stockpile of the metal. World production of the oxide amounts to only a few tonnes per year of which only 1 percent is converted to the metal itself. This is produced by heating scandium fluoride (ScF_3) with calcium metal and it costs \$8–10 per gram depending on its purity; in bulk it costs around \$5,000 per kilogram.

Scandium is a superior kind of aluminium, being equally lightweight, although much more costly. It might one day have industrial implications because it scores over aluminium in having a melting point that is 900°C higher; this has led to its being used in aircraft. Adding around 0.5 percent scandium to aluminium increases its strength dramatically by raising the melting point of the metal by 800°C so that it can even be welded, which normal aluminium cannot. Soviet Russia developed this alloy in the 1960s for their advanced MiG-25 jet fighters and it is also part of the more recent MiG-29 and MiG-31 fighters.

There are a few, rather specialised, uses for scandium such as in neutron filters for nuclear reactors. The artificially produced radioactive isotope, scandium-46, has been used in oil refineries to monitor the movement of various fractions as the oil is refined. In a similar manner it can detect leaks in underground pipes carrying liquids.

In the USA, scandium alloy is used to make sports goods such as baseball bats and lacrosse sticks. A cricket bat made from it gave dramatically improved performance but it was deemed unsporting and outlawed, but lightweight bicycle frames are not so penalised. Another use of scandium is as a seed germinating agent and, when it is applied as a dilute solution of scandium sulfate to corn, peas and wheat, it increases the number of seeds successfully germinating.

Scandium oxide provides a specialised optical coating for detectors because it is transparent to UV light down to 225 nm.

Environmental element

Scandium in the environment	
EARTH'S CRUST	16 p.p.m.
	Scandium is the 35th most abundant element.
SOILS	average 7 p.p.m., but there is a wide range from 0.5–45 p.p.m.
SEAWATER	0.6 p.p.t. (barely detectable).
ATMOSPHERE	nil.

Scandium precipitates from water as the insoluble hydroxide, $Sc(OH)_3$, and will do this even from neutral water of pH 7, consequently there has been no geological process to concentrate this element to any extent and what there is tends to be widely distributed in many types of rocks. It has been detected in more than 800 minerals so far. In the aquamarine variety of the gemstone beryl, the blue colour is thought to be due to traces of scandium, although normally the compounds of this element are colourless.

Chemical element

Data file	
CHEMICAL SYMBOL	Sc
ATOMIC NUMBER	21
ATOMIC WEIGHT	44.955910
MELTING POINT	1,541°C
BOILING POINT	2,831°C
DENSITY	3.0 g/cm³
OXIDE	Sc_2O_3

Scandium is a soft, silvery, yellow-white metal and a member of group 3 of the periodic table. It tarnishes in air, and burns easily once it has been ignited. It reacts with water to form hydrogen gas, and will dissolve in many acids. Its preferred oxidation state is +3 as in $ScCl_3$ and the ion Sc^{3+} in solution.

Scandium has 25 known isotopes with mass numbers ranging from 36 to 60, with 12 known isomers. There is only one naturally occurring isotope: scadium-45 and it is stable. Of the radioactive isotopes, the longest-lived is scandium-46 with a half-life of 83.8 days and it emits beta-rays.

Element of surprises

1. When scandium iodide (ScI_3) is added to mercury vapour lights it turns their intense but harsh glare into something more akin to natural sunlight and consequently such lights are used in filming and for the floodlighting of outdoor sports arenas. Annually around 100 kg of scandium is used this way.

2. Scandium forms a hydride, ScH_2, which is a good conductor of electricity for reasons that have yet to be deduced.

106
Sg

Seaborgium

Pronounced see-borg-ee-um, the element is named after Glenn T. Seaborg (1912–99) who was involved in producing several new transuranium elements.

French *seaborgium*; German *Seaborgium*; Italian *seaborgio*; Spanish *seaborgio*; Portuguese *seabórgio*.

Essential element

The few atoms that have been made of this element have half-lives of less than two minutes, but these nevertheless gave rise to disputes lasting many years.

Element of creation and contention

In 1970, a team led by Albert Ghiorso (1915–) at the Californian Lawrence Berkeley National Laboratory (LBNL) bombarded californium-249 with oxygen-16 and was successful in producing isotope 263. However, they did not analyse the data carefully enough at the time. That it really had been produced was revealed the following year by detailed analysis of the data, but by then the experiment could not be re-run to confirm the discovery because the Heavy Ion Linear Accelerator (HILAC) in which it had been made was being up-graded to the super-HILAC. The LBNL physicists made no announcement of the discovery and never filed a discovery claim.

In June 1974, a team led by Georgy Flerov and Yuri Oganessian at the Russian Joint Institute for Nuclear Research (JINR) tried five target/projective combinations in attempts to produce element 106, being successful with two of them. They bombarded lead-208 with chromium-54 and got isotope 260, and lead-207 with chromium-54 and got the 259 isotope, although they were less sure about this one. They also made three unsuccessful attempts, bombarding lead-206 with chromium-54, lead-208 with chromium-52, and bismuth-209 with vanadium-51. This was the first use of the so-called cold fusion method as a way of discovering an element. Their results, however, could not be confirmed.

In September 1974, a team led by Albert Ghiorso at LBNL obtained isotope-263, with a half-life of 0.8 seconds, which they made in the super-HILAC by bombarding californium-249 with oxygen-18 ions. This experiment was repeated in 1993 using the 88-inch-diameter cyclotron, and the result confirmed the earlier findings. Several atoms of seaborgium have since been made by this method which produces one seaborgium atom per hour.

Neither the physicists working at JINR nor those at LBNL proposed a name for element 106. The years of controversy over elements 104 and 105 had cautioned them to wait until their respective claims had been adjudicated. Meanwhile in 1979, IUPAC introduced a temporary naming system based on the atomic number of an element, so element 106 became un-nil-hexium (Unh). For more about this method of naming elements see un-named elements.

In 1992, the Transfermium Working Group, set up by IUPAC and IUPAP—see Transfermium elements for details—issued its report which suggested that although the JINR group had discovered the element first, the LBNL work was the more convincing and that the discovery should be shared. The LBNL people refused to accept this, wanting sole recognition, and this they were accorded by IUPAC in 1993 when they submitted their confirming results, which reproduced the results of the 1974 experiment.

Then in 1994, the LBNL caused a stir by proposing the name seaborgium which IUPAC rejected on the grounds that it was wrong to name an element after a living scientist. The LBNL and the American Chemical Society (which approved the name in 1995 and decided that it should be used in all its publications) refused to relent. In the same year, IUPAC issued a list of suggested names for elements 104 to 109, but this caused controversy and there followed acrimonious negotiations and compromises that included three sets of suggested names for these elements. Ultimately, the name seaborgium was approved in 1997 along with name for elements 104 to 109. The story is told more fully in the Transfermium Elements section on page 566.

Chemical element

Data file	
CHEMICAL SYMBOL	Sg
ATOMIC NUMBER	106
ATOMIC WEIGHT	271 (longest-lived known isotope)
MELTING POINT	1,966°C
Properties such as melting point, boiling point, and density, are likely never to be measured because there will never be more than a few atoms produced at any one time.	
OXIDE	not known, but likely to be SgO_3

Seaborgium is a radioactive element and a member of group 6 (row 6d) of the periodic table, coming below tungsten. So far not enough of this element has been made at one time to be visible by the naked eye, but predictions are that it would be a silvery metal susceptible to attack by air, steam, and acids. By analogy with other members of group 6 its preferred oxidation state should be +6 although +5 and +4 might also be stable.

Some reactions have been carried out on a few atoms and these produced the following compounds: SgO, SgO_2F_2, and SgO_2Cl_2. Research has also shown that $SgO_2(OH)_2$ and $[Sg(H_2O)_6]^{6+}$ would be formed in solution. The work was done using only seven atoms of the element to study its chemistry in the gas phase and in solution.

There are 11 known isotopes of seaborgium with mass numbers 258 to 267 and 271, and three known isomers. The longest-lived is seaborgium-271 with a half-life of 1.9 minutes. (Seaborgium-272 is theoretically calculated to have a half-life of one hour.)

Element of surprise

Had the physicists at the Lawrence Berkeley National Laboratory checked their data more carefully after their successful discovery attempt in 1970, they would have submitted a discovery claim and element 106 would then have been given a different name. LBNL's later re-evaluation of the data that they had originally thought was inadequate, indicated that they did have complete and credible information for a discovery-claim submission, but they just hadn't realised it.

Selenium

Pronounced sel-ee-nee-um, the name is derived from *selene*, the Greek name for the Moon. French *sélénium*; German *Selen*; Italian *selenio*; Spanish *selenio*; Portuguese *selénio*.

Essential element

Sperm, semiconductors, soil, and special types of glass all need selenium.

Element of life

Selenium in the human body	
BLOOD	levels vary 70–150 p.p.b.
BONE	levels vary 1–9 p.p.m.
TISSUE	about 0.1 p.p.m.
TOTAL AMOUNT IN BODY	about 14 mg (14 000 µg).

In 1975, selenium was proved to be an essential element for humans. Yogesh Awasthi, based at Galveston in Texas, discovered that it was part of the antioxidant enzyme, glutathione peroxidase, which eliminates peroxides before they can form dangerous free-radicals. Then, in 1991, Professor Dietrich Behne at the Hahn-Meitner Institute in Berlin found selenium in another enzyme, deiodinase, which promotes hormone production in the thyroid gland.

It is now known that every cell of the body contains more than a million atoms of selenium. The parts with the highest levels are hair, kidneys and testicles. The dilemma of selenium, as with many other elements, is that if a person has either too little or too much of it in their diet, their health will suffer.

Methylselenosysteine is the natural selenium-containing amino acid that Nature synthesises. It is present in brassica vegetables and in garlic and other *Alliums*. Nature has programmed part of messenger RNA to be able to make this molecule which is then incorporated into proteins. So far around 14 seleno-proteins have been identified, and while they play a part in bodily tissues, such as that of the kidneys, their most vital role is in sperm.

Dangerous element

Hydrogen selenide gas (H_2Se) is extremely toxic and almost killed the chemist who first made it—see below.

Even though selenium is essential, it is possible to overdose on it and 5mg would produce a highly toxic reaction. Prolonged low level exposure to selenium, as occurred in some industries, caused weight loss, anaemia, and dermatitis as well as social isolation due to the intolerable smell its victims exuded. The recommended maximum daily intake is 0.45 mg (450 μg). Above this we risk selenium poisoning, the most obvious symptom of which is extremely foul breath and body odour. The smell is caused by volatile methyl selenium molecules which the body produces in an effort to rid itself of the unwanted excess.

In April 2009, at the prestigious US polo championships, many horses were given a supplement called Biodyl which is designed to boost their intake of certain vitamins and minerals, namely selenium, magnesium, potassium, and the B vitamins. However, this batch had been made up incorrectly with an excess of selenium and within three hours of ingesting the supplement the horses began to die of selenium poisoning. Soon all 21 horses that had been given it were dead.

Food element

Selenium in foods comes in the form of selenocysteine, which occurs in things like broccoli and garlic, or as selenomethionine, which is how it occurs in meat and grain. Selenocysteine has one carbon atom fewer that selenomethionine and it does not get incorporated into protein, as does selenomethionine.

The daily intake of selenium varies between 6 and 200 micrograms (μg), depending on the type of foods that are eaten. The average Westerner takes in about 65 μg per day which should be adequate to prevent selenium deficiency, although it is less than the recommended intake of 75 μg for men, but adequate to meet the recommended 60 μg intake for women. Some days the body may lose more selenium than it absorbs from food but this does not pose a threat because the body can draw upon its store of selenium in the bones. As a dietary supplement, selenium is usually taken in the form of sodium selenite (Na_2SeO_3) which is a white crystalline material soluble in water, and the daily dose is 50 μg. An alternative way is to take brewer's yeast that has been grown on a selenium-enhanced culture medium.

Most people get their selenium from breakfast cereals and bread, especially wholemeal bread, two slices of which will provide 30 μg although in the case of wheat and meat products the level depends upon the soil of the farm from which they came. Foods particularly rich in selenium are Brazil nuts and molasses (black treacle), both having more than 100 μg per 100 g.[171] Other foods with levels in excess or 30 μg per 100 grams are seafoods such as tuna, cod and salmon, and offal such as liver and kidney. Many nuts, and especially peanuts and cashew nuts, have similar amounts and a 100-gram bag of the latter can contain 65 μg. Other foods rich in selenium are wheat germ, bran, and Brewer's yeast.

A particularly rich source of selenium is the edible mushroom *Albatrellus pes-caprae* which is popular in Italy. This has 3,700 μg per 100 grams fresh weight, and a dish containing this amount would provide the eater with eight times the recommended

---

[171] Eating lots of Brazil nuts is not the best way of taking in selenium because they also contain radium and barium which are toxic metals.

daily maximum. Nevertheless, these are safe to eat because much of the selenium in mushrooms is not easily digested.

Medical element

Selenium is essential for human health, and problems associated with low selenium levels are cancer, heart disease, and infertility. A deficiency of this element is said to be linked to other conditions: anaemia, high blood pressure, arthritis, premature ageing, muscular dystrophy, multiple sclerosis, and AIDS. Selenium has a marked effect on the production of T-lymphocytes which are the naturally killer cells of the immune system and which protect the body and can even destroy tumour cells.

The link to cancer has been much investigated and there is epidemiological evidence in support of selenium helping to prevent cancer, although this has been refuted by other surveys. In one US test involving 1,300 elderly people over four years, there was a 30 percent decrease in new lung, bowel, and prostrate cancers among those who took a selenium supplement.

In some illnesses the evidence of selenium deficiency is compelling, as in Keshan disease. Large areas of China, such as the Keshan and Linxian regions, have selenium-depleted soils, and children of the Keshan region were particularly prone to congestive heart failure known as Keshan disease; those living in the north central province of Linxian had a high incidence of stomach cancer. The incidence of both diseases was reduced when selenium supplements were issued to the population.

According to some researchers, a declining sperm count among European men can be explained by the decrease in selenium in the diet. The average selenium intake fell by half in the last century, due in part to the decline in popularity of selenium-rich foods such as kidney and liver and the low level of selenium in wheat grown in Europe—see *Environmental element*. The link between selenium and male fertility was demonstrated by a Scottish researcher, Alan MacPherson, in 1993. He reported on a double-blind trial in which selenium supplements were given to a group of men and their sperm counts monitored. The men who were given a placebo continued to have low sperm counts, whereas the men who took the selenium pills soon doubled their output of viable sperm.

Element of history

Selenium was discovered by Jöns Jacob Berzelius (1779–1848) in Stockholm in 1817. He had shares in a sulfuric acid works at Gripsholm about a hundred miles northwest of the Swedish capital at whose Medical School he was professor of Chemistry and Medicine. In the summer of that year he spent several weeks at the works as a consultant, and he was intrigued by a red-brown sediment which collected at the bottom of the chambers in which the acid was made. He took some of the red deposit back to Stockholm for further study.

At first he thought it was the element tellurium because it gave off a strong smell of radishes when heated, but he came to realise that, although it was chemically similar to tellurium, it was in fact a new element. He named it selenium, from the Greek *selene* meaning Moon, to match the name tellurium, from the Latin *tellus* meaning

Earth. Berzelius noted that the new element was like sulfur in that it would vaporise on heating and re-deposit as 'flowers' on a cool surface, but it differed in that, whereas flowers of sulfur are bright yellow, those of selenium are red. This is the non-metallic form of the element.

Berzelius found that selenium was also present in samples of tellurium and that it was the selenium that made tellurium smell of radishes. He also became affected by it personally and his housekeeper one day admonished him for his bad breath, saying he'd been eating too much garlic. On another occasion he was overcome by breathing in hydrogen selenide gas (H_2Se) which he had produced without realising how deadly it is and he was ill for two weeks.

Within a few years of its discovery, selenium turned up in many more minerals and the metallic form of the element was refined and cast into medallions, some of which depicted Berzelius.

Economic element

Native selenium is occasionally found and there are around 40 known selenium-containing minerals, some of which can have as much as 30 percent selenium. All are rare, and generally they occur together with sulfides of metals such as copper, zinc and lead. Selenium is nowadays extracted from the slime that settles at the bottom of tanks in which copper is refined electrolytically and this slime may contain up to 5 percent selenium.

Production of selenium amounts to about 1,500 tonnes a year, of which 10 percent is recycled from industrial waste or reclaimed from old photocopiers. The main countries producing selenium are Japan (800 tonnes), Belgium (300 tonnes), Sweden (130 tonnes) and Russia (100 tonnes). Several other countries produce smaller amounts. Production in the USA is not reported for strategic reasons but is thought to exceed 200 tonnes. World reserves of selenium are estimated to be around 170,000 tonnes.

Consumption is as follows: USA 30 percent; Europe 35 percent; Japan 25 percent; and the rest of the world 10 percent. The final destination of the element is as follows: glass manufacture 40 percent; metallurgy 30 percent and especially in manganese production; agriculture 15 percent; electronics 10 percent such as the photovoltaic materials copper-indium-gallium-diselenide (CIGS); and pigments 5 percent.

The metallic form of selenium has the curious property of increasing its electrical conductivity by a thousand fold when light falls on its surface. This was discovered in 1873 by the English telegraph engineer Willoughby Smith (1828–91). Selenium behaves as a semiconductor because its band gap, which is the energy needed to promote an electron to the conduction band is only 2 eV when it can move freely. This is the kind of energy which light waves can provide. As a consequence, selenium found use in light-sensitive photoelectric cells, light meters, solar cells, and photocopiers. These uses once accounted for about a third of all selenium production and required high purity (99.99 percent) selenium, but they have been in decline for many years.

The largest use of selenium is the glass industry which adds sodium selenate (Na_2SeO_4) and sodium selenite (Na_2SeO_3) to glass to decolourise it. On the other hand, cadmium selenide is added to glass in larger amounts to produce a magnificent ruby colour. Some is used to make bronze architectural glass for office building because it screens out the sun's rays. Selenium pigments provide ceramics, paint, and plastics with a range of orange, red, and maroon colours. Cerium selenide makes a striking and permanent red pigment and is used in place of the once popular cadmium selenide, which is now banned on account of the cadmium.

Other uses of selenium are in metal alloys, such as the lead plates used in storage batteries, and in rectifiers which convert AC to DC current. Selenium has been used to vulcanise rubber as it produced a much more durable material that was needed for industrial belts. Some selenium compounds are added to anti-dandruff shampoos because it is toxic to the scalp fungus *Malassezia* which causes this condition.

Were the commercial sources of selenium ever to become exhausted the element could be harvested from selenium-rich soils by growing crops of selenium-scavenging plants, so-called hyper-accumulators, such as milk-vetch. In this way 7 kilograms per hectare (3 kg per acre) could be extracted. The current world needs for selenium would require about 250,000 hectares to be farmed this way, but reserves of selenium in known ore deposits amount to over 100,000 tonnes, so it will be quite some time before the farming of selenium crops is introduced.

Environmental element

Selenium in the environment	
EARTH'S CRUST	50 p.p.b.
	Selenium is the 67th most abundant element.
SOILS	av. 5 p.p.m.
SEAWATER	0.2 p.p.b.
ATMOSPHERE	*ca.* 1 nanogram (ng) per cubic metre.

Selenium is among the rarer elements on the surface of this planet, and rarer than silver. Nevertheless, some soils have too much.

The amount in the atmosphere is only about 1 ng (a billionth of a gram) per cubic metre and comes as methyl selenide and dimethyl selenide, which are released from soils, lakes, and sewage by anaerobic bacteria. Just how important this is came to be realised in 1989, when the oil company Chevron stopped discharging selenium-polluted wastewater into San Francisco Bay. They pumped it instead to a 35-hectare wetland area that had been specially created to deal with the toxic waste, where it was thought that the selenium would end up in the sediment and from which it could be removed. Somewhat surprisingly, when the company scientists monitored the sediment they found only half the selenium that they expected. The rest had simply been digested by microbes, turned into gases, and carried away on the wind.

Selenium is present in the atmosphere as its methyl derivatives; these being the counterparts to the sulfur compounds which are an important part of the sulfur cycle in Nature (see page 511). There is a similar selenium cycle and about 20,000 tonnes of selenium enters the atmosphere every year, of which two-thirds comes from human activity. This may even be beneficial if it washed down by rain on to selenium-deficient land.

Selenium pollutes the environment around certain facilities, such as coal burning power stations, because this fossil fuel can have as much as 10 p.p.m. Metal smelters which use sulfide ores, and municipal incinerators also release small amounts of selenium.

Well-fertilized agricultural soil generally has about 400 mg per ton of selenium since the element is naturally present in phosphate fertilizers, and is often added as a trace nutrient. In the UK farmers have been giving their animals selenium supplements for years to keep them healthy, and in Finland there has been a programme for selenium enrichment of the soil by adding the element to fertilizers for many years. When lambs and calves are raised on selenium-deficient soil they are prone to white muscle disease which results in heart weakness. This can be prevented by top-dressing of pasture with selenised fertilisers. In 1984, the Finnish authorities decided to put selenium in all their fertiliaers because the average diet in that country was only delivering 25 micrograms per person per day, barely above the 20 microgram level needed to prevent Keshan disease.

Some parts of the world have soils with high levels of selenium. This can be taken up by the roots of some plants to levels that make them toxic to grazing animals, which suffer a condition known as blind staggers. The famous traveller Marco Polo (1254–1324) observed that the animals of Turkestan behaved in a curious way as though they were drunk. The plant generally responsible for the blind staggers was the milk-vetch (*Astragalus*) which can concentrate selenium up to 1.4 percent of its dry weight. The Great Plains of the USA and Canada are also rich in selenium, and cowboys of the Wild West knew that their herds could be affected it they ate vetch, which they called locoweed from the Spanish word *loco* meaning insane. In 1934, the American Orville Beath (*c.*1900–88) proved that the staggers was caused by excess selenium in an animal's diet.

Chemical element

Data file	
CHEMICAL SYMBOL	Se
ATOMIC NUMBER	34
ATOMIC WEIGHT	78.96
MELTING POINT	217°C
BOILING POINT	685°C
DENSITY	4.8 g/cm³
OXIDES	SeO_2 and SeO_3

Selenium is a member of group 16 (row 4p) of the periodic table. It exists in two forms: a silvery metal or a red powder; the former is produced from the latter by heat. There are other forms as well. Selenium's ability to give and take electrons explains why it is different from sulfur. When it gives electrons it is acting as a metal, when it accepts them it is acting as a non-metal. The metal consists of spirals of linked selenium atoms while the red form had stacked rings each made up of eight selenium atoms (Se_8).

Selenium burns in air, is unaffected by water, but dissolves in concentrated HNO_3 and alkalis. Selenium exhibits various oxidation states: -2 as in H_2Se and sulfides such as ZnSe, $+2$ as in $SeCl_2$, $+4$ as in $SeCl_4$, and soluble selenites SeO_3^{2-}, and $+6$ as in SeF_6, and selenates SeO_4^{2-}.

There are 31 known isotopes of selenium with masses 65 to 95, and nine known isomers. There seven naturally occurring isotopes: selenium-80 which accounts for 50 percent and which is radioactive with a half-life of 10^{18} years; selenium-78 (23.5 percent) which is stable; selenium-76 (9.5 percent) which is also stable; selenium-82 (8.5 percent) which is radioactive with a half-life of 1.08×10^{20} years; selenium-77 (7.5 percent) which is stable; selenium-74 (1 percent) which is stable; and selenium-79 (trace amounts) which is radioactive with a half-life of 327,000 years and produced as a fission product. For research and medical purposes, the radioactive isotope selenium-75 is employed and this has a half-life of about 17 weeks.

Element of surprises

1. Although selenium is itself toxic in low doses, it actually counteracts the effects of other metal toxins, especially cadmium, mercury, arsenic, and thallium, and it is known as an *antagonist* for these metals. Tuna fish, which accumulates higher than expected levels of mercury, is thought to be safe to eat because this fish protects itself by taking in an atom of selenium for every mercury atom it absorbs.

   Selenium protects against mercury poisoning by removing it from the enzyme sites which it is blocking. Mercury is chemically attracted to selenium much stronger than it is to sulfur, the element to which it is bonding in the poisoned enzymes. With the mercury removed, the enzymes can then function as normal.

2. In the 1860s the idea of television was first proposed and in the 1880s experiments with selenium were carried out as the substance that would make it possible to transmit motion pictures over wires. It was thought that because selenium increases its electrical conductivity when light falls on it surface, this might be a way to transmit a rapid sequence of impulses corresponding to light and dark down a wire, and to reconstruct them into moving pictures at the other end by stimulating patterns in selenium. The experiment was not very successful.

Silicon

Pronounced sil-i-kon, the name is derived from the Latin *silex* or *silicis* meaning flint. Readers should note that silicon is not the same as *silicone* although the latter term is sometimes mistakenly used in place of the former. Silicates are salts of silicic acid H_4SiO_4, aka orthosilicic acid, more correctly written as $Si(OH)_4$.

French *silicium*; German *Silicium*; Italian *silicio*; Spanish *silicio*; Portuguese *silício*.

Essential element

Mineral silicates are what the surface of the Earth mainly consists of, while silica (silicon dioxide) in the form of quartz is the world's accurate timekeeper. Silicon itself is part of most semiconductors and solar panels, while silicones have hundreds of uses.

Cosmic element

Silicon falls to Earth in the form of certain meteorites, known as aerolites, which are predominantly silicon dioxide, and in the form of glass-like tektites (which are also silicon dioxide) although these are of disputed origin.

Element of life

Silicon in the human body	
BLOOD	4 p.p.m.
BONE	17 p.p.m.
TISSUE	100–200 p.p.m..
TOTAL AMOUNT IN BODY	av. 1g.

Silicon as silicon dioxide (aka silica, formula SiO_2) is essential to some species such as diatoms and sponges which use it to make their cell walls and skeletons.[172] It is still not understood how these organisms, of which there are tens of thousands of different kinds, produce their remarkable and intricate glass-like structures using only the low levels of orthosilicic acid which are present in sea water.

Nettle stings are tiny hypodermic needles of silica.

Research in 1972 proved silicon to be necessary for bone growth, at least in chickens and rats, and it now seems likely that this element is an essential nutrient for humans

---

[172] Vast deposits of the mineral kieselguhr were laid down millions of years ago as the silica skeletons of diatoms.

as well. It concentrates in no particular organ of the body but is found mainly in connective tissue and skin. The daily dietary intake of silicon varies quite markedly from as little as 20 mg to as much as 1,200 mg, with cereals being the main source.

Dangerous element

The natural mineral forms of silicon are not poisonous as such but as dust they can cause lung diseases, namely silicosis and asbestosis, and the latter can lead to cancer.

Miners, stone-cutters, and sand-blasters develop silicosis of the lung which is a recognised occupational disorder caused by the inhalation of minute particles of silica. Silicosis is one form of the condition known as pneumoconiosis, the symptoms of which are wheezing, coughing, and shortness of breath.

Asbestos miners, shipyard workers, and those making insulation and brake-linings from asbestos, developed the more virulent lung disease asbestosis, although certain forms of asbestos were more dangerous than others in causing it. The first case of asbestosis was diagnosed in the UK in 1899 and the man died the following year. Despite the risks, and the introduction of safety regulations in the 1930s, asbestos continued to be widely used as a fire-retardant building material, as cheap roofing, and to insulate hot pipes etc., until it became obvious in the 1960s and 1970s that it really was a delayed action cancer-forming agent. Wherever it has been used, then special precautions have to be taken when it is removed and disposed of to special sites.

Asbestos became the subject of extensive litigation in the USA and consequently it is no longer permitted except in a few industries where it is needed because it can survive contact with extremely corrosive chemicals. Even those industries, such as chlorine production, have now turned to using alternative materials such as PTFE (polytetrafluoroethylene).

Asbestosis can lead to lung cancer many years after exposure. The reason why tiny asbestos fibres in the lungs can cause this appears to be due to the human immune system. This tries to protect cells around the fibre as it seeks to attack the fibre itself, which in fact it is unable to dislodge. In so doing it releases a protein that prevents cell death, but ultimately this protein also protects any cancerous cells which form.

In the USA in 2006, a large number of fraudulent lawsuits for legal compensation for silicosis were discovered mainly from those who had earlier claimed for compensation for asbestosis. Because it is very unlikely that someone would suffer from both conditions, almost all of these 100,000 cases were dismissed. The ill-qualified doctors who were employed by lawyers to certify the disease were found to be relying on evidence that a US Senate Committee decided was worthless.

Another kind of silicon material, the polymer known as silicone, also fell foul of US law. The silicones had been widely used as breast implants, most of which had been inserted purely for cosmetic reasons to enhance, rather than replace lost tissue. They were introduced in the 1960s, and several million women—and a few former men—have had them inserted.

In the USA during the 1980s there were some well publicised court cases brought by women who claimed that silicone leaking from their breast implants had adversely affected their health, causing auto-immune disease and even cancer. The Food and

Drugs Administration (FDA) had previously not investigated silicone implants because they were introduced before a 1976 US law requiring new medical devices to be approved, and in any case silicones had been thoroughly investigated in the 1940s and 1950s and found to have no adverse effect on living things. Nevertheless, successful law suits forced the leading silicone manufacturer, Dow Corning, to seek refuge in bankruptcy and offer a $4 billion compensation package. The FDA placed a moratorium on their use in 1992 as a result of medical/legal reports about their adverse effects. However, no reliable scientific evidence has come to light showing that these compounds present any hazard to humans. The American Academy of Neurology in 1997, and the prestigious Karolinska Institute in Sweden in 1998, reported that women with silicone breast implants were no more likely to suffer adverse health effects than women without them. The FDA finally lifted the ban on silicone breast implants in 2006.

Medical element

Louis Pasteur (1822–95) said that silicon would prove to be a treatment for many diseases and in the first quarter of the last century there were numerous reports of French and German doctors using sodium silicate successfully to treat conditions such as high blood pressure and dermatitis.[173] By 1930, such treatments were seen to have been in vain and this medication fell out of favour. So things remained, until the discovery that silicon might have a role to play in human metabolism, and thence followed suggestions that it could have a role in conditions such as arthritis and Alzheimer's disease, but no new treatment based on these assumptions has yet emerged.

Silicones appear to be bio-inert, in other words they are tolerated by the body, possibly because enzymes cannot bind to them and so decompose them. For this reason they are used for intravenous tubing, breathing tubing, and for dialysis. Silicone lenses are uses as replacement corneas in cataract surgery. Professor Henry Reid, at Canniesburn Hospital, Glasgow, Scotland, discovered the remarkable power of silicones to reduce scarring in badly burned patients. Applying a dressing coated with silicone gel to badly burned skin speeds up the body's ability to replace scar tissue with new skin, and the process of healing then takes weeks instead of years. The effect of the silicone gel is to increase the water content of the affected skin and so speed up its regeneration.

Other medical uses of silicones include indigestion tablets to relieve colic or gastric wind in humans, which can be especially painful after surgery, and to relieve bloat in cows. The silicone acts as a defrothing agent so allowing air or gas to escape.

Element of history

Silica, in the form of sharp flints were the among first tools made by humans. The ancient civilisations used other forms of silica such as rock crystal, and knew how to turn sand into glass, developing sophisticated technology for its production,

---

[173] Those who preferred a more natural, i.e. herbal, remedy were advised to drink tea made from the silicon-rich horsetail plant.

colouring, and shaping. Considering silicon's abundance, it is somewhat surprising that silicon aroused little curiosity among early chemists, even though it was clearly important to plants, and it had long been known that the ash from straw burning was capable of being turned into glass.

In 1800, Humphry Davy (1778–1829) said that silica was a compound not an element, as Lavoisier had assumed, but he was in no position to say what its composition really was, because attempts to reduce it to its components by electrolysis had failed. In 1811, Joseph Gay Lussac (1778–1850) and Louis Jacques Thénard (1777–1857) came near to isolating it by adding potassium metal to silicon tetrachloride. These chemicals reacted violently together and produced an impure form of silicon which, strangely, they did not attempt to purify. For this reason the credit for discovering silicon usually goes to the Swedish chemist Jöns Jacob Berzelius (1779–1848) of Stockholm, who in 1824 obtained silicon by heating potassium fluorosilicate with potassium metal. Although his product was contaminated with potassium silicide, he removed this by stirring it with water, with which it reacts, and thereby obtained relatively pure silicon powder. This was amorphous silicon. Berzelius called the new element silicium, but in 1831 Professor Thomas Thomson (1773–1852) of the University of Glasgow argued that it was a non-metal element and should be named silicon to accord with similar non-metals, boron and carbon, and this became its name in English.

The other form of silicon is crystalline silicon, which was first produced accidentally in 1854 by Henri Deville who electrolysed an impure melt of sodium aluminium chloride and produced aluminium silicide. When this was treated with water the aluminium dissolved leaving shiny platelets of silicon.

Economic element

Silica as sand is to be found in abundance in all parts of the world and is used as such for many purposes, as well as being the source of the silicon produced commercially. Industrially produced silica is a multibillion dollar industry and it ends up in a large variety of products ranging from construction fillers to toothpaste.

A few silicate minerals are mined, for example, talc (magnesium silicate hydrate) in Austria, Italy, India, South Africa, and Australia; and mica (sodium potassium aluminium silicate of variable composition) in Canada, USA, India, and Brazil. At the other extreme there are forms of silica so rare that they are desirable for this reason alone, and they are better known by the names of the precious and semi-precious gems that they represent. The gem-stone opal, which exhibits an attractive iridescence, is basically silica, as are rock crystal, agate, and rhinestone. The more precious gemstone amethyst is essentially a purple variety of silica, coloured by traces of manganese.

Quartz is pure silicon dioxide, and while some of this occurs naturally as rock crystal, that which is used commercially is manufactured. Quartz crystals have a peculiar property of resonating (vibrating) at a very precise frequency when subjected to the pulsation of an electric current, and this is the technology behind the quartz of clocks and watches, which thereby keep exact time. Quartz crystals are also used in radio and television transmitters.

Silicon, as the element, is the basis of several industries. Sand is reduced to silicon as a raw material for the steel, chemical, and electronics industries by heating it with coke (carbon) in an electric furnace with carbon electrodes, operating at 2,000°C.

Modern kinds of glass

Humans have been making glass for almost 3,000 years. It is made from silica and sodium carbonate and is basically a cross-linked Si–O polymer. It is transparent because it has no internal boundaries to reflect the light which enters it.

Over the past century, glass technologists and chemists have produced some remarkable versions of glass, some that is virtually unbreakable (made by adding boron), some which will even stop a bullet (by being layered), some which stop heat escaping from a building in winter and prevent heat entering in summer, and even some which will clean itself.

Sageglass has a coating which darkens when a low voltage is applied across it and it does this by converting lithium ions (Li^+) to form a film of lithium metal which reflects 90 percent the heat of the sun's heat while still allowing some visible light to pass through.

PowerGlaz is architectural glass designed for facades and roofs and it contains an array of silicon-based solar cells laminated between two layers of glass. These can deliver around 900 watts of power when exposed to the midday sun and the consist of 12.5 cm (6 inch) square cells in panels that are are 3.3×2.2 m (11×7 ft) in size.

Activ glass is self-cleaning and this is made by exposing the surface of molten glass to titanium tetrachloride ($TiCl_4$) and steam, which react chemically to form a layer of titanium dioxide (TiO_2) permanently bonded to the glass surface. The TiO_2 absorbs energy from the sun and creates superoxide free radicals from the oxygen of the air. These convert organic dirt like grease on the glass to CO_2 gas. Any inorganic dirt left behind is then easily washed away because the surface layer of titanium dioxide retains a film of water, unlike ordinary glass on which rain gathers as droplets.

Most silicon (50 percent) goes into making metal alloys such as aluminium-silicon and ferrosilicon (iron-silicon), which are used in dynamo and transformer plates, springs, and machine tools. Almost 4 million tonnes of ferrosilicon are made each year, and it is used to deoxidise steel. The silicon that is needed for making ferrosilicon steels needs only to be around 96 percent pure, and these alloys contain between 8 and 13 percent silicon. For other alloys, such as that with aluminium, the silicon needs to be at least 98 percent pure—and preferably 99 percent—and this grade is known as metallurgical grade silicon; 500,000 tonnes of this are made each year. Adding silicon to aluminium improves its castability and such alloys are used for engine blocks and cylinder heads.

The second largest use of silicon (45 percent) is to manufacture silicones (aka siloxane, dimethicone, and various trade names). These consist of chains or rings of alternate silicon and oxygen atoms with every silicon having two methyl (CH_3) groups attached.[174] These keep the chains stable and allow them to move smoothly over one another, which is why silicone oil is such a good lubricant. Depending on the lengths of the silicon-oxygen chains, it is possible to have a light mobile fluid with very short chains, or a viscous oil with long chains. Interconnecting ('cross-linking') these chains produces silicone rubber and resins, which are noted for their water-repellency and resistance to oxidation and chemical attack.

The first silicones were made by the British chemist Frederic Stanley Kipping (1863–1949) in the 1920s, but they were so difficult to produce it seemed unlikely they would ever find a use outside the laboratory. Then, in 1940, Eugene Rochow (1910–2002), a chemist working for the Corning Glass Works in New York State found a cheap and easy way of making them by heating elemental silicon (metallurgical grade) with organochlorine compounds under pressure at 350°C, and treating with water the organo-chlorosilanes which formed. Today, around 2 million tonnes of silicones are made each year, of which China produces around 1,000,000 tonnes, and European firms produce about 500,000 tonnes.

Silicones remain stable when hot and are extremely versatile, finding use as high temperature lubricants, insulating oils, waterproof sealants, rubber hose, and plastic components for electrical goods, medical equipment, and car engines. Many silicone oils are used in cosmetics and hair conditioners because they not only act as a protective layer, but leave skin and hair feeling silky smooth. The liquid silicone decamethylcyclopentasiloxane, which is a ring compound of five Si–O units, is used both in personal care products for skin and hair, and as a dry-cleaning fluid.

Silicone rubber conducts heat rapidly so it is used for finger-touch contact switches. It also makes ideal lightweight insulation for high-voltage equipment. Silicone putty is used for sealing around roofs, windows, pipes, and especially bathroom and kitchen fittings because it repels water. Silicones can waterproof fabrics such as canvas.[175]

The imprint of silicone rubber boots are to be found in the dust on the surface of the Moon, left there by the *Apollo 11* astronaut Neil Armstrong in 1969.[176]

Another key use of silicon is for semiconductors, for which the silicon must be pure, and this is achieved in various ways, starting with a volatile compound such as silicon tetrachloride, $SiCl_4$, or trichlorosilane, $HSiCl_3$, which can be easily purified and then decomposed to silicon itself by reaction with a metal like zinc or by strong heating. Even pure silicon has to be made ultra pure when destined for semiconductor use and this is achieved by zone refining. In this technique, a layer of molten silicon moves along an ingot of the element with impurities collecting in the molten zone. Silicon crystals have to be grown which have perfect lattice alignment of the

---

[174] Other organic groups can also be attached but the methyl silicones are the predominant ones.

[175] However, when they were used in the 1960s to waterproof and protect ancient stone monuments against urban pollution they proved to be a disaster. The silicone film slowly peeled away, sometimes removing the surface of the carvings it was meant to be protecting.

[176] They also left a permanent memorial of their visit in the form of a silicon (note, not silicone) disc inscribed with goodwill messages from Planet Earth.

atoms and again special techniques are available for producing these, and for including tiny, regulated, amounts of other elements such as boron and phosphorus within their structure.

Silicon carbide (SiC), better known as carborundum, is almost as hard as diamond and is widely used as an abrasive in powders, pastes, and sandpapers. It is produced by heating silicon dioxide and carbon in an electric furnace at 3,000°C but without the silica being in excess. It was discovered accidentally in 1891 by Edward Goodrich Acheson (1856–1931). Pure silicon carbide is transparent and is used in lasers and gives coherent light of wavelength 456 nm, while newer methods of synthesis produce single crystals that may be used for sophisticated equipment such as X-ray mirrors, solar cells, and high temperature transistors.

There are several forms of sodium silicate, which are produced by heating soda (sodium carbonate) and sand to 1,500°C. The product, a glass-like solid, is then dissolved in water under pressure and from the resultant solution sodium silicates can be crystallised. These are used in detergents, adhesives, textile bleaching, and in enhanced oil recovery. By pumping a solution of sodium silicate down an exhausted oil well it is possible to force more oil out of a nearby production well. The alkalinity of the solution neutralises acidic compounds in the oil-bearing strata and this releases more oil.

Environmental element

Silicon in the environment	
EARTH'S CRUST	28%
	Silicon is the 2nd most abundant element after oxygen.
SOILS	the key component.
SEAWATER	30 p.p.b at the surface, 2 p.p.m. in the deeper layers.
ATMOSPHERE	traces as dust.

The Earth's crust is composed primarily of silicate minerals, which explains why silicon is second only to oxygen in its abundance. The great variety of silicate rocks comes about as a result of the chemical versatility of the silicate ion, SiO_4^{4-}, which consists of a silicon atom attached to four oxygen atoms—see *Chemical element*. Two of these ions can link together sharing oxygens to give disilicates, or several can join together in a chain to give metasilicates, which can take many forms. They may be chain-like as in amphibole minerals, or even share three oxygens to form sheets of layered silicates. The last type of mineral often has a flaky appearance such as mica, vermiculite, or talc, and the flakiness explains the smooth feel of talc which accounted for its former widespread use in skin care. (For all these silicates, there needs to be counterbalancing positive metal ions.)

Silicates dissolve slightly in water to form silicic acid, $Si(OH)_4$, and in this way 200 million tonnes of silicates move around the environment in lakes and seas each year. In addition, 4 billion tonnes of silica particles are annually carried by rivers to the seas and oceans.

Solar panels promise great environmental benefits as a benign way of producing photovoltaic (PV) electricity—see box. There are some mega arrays of silicon-based solar PVs such as that in Portugal which is known as Girassol ('sunflower') at Moura. This consists of 350,000 solar panels, covering an area of 112 hectares (equivalent to 168 football pitches), and it generates 62 MW of power capable of serving the needs of 20,000 homes.

Silicon and photovoltaic (PV) power

The first solar cells appeared in the 1950s and they were fitted to satellites put into orbit round the Earth. They were made of pure polycrystalline silicon and they could convert 4 percent of the sun's rays to electricity. Although alternative materials are now being used to make second and third generation solar panels, silicon nevertheless accounts for around 90 percent of the solar cells in use around the world, and they have conversion efficiencies of about 15 percent. Japan produces 95 percent of suitable grade silicon.

The PV industry used to take the silicon which was not wanted by the microelectronics industry, but it now uses so-called metallurgical silicon which is cheap but then has to be purified. Tiny amounts of phosphorus or boron also have to be incorporated into the silicon before the whole is cast into large ingots weighing 240 kg. These ingots are cut into 25 brick-shaped blocks and each is sliced into 400 wafers (250 microns thick) using a fine cutting wire so that each ingot eventually produces 10,000 of them.

Thin film PVs use amorphous silicon which is radically different from crystalline silicon in that it does not need to be cast into ingots and it can be applied to a substrate such as glass or, better still, a flexible material so that the completed panels can be moulded to any surface.

Chemical element

Data file

CHEMICAL SYMBOL	Si
ATOMIC NUMBER	14
ATOMIC WEIGHT	28.0855
MELTING POINT	1,410°C
BOILING POINT	2,355°C
DENSITY	2.3 g/cm³
OXIDE	SiO_2

Ultrapure crystals of silicon have a blue-grey metallic sheen. The element is a member of group 14 (row 3p) of the periodic table. Bulk silicon is unreactive towards oxygen because a surface layer of silicon dioxide forms, which protects the rest of the material. It is also unreactive towards water and acids, except hydrofluoric acid, but dissolves in hot alkali. The preferred oxidation state of silicon is +4 which it exhibits as $SiCl_4$ and as silicates which are salts of orthosilicic acid, $Si(OH)_4$, although this molecule is not a separable chemical entity as such. Silicates form a series of highly complex compounds which are to be found in various minerals. There are other oxidation states such as −4 as in SiH_4 and various metal silicides such as magnesium silicide, Mg_2Si. There is a series of hydrogen compounds known as silanes which are comparable to the hydrocarbon alkanes, and these have the same general formula: Si_nH_{2n+2}, although in the silanes they only exist up to n = 8. The silanes are spontaneously flammable.

When all four oxygens of the silicate ion (SiO_4^{4-}) are shared, the result is silica itself, SiO_2, and there is no residual negative charge. Silica can take several forms, for example, quartz has a spiral arrangement, whereas vitreous silica has no long-range order at all. With a tiny amount of aluminium is present, silicates form even more remarkable structures called zeolites, which are three-dimensional networks with large channels and cavities into which small molecules can enter and become trapped. Some zeolites act as catalysts turning gases into liquid fuels. Others are known as molecular sieves, and these are used for drying solvents and gases.

There are 23 known isotopes of silicon with masses ranging from 22 to 44, and no known isomers. There are three main naturally occurring isotopes: silicon-28 which accounts for 92.2 percent, silicon-29 (4.7 percent), and silicon-30 (3.1 percent). None is radioactive. Also present are two radioactive isotopes, silicon-32 and silicon-34 which are produced in minute amounts and these are emitted from the decay of naturally occurring isotopes of plutonium, americium, and curium. Silicon-32 is also produced in the atmosphere by cosmic ray bombardment of argon which knocks chunks from the nucleus, a process known as spallation.

The longest-lived radioactive isotope is silicon-32 with a half-life of 170 years.

Element of surprises

1. The most accurate measurement of length is based on the lattice constant of ultrapure crystalline silicon, held under strict temperature and pressure conditions. By calibrating gamma-rays emitted from an excited atom of the isotope iron-57, against the almost exact Bragg back scattering from a reference silicon crystal, the wavelength of the gamma-ray can be measured to an accuracy of 0.08625474 nanometres, in other words accurate to 0.0000000008625474 of a metre.

2. The beautiful and hard-wearing flooring of many public buildings such as airports and shopping malls consists of artificial stone manufactured by heating together ground-up quartz and a silicon chemical such as vinyltriethoxysilane, $H_2C=CHSi(OCH_2CH_3)_3$ which reacts chemically to form a strong bond with the quartz. The resulting solid can be cut and polished like marble.

3. A perfect sphere of silicon consisting entirely of the isotope silicon-28 has been produced and will be used to define the kilogram. The pure isotope was made in Russia by centrifugally separating silicon tetrafluoride, SiF_4 to produce $^{28}SiF_4$, and then converting this to silane SiH_4 and finally to pure silicon. This was then

taken to the Institute of Crystal Growth in Berlin where it was melted and solidified several times over a period of six months to remove trace contaminants. The resulting 5 kg mass was then sent to the Australian Centre for Precision Optics in Sydney where it was shaped into a perfect sphere, a process that took 12 weeks. The number of atoms of silicon in it will be counted and from this its true weight will be calculated. It is then up to the International Committee for Weights and Measures to decide whether it can be used as the standard of weight instead of the platinum-iridium bar held in the vaults of the International Bureau of Weights and Measures in Paris, and whose weight appears to have changed slightly over the years.

4. Silly putty, aka bouncing putty, consists of silicone polymers to which boric acid, $B(OH)_3$, has been added along with other ingredients to give it body, colour, and texture.

Silver

Pronounced sil-ver, the name is derived from *siolfur*, the Anglo-Saxon word for this metal or the old German word *silubr*. The chemical symbol Ag comes from the Latin word *argentum*, meaning white, shining.

French *argent*; German *Silber*; Italian *argento*; Spanish *plata*; Portuguese *prata*.

Some elements take their name from the country where they originated, such as americium, or the homeland of their discoverer, such as polonium (Poland), and one has even been named after a river: rhenium (Rhine). Silver is unique in that it has given its name to a country and a river: the country is Argentina and the river is the River Plate. The Spanish name for the latter was *Rio de la Plata*, river of silver, so called because it was a rich source of silver plundered from the native population in the 1500s and 1600s. When the people of that region threw off Spanish rule in the 1800s they changed the name of their country to Argentina, thereby severing the links with its colonial past but preserving the links with its mineral wealth.

Essential element

Its unique ability to conduct heat and electricity make it invaluable in several applications, and its ability to kill microbes may well mean it has a remarkable future.

Element of life

Silver in the human body	
BLOOD	3 p.p.b.
BONE	varies 10–40 p.p.b.
TISSUE	varies 10–250 p.p.b.
TOTAL AMOUNT IN BODY	av. 2 mg.

Silver has no biological role and indeed is especially toxic to lower organisms.

Food element

Silver is present in many of the things that we eat. For example flour contains about 0.3 p.p.m. and bran about three times this level. Milk has between 25 and 50 p.p.b., beef, pork, and mutton around 40 p.p.b. and fish can have as much as 10 p.p.m. The daily intake of silver is between 20 and 80 micrograms (µg), depending on the food that is eaten, but such tiny amounts pose no threat to health. The hydrochloric acid in the stomach protects the body by precipitating the silver as insoluble silver chloride so that as much as 90 percent of ingested silver passes through the gut unabsorbed. That which is absorbed, lodges in the liver and skin.

In Europe, silver as a food additive has the code E174, signifying it is safe to use. Silver-coated objects can be eaten, and they often adorn wedding cakes. Small sugar balls covered in silver may be used to decorate birthday cakes and other confections. In India, ultra-thin silver foil is called *vark* and put on pastries to give them a shiny appearance.

There are several anecdotes which indicate that silver has long been known to protect drinking water against disease and may explain why silver coins were to be found at the bottom of many wells. It was said that the Persian King, Cyrus the Great, (*ca* 600–530 BC) travelled with his own supply of water in silver vessels. The ancient Phoenicians stored water, wine, and vinegar in silver containers, and even in the 1800s, seamen would drop silver coins in water and wine butts when going on long ocean voyages. In the USA in the early 1900s, people put silver dollar coins in milk to retard spoilage and retain freshness.

Medical element

The Medieval pharmacists sold silver nitrate, which they called lunar caustic, as a remedy for rubbing on warts to remove them. It worked, and was so used for hundreds of years. In 1881, the German obstetrician, Dr Carl Franz Credé (1819–92) showed that it was possible to prevent the blindness which afflicted some babies shortly after birth, by using eye drops containing 2 percent of silver nitrate. This solution killed the virus that caused the disease. Silver nitrate was also used to treat severe burns, which it did by preventing infection.

Silver acetate was once prescribed as anti-smoking tablets, effective because of the bitter taste they produced when a cigarette was smoked. Treatment with silver medicines had its drawbacks. A high intake into the body caused the skin, hair, and eyes to take on a grey tinge, a condition known as argyria, but now rarely met with.

Hip and knee replacement joints can be treated with a surface layer of silver to provide protection against the bacterial contamination which accounts for around 10 percent of replacement patients having to undergo repeat surgery.

The silver ion, Ag^+, is deadly to both bacteria and viruses and it looks as if silver preparations may return once again to hospitals because they kill antibiotic-resistant strains of infection. When bacteria, viruses, and fungi are exposed to silver solutions the silver is attracted to the sulfur-sulfur bonds in their enzymes and denatures these essential components of cell activity and thereby killing the microbe. (Nanosized particles of silver metal atoms can also be effective.) A polymer-protected form of silver has been developed, called Argleas, which releases its silver at a controlled rate. This new form of silver has received approval from the US Food and Drug Administration (FDA) for use in antimicrobial dressings and with catheters.

Colloidal silver is marketed as an over-the-counter treatment for acne. Colloidal silver was approved by the US FDA in 1999 for sale as a dietary supplement as long as no medical or health claims were made on its behalf. Nevertheless it is used by some as an alternative treatment, or as a preventative, of various illnesses although there is no support for its being effective.

Element of history

Slag heaps near ancient mine workings in Turkey and Greece reveal that silver deposits were first mined as long ago as 3000 BC. Silver was known to all the ancients but, unlike gold, it is rarely found as the natural metal, so that even though silver ores are much more abundant, the metal did not come into use until later. Indeed, when it first appeared in ancient Egypt it was more valued than gold.

Silver was refined by cupellation, a process invented by the Chaldeans, who lived in what is now southern Iraq, around 2500 BC and described in the *Bible* (Ezekiel, 22: 17–22). The method consisted of heating the molten metal in a shallow, porous cup known as a cupel, over which blew a strong draft of air. This oxidised the other metals, such as lead, copper, and iron, leaving only silver (and any gold that was present) as a globule of molten metal.

The rise of Athens, and its remarkable civilisation, was made possible partly through their exploitation of local silver mines at Laurium, which they operated from 600–300 BC, producing about 30 tonnes a year of the metal. The ore was mainly galena (lead sulfide) but with a silver content of several per cent and extracted by cupellation. These mines continued to be operated right through the Roman Empire as well, although most Roman silver came from Spain. In the Middle Ages, German mines became important as the main supplier of silver to Europe.

Silver was also mined by the ancient civilisations of Central and South America, but trade in this metal really opened up after the Spanish conquest. Rich deposits were discovered in 1535 at Charcas, Peru, then at Potosí in Bolivia in 1545, and Zacatecas, Mexico in 1548. Production from these sources totalled more than 500 tonnes per year.

In the 1800s, silver was found in the USA, especially the Comstock Lode in Nevada, and by 1870 production world wide amounted to 3,000 tonnes a year. It was to continue to grow.

Economic element

Metallic silver occurs naturally as crystals, and also as a native alloy with gold, called electrum. More generally metallic silver is found as a compact mass, and there are small deposits in Norway, Germany, and Mexico. The chief silver ores are acanthite (silver sulfide, Ag_2S) and stephanite (silver antimony sulfide, Ag_5SbS_4) of which the former is more important and is mined in Mexico, Bolivia, and Honduras, while stephanite deposits are worked in Canada. Other silver producing countries are Peru, USA, and Australia all of which produce more than 1,000 tonnes per annum. Silver is also found as the mineral horn silver (silver chloride, AgCl) and silver arsenide (Ag_3As). However, silver is mostly obtained as a by-product in the refining of other metals such as copper, and techniques for its separation include froth flotation and electro-refining.

World production of new silver is around 20,000 tonnes per year, of which only about a quarter comes from dedicated silver mines, while the rest is the by-product of other metals; 40 percent comes from lead-zinc mining, 22 percent from copper, and

13 percent from gold. Reserves amount to around 550,000 tonnes. In addition to new silver, about 10,000 tonnes per year of silver is reclaimed.

Silver is no longer used in coins although it is still used for medals. Silver coins were first minted by the Lydians (present-day western Turkey) in the seventh century BC and silver was an essential coinage metal until it was replaced in the last century by a harder-wearing copper-nickel alloy. For thousands of years silver coins were in circulation but had the disadvantage of being easily worn away, despite the hardening effect of alloying it with a little copper. There are no circulating currencies which use silver coins today, although some countries continue to mint silver coins as commemorative pieces; the USA annually mints both 50 cent and one dollar denominations. (The minting of silver coins ceased when the value of the metal exceeded the face value of the coin.)

Silver is unique in three ways: it has the highest electrical conductivity, it has the highest heat conductivity, and it has the highest reflectivity of any metal. These features are exploited commercially in electrics, grinding wheels, and mirrors. Silver is widely used for electrical and electronic devices, particularly because it makes and breaks electric circuits cleanly and so is used in all electro-mechanical keyboards. Even the primitive telegraph devices introduced in the 1830s had silver contacts. Silver solder is used to stick industrial diamonds to grinding wheels, and has the added advantage of dissipating more effectively the heat generated during their operation.

Silver provides an attractive shine for trophies, jewellery, as well as mirrors. Tableware and trophies are made of the alloy Sterling silver which is 93 percent silver and 7 percent copper. Jewellery is more likely to be 80 percent silver, 20 percent copper. Electroplated nickel silver, known as EPNS, was once popular as cutlery and, as its name implies, was made by depositing a thin surface layer of silver by means of an electrolytic cell.

Silver was important to the development of photography and some is still used for this method of image making—see box. Silver bromide and iodide were needed because of their sensitivity to light. This action causes some of the silver ions to revert to the metal in the form of tiny nuclei. The film is then developed with a reducing agent which causes more silver to deposit on these nuclei. When the image (negative) has the desired intensity the unaffected silver bromide or iodide is removed by dissolving in a fixing agent leaving the image behind. Silver salts continue to be a part of photography despite the rise in digital imaging, mainly because it produces high quality images and protects against illegal copying.

Light-sensitive glass works on the same principle except in this case the light-activated electrons convert colourless Ag^+ to metallic silver, and so darken the glass. When the sunlight fades they return to copper ions which are also part of the glass, and it is these which donate and accept the electrons as required.

Silver and the early days of photography

In the 1790s, Thomas Wedgwood (1771–1805) who was the son of the potter Josiah Wedgwood, investigated light-sensitive substances. He asked Humphry Davy to assist him and together they experimented with paper or leather coated with silver nitrate and silver chloride. They tried several approaches including the use of a camera and succeeded in making images, possible as early as 1799. However, the images faded because they knew of no way to make them permanent, which required the unaffected silver salts to be removed while leaving behind that silver which had been converted to metallic silver atoms by the action of light.

What they had overlooked was the discovery that silver chloride and bromide are soluble in ammonia, which Carl Wilhelm Scheele (1742–86) had reported in 1777. Had they known this, they could almost certainly have eventually made their images permanent.[177] For the next two years, Wedgwood and Davy experimented, making images of leaves and insect wings. Davy later making small silhouette images of individuals. In 1802, an article written by Davy detailing Wedgwood's experiments was published in *Journals of the Royal Institution of Great Britain*. He indicated that the images had to be kept in the dark and viewed by the light of a dim candle but even so they eventually faded.

In 1816, Joseph Nicéphore Niépce (1765–1833), using silver chloride and a camera, took the first photograph. Dissatisfied because it was only semi-permanent, he continued his research. The earliest known surviving photograph is from 1825 and this is Niépce's photo of a Flemish painting entitled 'Boy Leading a Horse'.

Brazing, which is the soldering and bonding of one metal to another, is best done with silver-tin alloys, sometimes with a little copper or titanium added as well, and these have generally replaced the silver-tin-lead alloys formerly used as solders. Such solders are not only used in domestic appliances they are also a key part of the micro-world of printed circuits and the macro-world of power lines. Silver nanowires can now be made and these are suitable as electrodes. They are transparent, flexible and robust and might one day be components of solar cells and other systems.

Silver can be used to destroy dangerous materials. An electrochemical cell has been devised by Accentus, a subsidiary of AEA Technology, which uses a solution of silver nitrate in nitric acid as the electrolyte. Under the influence of an electric current this creates silver(II) ions, Ag^{2+}, which are extremely strong oxidising agents, and capable of reacting chemically with all kinds of potentially dangerous chemicals such as nerve gases. When these are fed into the reaction vessel they are converted to CO_2, water, and harmless salts which can then be safely disposed of. The process can even be used by the pharmaceutical industry to recover radioactive carbon-14.

It requires only a little silver to sterilise water, and 10 p.p.b. will suffice. Swimming pools can be made safe by adding silver salts to the water as an alternative to chlorination.

Clothing such as socks, boxer shorts, and cycling pants can have silver fibres woven into them to prevent bacteria producing the sulfides and acid molecules which cause

---

[177] The discovery of what later became the universal 'fixer', sodium thiosulfate, and its ability to dissolve all silver salts, was made by the astronomer John Herschel (1892–71) who published his discovery in 1819.

unpleasant odours. Incorporating silver nanoparticles into clothing prevents bacteria on the skin from digesting the molecules in sweat and forming the obnoxious smells.

Silver has other uses. In dentistry there are silver-tin-mercury or silver-tin-copper-mercury amalgams for fillings, while in high capacity long life batteries there is silver oxide; and when it comes to increasing rainfall then silver iodide can be used. If clouds are seeded with this then it will initiate rainfall. One gram of silver iodide can provide a trillion seed crystals (10^9) around which water vapour will condense into droplets large enough to fall to earth as rain. An aircraft flying through a rain cloud dispersing this chemical as a fine smoke can produce a torrential downpour, sometimes with disastrous results. Secret tests carried out by Britain's Ministry of Defence in August 1952 in the West of England resulted in a flash-flood that engulfed the seaside resort of Lynmouth, sweeping away houses and sending 31 inhabitants to their deaths. (Rain-formation was being tested as a strategy for bogging down tanks and troops and thereby hindering their advance.)

Paint containing nanoparticles of silver prevents the growth of mould and is being used in hospitals. Tests have shown that it kills 99.999 percent of bacteria which settle on it. When used on outside surfaces, such paint reflects 90 percent of incident sunlight thereby keeping indoor temperatures lower.

Environmental element

Silver in the environment	
EARTH'S CRUST	70 p.p.b.
	Silver is the 66th most abundant element.
SOILS	av. 0.5 p.p.m.
SEAWATER	about 0.1 p.p.t. at the surface, about 2 p.p.t in the depths.
ATMOSPHERE	traces.

Silver levels in soil are not usually high except in mineral-rich areas when they can sometimes be as much as 44 p.p.m. near old workings. Plants can absorb silver and measured levels come in the range 0.03 to 0.5 p.p.m. (dry weight). Plant ash generally contains some silver but less than 5 p.p.m. Fungi and green algae can have as much as 200 p.p.m. and horsetail, lichen, mosses, and some trees will take in more than expected, although only in response to silver being in the soil at enhanced levels.

Chemical element

Data file	
CHEMICAL SYMBOL	Ag
ATOMIC NUMBER	47
ATOMIC WEIGHT	107.8682
MELTING POINT	962°C

Silver

BOILING POINT	2212°C
DENSITY	10.5 g/cm³
OXIDE	Ag_2O

Silver is a soft metal with a characteristic silver sheen when polished. It is a member of group 11 (row 4d) of the periodic table. Silver, like gold, is malleable and ductile. It can be beaten so thin as to be almost transparent, and 1 g of the metal can be drawn into a wire nearly 2 km long. Silver is stable to water and oxygen but is slowly attacked by sulfur compounds in the air to form a black sulfide layer, hence silver objects need regular cleaning.[178] Silver dissolves in sulfuric and nitric acids. Its preferred oxidation state is +1 as in the famously insoluble AgCl, and the highly soluble $AgNO_3$ when it is present in solution as Ag^+. There are higher oxidation states, namely +2 as in AgF_2, and +3 as in the rare complex ion AgF_4^-.

Silver occurs mainly as the stable isotopes Ag-107 (52 percent) and Ag-109 (48 percent). There are also a number of naturally occurring radioactive isotopes produced as fission fragments in uranium-bearing deposits. There are 38 known isotopes of silver with mass numbers 93 to 130, and 33 known isomers. The longest-lived radioactive isotope is silver-105 with a half-life of 41.3 days, but there is a long-lived isomer, silver-108m, whose half-life is 418 years.

Element of surprises

1. During World War II, the USA used silver in its atomic weapons programme at the Oak Ridge National Laboratory (known as the Clinton Laboratories until 1948). In 1943 a nuclear reactor was built to study ways of producing plutonium but the copper needed for its electrical connectors was in short supply so silver was used instead and this was borrowed from the US Treasury. After the war, copper replaced the silver which was melted down and returned.

2. Not all microbes are killed by silver. The bacterium *Pseudomonas stutzeri*, which was discovered in a Canadian silver mine, protects itself by depositing the unwanted metal in tiny cavities in its cell walls.

3. Another unique property of silver is its compound with hexacyanocobalt(III), which has the formula $Ag_3[Co(CN)_6]$. When crystals of this are heated they expand in one direction while contracting in another.

---

[178] This is the reason why silverware should never come into contact with high-sulfur foods like egg, onion, garlic, or mustard. A sulfide layer does not protect the rest of the silver, so staining can go deep. Some electroplated silverware is protected with a coating of rhodium.

Sodium

Pronounced soh-dee-um, the name is derived from the English word *soda*, and its name has been linked to sodanum, a medieval headache cure based on soda, which is sodium carbonate. In fact Humphry Davy, the first person to isolate it, decided on the name sodium after consulting his contemporaries. The chemical symbol Na comes from the Latin *natrium* meaning soda.

French *sodium*; German *Natrium*; Italian *sodio*; Spanish *sodio*; Portuguese *sódio*.

Essential element

Sodium as salt[179] (sodium chloride, NaCl) is both essential for life and is a primary feedstock of the chemical industry, while the metal itself is needed in the production of airbags and biofuel.

Cosmic element

The Sun and stars shine with visible yellow light given out by sodium atoms in a high energy state. In 2002, sodium was observed to be present in the atmosphere of a giant planet that circles the star HD 209458 which is 150 light years from Earth.

Element of life

Sodium in the human body	
BLOOD	plasma 3,500 p.p.m. (0.35%); red cells 250 p.p.m.
BONE	10,000 p.p.m. (1%)
TISSUE	varies between 2,000–8,000 p.p.m. (0.2–0.8%)
TOTAL AMOUNT IN BODY	100 g.

Sodium is essential for animals, less so for plants, and indeed these generally contain relatively little. Sodium in the body is mainly in the fluid outside cells, as the table above shows for blood, the situation being just the opposite to that which pertains with potassium. Blood needs a lot of sodium to regulate osmotic pressure and blood pressure, as well as to help solubilise proteins and organic acids. From the dawn of life, cells had to struggle to keep sodium out, in the face of an osmotic pressure from a marine environment in which sodium was abundant.

---

[179] In chemistry, the term 'salt' covers a whole group of compounds, namely those consisting of a combination of positive and negative ions, but in ordinary speech salt refers specifically to one kind of salt, that of sodium chloride, Na^+Cl^-, and this is how it will be used here.

Sodium's most important function is the movement of electrical impulses along nerve fibres, and this it does in conjunction with potassium, with both able to move through the cells of the nerve membrane. As these ions move, they generate a wave of electrical impulse that is equivalent to a current passing along the nerve. Sodium moves inwards and is then quickly expelled outwards again by the so-called 'sodium pump' while potassium moves out and is then carried back. The sodium pump accounts for as much as 40 percent of a body's energy needs.

It is necessary to take in a regular supply of sodium because it is continually lost from the blood stream as it is filtered out by the kidneys. Although a lot of sodium is recycled, some exits via urine, the faeces, and the sweat glands. Humans sweat more than any other animal and we can lose two litres of water per hour during extreme physical exercise. Along with the water, which is extracted from the blood by the sweat glands, salt is also lost and needs to be replaced, hence the need for salt to be included in sports drinks.

Dangerous element

Sodium compounds are not hazardous insofar as their sodium content is concerned, but excess salt can provoke a toxic response, such as vomiting. The positively charged sodium ion Na^+, is not technically a poison but it is a cause of concern because too much salt in the body puts unnecessary stress on those with a predisposition to high blood pressure and heart disease. When the body has too much salt it compensates by retaining more water, and it is this which causes increased pressure on the arteries, resulting in high blood pressure with all the risks to health this implies.

Salt has the power to harm, and was criminally used to this end as in the case of an Edinburgh woman who was sentenced to four years in jail for continually making her daughter ill by injecting salt solution into her daughter's feeding tube. On each occasion the child had to be hospitalised and doctors diagnosed an eating disorder, hence the need for a feeding tube. Eventually the child was given so much salt that she suffered a stroke that left her permanently brain damaged and the real cause of her illness was finally realised. The mother's behaviour was diagnosed as the syndrome known as Munchausen by Proxy.

Food element

Salt has been part of the human diet for thousands of years and was important as a way of preserving food. Few foods are preserved this way today. The amount of sodium a person consumes varies from individual to individual and from culture to culture; some people get as little as 2 g per day, some as much as 20 g.

The sodium which comes with a diet of fresh foods is equivalent to around 3 g of salt a day, and is all that a person requires. Foods which have more than 100 mg per 100 g of sodium are tuna, sardines, eggs, butter, cheese, and pickles. Vegetables generally have very little, but celery and peas have quite a lot. However, the average intake of salt in a Western style diet is about 9 g a day, with 3 g coming from fresh food, 3 g coming from processed foods, such as cereals and snacks, and 3 g from table salt. A Japanese person can have double this amount because of the prevalence of seafood in the diet.

Sodium chloride triggers a specific reaction on the tongue and is one of the five basic taste sensations.[180] This is rather puzzling because no other salt in the diet provokes such a response. Nor is it only humans who like salt; grazing animals such as cows will lick salt cakes if they are available, the reason being the relatively low level of sodium in grass.

Salt is called rock salt when it comes from mineral deposits, and sea salt when it is obtained by evaporating seawater. Both are refined by re-dissolving them in fresh water and evaporating the filtered solution until the salt re-crystallises as pure sodium chloride which is sold to food manufacturers, and as cooking salt. Table salt, on the other hand, has a little magnesium carbonate added to keep it free-flowing.

Salt is used in cooking to reduce bitterness which it does by blocking certain sensors on the tongue. It also has an effect also on the texture of the food at a molecular level, as revealed for proteins which curl up more tightly in its presence and this gives a better mouth-feel.

Medical element

A sudden excess of sodium, in the form of salt solution, acts as an emetic. Continued high levels of sodium in the diet lead to long-term health risks. On the other hand there are instances of sodium deficiency, and if this is not severe then in animals and humans it causes muscular spasms, popularly known as cramp, or charley horse in the USA. A severely reduced intake of salt can pose a real risk, as revealed by a certain detox 'hydration' diet promoted in the UK. This required an increased water intake of 3 litres per day while excluding as much salt as possible from the diet. It led to one woman in Oxfordshire suffering brain damage and epileptic seizures, and she received £800,000 in compensation in 2008 when she sued the promoters of the diet.

There are three grades of reduced salt diet which are referred to as low-salt, no-salt, and salt-free. A low-salt diet means no table salt or salt used for cooking. Foods high in salt, such as potato snacks and certain cheeses, are not allowed. Such a diet still provides about 6 g of salt a day, which comes from things like breakfast cereals, biscuits, cheese, and processed foods. A no-salt diet cuts out most of these and limits bread to three slices per day, so that salt intake is around 3 g. A salt-free diet as recommended by doctors excludes all salt except that which comes as a natural part of the food we eat, which may be less than 2 g per day; a patient on this diet eats mainly foods which have very little sodium anyway, such as boiled rice.

The World Health Organisation advocates population-wide measures to reduce salt intake and companies are trying to find ways to do this and yet retain salt's savoury taste. An alternative condiment is soy sauce whose umami (meaty) flavour compensates for a reduced salt level. The umami flavour comes from amino acids and other components.

---

[180] The others are sweet, sour (acid), bitter, and umami (savoury).

Salt and health

The links between salt intake and the health of the general public has proved to be a controversial subject with some believing too much salt causes illness, while others say it has little effect. In 1992, an analysis of several surveys, which together covered 47,000 people, was conducted by a group at St Bartholomew's Hospital, London, and they found a relationship between salt intake and high blood pressure. In the same year the *Journal of the American Medical Association* published a report from the Mount Sinai Hospital, Ontario, which came to a very different conclusion. It found that salt posed no threat at all for people whose blood pressure was normal. A report from the Oregon Health Sciences University came to the same conclusion in 1998, while another survey, chronicling the diets of 11,000 US citizens from 1970–90, even concluded that those who ate more salt tended to live longer, witness the longevity of the Japanese.

Salt is a blessing in tropical countries, where it saves millions of lives. Diarrhoea, and the resulting dehydration it causes, are reported to kill millions of babies and children every year. The answer to this potentially fatal illness is to drink a solution of glucose and salt and the United Nations children's organisation, UNICEF, distributes millions of sachets for making up such solutions. They contain 20 g of glucose, 2 g of salt, 3 g of sodium citrate, and 1.5 g of potassium chloride, to be dissolved in a litre of boiled water. The same treatment can also help those suffering from cholera, which kills its victim by blocking the channels in the intestines that absorb water from the food in the gut. The result is severe dehydration and often death.

Radioactive sodium-24, with a half-life of 15 hours, is used in medical research.

Historical element

Salt has been known from prehistoric times. In the era of the Roman Empire it was considered so important that it was given as part of the wages of soldiers, hence the word *salary*. Sodium carbonate (also known as soda, or industrially as soda ash, Na_2CO_3) is mentioned in the *Bible* as a cleaning agent (Jeremiah 2: 22) although there it is called nitre, a name that eventually came to mean another salt, potassium nitrate. In the ancient world, sodium carbonate was harvested in Egypt where it came from the Natron Valley in which flood water from the Nile gathered each year and which yielded a crop of soda crystals as it dried out. Later a surface deposit of trona, another form of sodium carbonate, was located in the Sahara from where it was transported to Tripoli and exported.

The demand for sodium carbonate came partly from its use as a washing aid ('washing soda') and partly because it was needed to manufacture glass. In many areas it was produced from plant ash, especially that from certain types of fern and seaweed, which have high sodium levels.[181] However, even these could not meet the demand created by the Industrial Revolution and in 1783 Nicolas Leblanc (1742–1806) devised a method for making soda from salt; he was encouraged to do this as a result of the French Royal Academy of Science offering a prize for such a process in 1775. Earlier that century, Henri-Louis Duhamel du Moncease (1700–82) had shown that salt could be turned into soda by dissolving it in nitric acid, boiling to dryness, and then heating with charcoal. Leblanc's process used cheaper sulfuric acid to form sodium sulfate which was then heated with limestone (calcium carbonate) and charcoal.

Even at the start of the 1800s, the elemental composition of salt and soda remained a mystery. The great chemist Antoine Lavoisier (1743–94) even thought salt might contain nitrogen. The answer came from the Royal Institution in London in October 1807 when Humphry Davy (1778–1829) reduced caustic soda (sodium hydroxide, NaOH) to sodium. He had recently produced metallic potassium by the electrolysis of potassium hydroxide and he next turned his attention to caustic soda from which he obtained metallic globules of sodium. He reported that the electrolysis of caustic soda required a much higher current from his battery than that to generate potassium.

The following year, Louis-Josef Gay-Lussac (1778–1850) and Louis-Jacques Thénard (1777–1857) made sodium by heating caustic soda and iron filings to red heat. Later that century, sodium was produced from sodium carbonate and carbon at 1,100°C. However, the sodium metal made commercially is produced by the electrolysis of sodium hydroxide, the method used by Davy.

Economic element

Sodium occurs in many silicate minerals but none is exploited for the sodium it contains. Those deposits which are extracted with this end in view are halite (aka rock salt, NaCl) and trona, which is sodium carbonate bicarbonate, of composition $Na_3(CO_3)(HCO_3)$.[182] The main mining areas for halite are Germany, Poland, the USA, and the UK; and for trona are Kenya and the USA. A new mineral deposit in Colorado is being developed and this is nahcolite, which is almost pure sodium hydrogen carbonate.

World production of salt is around 250 million tonnes per year, and this huge amount is mainly extracted from salt deposits by pumping water down bore holes to dissolve it and pumping up brine. The main salt-producing countries are China (60 million tonnes), USA (40 million tonnes), India (20 million tonnes), and Canada (12 million tonnes) although almost all countries of the world produce some salt. A growing amount is now being extracted as a renewable resource from seawater, something that has been done on a small scale in many parts of the world for cen-

---

[181] The production of sodium carbonate from the ash of burnt seaweed was the chief industry of the UK's Scilly Isles for almost a century, beginning in 1684 [Colin Daly].

[182] The correct name for bicarbonate is hydrogen carbonate, and its formula is HCO_3^-. It will be referred to by its correct name hereafter.

turies. At Dampier, on the North West coast of Australia, there are vast evaporation ponds which yield 5 million tonnes of salt each year.

Salt is used mainly by the chemical industry—60 percent ends up there—where it is turned into chlorine gas, sodium hydroxide, and sodium carbonate, all of which are basic materials for the industry. Sodium hydroxide is the main alkali of the chemical industry—and it also finds domestic used in unblocking drains. About 20 percent of salt goes to the food industry where it is used as a preservative and flavouring agent, and the remaining 20 percent find various outlets, such as de-icing roads in winter.

Sodium carbonate production amounts to around 30 million tonnes per year, and whereas formerly this was produced from salt and limestone ($CaCO_3$) in the Solvay process, developed by Ernest Solvay in Belgium in 1861, much now comes from the vast natural deposits found in the USA, which represent at reserve of almost a billion tonnes.

Solid sodium carbonate is needed to make glass, and almost half ends up this way. A solution of sodium carbonate is the main acid-neutralising agent in water treatment, and it is added to fizzy drinks to boost the solubility of carbon dioxide. Sodium hydrogen carbonate also acts as a leavening agent in baking, as well as being used in the textile and leather industries, and for making soaps and detergents. Fire extinguishers which produce a jet of water or foam do so by generating carbon dioxide gas from the action of sulfuric acid on sodium hydrogen carbonate.

Sodium metal is commercially important and is produced in so-called Downs Cells (developed in 1924) wherein molten sodium chloride, mixed with calcium and barium chlorides, undergoes electrolysis. The other chlorides are there to lower the melting point of sodium chloride from 808°C to 600°C.[183] The molten salts are contained in a ceramic-lined vessel fitted with a graphite anode and a steel cathode, from which the molten sodium rises and can be skimmed off. As the molten metal cools it releases a few percent of calcium which crystallises out and is filtered off, leaving molten sodium of 99.9 percent purity. World production of sodium metal is about 90,000 tonnes per year, and is produced in France (30,000 tonnes), the USA (30,000 tonnes), and China (30,000 tonnes).

Sodium metal is used by industry in the manufacture of various chemicals and for extracting metals, namely beryllium, thorium, titanium, and zirconium which it does when it is heated with their halide salts. Other chemicals are produced from it such as sodium borohydride ($NaBH_4$), which is used for bleaching paper pulp, sodium azide (NaN_3), which is the explosive in air-bags of vehicles, and sodamide ($NaNH_2$), which is used in dye manufacture. Other products made from sodium metal are sodium methylate ($NaOCH_3$), which is used to manufacture biodiesel from plant oils, and polysilicon, which is needed for solar cells.

Sodium metal is also used in street lighting, where a tiny piece is sealed in neon-containing lamps. Such bulbs use less electricity and their light penetrates mist and fog better than other forms of street lighting. When the bulb warms up, the sodium vapour is excited and emits a characteristic yellow light of wavelength 589 nm.

---

[183] The composition of the molten salt electrolyte is 28% NaCl, 26% $CaCl_2$ and 46% $BaCl_2$.

Environmental element

Sodium in the environment	
EARTH'S CRUST	23,000 p.p.m. (2.3%)
	Sodium is the 6th most abundant element.
SOILS	major component.
SEAWATER	10,500 p.p.m. (1.05%)
ATMOSPHERE	traces.

Sodium has been leaching from the Earth's rocks and soils to its oceans for billions of years, and once there it remains for hundreds of millions of years because it is so soluble. The sea has a total of 3.5 percent dissolved salts of which sodium chloride is by far the major component. Sodium which drained to inland seas that eventually dried up, accounts for the vast deposits of sodium minerals which are to be found around the world.

The largest extinction of life on Earth occurred at the end of the Triassic era, around 250 million years ago, when almost 80 percent of creatures became extinct. What caused this to happen is still unknown and there may have been several factors, one of which might have been emissions from the Zechstein Sea. This covered $600,000 \, km^2$ and stretched from land that is now England to Western Russia. It became exceedingly salty as it was exposed to a dry continental climate and it then it began to give off millions of tonnes of organochlorine compounds such as chloroform ($CHCl_3$) and tetrachloroethane ($C_2H_2Cl_4$). In 2009, it was suggested that these contributed to the mass extinction.

Chemical element

Data file	
CHEMICAL SYMBOL	Na
ATOMIC NUMBER	11
ATOMIC WEIGHT	22.989770
MELTING POINT	98°C
BOILING POINT	883°C
DENSITY	0.97 g/cm³
OXIDE	Na_2O, also Na_2O_2 and NaO_2

Sodium is a soft, silvery-white metal which is a member of group 1 (row 3s), also known as the alkali metal group of the periodic table. It oxidises rapidly when cut and it has to be stored under paraffin. When it burns in air the main oxide it forms is sodium peroxide, Na_2O_2. The metal reacts vigorously with water to evolve hydrogen gas which sometimes ignites. On the other hand, sodium dissolves in liquid ammonia to form a deep blue solution which gradually becomes metallic looking as the amount of sodium increases and under such conditions it is believed that negative sodium ions, Na$^-$, are present. Normally the chemistry of sodium is that of the +1 oxidation state as manifested in the sodium ion Na$^+$, whose salts are almost all very soluble in water.

There are 20 known isotope of sodium of masses 18 to 37, and three known isomers. Naturally occurring sodium has only one main isotope, sodium-23, and it is not radioactive. However, there are traces of sodium-22 and sodium-24 produced by the cosmic ray spallation of argon in the upper atmosphere.

The longest-lived radioactive isotope is sodium-22 with a half-life of 2.6 years.

Element of surprises

1. Liquid sodium metal is used as the coolant in heat exchangers in some nuclear reactors. The liquid sodium coolant is contained in an inner loop round the reactor, but this tends to accumulate radioactive isotopes sodium-22 and sodium-24. Consequently, a second loop of liquid sodium is used to transfer the heat from the inner loop to the exterior of the reactor. It is potentially highly dangerous because sodium heated to above 125°C bursts into flames when exposed to the air, and the temperature quickly rises to 800°C whereupon it becomes almost impossible to extinguish. The French Superhenic breeder reactor is cooled with almost 6,000 tonnes of liquid sodium.[184] Researchers there have developed a special sodium fire extinguisher called Marcalina which is composed of lithium and sodium carbonates and graphite.

2. Salt crystals are tiny cubes and their surface can absorb moisture which sticks them together, hence the need to add a water-absorbing magnesium carbonate, $MgCO_3$ (food additive code E504i) or sodium aluminium silicate, $Na_2Al_2Si_{14}O_{32}$ (food additive code E544) to keep the salt from caking and to keep it free-flowing. If NaCl crystals are grown from water containing a little of the amino acid glycine, however, they form almost spherical crystals which don't stick together. As yet such salt has not become commercially available.

---

[184] France generates a larger percentage (77%) of its electricity from nuclear energy than any other country.

| 38 |
| Sr |

Strontium

Pronounced stron-tee-um it is named after the mineral strontianite which was first found in a mine at the small town of Strontian in Scotland.

French *strontium*; German *Strontium*; Italian *stronzio*; Spanish *estroncio*; Portuguese *estrôncio*.

Essential element

Flares and fireworks rely on strontium for strong red colours. This element is also added to some toothpastes to reduce the sensitivity of teeth to hot and cold.

Element of life

Strontium in the human body	
BLOOD	30 p.p.b.
BONE	35–140 p.p.m.
TISSUE	120–350 p.p.b.
TOTAL AMOUNT IN BODY	320 mg.

Some deep sea creatures incorporate strontium into their shells as strontium sulfate, and stony corals require it, which is why it needs to be added to the water of aquariums. Although strontium has no function in the human body it is present in sizeable amounts because of its chemical similarity to the essential mineral calcium, and like calcium a lot of it ends up in the skeleton. The analysis of bone for the ratio of its strontium isotopes ($^{87}Sr/^{86}Sr$) will reveal the region from which a person came. The analysis of birds' feathers for this ratio has been advocated as a method of tracking their migratory patterns.

Food element

While strontium is not an essential ingredient, nevertheless a lot that we eat contains it. Foods containing strontium range from the very low, for example, corn has only 0.4 p.p.m. and oranges 0.5 p.p.m., to the high, for example, cabbage has 45 p.p.m., onions 50 p.p.m. and lettuce 74 p.p.m. (all dry weights). Strontium-90 was found to concentrate in plants such as beans, but not in grain crops. We take in about 2 mg a day of strontium but it causes no harm.

Medical element

Strontium salts are given to treat osteoporosis because it stimulates bone growth and bone density. Strontium ranelate can be used to prevent bone loss in post-menopausal women. Those treated with the drug have a much reduced risk of bone fracture later in life. It is a recognised treatment in many countries but not in the USA or the UK.

Strontium-87m (the 'm' stands for metastable and signifies that this is an isomer) is a radioactive isotope with a half-life of 2.8 hours, which is used for diagnostic purposes in medical research. It quickly disappears from the body, and within a day less than 0.3 percent of the dose remains. Radioactive strontium-89 (half-life 50 days) is a beta-emitter and is employed as a radiopharmaceutical, and was approved for use in the USA in 1995. It offers a treatment for bone cancer because it concentrates in the active area where the cancer is growing and destroys it.

Dangerous element

Strontium is not toxic. Nevertheless it is dangerous as the radioactive isotope strontium-90 (half-life 28.8 years) which became a major worldwide pollution concern in the middle of the last century, when it was produced by above-ground nuclear bomb tests which took place from 1945 to 1980. They contaminated the whole planet with it. Strontium-90 is a beta-emitter and is a serious threat because it is one of the most powerful emitters of ionising radiation and therefore capable of causing serious damage to dividing cells. Its presence was detected in the milk teeth of infants in the 1950s, showing how prevalent it had become, having been washed out of the atmosphere on to grassland, to be eaten by cows, and so ending up in milk and other dairy products. Those children have carried it as a burden for the rest of their lives.

The accident at Chernobyl the nuclear power plant in the former Soviet Union in 1986 released this isotope into the surrounding areas and across northern Europe.

Element of history

In 1787, a dealer in mineral specimens in Edinburgh, Scotland, was offered a new stone that had been found in a lead mine at Strontian, on the west coast. He thought it was a barium mineral and told a local doctor, Adair Crawford (1748–95), whom he knew was interested in baryta (barium oxide) as a possible medicament. Crawford eventually realised that this was a new 'earth' which he named strontia (strontium oxide). He published a paper on his research in 1790 and this prompted a Thomas Charles Hope of Edinburgh to begin a fuller investigation the following year. Hope proved conclusively that it was a new element and noted that it caused the flame of a candle to burn red, whereas baryta gave a green colour.[185] Meanwhile Martin Heinrich Klaproth (1743–1817) in Germany was also working with strontianite and he had made both strontium oxide and strontium hydroxide.

---

[185] In fact this red colour had been noted by others soon after the mineral was discovered.

In 1799, another strontium mineral was discovered in Sodbury, Gloucestershire, England, where the locals were using it as gravel for paths in ornamental gardens. This was strontium sulfate and the mineral was named celestite.

Strontium metal itself was isolated in 1808 at the Royal Institution in London by the English chemist Humphry Davy (1778–1829). He generated it by the electrolysis of a mixture of strontium chloride and mercury oxide, the method he had used to isolate sodium, potassium, and barium metals.

Economic element

The chief ores are celestite (aka celestine, $SrSO_4$) and strontianite (strontium carbonate, $SrCO_3$). The celestite deposits around Yate in Gloucestershire were extensive and these supplied almost all the world's needs for strontium until World War II. The last mine there closed in 1994, not because the ore had been exhausted, far from it, but because China could mine it cheaper and it is now the leading producer of strontium. Other mining areas are Spain and Mexico. World production of strontium ores is about 1,000,000 tonnes per year, 70 percent of which is mined in China.

Strontium metal is produced by heating strontium oxide and aluminium under vacuum conditions. These react to form strontium and aluminium oxide and the strontium distils off and is collected. Demand for the metal is limited, a little being used as a 'getter' in vacuum tubes to remove the last traces of air. Strontium carbonate was formerly used to make the special glass for television screens and visual display units. It was added to the glass to absorb the secondary X-rays emitted by cathode ray tubes.

The brilliant reds of fireworks and warning flares is produced by strontium nitrate, $Sr(NO_3)_2$, and in this respect strontium is unique and has never been bettered.

Strontium-90 is a by-product of nuclear reactors from whose spent fuel it is extracted. Its high energy radiation can be used to generate an electric current, and for this reason it is used in space vehicles, remote weather stations, and navigation buoys. Some goes to make thickness gauges for measuring paper, plastic film, and fabrics. The high electron flux it emits can also be used to neutralise static electric charges of the kind that build up on machinery which handles paper, plastics, and fibres.

Other uses of strontium included an early method for separating sugar from sugar beet molasses. The sugar was precipitated as strontium disaccharide which was then converted back to strontium carbonate using CO_2 and the sugar was released. Even today, strontium chloride hexahydrate ($SrCl_2.6H_2O$) is an ingredient in some toothpastes that are formulated to desensitise teeth. It plugs the tiny holes in the dentine which give access to the nerve making hot and cold drinks painful. Modern luminous paints, which glow in the dark without the need for radioactive isotopes, rely on strontium aluminate. This absorbs light during the day and releases it slowly for hours afterwards.

Environmental element

Strontium in the environment	
EARTH'S CRUST	370 p.p.m.
	Strontium is the 16th most abundant element.

SOILS	av. 200 p.p.m., range 18–3,500 p.p.m. (i.e. 0.35%)
SEAWATER	8 p.p.m.
ATMOSPHERE	trace.

Generally, strontium becomes immobilised in the environment by precipitation as strontium carbonate, or incorporated into invertebrate shell material. Strontium is widespread, and some soils have as much as 3,500 p.p.m. (0.35%). The highest levels of strontium are found in deserts and the soils of forests.

Chemical element

Data file

CHEMICAL SYMBOL	Sr
ATOMIC NUMBER	38
ATOMIC WEIGHT	87.62
MELTING POINT	769°C
BOILING POINT	1,384°C
DENSITY	2.5 g/cm³
OXIDE	SrO

Strontium is a silvery-white, relatively soft metal, which is a member of group 2 (row 5s), also known as the alkaline earth group, of the periodic table. The bulk metal is protected to a certain degree by an oxide film, but even so it has to be stored under paraffin. The metal will burn in air if ignited, and it is attacked by water. Its chemically most stable form is in oxidation state +2 as the ion Sr^{2+} in salts and in solution.

Strontium has 33 known isotopes ranging from 73 to 105, and six known isomers. Four are naturally occurring: strontium-88 which comprises 82.5 percent, strontium-86 (10 percent), strontium-87 (7 percent), and strontium-84 (0.5 percent). None is radioactive. Strontium-87 is the product of radioactive decay and derives from rubidium-87, so that the ratio of strontium-87 to the other isotopes varies slightly with location. There are also traces of several radioactive isotopes, among them strontium-89 with a half-life of 50.53 days, and strontium-90, which is the longest-lived radioactive isotope with a half-life of 28.79 years. These occur as uranium fission products in uranium ores. The next longest-lived isotope is strontium-85 with a half-life of 64.8 days.

Element of surprise

Crystals of strontium titanate ($SrTiO_3$) outsparkle even diamond because they have a higher refractive index. However, unlike diamond, the surface is easily scratched and it has not been widely used as a gemstone, although it has been used in optical devices.

Sulfur

or Sulphur

Pronounced sul-fer, the name is derived from two possible words, the Sanskrit *sulvere*, or the Latin *sulfurium*, both ancient names for sulfur. In English there are two spellings of the word, and indeed both sulfur and sulphur were both used in the Middle Ages, but the International Union of Pure and Applied Chemistry (IUPAC) decided that sulfur is to be the name used for the element, and this is how it will be spelt here, along with sulfite and sulfate. Chemists also use the prefix 'thio' to denote the presence of sulfur in a molecule, for example, thioether is ether with sulfur instead of oxygen, that is, $C_2H_5SC_2H_5$. Thio comes from the Greek *theo* meaning 'of god', and this link is thought to arise because sulfur was a component of incense.

French *soufre*; German *Schwefel*; Italian *solfo*; Spanish *azufre*; Portuguese *enxofre*.

Because sulfur has been known for so long, there is a word for it in almost every language and one that is usually very different from its chemical name, such as brimstone (meaning burning stone) in English, *rikki* in Finnish, *ken* in Hungarian, *iwo* in Japanese, *liu huang* in Chinese, *gundhuk* in Hindi, *whanariki* in Maori, and *isibabule* in Zulu.

Essential element

Sulfur-containing amino acids are essential components of living cells, and sulfuric acid (H_2SO_4) is an essential raw material for many industries, and produced on a large scale.

Cosmic element

Elemental sulfur is found in some meteorites which can have up to 6 percent in the case of chondrites, and there appears to be a deposit of sulfur on the surface of the Moon near the crater Aristarchus. The Earth's iron core is thought to contain about 5 percent sulfur. Io, one of the moons of Jupiter, has a surface covered with sulfur and has active volcanoes that spew it forth. It has been suggested that sulfur-using bacteria might live there, of the kind that survive in sulfur-rich environments on Earth. In fact life on this planet may owe a lot to sulfur.

Conditions in the early seas were such that simple chemical reactions could have generated the range of amino acids that are the building blocks of life. Experiments have shown that reactions between hydrogen sulfide, ammonia, and simple carbon-containing molecules such as formaldehyde, will yield a wide range of amino acids. Some micro-organisms, which live near hydrothermal vents in the sea, still rely on reduced forms of sulfur, such as H_2S, as an energy source.

Sulfur tells us about the development of the Earth's atmosphere. Sedimentary rocks of the precambrian period reveal the change that occurred around 2.4 billion years ago, before which there were only traces of oxygen. By 2.1 billion years ago the levels of oxygen were substantial and this is shown by the sulfites and sulfates in the rocks laid down during this period. The ratio of sulfur isotopes reveals the geological processes and biological fractionations that were then in operation.

Element of life

Sulfur in the human body	
BLOOD	1,800 p.p.m. (0.18%)
BONE	varies 500–2,400 p.p.m. (i.e. 0.05–0.24%)
TISSUE	varies 5,000–12,000 p.p.m. (0.5–1.2%)
TOTAL AMOUNT IN BODY	140 g.

All living things need sulfur and it is especially important for humans because it is part of the amino acid methionine which is an absolute dietary requirement—see below. Other sulfur-containing amino acids such as cysteine can be made within the body as required. Plant roots extract sulfate ions, SO_4^{2-}, from the soil and reduce this to sulfide, S^{2-}, before incorporating it into amino acids. Cysteine is needed to link peptide chains by forming sulfur-to-sulfur bonds between them. It is also abundant in the single-stranded protein keratin which is the main constituent of hair and nails. Eggs contain a lot of cysteine because feathers particularly need this amino acid.

Metalloproteins, which bind metals, rely on sulfur atoms to attach themselves to essential metals such as iron, copper, zinc, and molybdenum. They also bind to harmful metals such as cadmium which can then be safely locked away in the liver to await disposal.

Food element

The average person takes in around 900 mg of sulfur per day, mainly in the form of protein. Not all protein foods contain the essential amino acid, methionine, although all fish and meats do, and these can contain up to 4 percent. Plant protein can have as much as 2 percent, but many plant proteins may be deficient in it, such as those of peanuts, lentils, and kidney beans. However, soybeans contain it.

Some sulfur in food and drink comes in the form of preservatives based on sulfur dioxide (SO_2) or its derivatives, the sulfites (SO_3^{2-}) which kill bacteria and prevent food from going brown; in particular SO_2 in solution is widely used with fruits and vegetables because it preserves their natural colour. For example, peeled potatoes stay white.

When grapes are crushed, the juice will spontaneously ferment due to the action of wild yeasts on the surface of the fruit. As little as 100 p.p.m. of SO_2 is enough to prevent undesirable yeasts from multiplying while allowing desirable yeasts to flourish. These are often added as specially cultured yeasts and they can survive high levels of sulfur dioxide.

Wines may also be treated with more SO_2 just before bottling, to prevent further fermentation, and this can result in levels as high as 350 mg per bottle. Most of this sulfur dioxide eventually reacts with other components in the wine and disappears, but some young white wines have noticeable amounts of up to 50 mg per bottle, and this can cause problems—see *Medical element.*

Sulfur plays a part in other dietary components. Vitamin B_1, also known as thiamine, has a ring structure incorporating a sulfur atom as part of its molecule. A deficiency of this vitamin leads to the disease beriberi. The co-enzyme biotin also contains sulfur, as do natural flavour agents that give a characteristic aroma to cooked lamb, passion fruit, roasted coffee, and peanuts. The flavour and smell of onions, and garlic arises from the sulfur molecules di-2-propenyldisulfide ($CH_2=CHCH_2SSCH_2CH=CH_2$) which is in garlic, and propanethial S-oxide ($CH_3CH_2C=S=O$) which is in onions.

The synthetic sweeteners, saccharin, cyclamate, and acesulfam are compounds in which sulfur is a component. In each case it is bonded to a nitrogen atom, a combination that is critical to activating the sweetness receptors of the tongue. These molecules pass through the body undigested.

Dangerous element

Elemental sulfur is not toxic, but the gases sulfur dioxide and hydrogen sulfide, H_2S, are toxic, with the latter being dangerous even in small doses. It is particularly insidious because it deadens the sense of smell so that a person might be unaware of the danger it poses. Sulfur dioxide on the other hand is easily recognised by its choking smell which irritates the nose and lungs. It can trigger an attack of asthma in those who are sensitive to it, and their bodies respond by releasing histamines which are responsible for triggering the various unpleasant symptoms such as nausea, diarrhoea, headaches, and skin rashes.

Almost all volatile sulfur molecules are unpleasant, in some cases notably so—see box.

Element of bad smells

If something smells truly awful then it probably contains sulfur and this is exemplified by skunk odour, which arises from several sulfur compounds, namely 2-butene-1-thiol, $CH_3CH=CHCH_2SH$, 3-methyl-1-butanethiol $(CH_3)_2CHCH_2CH_2SH$, 2-butenyl thioacetate $CH_3CH=CHCH_2SCOCH_3$, 3-methyl-1-butanethioacetate $(CH_3)_2CHCH_2CH_2SCOCH_3$, 2-phenylethanethiol $C_6H_5CH_2CH_2SH$, and 2-quinolinemethanethiol $C_9H_6NCH_2SH$.

Less repulsive, but socially dissociating, are the sulfur chemicals of halitosis (bad breath) and this is also caused by three sulfur-containing molecules: hydrogen sulfide H_2S, methyl mercaptan CH_3SH, and dimethyl sulfide $(CH_3)_2S$. The amounts of these gases which we can detect are tiny: hydrogen sulfide can be detected at only 5 p.p.m in air;[186] dimethyl sulfide at even less, 1 p.p.m; but the world's smelliest

[186] Hydrogen sulfide is said to smell of rotten eggs and indeed these give off this gas, which is formed by bacteria decomposing the methionine in the yolks.

molecule is methyl mercaptan which requires only 0.02 p.p.m to register with the nose. These gases are produced in the mouth from gram negative bacteria acting on cysteine and methionine in food particles or even attacking the lining of the mouth itself. Hydrogen sulfide gas is also released from proteins by bacteria in the intestines.

The titan ariam lily (*Amorphophallus titanium*) only flowers once every four years. It produces an attractive bloom but a repulsive smell, at least as far as humans are concerned. However, a species of bee finds it irresistible and the plant relies on this to carry out its pollination. The smell comes from dimethyl disulfide, CH_3SSCH_3, and dimethyl trisulfide, CH_3SSSCH_3, and this has been described as the aroma of rotting fish. In Indonesia the lily is known as the corpse plant.

Medical element

The use of sulfur in medicine goes back to prehistoric times and it continued to be prescribed for nearly 3,000 years. A popular remedy for digestive upsets was brimstone and treacle. This was a laxative which worked by the sulfur being converted to alkali sulfides in the intestine, and these acting as an irritant. Sulfur could also be taken in tablet form, the dose being 1–4 g. Other sulfur preparations included ointments mixed with wool fat, for external application to treat acne or scabies, a parasitic infection. It was also applied as a solution in organic solvents, such as chloroform or ether, in which it is soluble.

Sulfur medicaments are no longer used, although sulfur atoms are incorporated into the molecules of modern drugs, such as penicillin, which is a natural sulfur compound. Its forerunners, the sulfa drugs, started life as dyes but were found to have antibiotic properties when tested on rabbits and were then used on humans.

Element of history

Sulfur is mentioned 15 times in the *Bible*, and was best known for destroying the twin cities of Sodom and Gomorrah. It was also known to the ancient Greeks, and in Homer's *Odyssey* the hero Odysseus says: 'bring me sulfur, old nurse, that I may purify the house.' This is a reference to sulfur which was burnt as a fumigant.

The Romans mined sulfur in Sicily and used it for sulfur matches (which are easily lighted by touching them against a glowing ember). It was also used for bleaching cloth and making wine, both of which involved burning it to form sulfur dioxide, and allowing this to be absorbed by wet clothes or the grape juice.

Sulfur, along with mercury, was believed to be a component of all metals, a theory put forward by the famous alchemist Jabir ibn Hayyan, aka Gerber (*ca.* 721–815). It followed therefore that, given the right conditions, it should be possible to transmute one metal into another, the result being centuries of futile attempts by alchemists to find the Philosopher's Stone that would turn base metals into gold.

In the 1700s, Antoine Lavoisier (1743–94) thought that sulfur was an element, but in 1808 Humphry Davy (1778–1829) investigated it and found it contained hydrogen, which appeared to put paid to the idea. However, his sample was impure and, when Louis-Josef Gay-Lussac (1778–1850) and Louis-Jacques Thénard (1777–1857) proved it to be an element the following year, Davy obtained a better sample and convinced himself that it truly was an element.

Element of war

Sulfur was probably a component of Greek fire, the liquid combustible agent which the Byzantine Empire[187] used successfully for hundreds of years and which terrorised opponents. It was said to set the sea on fire and was usually projected at attacking ships like a flame thrower. It was reputed not to be extinguishable once it had started to burn. The exact composition of Greek fire remained a closely guarded secret, and no record now exists of how it was made. It is thought to have been sulfur dissolved in a mixture of hydrocarbons, these being distilled from rock-oil, the black tarry material that oozed from the ground in the region between the Black and Caspian seas.

Gunpowder, likewise was a closely guarded Chinese secret when it was discovered around 950 AD, and it too was made from sulfur. However, within 250 years it was known in Europe, and its recipe was revealed by Roger Bacon (1214–92) an English Franciscan monk. His recipe had charcoal, sulfur and potassium nitrate in the weight ratio of five parts to five parts to seven parts (approximately 30 percent to 30 percent to 40 percent), and these had to be finely ground together. Today the blend is charcoal (15 percent), sulfur (10 percent) and potassium nitrate (75 percent). Gunpowder only explodes if it is ignited in a confined space where the carbon dioxide and sulfur dioxide gases produced in the chemical reaction expand rapidly and can be used to project a bullet or cannon ball. Gunpowder is still the basis of firecrackers.

The mustard gas of World War I was the sulfur compound bis(2-chloroethyl) sulfide, $ClCH_2CH_2SCH_2CH_2Cl$, which was also a blistering agent as well as giving off a highly toxic vapour. Curiously, the antidote developed to protect troops against arsenical war gases is also a sulfur compound known as British Anti-Lewisite, aka BAL, $HSCH_2CH(SH)CH_2OH$. This is still used to treat people suffering from heavy metal poisoning because its sulfur atoms will bind to such metals and the resulting molecule is easily excreted.

Economic element

Sulfur occurs naturally near volcanoes. This is why Sicily supplied much of the sulfur for the ancient world. Its sulfur mines were still employing more than 50,000 men in the 1800s. The earliest way of processing it was to spread the mineral on sloping slabs and set fire to it at the top. As it burned it melted the sulfur below which then ran off and was collected, but the process wasted about half the sulfur.

Native sulfur occurs naturally as massive deposits in Texas and Louisiana in the USA. The Louisiana deposit was discovered in 1865, during drilling for oil. A thick layer occurs at a depth of 150 metres and it extended into Texas. Extraction proved difficult until Herman Frasch (1851–1914) studied the problem in the 1890s. He decided

---

[187] This lasted for almost a thousand years from 476 to 1461 AD.

that the only way to get it out was to melt it and pump it to the surface as a liquid. Sulfur melts at 113°C and Frasch decided to use superheated steam for the purpose. This was pumped down into the seam of sulfur and up came molten sulfur. This method of extraction became known as the Frasch process.

There are many sulfide minerals: pyrites and marcasite are iron sulfides; stibnite is antimony sulfide; galena is lead sulfide; cinnabar is mercury sulfide; and sphalerite is zinc sulfide. A lot of sulfur occurs as sulfates, the most abundant ones being anhydrite and gypsum, which are calcium sulfates, and epsomite and kiesterite which are magnesium sulfates. Other sulfate minerals are barite (barium sulfate) and celestite (strontium sulfate).

The chief source of sulfur for industry is the hydrogen sulfide of natural gas, oil, and tar sands such as the Athabasca Oil Sands of Alberta, Canada. It is imperative that the H_2S be removed before use. Removal from natural gas is achieved by scrubbing the gas with a solvent, such as ethanolamine, in which H_2S is soluble and then reclaiming this, before reacting it with a measured supply of oxygen. This reacts to form water and leaves the sulfur behind, which can then be used as chemical feedstock.

Sulfur comes also from mined sulfur, and from pyrites (FeS), the roasting of which at 1,200°C in the absence of air releases about half its sulfur. World production of sulfur exceeds 70 million tonnes a year, and reserves are of the order of 2.5 billion tonnes. Most sulfur, 65 million tonnes, is recovered from fossil fuels and the main producers are the USA (8.2 million tonnes), Canada (7.8 million tonnes), Russia (6.4 million tonnes), China (4.3 million tonnes), Saudi Arabia (2.9 million tonnes), and South Korea (2.8 million tonnes).

Applications for sulfur are very varied, as in vulcanising rubber; in corrosion resistant concrete which has greater strength and is frost resistant; and in gunpowder for fireworks. It can also end up in solvents such as carbon disulfide and dimethylsulfoxide (aka DMSO, $(CH_3)_2SO$) and in a host of other products of the chemical and pharmaceutical industries.

Charles Goodyear (1800–60) accidentally discovered the benefits of reacting sulfur with natural rubber, thereby converting this latex polymer from a sticky, soft material of limited use, to something of wide application suitable for hundreds of uses but especially tyres for vehicles.[188]

Most sulfur is converted to sulfur dioxide gas and most sulfur dioxide ends up as sulfuric acid, a key industrial chemical. The former gas is made in various ways: by burning sulfur or hydrogen sulfide, roasting metal sulfides in air, and some is even reclaimed from waste sulfuric acid and sulfates. Most sulfur dioxide goes into making sulfur trioxide, by heating it at 420°C with oxygen in the presence of a catalyst such as vanadium(V) oxide, and reaction of this with water produces sulfuric acid.

Sulfuric acid is the world's number 1 industrial chemical. It is needed for the making of many other products, the major ones being phosphate fertilizers, and the removal of rust from iron and steel (known as 'pickling') prior to their being painted.

---

[188] The term vulcanisation was suggested because Goodyear made his discovery when he placed a sample of rubber mixed with sulfur on a hot stove and his friend William Brockedon suggested he called the new material after Vulcan, the Roman god of fire.

The chemical industry uses sulfuric acid in the production of explosives, fuels, rayon, cellophane, paints, enamels, paper, detergents, dyestuffs, antifreeze, and detergents. The original surfactants that were introduced into detergents in the 1940s were derivatives of sulfuric acid in which one of the acid groups of H_2SO_4 was replaced by an organic group and the other acid group neutralised as the sodium salt. They made much better cleaning agents than traditional soaps because they did not form soap scum in hard water.

Most sulfuric acid (60 percent) is used to manufacture phosphoric acid, and most of that is used to make fertilizer phosphates. However, the use of sulfuric acid is changing. In some chemical plants it is manufactured from the sulfur dioxide released when copper sulfide ores as refined, whereas in other mining plants sulfuric acid is used to extract copper from its oxide ores.

Methyl mercaptan (aka methanethiol, CH_3SH) is the simplest member of a series of compounds in which there can be chains of up to 20 carbon atoms attached to sulfur. Mercaptans with three and four carbons are those we encounter when we smell a leak of gas because this is added purposely to odorise the gas. The one with 18 carbons is a wax used in silver polishes. Methyl mercaptan is used in the production of pesticides and herbicides and also to make methionine for fortifying animal feeds to increase the amount of this amino acid in their meat and milk.

Another gaseous sulfur compound is sulfur hexafluoride, SF_6, which is very stable and can be heated at 500°C without any decomposition, and is not even affected by molten sodium.[189] Its stability is the reason the gas is used in high voltage apparatus as an insulating gas because it does not break down even when exposed to electric sparks—but see *Environmental element* below.

Environmental element

Sulfur in the environment	
EARTH'S CRUST	260 p.p.m.
	Sulfur is the 17th most abundant element.
SOILS	major component.
SEAWATER	870 p.p.m.
ATMOSPHERE	various gases, totalling about 1 p.p.b.

In the main compartments of the environment the amounts of sulfur are as follows: in the soil on land there are 3 billion tonnes as sulfate, and 600 million tonnes in things living in it and on it. In the seas there are 1.3 million billion tonnes as dissolved sulfate, and 24 million tonnes in living things. In the atmosphere there are around 3.5 billion tonnes of gaseous sulfur compounds and these are continually being added to and removed.

There are many sulfur-containing compounds in the air. The most abundant one is carbon oxygen sulfide (COS) which comes mainly from swamps and resists oxi-

---

[189] The purity of SF_6 is checked by exposing albino mice to an atmosphere made up of 21% oxygen and 79% SF_6 for 24 hours. If the gas is pure they come to no harm.

dation. Its concentration is 0.5 p.p.b. Sulfur dioxide, on the other hand, is emitted in vast quantities but is rapidly washed out of the atmosphere by rain and its concentration is 0.2 p.p.b. Other volatiles are hydrogen sulfide and carbon disulfide, which also come from swamps, and dimethyl sulfide, which comes from the oceans. These are easily oxidised and washed out of the air so their levels are only about 0.01 p.p.b. each.

Sulfur as sulfur dioxide is emitted in large volumes from volcanoes, but this is now matched by that emitted by human activity, mainly from the burning of coal and heating oil, and these are responsible for acid rain. Up to 5 percent of the weight of a fossil fuel may be sulfur, and coal contains the most. Over 300 million tonnes of sulfur dioxide are released to the Earth's atmosphere each year from this source.

Sulfur dioxide from power stations can be removed by 'scrubbing' the flue gases with sprays of water and neutralising the acid with lime, but this is an expensive answer to the problem. As an alternative, the coal can be converted to synthesis gas, which is a mixture of carbon monoxide and hydrogen, leaving the sulfur behind, and these are burned separately to release their energy. Alternatively, powdered coal can be combusted on a fluidised-bed along with powdered limestone ($CaCO_3$) which traps the sulfur as calcium sulfate. However, there have been concerns expressed that in parts of Europe where soils were once replenished with sulfur via acid rain, that sulfur is being depleted and might eventually have to be added via sulfur-enhanced fertilizers.

Sulfur in various forms passes endlessly between soil, the atmosphere, and the sea. It is absorbed and released by all living things. Most organisms take in sulfur as sulfate and reduce this to sulfide so that it can be incorporated into the amino acids methionine and cysteine needed for protein.

Marine algae take in sulfate and convert it to dimethylsulfonium propionate, $(CH_3)_2S^+CH_2CH_2CO_2^-$, which they use to maintain their osmotic balance with seawater. When the algae die, or are eaten, this compound is converted to dimethyl sulfide which is released from the oceans in vast quantities (transferring as much as 50 million tonnes of sulfur a year to the atmosphere). Phytoplankton release dimethylsulfonium propionate and this attracts other sea creatures such as bacteria and zooplankton which consume it as a source of carbon and sulfur, and it is these which produce dimethyl sulfide as a by-product. In the atmosphere this is converted to dimethyl sulfonic acid by reacting with hydroxy radicals (OH) and then to sulfur dioxide and eventually to sulfuric acid, which may act as a planetary regulator by promoting cloud formation.

The bacteria *Pyrobaculum islandicum* live in hot springs and it is these sulfur-loving bacterium which laid down the vast deposits of uncombined sulfur during the Permian and Jurassic periods between 280 and 130 billion years ago. At the bottom of shallow seas conditions favoured these anaerobic bacteria which convert metal sulfates to metals sulfides and then to elemental sulfur.

A recent addition to the atmosphere has been trifluoromethyl sulfur pentafluoride, CF_3SF_5, which was first detected in 2000 and is a potent greenhouse gas, as is SF_6. Its source is unknown but probably produced from SF_6 by the electronics industries. Both have potentially long life-times in the atmosphere because they are so stable. The level of SF_6 is of the order of 5 p.p.t and of CF_3SF_5 is about 0.25 p.p.t.

Chemical element

Data file	
CHEMICAL SYMBOL	S
ATOMIC NUMBER	16
ATOMIC WEIGHT	32.066
MELTING POINT	113°C
BOILING POINT	444°C
DENSITY	2.1 g/cm³
OXIDES	SO_2 and SO_3

Sulfur is a non-metal and comes below oxygen in group 16 (row 3p) of the periodic table. This element is notable for its ability to bond to itself to form long chains and rings of up to 20 atoms. The most common form of sulfur consists of rings of eight atoms (S_8). When this is melted and then heated to around 175°C the rings open and join together to form long chains making the liquid very viscous and the solid obtained from it on cooling is like rubber and known as plastic sulfur. When sulfur boils it deposits as fine yellow orthorhombic crystals of S_8 rings and this is the most recognisable form of the element, and these are called flowers of sulfur. Sulfur is stable to air and water, but burns if heated. It is attacked by oxidising acids.

The chemistry of sulfur is complex because of its several stable oxidation states in which it can exist, extending from −2, as in H_2S and the S^{2-} ions of metal sulfides, to +6 as in H_2SO_4, sulfates, and SF_6. Sulfur dioxide and sulfites are in oxidation state +4 and they can act as reducing agents thereby becoming oxidised to sulfate.

There are 24 known isotopes of sulfur with masses 26 to 49, and one known isomer. There are four naturally occurring main isotopes of sulfur: sulfur-32, which makes up 95 percent, sulfur-33 (0.8 percent), sulfur-34 (4.2 percent), and sulfur-36 (0.02 percent). None is radioactive. There are also traces of radioactive sulfur-35 produced in the upper atmosphere by the spallation of argon by cosmic rays.

The longest-lived radioactive isotope is sulfur-35 which has a half-life of 87.5 days.

Element of surprises

1. A Venezuelan plant related to the mimosa produces carbon disulfide, CS_2, which it releases from its roots to protect them against nematodes (parasitic worms) and fungi.

2. When mice were placed in a chamber in which the air contained 80 p.p.m. of hydrogen sulfide, their oxygen consumption dropped dramatically and within hours their metabolic rate had fallen by 90 percent and their body temperature dropped to 15°C, just as in hibernating animals. When exposed to normal air,

the mice quickly returned to normal and were unaffected by the experience. The H_2S blocks the same receptors that O_2 uses and by which the body produces energy.

3. The release and tracking of SF_6 gas was used in 2007 at St John's Wood London Underground station to simulate what would happen were a nerve gas to be released during a terrorist attack.

Tantalum

Pronounced tan-ta-lum, the name is derived from the name of the legendary Greek figure, Tantalus, King of Sisyphus in Lydia.

Tantalus stole the food of the gods, thereby becoming immortal, and he then invited them to a feast at which he served up the flesh of his son, Pelops. The gods were not amused and as a punishment they decreed that Tantalus was to be tormented forever by thirst and hunger. In the Underworld he was to stand up to his neck in water, which receded whenever he tried to drink, and to be surrounded by the branches of vines heavy with ripe grapes, that swayed out of reach as he stretched to pick them. The discoverer of this element evidently felt similarly tormented by its elusive nature and named it accordingly.

French *tantale*; German *Tantal*; Italian *tantalio*; Spanish *tántalo*; Portuguese *tântalo*.

Essential element

Despite its difficulty of extraction and its high cost, tantalum has several niche applications, mostly industrial, but some very personal such as surgery implants. And mobile phones became less bulky thanks to tantalum.

Element of life

Tantalum in the human body	
BLOOD	not known, but low.
BONE	av. 30 p.p.b.
TISSUE	not known, but low.
TOTAL AMOUNT IN BODY	about. 0.2 mg.

Tantalum has no biological role to play, and it appears to be non-toxic. When rats were given doses of tantalum oxide (Ta_2O_5) it passed through them completely without effect and without any being absorbed. Radioactive tantalum-182 (half-life 114 days) showed that when the element was injected into the body it concentrated in the bones and kidneys. There are only a few cases of industrial injury caused by tantalum or its compounds and these were mainly skin rashes.

Medical element

Tantalum metal is well tolerated by the body and consequently has been used in many ways during surgery: as plates in the repair of skull fractures; as bolts to fasten broken bones; and as wire for sutures to mend torn ligaments. A few people who had tantalum implants have suffered skin rashes, but these cleared up when the implants were removed.

Tantalum is used for equipment in the pharmaceutical industries because it does not contaminate the materials it comes into contact with, and it allows the same apparatus to be used for many applications.

Some tantalum compounds are used in diagnostic medicine because it is more visible on X-rays than other agents, especially when examining the lungs.

Element of history

Tantalum was reported as a new metal in 1802 by the Swedish chemist Anders Gustav Ekeberg (1767–1813), who was Professor of Chemistry at Uppsala University. However, when William Wollaston (1766–1828), the eminent British chemist, analysed minerals from which it had been extracted he declared it was identical to the new metal niobium (then known as columbium) which has been discovered the year previously. These two elements often occur together and are chemically very similar, which makes them difficult to separate even today, and especially so by the methods available two centuries ago. It was as a result of their similarity that there was confusion regarding their identification.

More than 40 years were to pass before Heinrich Rose (1795–1864) separated tantalum and niobium and proved conclusively that they were different. That was in 1846. It was a remarkable achievement, and yet *pure* tantalum was not produced until Werner von Bolton of Charlottenburg finally accomplished it in 1903.

No one now doubts that Ekeberg had isolated tantalum and so was the discoverer of a new element, thanks to his exceptional skills as a chemist, yet even his samples must have contained quite a lot of niobium.

Economic element

Occasionally, tantalum is found as the uncombined metal. The chief tantalum ores are tantalite,[190] which also contains iron, manganese and niobium, and samarskite, which contains seven metals. Another mineral which contains both tantalum and niobium is pyrochlore which occurs as large deposits in Canada and Brazil.

The amount of tantalum in ores that are mined is calculated to be as high as 70,000 tonnes per year, most of which is from Brazil (90 percent), Canada, and Australia, with smaller amounts being produced in China and Rwanda. A lot is obtained as a by-product of tin extraction, and some tin slag can contain as much as 10 percent of tantalum. The actual demand for tantalum metal itself is around 1,200 tonnes a year, and reserves total around 150,000 tonnes which may be an underestimate because deposits in Australia are exceptionally large.

The difficulty in refining tantalum is separating it from niobium, and this is done by converting both metals to their potassium fluoro-derivatives which can be separated by solvent extraction or ion exchange. Tantalum is then reclaimed by the electrolysis of the potassium fluorotantalate, or by reacting this salt with sodium metal.

Tantalum finds use in four areas: in high temperature applications; in electrical devices; as surgical implants; and for handling corrosive chemicals. More would

---

[190] If the amount of niobium exceeds that of tantalum the ore is generally known as columbite.

be used but for its high cost. It is rarely used as an alloying agent because it tends to make metals brittle, although with some metals it has a dramatic strengthening effects and it is used in turbine blades, rocket nozzles, heat shields, and nose caps for supersonic aeroplanes.

The metal is easily worked cold by rolling, forging, or drawing into wire. The wire it makes is very strong. The metal became available commercially in 1922, and it was soon being used to make all kinds of things, from corrosion-resistant laboratory apparatus to jewellery. Accurate weights for analytical chemistry were made from tantalum, which was seen as a suitable replacement for platinum.

Tantalum resists corrosion and is almost impervious to chemical attack. For this reason it has been employed in the chemical industry, for example, for heat exchangers in boilers where strong acids are vaporised. It is ideal for making parts for vacuum furnaces. It is possible to coat other metals with tantalum and even a layer as thin as 100 microns will provide excellent corrosion resistance. The oxide layer which forms on the surface of the metal consists of tightly packed atoms, not only protecting the bulk of the metal, but also acting as an insulating (dielectric) layer, which is why a major use of tantalum is as very thin capacitors, which are a part of personal equipment like mobile phones.

Other ways in which tantalum has been employed are as the electrodes in neon lights, in AC/DC rectifiers, for the glass of special lenses, and for the spinneret dies through which cellulose solutions are forced in the process of making rayon fibres.

Tantalum oxide makes glass with a high refractive index and this is ideal for camera lenses.

Environmental element

Tantalum in the environment	
EARTH'S CRUST	2 p.p.m.
	Tantalum is the 51st most abundant element.
SOILS	*av.* 1–2 p.p.b.
SEAWATER	2 p.p.t.
ATMOSPHERE	nil.

Because tantalum oxide is so insoluble there is almost none of this element to be found in natural waters. Few attempts have been made to measure its level in soils and what work has been done suggests a range of from 0.1 to 3 p.p.m., with most soils being at the lower end. Only tiny amounts of tantalum are taken up by plants and vegetation rarely exceeds 5 p.p.b., and the same goes for food plants, with *ca.*, 0.5 p.p.b. (fresh weight).

Chemical element

Data file	
CHEMICAL SYMBOL	Ta
ATOMIC NUMBER	73
ATOMIC WEIGHT	180.9479
MELTING POINT	2,996°C
BOILING POINT	5,425°C
DENSITY	16.7 g/cm^3
OXIDES	TaO_2 and Ta_2O_5

Tantalum is a shiny, silvery metal which is soft when pure and it is a member of group 5 (row 5d) of the periodic table. Tantalum is unaffected by aqua regia, but it is attacked by hydrofluoric acid and molten alkalis. Its preferred oxidation state is +5 as in TaF_5 and tantalate salts such as $LaTaO_4$ in which there is the TaO_4^{3-} ion, but other oxidation states are known such as +3 as in $TaCl_3$, and +4 as in the lower oxide TaO_2 and in compounds like $TaCl_4$.

There are 32 known isotopes of tantalum with mass numbers 155 to 186, and 37 known isomers. Tantalum is predominantly the isotope tantalum-181, which accounts for 99.99 percent, the remainder being the isomer tantalum-180m which is weakly radioactive with a half-life of more than 1.2 million billion years (1.2×10^{15})—see below. The ground-state isomer, tantalum-180, has a half-life of only 8.1 days. The next longest-lived radioactive isotope is tantalum-179 with a half-life of 1.82 years.

The existence of tantalum-180m as a naturally occurring primordial isomer has stimulated much enthusiasm among some physicists who theorise that deformed long-lived isomers of transfermium elements might occur naturally on planet Earth—see roentgenium and un-named elements for details.

Element of surprises

1. Wars have been fought over tantalum. Demand for the metal increased with the use of mobile phones to such an extent that, in the late 1990s and early 2000s, there was conflict between states in central Africa over the ore deposits which straddled their borders. Eventually the UN intervened and suggested an embargo on ore from the region. However, the Security Council decided not to enforce it because of the hardship it would cause to local miners and their families.

2. A form of tantalum carbide, TaC, has been made at the Los Alamos National Laboratories, in the USA, which is even harder than diamond. TaC has a melting point of 3,738°C and it is used in special cutting tools.

Technetium

Pronounced tek-nee-shee-um, it is named from the Greek word *tekhnetos* meaning artificial. It might well have been called masurium—see *Element of history* below.

French *technétium*; German *Technetium*; Italian *tecneto*; Spanish *tecnecio*; Portuguese *tecnécio*.

Essential element

The medical diagnosis of difficult-to-detect cancer is carried out using the radioactive isomer technetium-99m, and this technique is performed on tens of millions of patients around the world every year.

Cosmic element

The longest-lived isotope of technetium has a half-life of 4.2 million years, which means than any technetium around when the Earth formed 4.57 billion years ago would be long gone. All the more surprising when, in 1952, Paul Merrill (1887–1961), an astronomer at the Mount Wilson and Palomar Observatories in California examined the light given off by stars known as red giants, and found that the spectra contained lines showing that the stars were rich in technetium. Because all isotopes of technetium are relatively short-lived compared to the age of stars, and especially compared to the age of red giants which are stars at the end of their lives, it could only mean that this element was being formed within the star itself. This proved that the stars are the furnaces in which the chemical elements are being manufactured.

Element of life

The total amount of technetium naturally in the human body, and in all living cells, is zero. Indeed all the technetium present when the Earth formed would have disappeared before life emerged on this planet, so there can never have been a biological role for it to play.

Medical element

The isotope isomer technetium-99m is used in medical diagnosis. It is produced by the radioactive isotope molybdenum-99, which can be extracted from nuclear fission products, but it is mostly made by bombarding specially designed targets of highly enriched uranium-235 with neutrons, and this is how it is made at Chalk River, Ontario, Canada, which for more than 50 years has supplied two thirds of the global demand. It is also produced in other countries such as the Netherlands, South Africa, and Argentina, and in these last two countries it is made from less enriched uranium-235 and this method is now being encouraged and promoted in the USA.

Molybdenum-99 decays with a half-life of 67 hours to form technetium-99m, and the two elements are easily separated. Despite its short half-life, supplies of molybdenum-99 are sufficient to deal with a crisis in production of technetium-99m as happened in 2009 as a result of a power blackout in Ontario, which affected the Canadian nuclear facility for several weeks.

Technetium-99m has a half-life of six hours and emits only gamma-rays as it reverts to technetium-99. It is used in a medical diagnostic technique known as immunoscintigraphy, in which it is incorporated into a monoclonal antibody that has sites capable of binding to cancer cells. By mapping the gamma-ray emission it is possible to diagnose the disease within a few hours, and the method is especially useful is diagnosing difficult-to-detect cancers in the intestine, as well as cancers of the kidney, brain, and bone.

When technetium-99m, together with a tin compound, is injected into the body it binds to red blood cells and can be used to map disorders associated with the circulatory system. When it is accompanied by a diphosphate ion it binds strongly to the heart muscle and this can be used to assess the damage done by a heart attack.

The by-product of technetium-99m treatment is technetium-99 which is quickly eliminated from the human body, but in any case its long half-life of 212,000 years means that it adds only a little to the body's burden of naturally occurring radioactive isotopes.

Element of history

Technetium occupies the place below manganese in the periodic table, and it long tantalised chemists because it could not be found, despite reports to the contrary. Some scientists claimed to have discovered it and gave it names such as davyum, lucium and nipponium, but none of the claims was substantiated. A report in 1925 of its discovery by three German chemists, Ida Tacke (1896–1978), Walter Noddack, (1893–1960), and Otto Berg (1873–1939),[191] appeared to have more credibility than most. They named it masurium (symbol Ma) after that part of East Prussia which was Noddack's homeland. 'Masurium' was detected by bombarding a sample of the mineral columbite with a beam of electrons and from the faint X-rays that were emitted, the presence of element 43 was deduced. It has since been assumed that they were wrong in finding technetium, although it now appears they might have been right.

Their method of analysis was repeated in 1999 by David Curtis of the Los Alamos National Laboratory in the USA, using a sample of uranium from a Canadian ore deposit. A few of the uranium atoms that undergo spontaneous nuclear fission will produce technetium and Curtis was able to detect and estimate that there was about 1 nanogram (1×10^{-9} g) of technetium per kg of uranium. Moreover, he was able to show that the method used by the German workers would also have been sensitive enough to detect these small amounts. As columbite has been known to contain as

---

[191] Ida Tacke eventually became Mrs Noddack.

much as 10 percent uranium, there must also be some technetium atoms in this mineral. Tacke, Noddack and Berg were probably the first chemists to find technetium.

Technetium really only came to light in 1937 when Emilio Segrè (1905–89) and Carlo Perrier (1886–1948) at the University of Palermo in Sicily separated it from a sample of molybdenum. The discovery of element 43 resulted from an unusual sequence of events, with Ernest Lawrence unknowingly making it possible; it did not result from a planned experiment to discovery it. In 1936 Segrè visited the University of California and met Lawrence. He was shown the 37-inch cyclotron and while he was there, Segrè asked for and was given pieces of the irradiated scrap metal that were laying about so that he could examine them when he returned to Italy. After returning, in early 1937 he received another piece of irradiated metal which was the molybdenum foil that had been part of the cyclotron's deflector material. Lawrence sent it because he thought that Segrè would like to see what it contained. It had been a deflector during many experiments and had undergone a lot of deuteron bombardment. It was from this that Segrè first separated technetium.[192]

Economic element

Technetium is rarely encountered outside nuclear facilities or research laboratories, apart from that needed for medical diagnosis. World production of technetium runs to many tonnes because it is extracted from spent nuclear fuel rods. However, only tiny amounts of this have any commercial use. One isotope, technetium-95m, with a half-life of 61 days, is used as a source of gamma rays.

Environmental element

There used to be only traces technetium on the Earth, but today there is much more of this element which is produced in tonne quantities in nuclear reactors, so it is adding to the planetary burden of unwanted radioactive waste. So far, nuclear reactors have produced about 80 tonnes of the metal and above-ground nuclear weapons testing from 1945 to 1980 created and released about a tonne of technetium to the environment.

Chemical element

Data file	
CHEMICAL SYMBOL	Tc
ATOMIC NUMBER	43
ATOMIC WEIGHT	98
MELTING POINT	2,172°C
BOILING POINT	4,877°C
DENSITY	estimated to be 11.5 g/cm³
OXIDES	TcO_2 and Tc_2O_7

[192] Segrè was dismissed from his academic post in 1938 because he was opposed to Mussolini's fascist regime, so he emigrated to the USA.

Technetium is a radioactive metal that is unusual in being a relatively light element (atomic number 43), yet having no stable isotopes. It is a member of group 7 (row 4d) of the periodic table. As the bulk metal it is silvery, but it is more commonly obtained as a grey powder. Technetium resists oxidation but slowly tarnishes in moist air, and burns in oxygen when in powder form. It dissolves in nitric and sulfuric acids, but not in hydrochloric acid. It exhibits oxidation states +4, +5, and +7, the last of these as the pertechnetate ion TcO_4^-. Like its fellow element in group 7, rhenium, it also forms the curious hydride ion TcH_9^{2-}.

There are 36 known isotopes of technetium of mass numbers 85 to 120, and 22 known isomers. The longest-lived isotopes are technetium-98 with a half-life of 4.2 million years, and technetium-97 with a half-life of 2.6 million years. Technetium occurs naturally in uranium ores, where it is produced by spontaneous fission, but only in trace amounts. It was first identified in the uranium ore pitchblende from the Congo in 1962 where is was present at a concentration of 0.2 micrograms per tonne.

Element of surprise

The best way to protect steel that is in contact with aerated water is to dissolve a little potassium pertechnetate ($KTcO_4$) in the solution: 55 p.p.m. is sufficient to prevent any corrosion, even when the system is under pressure and the temperature is 250°C. Remarkable as this protection is, potassium pertechnetate can be used only in closed systems because all its isotopes are radioactive.

Tellurium

Pronounced tel-oor-ee-um, the name is derived from the Latin *tellus*, meaning Earth. French *tellure*; German *Tellur*; Italian *tellurio*; Spanish *teluro*; Portuguese *telúrio*.

Essential element

Tellurium is an element to avoid if at all possible—on account of its socially isolating effects—although there are some applications for which it might have a future.

Cosmic element

Tellurium is much rarer on Earth than it is in the universe in general. The reason appears to be that as this planet was forming, and there was much more hydrogen around, this formed tellurium hydride (H_2Te) gas which was lost into space.

Element of life

Tellurium in the human body	
BLOOD	6 p.p.b.
BONE	no data
TISSUE	av. 15 p.p.b.
TOTAL AMOUNT IN BODY	about 0.7 mg (700 µg).

There is no known biological role for tellurium. Micro-organisms can absorb tellurium and then emit it in a volatile form by methylating it to dimethyl telluride, $Te(CH_3)_2$, hence it is detectable in air, especially near copper refineries. The tellurium which is ingested is absorbed into the blood stream and slowly excreted in the urine although some is converted to the obnoxious and volatile dimethyl tellurium which is then expelled from the lungs or excreted via the sweat glands.

Food element

Plants can take in tellurium from the soil and have been found with levels as high as 6 p.p.m. although few vegetable plants have more than 0.5 p.p.m. and most have less than 0.05 p.p.m. (fresh weight), with apples having the least of all. Onion and garlic, however, can have up to 300 p.p.m. (in this case recorded as their dry weight).

The daily intake for the average person is judged to be around 600 µg, most of which is absorbed and excreted in the urine, with most of the remainder passing through the gut unabsorbed, apart from around 10 micrograms (µg) which is exhaled on the breath as volatile dimethyl telluride or via the sweat glands.

In 1884, tests were carried out on volunteers who were given 0.5 µg of tellurium oxide, TeO_2, by mouth. In an hour it was detectable on their breath and the smell did not disappear for 30 hours. Some men, given doses of 15 milligrams (mg), were still found to have 'tellurium breath' eight months later.

Dangerous element

Elemental tellurium has low toxicity but unpleasant side effects, producing extremely bad breath and body odour. Those who worked with it often developed what was called 'tellurium breath' even when there was as little as 10 µg of the element per cubic metre of air. Those who come into contact with tellurium are advised to take extra vitamin C which reduces considerably the smell of tellurium on the breath.

As little as 2 grams of sodium tellurite (Na_2TeO_3) can be fatal and this was discovered accidentally in 1946 when three soldiers were given it from a mislabelled bottle as part of their medical treatment. Two of them died within six hours. Acute tellurium poisoning results in vomiting, inflammation of the gut, internal bleeding, and respiratory failure. Chronic poisoning produces garlic breath, tiredness, and indigestion.

Element of history

Tellurium was discovered in 1783 by Franz Joseph Müller von Reichenstein (1740–1825) at Sibiu, Romania. Müller, as he was known, had amassed a collection of rare minerals during his trips around Europe and was particularly intrigued by one which came from a mine near Zalatna in central Transylvania, Romania. It had a metallic lustre and he thought it was a sample of uncombined antimony. Preliminary tests showed that it was not antimony, and Müller announced that it was bismuth sulfide. However, further investigation showed neither of these two elements to be present, and he finally concluded that it was a compound of gold with an unknown element.[193]

For three years Müller researched it, often calling it *metallum problematum* or *aurum paradoxum* because of its puzzling nature. Finally he was convinced that it was a new element, and he sent a sample of the mineral to other chemists asking them to confirm his findings, which they did. He published his discovery in an obscure journal in Vienna.

In 1796, Müller sent a sample of the mineral to the eminent German chemist Martin Klaproth (1743–1817) in Berlin who also investigated it, confirmed it was a new element, obtained a pure sample and decided to call it tellurium. Rather strangely, this was not the first sample of tellurium to pass through his hands. In 1789, he had been sent some by a Hungarian scientist, Paul Kitaibel (1757–1817) who also claimed to be its discoverer. There is no doubt that Kitaibel had independ-

---

[193] The mineral was gold telluride, $AuTe_2$, and it is one of the few naturally occurring compounds of gold.

ently discovered it, and naturally he was somewhat upset when he read Klaproth's account of tellurium in which Müller was mentioned but he was not. A pained correspondence followed.

There is no doubt that Müller should get the credit for discovering tellurium although he seemed reluctant to have it widely known, hence his decision to publish news of it in a little-read journal. Because he had already once wrongly identified the mineral from which it came, this is perhaps understandable.

Economic element

Samples of uncombined tellurium are sometimes found, but they are extremely rare. There are some tellurium minerals, such as calaverite ($AuTe_2$), sylvanite ($AgAuTe_4$), and tellurite (TeO_2) but none is mined as a source of the element. Commercial tellurium is obtained from the anode slime of copper refining, which can contain as much as 8 percent. World production varies between 150 and 200 tonnes per year, coming from the USA (50 tonnes), Peru, Japan, and Canada. The reserves of this element have not been assessed but since demand is small it is unlikely there would ever be a shortage.

Tellurium is a semi-metal usually obtained as a grey powder by the electrolytic reduction of sodium tellurite solution. It is used in alloys with copper and stainless steel to improve their machinability, and it is added to lead (0.05 percent) to make it harder and more acid-resistant, especially that which is used in batteries. At various times, tellurium has been used to vulcanise rubber, to tint glass and ceramics, in electronic devices, and as an industrial catalysts in oil refining.

When added to rubber, tellurium speeds up the curing process and makes the product less susceptible to ageing and less likely to be affected by oil, which softens normal rubber. Tellurium is also used in photoreceptors and other microelectronic devices, in the form of bismuth telluride.

Cadmium telluride is used to make solar panels and indeed it has the highest efficiency for solar cell electricity generation. Cadmium telluride, in which some of the cadmium is replaced by zinc, is used in X-ray detectors.

Tellurium suboxide (TeO) is used in some types of rewritable CDs and DVDs.

Environmental element

Tellurium in the environment	
EARTH'S CRUST	5 p.p.b.
	Tellurium is the 72nd most abundant element.
SOILS	generally around 0.05 p.p.m., but can be as high as 30 p.p.m.
SEAWATER	0.15 p.p.t.
ATMOSPHERE	traces.

Tellurium is present in coals at up to 2 p.p.m. and this is probably the major release of this element to the environment.

Chemical element

Data file	
CHEMICAL SYMBOL	Te
ATOMIC NUMBER	52
ATOMIC WEIGHT	127.60
MELTING POINT	450°C
BOILING POINT	990°C
DENSITY	6.2 g/cm³
OXIDES	TeO, TeO$_2$, TeO$_3$

Bulk tellurium is a semi-metal, silvery-white and metallic-looking, but is usually available as a dark grey powder. It is a member of group 16 (row 5p) of the periodic table, where it lies on the boundary of metals and non-metals, and has properties of both. (It even occurs as a native metal.) As a metal it is brittle and easily ground to powder. Tellurium burns in air or oxygen, is unaffected by water or hydrochloric acid, but dissolves in nitric acid. Tellurium exhibits oxidation states −2 as in H_2Te, +2 which is rare, for example, $TeCl_2$, +4 as in $TeCl_4$, and the tellurite ion TeO_3^{2-}, which is the form encountered in solution, and +6 as in TeF_6, and the tellurate ion TeO_4^{2-} in solution. These last two oxidation states are the more usual ones.

There are 38 known isotopes of tellurium with mass numbers 105 to 143, and 17 known isomers. Tellurium has eight main naturally occurring isotopes: tellurium-130 is the most abundant accounting for 34.5 percent; it is radioactive with a half-life of 7.9×10^{20} years. The second most abundant is tellurium-128 (31.8 percent), and this too is radioactive and has the longest measured half-life of any element's radioactive isotope at 2.2×10^{24} years. The abundances of the other isotopes are tellurium-126 (18.7 percent), tellurium-125 (6.9 percent), tellurium-124 (4.6 percent), tellurium-122 (2.5 percent), tellurium-120 (0.89 percent), which has a half-life of 10^{16} years, and tellurium-123 (0.87 percent) which has a half-life of 10^{13} years.

There are also traces of tellurium-127, -129, -131, and -132 that occur as fission products in uranium-bearing deposits.

Element of surprises

Dimitri Mendeleyev found tellurium rather vexing when he was drawing up his first periodic table of the elements in 1869. The list of atomic weights showed tellurium to be heavier than the element that followed it, iodine. Tellurium was 128 whereas iodine was 127. Mendeleyev concluded that one of them must be wrong because tellurium clearly preceded iodine in his periodic table. Despite intensive work to reassess their relative atomic weights, they remained as 128 and 127 respectively.

The explanation of the anomaly had to wait almost 50 years until it was realised that elements could have numerous isotopes. The dominant isotopes of tellurium are 130 and 128, which together account for two thirds of the mass and give an overall atomic weight of 127.6, whereas for iodine there is only one dominate isotope, iodine-127; its atomic weight is 126.9

Terbium

Pronounced ter-bee-um, it is named after Ytterby, Sweden—see 'The village of the four elements' under yttrium.

French *terbiuim*; German *Terbium*; Italian *terbio*; Spanish *terbio*; Portuguese *térbio*.

Terbium is one of 15 chemically similar elements referred to as the lanthanoid or rare earth elements, which extend from element atomic number 57 (lanthanum) to element atomic number 71 (lutetium). The term rare earth is a misnomer because some are not rare. The minerals from which they are extracted, and the properties and uses they have in common, are discussed under Lanthanoids.

Essential element

Alloys containing terbium have unusual applications such as in audio equipment, and phosphors with terbium have led to improved X-ray screens.

Element of life

The amount of terbium in the human body is not known, but is very small. Terbium has no biological role. It is present in humans but it is difficult to know how much there is, because little work has been done to separate out the various amounts of the different lanthanoids in the body. The levels are highest in bone, with smaller amounts being present in the liver and kidneys. No one has monitored diet for terbium content, so it is difficult to judge how much we take in every day but it is probably only a few micrograms or so per year. Very little terbium is taken up by plant roots so it does not get into the human food chain, and those vegetables that have been analysed had less than 1 p.p.b. (dry weight).

Terbium may be mildly toxic by ingestion, but insoluble salts are non-toxic. Terbium compounds are very irritating if they come into contact with the skin and eyes.

Element of history

Terbium was first isolated in 1843 by the Swedish chemist Carl Gustav Mosander (1797–1858) at Stockholm. He suspected that the earth yttria (yttrium oxide), which had been discovered in 1794, might harbour other elements, which is what had happened when he had investigated ceria (cerium oxide) a few years earlier and extracted lanthanum from it. So it proved to be, and Mosander was able to show that although yttria was mainly yttrium oxide, its colour was due to two other oxides: terbium oxide (yellow) and erbium oxide (rose pink). There is more of this story in the section on yttrium.

Economic element

The most important lanthanoid ores are monazite and bastnäsite, but neither contains more than a fraction of a per cent of terbium—monazite contains just 0.05 percent and bastnasite even less at 0.02 percent—although enough is present for it to be extractable. Terbium is also found in the minerals cerite, gadolinite, xenotime, while the ore euxenite contains as much as 1 percent.

The main mining area is China, and there are deposits in the USA, Brazil, India, Australia, Greenland, and Tanzania although these are not generally worked. The reserves of terbium are estimated to be around 300,000 tonnes. World production of terbium is around 450 tonnes per year. The metal itself is produced by heating terbium fluoride and calcium in a tantalum crucible under vacuum conditions. However, in 2006 a group of chemists at Wuhan University, China, proved it was possible to produce the metal by the electrolysis of terbium oxide in molten calcium chloride, and this process used less energy.

Terbium has few commercial uses, partly because it is such a rare element and therefore costly, being four times more expensive than platinum. An unusual property of terbium alloys is that they will lengthen or shorten when exposed to a magnetic field; one such is an alloy of terbium, dysprosium, and iron of composition $Tb_{0.3}Dy_{0.5}Fe_2$. A rod of this material called Terfenol–D will grow longer or shorter by a tiny amount, and if the magnetic field is linked to an audio signal then any surface in contact with the rod will then act as a loudspeaker. The SoundBug is based on Terfenol-D and this can link devices like personal stereos to a flat surface such as a window which then becomes the speaker. These so-called magnetostrictive alloys can store a lot of strain energy and research is currently going on into how this might be used to work tiny motors, pumps, and injection systems (see also dysprosium).

Terbium is used in lasers (sodium terbium borate emits green laser light at 546 nanometres), low-energy light bulbs (see also europium), and mercury lamps. Terbium has led to increased safety in X-ray diagnosis by reducing the time a patient is exposed to these dangerous rays. This has been achieved by having the more responsive terbium-doped phosphors for the imaging screens. These allow the same quality image to be produced in a quarter of the time previously required.

Environmental element

Terbium in the environment	
EARTH'S CRUST	1 p.p.m.
	Terbium is the 57th most abundant element.
SOILS	approx. 0.7 p.p.b., range 0.5–1.8 p.p.b.
SEAWATER	0.2 p.p.t.
ATMOSPHERE	virtually nil.

Terbium is one of the rarer lanthaoid elements although it is still twice as common in the Earth's crust as silver. It poses no environmental threat to plants or animals.

Chemical element

Data file	
CHEMICAL SYMBOL	Tb
ATOMIC NUMBER	65
ATOMIC WEIGHT	158.92534
MELTING POINT	1,356°C
BOILING POINT	3,120°C
DENSITY	8.23 g/cm³
OXIDE	Tb_4O_7, which is a mixture of Tb_2O_3 and TbO_2.

Terbium is a silvery metal, soft enough to be cut with a knife, and a member of the lanthanoid group (row 4f) of the periodic table. It is reasonably stable in air, but is slowly oxidised and reacts with cold water. Its preferred oxidation state it +3 as in TbF_3, but there are some compounds known in which it is in oxidation state +4 such as TbF_4, as well as in the mixed oxide Tb_4O_7. When TbF_4 is heated it gives off pure fluorine gas (F_2). In solution terbium exists as Tb^{3+} ions.

There are 36 known isotopes of terbium with mass numbers 136 to 171, including 23 known isomers. There is only one naturally occurring isotope in ores and that is terbium-159, which is not radioactive. There are also minute traces of some radioactive isotopes in uranium ores, where they have been formed by nuclear fission. The longest-lived radioactive isotope is terbium-158 which has a half-life of 180 years.

Element of surprises

1. Chemists in Ireland have discovered a terbium complex which can selectively identify aspirin and when the two interact, the terbium will fluoresce to show that it has attached an aspirin molecule to itself.

2. There was once a plan to use a ceramic containing terbium for false teeth, because it mimics the gleam of real teeth but nothing came of it.

Thallium

Pronounced thal-ee-um, the name is derived from the Greek *thallos*, meaning a green shoot.

French *thallium*; German *Thallium*; Italian *tallio*; Spanish *talio*; Portuguese *tálio*.

Essential element

Special types of glass with a high refractive index need thallium and when so used this infamously murderous element is safe.

Element of life

Thallium in the human body	
BLOOD	0.5 p.p.b.
BONE	2 p.p.b.
TISSUE	4–70 p.p.b.
TOTAL AMOUNT IN BODY:	*ca* 0.5 mg.

Thallium is widespread and finds its way into most plants and thence into the food chain. Most roots have no difficulty in absorbing thallium from the soil in which they grow and the more thallium that is present the more they absorb. Vegetables and meats have 0.02–0.12 p.p.m., amounts too small to affect health. Sometimes quite high levels of thallium in vegetation have been measured; pine trees can have 100 p.p.m. while some flowers have reached 17,000 p.p.m. (1.7 percent). Thallium has been detected in spring waters and wine.

There is no biological role for thallium although some marine organisms appear to be able to concentrate it in their tissues. Nevertheless, the average person takes in about 2 micrograms of thallium a day, as part of the diet, and this accumulates in the body over time, with the highest concentrations in the kidneys and liver.

Medical element

Thallium was once part of the medical pharmacopoeia being prescribed by doctors as a depilatory (hair remover) and given as a pre-treatment for ringworm of the scalp. (This is a fungus disease that is highly contagious and often transmitted by contact with animals.) The thallium caused the patient's hair to drop out, so that the condition could be more easily treated. Its ability to do this was discovered in 1898 by a Dr R. Sabourand, the chief dermatologist at the St Louis Hospital in Paris, when he was searching for something that might cure the night sweats experienced by tuberculosis patients. Thallium had no effect on those symptoms, but it caused hair loss so it became the standard treatment for hair removal for almost 50 years. In the 1930s, thallium acetate was even sold as Koremlou cream for removing unwanted facial hair.

The medicinal dose for hair removal was 8 mg per kg body weight and an adult would be given 500 mg. This treatment was reported to cause 40 percent of patients to experience other side effects, albeit mild ones, and these were reported to disappear after three weeks. Hair loss would begin after 10 days or so, but such hair loss would today be taken as an indication that a person was suffering from near-lethal thallium poisoning. Accidental overdoses of thallium acetate led to several fatalities in the 1920s and 1930s and as a result thallium treatments were phased out and ceased being used after the 1950s. In one incident a group of boys in Budapest who had ringworm were given 5,000 mg doses instead of 500 mg and they all died. The same thing happened at an orphanage in Granada, Spain, when 14 out of 16 children died after being given overdoses due to the pharmacist's weigh scales being faulty.

A radioactive isotope, thallium-201 which has a half-life of 73 hours, is used in the diagnosis of heart disease. This isotope decays by capturing an electron and emitting low energy gamma-rays so it gives good images without putting the patient at risk. Thallium-201 will displace some of the potassium in the heart muscle, but only if there is an adequate supply of blood to carry it there. The patient is given an injection of the isotope and then scanned by a scintillation counter before and after physical exercise. The uptake of thallium-201 by the heart, and its distribution within the heart, will reveal the extent of the damage. The isotope is generated on site from lead-201. This has a half-life of 9.33 hours and this is produced in a cyclotron by bombarding ordinary thallium atoms with protons or deuterons.

Dangerous element

Thallium mimics potassium so effectively that it can displace it at all sites around the body but most damagingly along the central nervous system which soon begins to malfunction. The symptoms that follow include lethargy, numbness, tingling of the hands and feet, slurred speech, insomnia, and blackouts. Thallium upsets the sugar metabolism of the body and produces symptoms of diabetes. In addition men are affected sexually and become impotent. Thallium eventually affects hair follicles and these become unable to produce more hair. The existing hair drops off, and this occurs all over the body. (When all the thallium has been excreted from the body, the hair will begin to grow again and return to its normal state.)

Studies using the radioactive isotope thallium-204 (half-life of 3.78 years) have traced the movement of thallium in the body, showing that it accumulates in the bones, the kidneys, walls of the stomach, and the intestines, the pancreas, and the salivary glands. The studies showed that it takes at least a month to remove half of a given dose, but some is released so slowly that it can still be detected in the urine after three months. Excretion is mainly through the faeces and urine with the former predominating. Once it has penetrated the body, thallium tends to form thallium chloride, which is much less soluble than potassium chloride and consequently it takes the body a long time to eliminate thallium.

A fatal amount of a thallium salt such as the acetate or the sulfate is around 800 milligrams for an adult, which is less than a quarter of a teaspoonful. Such a dose acts as follows: on day 1 there may be no symptoms at all or only mild symptoms typical of a cold; on day 2 the victim will suffer the symptoms of gastroenteritis, such as vomiting and diarrhoea, as well as a sensation of pins and needles in the feet; on day 3 there will be band-like pain around the body, joint pains, the feet become very sensitive to touch and sleep is almost impossible. As the days continue these conditions worsen and eventually death intervenes.

The body tries to rid itself of the thallium, and does so by excreting it into the intestines, but there it may be reabsorbed again in mistake for potassium. Whether a victim of thallium poisoning survived depended primarily on thallium poisoning being recognised as such, and quickly. If it went undetected then even the most intensive care could not succeed in saving life because there was no antidote for the poison until the early 1970s when a German pharmacologist, Horst Heydlauf of Karlsruhe, discovered that Prussian blue, the dye of blue ink, can break this cycle of excretion and re-absorption by exchanging its potassium for thallium to which it clings to more strongly and carries it from the body.

One of the first persons to be treated with Prussian blue was a woman aged 26 who had been admitted to a South African hospital in 1972 with thallium poisoning, having taken an estimated 700 mg. She was treated with four daily doses of 3.75 grams of the dye for 13 days and successfully recovered. Cases of thallium poisoning in which people have taken a more-than-lethal dose of thallium sulfate have been dealt with very successfully by the administration of Prussian blue. This can be given by mouth but is best administered via a tube directly into the duodenum because thallium poisoning hampers the progress of material from the stomach downwards by closing the pylorus, the sphincter at the base of the stomach.

The crime writer Agatha Christie is often blamed for bringing thallium sulphate to the attention of would-be-poisoners. Her detective story, *The Pale Horse*, which was published in 1961, described the symptoms of thallium poisoning. It showed how these could easily be attributed to other causes—in this case to black magic curses.

Her fictional tale may also have suggested thallium as a suitable poison to Graham Young who in 1971 put thallium sulfate into his workmates' coffee at a factory in Bovingdon in Hertfordshire, England. Several workers were taken ill and two died of a mysterious illness. Only one of the 43 doctors who examined the victims correctly diagnosed thallium poisoning and indeed it was only when Young himself suggested that the cause might be thallium, that he was caught. Young was found guilty of murder in 1972 and sentenced to life imprisonment. He committed suicide in 1990.

In Holland a woman used hair remover cream to dispose of members of her family. She had killed seven of them before she was arrested. Six other victims recovered from the doses she had given them and it was one of these who was correctly diagnosed as suffering from thallium poisoning which led to the woman's arrest. Her other victims had been diagnosed as having encephalitis, brain tumour, alcoholic neuritis, typhus, pneumonia, and epilepsy.

Murderers who poison people may think they are in the clear once their victim has been cremated. However, in the case of Graham Young it provided him with no such protection. His case became a milestone in forensic detection because the ashes of Bob Egle, one of his victims, were reclaimed and analysed by a technique known as atomic absorption spectrometry. This revealed a level of 5 parts per million of thallium which was well above any natural level of this element in cremated remains, and proof that he had been poisoned with it.

In the 1980s and 1990s the Iraqi regime of the dictator Saddam Hussein disposed of its opponents with thallium sulfate, both inside and outside Iraq. An agent poisoned several prominent Iraqis in London and in particular members of the Patriotic Union of Kurdistan a militant anti-Saddam group. Nor has thallium's lethal toxicity been forgotten in Iraq. There was an attempted mass poisoning at a Baghdad sports club in 2008 when someone added thallium to cakes which were supplied to the club. Nine people were taken ill and two children died.

Not all thallium deaths are malicious, some are accidental, such as the incident in Guyana in 1987 which affected hundreds of people, of whom 44 died, when they drank the milk from cows that had eaten molasses which had been poisoned with thallium sulfate in order to kill sugar-cane rats.

Element of history

Thallium was controversial from the time it was discovered, including who had priority regarding its discovery. Was it William Crookes (1832–1919) of the Royal College of Science in London, or Claude-Auguste Lamy (1820–78), a physicist of Lille, France? Crookes first observed a green line in the spectrum of some impure sulfuric acid, and realised that it meant a new element was present. That was in 1861. He confirmed that it really was a new element by measuring the green light using a spectrometer and he could see that the green was in a different part of the spectrum to the green lines that other elements like barium display. (We now know that the actual wavelength of this line is at 535 nm.) Crookes rushed to announce his discovery in the 30 March 1861 issue of *Chemical News*, a weekly publication that he edited, and he gave the new element the name by which it was to be known. He began to investigate thallium (which he so named because the colour of its spectral lines reminded him of a fresh green shoot), but over the next year he only managed to make small amounts of a few simple compounds.

Meanwhile, in 1862 Lamy began to research thallium more thoroughly even to the extent of extracting the metal itself and casting a small ingot of it. The French Academy consequently credited him with the discovery. He sent the ingot to the London International Exhibition of that year, where it was acclaimed as a new metal and he was awarded a medal. Crookes was furious when he learned what had happened and he campaigned to have the award given to him, on the grounds that he was the true discoverer. Accusations and counter-accusations followed and eventually the exhibition committee felt obliged to award Crookes a medal also.

A thallium mineral was found in Sweden in 1866, and named crookesite in honour of Crookes but it is rare. In 1867, thallium oxide was found to be a component of the flue dust of pyrite-roasting kilns.

Economic element

Thallium minerals are rare, but a few are known such as crookesite (copper thallium selenide, Cu_7TlSe_4) and lorandite (thallium arsenic sulfide, $TlAsS_2$). Thallium is mainly obtained as the by-product of copper, zinc, and lead smelting, but some is extracted from pyrites when the sulfur from this ore is to be used to make sulfuric acid. World production of thallium compounds is around 30 tonnes per year, although only a tiny fraction of this is converted, electrolytically, to the metal itself. World reserves of thallium are estimated to be around 600,000 tonnes.

Thallium sulfate is still used in developing countries and is particularly effective as a sugar syrup against rats, cockroaches, and ants. However, thallium pesticides have been banned in most developed countries.

Most thallium is used by the electronics industry as thallium sulfide, thallium selenide, and thallium arsenide for photoelectric cells. Thallium is also used to make special glass for high-refractive lenses. Some ends up as thallium bromide-iodide crystals for infrared detectors because these are transparent to this kind of radiation. Thallium salts are used as reagents in chemical research; for example, thallium(III) nitrate is a selective oxidising agent.

Thallium metal is not used as an alloying agent although the addition of 8 percent of thallium to mercury enabled thermometers to record temperatures down to −58°C, which is 20 degrees lower than with mercury alone.

Environmental element

Thallium in the environment	
EARTH'S CRUST	0.6 p.p.m.
	Thallium is the 59th most abundant element.
SOILS	0.1–0.3 p.p.m. but can be as high as 2.8 p.p.m.
SEAWATER	av. 10 p.p.t.
ATMOSPHERE	the presence of thallium in the layers of a peat bog in Switzerland which go back thousands of years shows that some thallium is deposited naturally from in the atmosphere but only in tiny amounts.

Thallium is not a rare element; it is 10 times more abundant than silver. The element is widely dispersed, mainly in potassium minerals such as sylvite. There has been no significant contamination of the environment by thallium from industry, unlike that caused by its neighbours in the periodic table, mercury and lead. It has been estimated that around 600 tonnes a year is emitted by smelters and metal processing industries, and a similar amount from coal-fired power stations.

In 1980, vegetation growing around a cement works in Germany was found to contain raised levels of thallium, this being emitted from the cement kilns in which pyrite, with unsuspected high levels of thallium, was being used to make a new type of cement. Cabbages were found to have the highest levels of thallium, with up to 45 p.p.m. (fresh weight), and grapes also had a high level with 25 p.p.m. Normally these plants would have less than 0.1 p.p.m. Even the eggs of local hens had more than 1 p.p.m. although most of this was concentrated in their shells.

Chemical element

Data file	
CHEMICAL SYMBOL	Tl
ATOMIC NUMBER	81
ATOMIC WEIGHT	204.3833
MELTING POINT	304°C
BOILING POINT	1,457°C
DENSITY	11.9 g/cm³
OXIDES	Tl_2O and Tl_2O_3

Thallium is a soft, silvery-white metal, and a member of group 13 (row 6p) of the periodic table. It tarnishes readily in moist air and reacts with steam to form the hydroxide (TlOH). The metal is attacked by acids. Thallium can exist in two oxidation states: +1 and +3. In the lower state, as the ion Tl^+, it resembles potassium K^+. The higher oxide, Tl_2O_3, is formed by heating the lower oxide in the presence of oxygen, and when treated with concentrated nitric acid it forms the thallium(III) nitrate, $Tl(NO_3)_3$, used in research.

Thallium has 37 known isotopes of mass numbers 176 to 212, and 43 known isomers. Two isotopes occur naturally in thallium ores and these are thallium-205, which accounts for 70 percent, and thallium-203 (30 percent). These are considered to be stable although it has been suggested they may be radioactive, but if so they have immeasurably long half-lives. In fact thallium appears to be the element of highest atomic number that is stable, with thallium-205 having the highest stable atomic mass.

There are minute amounts of five other naturally occurring radioactive isotopes, thallium-206 to 210, in uranium and thorium minerals where they are part of the decay chain of these elements; all have short half-lives. The longest-lived radioactive isotope is thallium-204 which has a half-life of 3.78 years; the next longest-lived is thallium-202 with a half life of 12.23 days.

Element of surprises

1. Thallium has the curious ability to be absorbed through the skin and handling its compounds directly can lead to the loss of the fingernails.

2. One way of measuring the density of a heavy mineral was by the flotation method. For some minerals a particularly dense liquid was required and a solution of thallium(I) formate was used. This can have a density as high as 4.25 g/cm³ which is more than four times denser than ordinary water.

3. A case of misdiagnosed poisoning with thallium sulfate was reported in 1977, and the victim was a young girl from Qatar. Her parents took her to London for treatment but the doctors at the Royal Postgraduate Medical School at Hammersmith Hospital could find nothing that would account for her condition. There seemed no hope for her until a nurse suggested that she might be suffering from thallium poisoning. The nurse drew their attention to the book she had been reading, which was *The Pale Horse*, and the doctors realised they were dealing with a case of thallium poisoning—and the child's hair was already falling out. She was immediately treated with the antidote potassium ferric ferrocyanide and her life was saved.

Thorium

Pronounced thor-ee-um, it was named after Thor, the Scandinavian god of war.
French *thorium*; German *Thorium*; Italian *torio*; Spanish *torio*; Portuguese *tório*.

Essential element

Thorium has the potential to solve the world's need for electricity for the next thousand years or more.

Cosmic element

Much of the internal heat of the Earth is thought to come from the radioactive decay of thorium.

Element of life

Thorium in the human body	
BLOOD	0.2 p.p.b.
BONE	2–12 p.p.b.
TISSUE	no data.
TOTAL AMOUNT IN BODY	40 mg.

There is no biological role for thorium, nor would we expect there to be, because this element is radioactive in all its forms. We take some in with our diet and the average amount is estimated to be around 3 micrograms a day. Plant roots can absorb thorium, but it is only present at p.p.b. levels in most cases, although concentrations of over 1 p.p.m. have sometimes been recorded (based on dry weight). The levels in vegetables are in the range 5–20 p.p.b. (again dry weight).

Research has shown that 99.98 percent of ingested thorium passes through the body unabsorbed, and of the 0.02 percent which is absorbed, three-quarters deposits in the skeleton. Thorium compounds are thought to be moderately toxic chemically and those who worked with its compounds sometimes suffered from a form of dermatitis, while prolonged exposure was likely to increase the risk of cancer.

Deadly medical element

In the early years of X-ray diagnosis, patients were injected with a colloidal suspension of thorium dioxide, ThO_2, to highlight the bloodstream and organs of the body. Thorotrast was its commercial name and it was introduced in the 1920s and given to highlight the gall bladder and bile duct prior to surgery. It was also used to examine the lymphatic system and spleen. In 1954, its use was discontinued in the face of overwhelming evidence by then that it caused major damage. The 10,000 individuals who underwent this treatment were later found to have higher incidences of leukaemia, bone cancers, and abnormal chromosomes due to radiation, while some suffered liver damage which included cirrhosis and the rare malignant disease haemangiosarcoma. The reason it took so long for thorium's baleful effects to be realised was the long latent period, of up to 30 years, before these conditions manifested themselves.

Element of history

In 1815, Jöns Jakob Berzelius (1779–1848), who was Professor of Chemistry at the Royal Karolinska Institute, Stockholm, thought he had extracted a new element from a recently discovered mineral, and he named it thorium. The mineral later turned out to be the phosphate of a metal that was already known: yttrium. Nevertheless, Berzelius so liked the new name that he used it again when, in 1829, he really did discover a new element. This he had extracted from a curious rock specimen sent to him by Hans Esmark, a Norwegian pastor and amateur mineralogist, who had found it near Brevig. Esmark was the son of the Professor of Mineralogy at the University of Christiania, Oslo, and it was he who realised that the mineral had not previously been recorded, so he suggested his son should send it to Berzelius. The mineral was thorium silicate, and is now known as thorite. Berzelius even produced a sample of metallic thorium by heating thorium fluoride with potassium, and confirmed it as a new metal.

The radioactivity of thorium was first demonstrated in 1898 by Gerhard Schmidt (1865–1949), and confirmed by Marie Curie later that same year. It was the second element, after uranium in 1896, to have its radioactivity revealed.

Economic element

Thorium occurs naturally as the minerals thorite ($ThSiO_4$), uranothorite ((Th,U)SiO_4), and thorianite (ThO_2). It is a major component of the lanthanoid ore monazite which can contain up to 12 percent thorium oxide, and this is currently the source of the world's supply. Thorium is also present in significant amounts in the minerals zircon (see zirconium), titanite (see titanium), gadolinite (see yttrium), and betafite (see uranium). World production of thorium is in excess of 5,000 tonnes per year. Known reserves exceed 3 million tonnes and the countries with the largest deposits are Brazil, India, Norway, Turkey, and the USA. The USA has a reserve of 300,000 tonnes which is about 20 percent of the world's supply, and most of it is in Idaho's Lemhi Pass.

Until the inherent dangers associated with its radioactivity were realised, thorium and its compounds found some important retail outlets, the best known of which was the gas mantle. These were made in their hundreds of millions and consisted of thorium oxide (plus 1 percent cerium oxide) which, when heated, emitted a bright white light. It was Karl Auer's[194] invention of these which enabled town gas to compete successfully with electricity for domestic and street lighting well into the last century.

[194] There is more information about Auer under neodymium and lutetium.

While these uses of thorium have now been discontinued, the metal and its compounds are still employed in other ways, although not ones which expose the general public to its radiation. Today thorium oxide is used in the manufacture of refractory materials for the metallurgical industries. It is also employed as a catalyst in the process which converts syngas (which is a mixture of carbon monoxide and hydrogen) into hydrocarbons. It is also used as a catalyst for converting ammonia to nitric acid. Thermionic emitting devices, and photoelectric cells that measure the wavelength of ultraviolet light, both contain thorium.

Thorium metal can be extracted by heating thorium oxide with calcium, or reacting thorium tetrafluoride with magnesium, the process being carried out under an atmosphere of argon. Thorium reacts rapidly with the oxygen of the air and becomes coated with a protective layer of thorium oxide, and this is why thorium crucibles are used in high temperature research. Thorium was formerly used to control the grain size of the tungsten that went into the filaments of the once common incandescent light bulbs. As a metal, it is soft and ductile, but its alloys can be very strong, and adding thorium to magnesium enables this metal to be fabricated into components that can withstand high temperatures. It can also be added to nickel to improve that metal's creep resistance.

Like uranium, thorium could be a source of nuclear fuel. There is probably more energy locked up in the world's thorium than in all known fossil fuels deposits and in theory thorium could supply all the world's energy needs for several thousand years. When thorium-232 is bombarded with neutrons it transforms first to thorium-233, which has a half-life of only 22 minutes, decaying to uranium-233. This then undergoes fission in the same way as uranium-235, liberating lots of energy plus enough neutrons to start a controlled chain reaction to keep the process going.

Thorium needs to be 99.999 percent pure it if is to be used as nuclear fuel, but the problems of producing such high grade material have been solved. Thorium can be 'burnt' in a nuclear reactor as either the metal itself or as the oxide, nitrate, fluoride, or chloride. One of the advantages of a thorium-powered nuclear reactor would be that it would not generate plutonium. An experimental thorium-fuelled reactor was built in the USA in the 1960s at the Oak Ridge National Laboratory, with the thorium present as liquid thorium fluoride (TlF_4) and it operated for five years, but was discontinued due to lack of funding.

Renewed research into nuclear energy based on thorium is being undertaken in India and Russia. Thorium does not require isotope enrichment as does uranium. It is four times more abundant than uranium and much cheaper to extract because it consists of a single isotope, whereas uranium has to be processed to extract the 0.7 percent of it which is uranium-235, the isotope which is fissile and which can be used as fuel. Thorium reactors will be of the liquid fluoride type known as LFTRs (liquid fluoride thorium reactors) which use molten thorium fluoride dissolved in a mixture of molten lithium and beryllium fluoride at 630 °C. The heat which the reactor produces is used to power a turbine to generate electricity. India's thorium-fuelled reactor should come on-stream in 2011 and plans are to have 30 percent of that country's electricity generated by such reactors by 2030.

Environmental element

Thorium in the environment	
EARTH'S CRUST	12 p.p.m.
	Thorium is the 37th most abundant element.
SOILS	range 0.8–11 p.p.m.
SEAWATER	*approx.* 10 p.p.t.
ATMOSPHERE	nil.

Thorium is surprisingly abundant in the Earth's crust, being almost as abundant as lead. It is three times more abundant than uranium, five times more abundant than tin, and 200 times more abundant than silver. Granite contains up to 80 p.p.m. of thorium. Because thorium oxide is highly insoluble, very little of this element circulates through the environment, although the thorium ion, Th^{4+}, is soluble, especially in the presence of organic acids in soils, and the level of thorium around coal-burning and oil-burning power plants can be relatively high due to its release from these fossil fuels. Near one such Italian power plant, soils were found to have 40 p.p.m.

Thorium is an alpha emitter, albeit only weakly so and it decays eventually to lead-208, passing through a series of short-lived isotopes, although one of them is the gas radon, and for this reason thorium materials need proper ventilation. However, this isotope of radon is radon-220 with a half-life of only 55 seconds, which means that it does not have time to escape from subsoil and into houses so it is not an environmental hazard.

Chemical element

Data file	
CHEMICAL SYMBOL	Th
ATOMIC NUMBER	90
ATOMIC WEIGHT	232.0381
MELTING POINT	1,750°C
BOILING POINT	4,788°C (Thorium has the longest liquid range of all the elements.)
DENSITY	11.7 g/cm³
OXIDE	ThO_2 which has the highest melting point of any oxide at 3300°C

Thorium is a radioactive, silvery metal, and the most abundant of the actinide series of elements. Its most stable oxidation state is +4. As a bulk metal it is protected by an oxide coating, but this can be attacked by steam, and will slowly dissolve in acids. The powdered metal has been known to ignite spontaneously. Thorium's preferred oxidation state is +4, as in ThF_4 and $ThCl_4$, although there are some compounds in oxidation state +3 such as the iodide, ThI_3.

Recently thorium has shown interesting chemical behaviour such as activating C–H bonds in the formation of some unusual compounds. This work has been done at the Los Alamos National Laboratory in the USA.

There are 30 known isotopes of thorium with mass numbers ranging from 209 to 238, and three known isomers. There are nine thorium isotopes which occur naturally: thorium-226 to thorium-232, thorium-234, and thorium-235. Thorium-232 is the predominant one, and with a half-life of 14 billion years; the amount on Earth is still about 85 percent of that which was present when the planet formed. The next longest-lived isotope is thorium-230, with a half-life of 75,690 years.

The other naturally occurring thorium isotopes have short half-lives and are produced by the decay of the radioactive elements plutonium, uranium, actinium, and even thorium itself. For example, uranium-235 decays via thorium-231 (half-life 26 hours) and thorium-227 (19 days), while uranium-238 decays via thorium-234 (24 days) and thorium-230 (75,380 years). Thorium-232 itself also decays via thorium-228 which has a half-life of two years.

Element of surprises

1. Thorium compounds used to be ingredients in a toothpaste which was first marketed in the early 1900s by the Berlin Auer company, and sold in Germany for many years. In World War II, another thorium-based toothpaste called Doramad also came on the market there and its radioactivity was claimed to be beneficial because it attacked bacteria. During World War II, the allies discovered that a military contractor in Germany had acquired a large stockpile of thorium which by then they knew might be used for atomic weapons. It transpired he was planning to launch another thorium-based toothpaste after the war.

2. Yet more thorium isotopes have been announced. In 2007, a paper was published which claimed four other naturally occurring ones: thorium-211, thorium-213, thorium-217, and thorium-218 saying these were also primordial and were long-lived neutron deficient, super-deformed, and hyper-deformed isomers with half-lives longer than 100 million years. These are new isomers of nuclei which because of their extreme shape-distortion are highly resistant to decay. They exhibit low instability and extremely long half-lives. They would have been produced by the extreme temperatures and pressures in a Type II supernova. As yet these findings have not been widely accepted.

3. Thorium has been shown to be capable of forming more chemical bonds than any other element. Thorium tetrakisaminodiborane, formula $Th[H_3BN(CH_3)_2BH_3]_4$, has 15 hydrogen atoms bonded to the central thorium atom. The molecule was reported in 2010 and made by Gregory S. Girolami and colleagues at the University of Illinois, Urbana-Champaign.

Thulium

Pronounced thyoo-lee-um, it is named after Thule, the ancient name for Scandinavia.

French *thulium*; German *Thulium*; Italian *tulio*; Spanish *tulio*; Portuguese *túlio*.

Thulium is one of 15 chemically similar elements referred to as the lanthanoid or rare earth elements, which extend from element atomic number 57 (lanthanum) to element atomic number 71 (lutetium). The term rare earth is a misnomer because some are not rare. The minerals from which they are extracted, and the properties and uses they have in common, are discussed under Lanthanoids.

Essential element

Thulium finds use in stage and studio lighting and in lasers that are safe to use for surgical procedures.

Element of life

The total amount of thulium in the human body in not known, but is tiny. Thulium has no biological role, although it has been noted that thulium salts stimulate metabolism. They are regarded as slightly toxic if they are soluble and taken in large amounts, but non-toxic if they are insoluble. It is difficult to separate out the various amounts of individual lanthanoids in the human body, but levels are highest in bone, liver and kidneys, and this is no doubt how thulium distributes itself in the body, but in any case it is reasonable to assume that actual amounts are very low. Thulium is not taken up by plant roots to any extent and so does not get into the human food chain. Vegetables typically have only 1 p.p.b. (dry weight) of thulium. The thulium we take in with our food is probably only a few micrograms a year.

Element of history

Thulium was first isolated in 1879 as its oxide by Per Teodor Cleve (1840–1905) at the University of Uppsala. The discovery of several lanthanoid elements began with yttrium which, unbeknown at the time it was discovered, was contaminated with traces of other rare earth elements. In 1843, erbium and terbium were extracted from it, and then, in 1874, Cleve looked more closely at erbium and realised that it must contain yet other elements because he observed that its atomic weight varied slightly depending on the source from which it came. He extracted holmium from erbium in 1878 and thulium a year later in 1879.

In 1911, the American chemist Theodore William Richards (1868–1928) performed 15,000 recrystallisations of thulium bromate in order to obtain a pure sample of the element and so determine exactly its atomic weight. Richards won the 1914 Nobel Prize for chemistry.

Economic element

The most important thulium-containing ores are monazite, which contains 20 p.p.m. (0.002 percent), and bastnäsite, which contains even less, at 8 p.p.m. (0.0008 percent). The chief ores from which thulium is obtained are mainly mined in China, but there are large ore bodies in the USA, Brazil, India, Australia, Greenland, and Tanzania. Reserves of thulium are estimated to be about 100,000 tonnes.

World production of thulium is around 50 tonnes per year as thulium oxide. The metal itself was first obtained by heating thulium oxide with lanthanum, although it is now produced by heating thulium fluoride with calcium. Because it is so expensive, and has little to offer that other rare earth elements cannot provide, thulium finds only a few applications.

Thulium iodide is important to lighting designers because it contributes a strong green emission line to high intensity discharge lights. Thulium irradiated in a nuclear reactor produces an isotope which emits X-rays and this has been used as a portable X-ray source for medical and dental use. Thulium-doped YAG (which stands for yttrium aluminium garnet) is used for lasers and it operates over a narrow range of wavelengths (1.93 to 2.40 microns) making it safe to use for laser-based surgery.

Environmental element

Thulium in the environment	
EARTH'S CRUST	0.5 p.p.m.
	Thulium is the 61st most abundant element.
SOILS	*approx.* 0.5 p.p.m., range 0.4–0.8 p.p.m.
SEAWATER	0.25 p.p.t.
ATMOSPHERE	nil.

Thulium is the second rarest lanthanoid element—promethium is less abundant but that does not exist in lanthanoid ores on Earth because all its isotopes are radioactive and have long since decayed away. Although it is rare, thulium is still twice as abundant in the Earth's crust as silver. It poses no environmental threat to plants or animals.

Chemical element

Data file	
Chemical symbol	Tm
Atomic number	69
Atomic weight	168.93421
Melting point	1,545°C
Boiling point	1,947°C
Density	9.32 g/cm³
Oxide	Tm₂O₃

Thulium is a bright silvery metal, which is soft enough to be cut with a knife, and which is a member of the lanthanoid group (row 4f) of the periodic table. It slowly tarnishes in air, but is more resistant to oxidation than most rare earth elements. It reacts with water. Its preferred oxidation state, indeed its only oxidation state, is +3 as in compounds such as TmF_3 and $TmCl_3$. In solution it is present as the Tm^{3+} ion, surrounded by nine water molecules.

There are 35 known isotopes of thulium with mass numbers 145 to 179, and 26 known isomers. There is only one naturally occurring isotope in minerals and that is thulium-169, which is not radioactive. The longest-lived radioactive isotope is thulium-171 with a half-life of 1.92 years.[195]

Element of surprise

Thulium-doped calcium sulfate has been used in personal radiation dosimeters because by its fluorescence it can register particularly low exposure levels.

---

[195] Nuclear theory suggests that thulium is a radioactive element and that its lone stable isotope is in fact radioactive but with an incredibly long half-life.

Tin

Pronounced tin, the name comes from the Anglo-Saxon *tin*, and the chemical symbol Sn comes from the Latin *stannum*, which may have derived from the Sanscrit word *stan*, meaning hard, or from the Cornish word for tin which was *stean*, and this was known to the Romans. It is also where the word stannery, the name for a tin mine, comes from. Some tin compounds are also called stannates.

French *etain*; German *Zinn*; Italian *stagno*; Spanish *estaño*; Portuguese *estanho*.

Essential element

The discovery that tin could harden copper to form bronze was a major step in human progress 5,000 years ago, as was the discovery in the 1950s that molten glass would float on liquid tin and produce the perfect window glass of modern city tower blocks.

Cosmic element

Tin is created in stars by the slow-neutron-capture process (s-process). It is produced from indium (element 49) and this begins when an indium-115 nucleus captures a neutron to become indium-116. This then undergoes beta decay to become tin-116. The process occurs in large stars but it takes thousands of years to convert one element to the other.

Element of life

Tin in the human body	
BLOOD	av. 0.4 p.p.m.
BONE	1.4 p.p.m.
TISSUE	0.3–2.4 p.p.m.
TOTAL AMOUNT IN BODY	30 mg.

There is no evidence of humans needing this metal although at one time it was thought to be part of the stomach hormone, gastrine. However, it may be essential to some creatures; rats fed on a tin-free diet failed to grow properly, but recovered when a tin supplement was given, suggesting it has some essential role.

Plants easily absorb tin, but it is not essential or even beneficial for them to do so, and most of that which is absorbed is retained in their roots. Nevertheless, tin in wheat grains is detectable at 7 p.p.m, and in maize (corn) is 3 p.p.m. Plants growing in uncontaminated soil can have up to 30 p.p.m. of tin, whereas those growing on contaminated soil can have very high levels. Sugar beet grown adjacent to a chemical plant had 1000 p.p.m. (0.1 percent) while vegetation near a tin smelter reached levels of 2000 p.p.m. (0.2 percent) measured as dry weight.

Food element

The average person takes in around 0.3 mg of tin a day, of which 0.2 mg comes naturally from food. Less than 3 percent of the tin in food is absorbed by the body, and that which is absorbed is mainly excreted in the urine but some tends to collect in the skeleton and liver. Over time the amount slowly builds up, although not to anything like a dangerous level. Tin has always been part of the human diet, but it greatly increased when canned food was introduced in the early 1800s. The lacquering of the inside of cans has considerably reduced the amount of tin that leaches into the food they contain. The USA has imposed limits of 300 p.p.m. for tin in canned food; the UK limit is 200 p.p.m.

The first containers for preserving food were made of glass and produced by the Frenchman Nicholas Appert in 1810. The jars were filled, made airtight, and then sterilised by heating. John Hall and Bryan Donkin in Deptford, England, copied the idea but used tinplate cans instead. Their canned meats were particularly popular with officers of the Royal Navy, and were an essential part of the food stores taken on long voyages of exploration, such as those undertaken to find the North West Passage around the north of Canada and Alaksa. Such voyages could end in disaster; this was attributable to the canned food, not because of the tin, but because of the lead solder which sealed the cans—see lead.

Dangerous element

Tin compounds can be poisonous by ingestion, but to what extent the tin they contain is responsible for this is debatable, and inorganic tin compounds are generally regarded as non-toxic. (Tin fluoride, SnF_2, is safe enough to be used in fluoride toothpaste.) However, organo-tin compounds are toxic, especially when the tine atoms have three organic groups attached. Such organo-tin compounds are able to penetrate biological membranes and once inside a cell they damage various metabolic processes. Trimethyl tin and triethyl tin are this type and are particularly toxic to humans, but larger organic groups were found to be much less toxic which is why some, such as tributyl tin (TBT), have been widely used as biocides, including putting it on clothes that were likely to become sweaty, such as T-shirts, to counteract the bacteria that breed on them and emit unpleasant odour molecules.

TBT in particular was used to protect the hulls of ships against marine encrustation. It became an environmental problem in the 1960s even though it was proving to be very profitable. It prevented the drag caused by marine growths on hulls. It reduced the time spent in dry-dock, to the extent that some ships needed repainting only once every five years. Although TBT saved energy and other resources, to an estimated $7 billion per year, by the 1980s it was clear that species such as oysters, marine snails and dog-whelks, which flourish in coastal waters, were undergoing strange sexual changes or becoming infertile. The cause was TBT and as little as 1 nanogram per litre of sea water (1 p.p.t.) was capable of causing the observed changes.

The use of TBT has now been outlawed in most countries. The EU banned it in 2003 and any ship entering an EU port after 2008 had to be certified as not using TBT.

Triphenyl-tin was thought not to be toxic to humans and has been used in fungicidal crop sprays, but research in 2002 showed that it could suppress the body's natural killer cells which are its first line of defence against cancer. Also worrying was the discovery that triphenylt-in was damaging to the central nervous system of tadpoles and so was a threat to frog populations.

The reason why these kinds of compounds are dangerous came to light in 2003 when it was shown that trimethyl tin targeted a particular protein found in the membranes of the central nervous system. That had two cysteine amino acid residues close together and it is to the sulfur atoms of these that the tin attaches itself thereby deforming the protein from its natural configuration.

In France in the 1960s and early 1970s, diethyl tin di-iodide was prescribed as the drug Stalinon for the treatment of staphylococcal skin infections. Diethyl tin has only two organic groups attached and was thought to be safe, but the treatment led to several deaths, which were probably due to contamination by triethyl tin iodide.

Element of history

Tin by itself was of little use but a tin ring and pilgrim bottle have been found in an Egyptian tomb of the eighteenth dynasty (1580–1350 BC) despite the fact that there was no tin ore to be found in Egypt itself. Pure tin has also been found at Machu Picchu, the mountain citadel of the Incas. The Chinese were mining tin around 700 BC in the province of Yunnan and where tin mines are still being worked.

When copper was alloyed with a small amount of tin—around 5 percent was the usual amount—it produced bronze, which not only melted at a lower temperature, so making it easier to work with, but it also produced a metal that was much harder, and ideal for tools and weapons. This discovery is celebrated in the term Bronze Age, which was a recognised stage in the development of civilisation. How bronze was discovered we do not know, but the peoples of Egypt, Mesopotamia, and the Indus valley started using bronze around 3000 BC.[196]

It was not known where these early bronze smiths got their tin until, in 1989, an ancient cassiterite mine was discovered in the Taurus Mountains of Turkey. Later, tin was traded around the Mediterranean by the Phoenicians who obtained it from Spain, Brittany, and Britain, especially from the Scilly Isles and Cornwall. Julius Caesar mentions British tin in his book *Commentaries on the Gallic War*. Tin from Cornwall was appreciated for its quality and was first mined around 2000 BC. Mining continued there until very recently, when it became uneconomic.

Throughout history, bronze was used not only for weapons, but to create great works of art. Among the most famous was the third century BC Colossus of Rhodes, a 30 metre high statue of the Sun God (destroyed in an earthquake in 224 BC). Other monumental bronzes were the Great Buddha of Kamakura, Japan, made in 1252 AD and the East Door of the Baptistry of St John, Florence Italy, made by Lorenzo

---

[196] The ancients tried to relate things on Earth to heavenly bodies. The Sun was clearly related to gold, the Moon to silver, Venus to copper, Mars to iron, Mercury to mercury, Saturn to lead, and Jupiter to tin.

Ghiberti in 1400s. The Benin Bronzes of Nigeria which were cast in the 1500s are another fine example of the art of the bronze smith.

Tinplate was first mentioned by Theophrastus in 320 BC. There was a large industry in the Middle Ages in Europe which was based on coating iron with tin to protect it. In the 1600s, Saxony was the centre for tin-plating, after tin mines were established in the area. Tin-working employed upwards of 80,000 men and the goods they made were exported all over Europe. By the mid-1700s, tin was being used to make plates, pans, tankards, teapots, and coffee-pots, as well as for plating lots of other domestic utensils such as caldrons and basins. About this time the tin smelters of Cornwall were already adding bismuth to make tin harder and give it a brilliant lustre.

Tin can be hammered into thin sheets, known as tin foil and this was used for hundreds of years until aluminium foil replaced it.

Economic element

There are a few tin-containing minerals, but only one ore is of commercial significance and that is cassiterite, which is tin(IV) oxide (SnO_2). The main mining areas are to be found in the 'tin belt' which stretches from China through Thailand, Burma, and Malaysia to the islands of Indonesia. Total mine output of tin is around 300,000 tonnes per year and the amounts produced by smelters total 10 percent more than this, the extra coming from the recycling to tin-plated metals like tin cans. The main producers of mined tin are China (136,000 tonnes), Indonesia (100,000 tonnes), Peru (40,000 tonnes), Bolivia (16,000 tonnes), and Brazil (13,000 tonnes). The Congo also produces tin and this may even be as much as 15,000 tonnes but from illegal mines that fund opposition forces. Malaysian mines, which once produced 40 percent of the world's tin, now only produce 2,000 tonnes, and the UK which was a key tin producer of the ancient world now produces none at all. There are around 11 million tonnes of tin in known deposits of which around half is economically recoverable. Most of this material is in China and Brazil.

Tin is produced by heating cassiterite with coke in a furnace. Half the tin produced ends up in solder, a fifth goes into tin plate, with a similar amount being used by the chemical industry. Tinplate is made by hot-dipping sheet steel into a solution of tin(II) chloride or tin(II) sulfate, or by electrolytic deposition. A tin can is coated with a layer of tin only about 15 microns thick. More than 100 billion cans of food are produced every year and about 30 percent of these are collected for recycling.

Tin alloys are employed in many ways: as solder (33 percent tin, 76 percent lead); pewter (85 percent tin, 7 percent copper, 6 percent bismuth, and 2 percent antimony); bell metal (76 percent copper, 24 percent tin); Babbitt metal (for example, 89 percent tin, 7 percent antimony, 4 percent copper) which is used for heavy bearings; and dental amalgams (a typical modern one being 60 percent silver, 27 percent tin and 13 percent copper).[197] The niobium-tin alloy that cor-

---

[197] Tin was sometimes used to adulterate silver, and in 1124 AD, in the reign of King Henry I of England, the men who produced silver coins at the Royal Mint were accused of doing just this and were sentenced to have their right arms amputated. However, forensic investigations in the 1900s failed to find any of the silver coins of that period had more than a tiny amount of tin.

responds to the formula Nb_3Sn is used for superconducting magnets. As little as 0.1 percent tin added to cast iron improves its strength and wear resistance. Tin is also added to many other castings such as engine cylinder blocks, crankshafts and brake drums.

Tin-lead solder was banned in the EU from 2006 except where no alternative is available. However, a new solder consisting of 95.5 percent tin, 3.9 percent silver, and 0.6 percent copper, and known as SAC after the initial letter of each element's chemical symbol, is just as good and may even be better for electronics, although it has the drawback of melting at a slightly higher temperature of 217°C compared to tin-lead solder which melts at 183°C.

A tank of molten tin is used to produce float glass by the Pilkington process. Tin is also deposited electrically onto glass as a thin film, which increases its resistance to breaking and allows thinner-walled containers to be manufactured. Applied as a thicker film it makes glass conducting, and such glass is used for panel lighting.

Some tin compounds find outlets. Tin(IV) oxide, SnO_2, is used for ceramics and has been for millennia; the ancient Babylonians produced wall tiles in the ninth century BC covered with a tin-based glaze. Some tin pigments give beautiful colours such as tin-vanadium which is yellow and cobalt-stannate which is sky blue (known as Cerulean Blue). Tin oxide is used in gas sensors because as it absorbs a gas its electrical conductivity increases and this can be monitored. Tin(II) chloride, $SnCl_2$, is used as a mordant in dyeing calico and natural silk. The tin becomes attached to the fabric and then the dye molecules attached themselve to the tin. In the case of silk dyes the mordant and dye can more than double the weight of the fabric.

Zinc stannate, Zn_2SnO_4, is the preferred fire-retardant for things like plastics because it also suppress smoke emission.

Environmental element

Tin in the environment	
EARTH'S CRUST	2 p.p.m.
	Tin is the 49th most abundant element.
SOILS	av. 1 p.p.m.
SEAWATER	4 p.p.t.
ATMOSPHERE	virtually nil.

Tin(IV) oxide is insoluble, and the ore strongly resists weathering, so the amount of tin in soils, and particularly in natural waters, is low. The concentration in soils is generally within the range 1–4 p.p.m. but some soils have less than 0.1 p.p.m. while peats can have as much as 300 p.p.m. Tin became an environmental problem when TBT was widely used in marine paints which were introduced in the 1960s—see *Dangerous element* section above.

Chemical element

Data file	
CHEMICAL SYMBOL	Sn
ATOMIC NUMBER	50
ATOMIC WEIGHT	118.710
MELTING POINT	232°C
BOILING POINT	2,270°C
DENSITY	7.3 g/cm³ (white tin) 5.8 g/cm³ (grey tin).
OXIDES	SnO and SnO_2.

Tin is a soft, pliable, silvery-white metal (white tin) and is a member of group 14 (row 5p) of the periodic table. It is unreactive because it is protected by an oxide film on the surface, but it dissolves in acids and bases. Pure tin undergoes a transition when cooled below 13°C from the metal form, known as beta-tin (white tin), to a powdery form, known as alpha tin (grey tin)—see below. The transition to grey tin can be prevented by the addition of small amounts of antimony or bismuth. Tin's preferred oxidation states are +2, as in $SnCl_2$, which is a strong reducing agent, and +4, as in $SnCl_4$ and stannates which refers to the ions SnO_3^{2-} and SnO_4^{4-}. One compound of tin important in chemical research is tributyltin hydride, $(C_4H_9)_3SnH$, which is used in reactions that involve free radicals. The tin-hydrogen bond is easily broken leaving $(C_4H_9)_3Sn^{•}$. The black dot signifies the loose electron which characterises all free radicals, and it this which allows these molecules to undergo all kinds of reactions.

There are 40 known isotopes of tin with masses 99 to 138, and 32 known isomers. Tin has 10 main isotopes: tin-120 accounts for 32.5 percent, tin-118 (24 percent), tin-116 (14.5 percent), tin-119 (8.5 percent), tin-117 (7.5 percent), tin-124 (5 percent), tin-122 (4.5 percent), tin-112 (1 percent), tin-114 (0.7 percent), and tin-115 (0.3 percent). Nine are stable, the most of any element. Previously it was thought that tin-124 was stable but it was found to be radioactive with a half-life of 10^{18} years. The second longest-lived radioactive isotope is tin-126 with a half-life of 230,000 years.

Tin has four radioactive isotopes that are found as the fission products in uranium deposits: tin-121, tin-123, tin-125, and tin -126.

Element of surprises

1. Tin can decay to dust but very slowly even at −10°C, although at −33°C it becomes much quicker, and such low temperatures are not unknown. Tin 'plague' as it was known came to public notice during the hard Russian winter of 1850 when the temperature dropped to record lows for weeks on end. The most noticeable effect was on organ pipes in churches which were generally made from tin (95 percent tin, 5 percent antimony) and which became covered with scaly patches that crumbled when touched

It has been suggested that many of Napoleon's Grand Army froze to death on its retreat from Moscow in 1812 because the buttons on their uniforms were made from tin which disintegrated. This is merely an entertaining myth, because at the time the buttons on a soldier's uniform were made of bone. Only officers had metal buttons, and those were made of brass.

The deaths of the members of the Scott expedition to the South Pole in 1912 have also been attributed to tin plague. The explorers left stores, including cans of paraffin, at various points as they trekked to the South Pole to be used on the way back. When these were needed, however, the cans were found to be empty, the fuel having drained away through tiny holes in the tin-soldered joints. The result was that Scott and his comrades eventually died of exposure. The cans of food in their stores were unaffected and were still found to be edible when some were opened more than 50 years later, but even if the tin of such cans had flaked off, the food inside them would have been preserved by the intense cold.

2. When a piece of pure tin is bent it gives out a supposedly plaintive cry, although to hear it, the bending has to be done near to one's ear.

3. Tin has two 'doubly magic' isotopes, these having 'magic numbers' for both the number of protons and the number of neutrons. Tin-100 has 50 protons and 50 neutrons, both of which are 'magic' numbers in nuclear science terms, while tin-132 has 50 protons and 82 neutrons, again a 'magic' number. Only one other element has doubly magic isotopes and that is calcium-40 and calcium 48. However, being 'magic'. doesn't mean these isotopes are particularly stable: tin-100 has a half-life of one second, while tin-132 has a half-life of 40 seconds.

Titanium

Pronounced tit-ay-nee-um, and named after the Titans, the sons of the Earth goddess of Greek mythology. The word titanate refers to the grouping TiO_3^{2-}.

French *titane*; German *Titan*; Italian *titanio*; Spanish *titanio*; Portuguese *titânio*.

Essential element

Titanium metal is the metal of the future because of its lightness, durability, and versatility. Titanium dioxide is the perfect ingredient for paint and sunscreen lotions.

Cosmic element

Rocks brought back from the Moon had 12 percent titanium dioxide, showing that it is as plentiful there as it is on Earth. Titanium oxide bands are prominent in the spectra of M-type stars. These are the coolest type of star, with a surface temperature of less than 3,200°C, and are red in colour.

When meteorites are in space they are bombarded with cosmic rays which create new isotopes such as titanium-44. Meteorites that have fallen to Earth in the past 240 years have been analysed for their titanium-44 content and this has allowed the Sun's activity to be plotted and has shown that it increased in the last century.

Element of life

Titanium in the human body	
BLOOD	*approx.* 50 p.p.b.
BONE	*approx.* 40 p.p.m.
TISSUE	1–2 p.p.m.
TOTAL AMOUNT IN BODY	*approx.* 700 mg

There is no known biological role for titanium. Most plants contain some, generally about 1 p.p.m. (dry weight), although a few have up to 80 p.p.m., such as horsetail and nettle. Food plants contain 2 p.p.m. or less. Titanium can have a beneficial effect on plants, stimulating the production of carbohydrates, so there may be some biological role for this element, but the mechanism by which it makes plants grow bigger has yet to be identified.

There is a detectable amount of titanium in the human body and it has been estimated that we take in about 0.8 mg a day, but most passes through us without being absorbed. Titanium is not a poisonous metal and the human body can tolerate titanium in large doses. If anything, titanium has a stimulatory effect.

Medical element

In the 1950s, surgeons discovered that titanium metal was ideal for pinning together broken bones. It resists corrosion, bonds well to bone, and is not rejected by the body. Hip and knee replacements, pace-makers, bone-plates, and cranial plates for skull fractures, can be made of titanium and remain in place for up to 20 years. Titanium implants have been used for attaching false teeth, by means of metal plugs inserted into the jaw bones using a technique developed by Professor P. I. Branemark of Gothenburg, Sweden. There are people with such implants who have had them for more than 30 years.

Before inserting a titanium device into the body it has to be prepared by exposing it to a high temperature plasma arc which strips off the surface atoms, thereby exposing a fresh layer of the metal which is instantly oxidised. It is to this oxide film that tissue bonds strongly.

Element of history

The first titanium mineral was discovered in 1791 in a remote village in Cornwall, southwest England, by the Reverend William Gregor (1761–1817) who was the vicar of Creed. He was intrigued by a black sand that he had come across at the side of a stream in the near-by parish of Menachan. The sand was odd because it was attracted to a magnet. He analysed it as best he could and deduced it was made up of two metal oxides: iron oxide, which explained the magnetic properties, and another one which was not magnetic and which he could not identify. He realised he had stumbled upon a new metal and reported it to the Royal Geological Society of Cornwall. He also announced his discovery in the 1791 edition of the German science journal, *Crell's Annalen*, in which he suggested calling the new mineral menachanite after the name of the village where it had been found. This mineral is now known as ilmenite ($FeTiO_3$). Had he been able to isolate the metal, and prove it was a new element, his intention was to call it menaccin.

Four years later, in 1795, the eminent German scientist Martin Heinrich Klaproth (1743–1817) of Berlin, re-discovered the same element and it was he who gave it the name titanium, derived from the Titans of Greek mythology, and he said he chose this name because in itself it had no meaning so 'could give no rise to any erroneous idea.' Klaproth produced the metal oxide from a specimen of a red ore known as Schörl, which he had been sent from Hungary. This mineral is a form of rutile (TiO_2). When he was told of Gregor's earlier discovery he investigated a sample of menachanite and confirmed it too yielded the same metal oxide.

Neither Gregor not Klaproth lived to see the metal itself. Although Klaproth tried hard to reduce the oxide, he did not succeed. Indeed, titanium is impossible to extract from its oxide ore by heating with carbon, and while this does react to remove the oxygen, it reacts further to produce titanium carbide which is also very intractable. Others in the 1800s who tried to isolate titanium and thought they had succeeded, were invariably disappointed when they discovered that they had obtained only titanium nitride. Those who produced samples of the metal in which there was only a few percent of impurities found it to be brittle and unworkable.

It was not until 1910 that M. A. Hunter, working for General Electric in the USA, made pure titanium by heating titanium tetrachloride and sodium metal under high pressure in a sealed vessel. This yielded 99.8 percent pure titanium and it now revealed itself to be a rather remarkable metal, easily worked, incredibly strong, and able to retain this property at high temperatures. It is very resistant to corrosion.

Economic element

Titanium minerals are rutile, brookite, anatase (which are all forms of titanium dioxide, TiO_2), ilmenite (iron titanate, $FeTiO_3$) and titanite (calcium titanate silicate, $CaTiSiO_5$). There is also perovskite (calcium titanate, $CaTiO_3$) and this can be rich in niobium, cerium, and other lanthanoid elements and may be a source of these metals rather than of titanium. The stars in star sapphires and star rubies are due to titanium dioxide impurities.

The chief mined ore is ilmenite and it comes from South Africa, Australia, Canada, China, Norway, India, Vietnam, Ukraine, and the USA. There are also worked deposits of rutile in Sierra Leone, India, and South Africa. World production of ilmenite is 12 million tonnes per year, of rutile is 640,000 tonnes, and of all forms of TiO_2 is 6 million tonnes. Reserves of titanium amount to more than 600 million tonnes. Titanium metal production is around 100,000 tonnes per year and increasing.

Titanium is used either as the metal or as its oxide. Each has become important in human affairs in the past 50 years. Some titanium is converted to the trichloride ($TiCl_3$) which is used as a catalyst in the manufacture of polypropylene. Another chloride is titanium tetrachloride ($TiCl_4$) which is a crystal clear, volatile liquid which boils at 136°C and is consequently easy to purify. It has been used to create smoke screens because when it reacts with water it produces dense clouds of titanium dioxide particles.

The metal itself is produced via $TiCl_4$ which is heated with magnesium metal at 1,300°C. The process was first shown to be viable by Justin Kroll in Luxembourg in 1932, although he used calcium metal as the reducing agent. In the modern process the chemical reaction between the titanium tetrachloride and magnesium generates enough heat to keep the process going and the slag of molten magnesium chloride is tapped from the bottom of the reactor and is recycled electrolytically to magnesium metal and chlorine gas, which goes to making more titanium tetrachloride.

Titanium metal is light and strong and used in products as varied as golf clubs and spacecraft. It is set to become the metal of this century and while most is used for transport, some finds its way into everyday life. It ends up in laptops, adventure bikes, crutches, and body jewellery such as tongue studs and eyebrow rings. Backpackers prefer the titanium versions of food utensils and other equipment because they are much lighter to carry. However, it is industry which uses most titanium. It is as strong as steel but 45 percent lighter and it is immune to metal fatigue. Power plant condensers worldwide contain millions of metres of titanium piping and it is claimed that none has ever failed due to corrosion.

Titanium metal is sold in four certified grades which vary according to their oxygen content: grade 1 has 0.18 percent; grade 2 has 0.25 percent; grade 3 has 0.35 percent; and grade 4, the strongest, has 0.40 percent. The titanium alloy most used is known as 6AL-4V and consists of 90 percent titanium, 6 percent aluminium, and 4 percent vanadium. It is employed in aircraft, helicopters, spacecraft, and missiles. The Boeing 777 contains almost 60 tonnes of titanium and the Airbus A380 has 145 tonnes of which around 25 tonnes is in the engines. Process engineers use titanium in chemical plants where corrosive acids, chlorine, and sulfur compounds are handled. Titanium electrodes, often coated with a precious metal, are used in chlorine production, electroplating, and other similar applications.

Titanium has its drawbacks, such as low thermal and electrical conductivity. It lacks the elasticity of steels, becomes brittle when exposed to hydrogen gas, and tends to stick when in contact with other metals. Despite all this, its uses are increasing because of its inherent benefits of excellent strength-to-weight ratio and high temperature properties, which make it the preferred metal for gas turbine engines and their housing.

Titanium metal can cope with all kinds of extreme conditions due to its impervious layer of titanium dioxide. This layer is only 1–2 nanometres thick to begin with, although it continues to grow slowly, reaching 25 nm after about four years. This protective layer can resist most forms of chemical attack and even when it is damaged it instantly repairs itself. If the oxide film on titanium is artificially enhanced, by anodic oxidation, it produces an iridescent surface and is then suitable for jewellery, particularly popular for earrings.

Titanium is impervious to seawater. Submarines built by the Soviet Union in the 1960s had hulls of titanium, as have modern submarines, and their high performance jet fighters of the Cold War era relied on it. (The Russians are still among the major titanium metal producers.) The US high-speed (Mach 3) research and reconnaissance aircraft were 85 percent titanium. Today, titanium is used in off-shore oil rigs, and it is used for propeller shafts, rigging, and other parts of ships exposed to sea spray and water. It is also used on board ship for fire pumps, heat exchangers, and piping. Titanium metal is ideal for heat transfer applications in which the coolant is seawater or even polluted waters. Desalination plants rely particularly on titanium components. Titanium pipe is used in oil exploration and its light weight and flexibility make it the preferred material for deep sea exploration.

Architects are beginning to use titanium cladding for buildings, an example of which is the Guggenheim Museum. This occupies a waterfront site in Bilbao, Spain, and is sheathed in 33,000 square metres of pure titanium sheet which is guaranteed to resist corrosion for more than 100 years. The roof of Hong Kong's central railway station consists of 6,500 square metres of titanium sheeting.

A more recent method of making titanium metal has been developed and that uses an electrical process in which titanium oxide is reduced to the metal in a bath of molten calcium chloride at 950°C employing a potential of 3 volts. This method was discovered by Derek Fray, Tom Farthing, and George Chen of Cambridge University, England. The cathode of the electrolytic cell is the container, the electrolyte for the titanium dioxide is molten calcium chloride, and the anode is carbon. The electric current then converts the titanium ions to titanium metal which collects at the cathode while the oxygen ions move to the anode and are there released as oxygen gas. The FFC process, named after its discoverers, takes only 24 hours to make the same amount of titanium metal that takes the Kroll method a week or more to produce. It could increase the output of titanium to a million tonnes a year. The method can also produce titanium alloys in a singe process.

Titanium dioxide is the most used titanium chemical. It is the modern equivalent of white lead, but much safer because titanium compounds are not poisonous. More than half the titanium dioxide manufactured goes into paints, a quarter goes into plastics, and the rest into paper, fibres, ceramics, enamels, food colouring (wedding cakes for example), printing inks, and laminates. Paper contains titanium dioxide to prevent the print from one side of the page from showing through to the other side.

The titanium oxide industry started in the 1930s when paint manufacturers were seeking a replacement for white lead and found titanium dioxide to have excellent covering properties. It is non-toxic, does not discolour, and has a very high refractive index. (Refractive index measures a substance's ability to scatter light, and titanium's refractive index of 2.7 is even greater than that of diamond at 2.4.) This accounts for the brilliant whiteness titanium dioxide imparts to domestic appliances. Titanium dioxide ointment is used as a sunscreen because it prevents UV rays from reaching the skin and applied as a lotion of nanosized particles it appears invisible.

Titanium dioxide is traditionally made by dissolving titanium ore in sulfuric acid, precipitating the wet oxide and heating it at 1,000°C. The more modern process uses chlorine gas to convert the ore to $TiCl_4$ which is then oxidised with oxygen at 1,000°C. This regenerates the chlorine gas which is recycled. In an improved chloride process, the titanium tetrachloride is oxidised in a plasma arc at 2,000°C. The titanium dioxide is milled to form crystals of the right size and sometimes they are coated with aluminium oxide to make them mix with liquids more easily.

Self-cleaning glass windows have a micro thin transparent layer of TiO_2 which is deposited on the glass using chemical vapour deposition. The glass is exposed to a mixture of $TiCl_4$ and a source of oxygen, such as methanol, and heated to around 600°C whereupon they react to form TiO_2 which bonds firmly to the surface of the glass. It is this layer which acts to remove sticky organic dirt by catalysing its oxidation to CO_2 using the UV rays of sunlight. (Inorganic dirt is washed off by rainwater.) The layer of TiO_2 on glass is only a tenth of a micron thick. Bathroom fittings coated with fine particles of TiO_2 are being made in Japan and when these are exposed to UV light they oxidise organic molecules and kill bacteria. A method of killing airborne bacteria has been devised which works by passing air over UV-activated TiO_2.

Apart from titanium metal and titanium dioxide, there are other commercial products, albeit of relatively small demand. Titanium carbide, TiC, is made by the action of carbon black on titanium dioxide at 2,000°C. It is the most important metallic hard material after tungsten carbide, and in fact is the hardest of all the metal carbides, but it is rarely used because it is brittle and it has to be mixed with the carbides of tungsten, tantalum and niobium. Titanium will also burn in nitrogen, forming titanium nitride. Titanium nitride (TiN) is as hard as carborundum and used to give added strength as a coating to drills and other cutting tools.

Environmental element

Titanium in the environment	
EARTH'S CRUST	6 p.p.m.
	Titanium is the 9th most abundant element.
SOILS	*approx.* 0.33% (3300 p.p.m.), range 0.02–2.40%
SEAWATER	*approx.* 1 p.p.b.
ATMOSPHERE	traces in dust.

Titanium is widespread in rocks and soils and the majority of minerals contain it, and especially silicate ones. Titanium dioxide and titanates are the most stable of all soil components and because they are resistant to weathering very little titanium is leached into rivers.

In itself, titanium poses no threat to the environment, although at times the manufacture of titanium dioxide has been responsible for large scale pollution of rivers and coasts by acidic effluent, although this is no longer permitted.

Chemical element

Data file	
CHEMICAL SYMBOL	Ti
ATOM NUMBER	22
ATOMIC WEIGHT	47.867
MELTING POINT	1,660°C
BOILING POINT	3,287°C
DENSITY	4.5 g/cm³
OXIDES	TiO_2 (the most stable), TiO, and Ti_2O_3

Titanium is hard, lustrous, silvery metal which heads group 4 (row 3d) of the periodic table, and it is one of the transition metals. Powdered titanium metal will burn if ignited and it is used in fireworks to produce brilliant sparks. Titanium is unaffected by many acids (except hydrofluoric acid, phosphoric acid, and concentrated sulfuric acid) and by alkalis. The dominant oxidation state is +4, as in $TiCl_4$, but compounds of oxidation state +3 are also stable, such as $TiCl_3$.

There are 24 known isotopes of titanium with mass numbers 38 to 61, and two known isomers. There are five naturally occurring isotopes: titanium-48 accounts for 74 percent, titanium-46 (8 percent), titanium-47 (7.5 percent), titanium-49 (5.5 percent), and titanium-50 (5 percent). None is radioactive. The longest-lived radioactive isotope is titanium-44 with a half-life of 63 years. The next longest-lived is titanium-45 whose half-life is 185 minutes.

Element of surprises

1. Titanium dioxide is added to lipstick on account of its superb covering power. A typical lipstick contains 10 percent, although this can vary. Lighter shades require more in order to dilute the intensity of the red dye.

2. In 1999, the shortest laser pulses were achieved used a titanium-doped sapphire laser, which produced pulses of only four femto-seconds (4 *million* billionths of a second, that is, 4×10^{-15}). Atto-second pulses (10^{-18} second) pulses have since been achieved using other substances.

3. Some horseshoes are now made of titanium because it is lighter and stronger than traditional iron shoes.

4. Paving sidewalks in cities with so-called Noxer paving block will reduce atmospheric pollution from vehicles. These were pioneered by Japan's Mitsubishi Materials Corporation and they have a surface layer of titanium dioxide and it is this which converts the nitrogen oxides, NO_x, from exhaust fumes to harmless nitrate. In Japan there are 30 towns and cities where these paving blocks are now in place, and streets in the London Borough of Westminster are having them installed.

5. In the USA, the chemical company Du Pont has developed a new process for making titanium metal which requires much less energy. In it the TiO_2 is partly reduced with carbon before being completely reduced by electrolysis to give titanium powder.

Transfermium elements

Element 100 is fermium, and elements of atomic number 101 and above are called the transfermium elements. They have short half-lives, sometimes as short as fractions of a second, with the result that only a few of their physical or chemical properties have been recorded. Indeed, for some of them all that is known is that they can be made, and some have yet to be named. The naming of these elements has caused controversy in the past, mainly because those who claimed to have been the first to make them have sometimes not always had those claims verified by the International Union of Pure and Applied Chemistry (IUPAC) whose job it is to approve the names and symbols of the chemical elements.

None of these elements has a human role, a biological role, an economic role, or an environmental role.[198] The chemical data about them are limited to atomic number, known isotopes and their half-lives, and sometimes a little information about one or two of their compounds.

The half-lives of the longest lived isotopes of the transfermium elements*

Atomic number	Name and symbol	Isotope	Half-life as measured**	Half-life as calculated
101	Mendelevium, Md	Md-258	52 days	n/a
102	Nobelium, No	No-259	58 minutes	No-261, 2.78 hours
103	Lawrencium, Lr	Lr-262	3.6 hours	Lr-264/265, 10 hours
104	Rutherfordium, Rf	Rf-267	1.3 hours	n/a
105	Dubnium, Db	Db-268	29 hours	n/a
106	Seaborgium, Sg	Sg-271	1.9 minutes	Sg-272, 1 hour
107	Bohrium, Bh	Bh-270	61 seconds	Bh-273/274, 90 minutes
108	Hassium, Hs	Hs-269	9.7 seconds	Hs-276, 1 hour
109	Meitnerium, Mt	Mt-278	8 seconds	Mt-279, 6 minutes
110	Darmstadtium, Ds	Ds-278	10 seconds	Ds-280, 11 seconds
111	Roentgenium, Rg	Rg-281	22.8 seconds	Rg-283, 10 minutes ***
112	Copernicium, Cn	Cn-285	34 seconds	Cn-285m, 8.9 minutes ***
Un-named elements				
113	Ununtrium, Uut	Uut-286	19.6 seconds	Uut-287, 20 minutes
114	Ununquadrium, Uuq	Uuq-289	2.6 seconds	n/a
115	Ununpentium, Uup	Uup-289	220 milliseconds	Uup-291, 1 minute
116	Ununhexium, Uuh	Uuh-293	61 milliseconds	n/a

[198] It is just conceivable that some long-lived isotopes of these elements might one day come to light and that these could have useful properties.

117	Ununseptium, Uus	Uus-294	78 milliseconds	n/a
118	Ununoctium, Uuo	Uuo-294	0.89 milliseconds	n/a
119	Ununennium, Uue	n/a	n/a	n/a
120	Unbinilium, Ubn	n/a	n/a	n/a
121	Unbitrium, Ubu	n/a	n/a	n/a
122	Unbibium, Ubb	n/a	u n/a	n/a ***
123	Unbitrium, Ubt	n/a	n/a	n/a
124	Unbiquadium, Ubq	n/a	n/a	n/a
125	Unbipentium, Ubp	n/a	n/a	n/a
126	Unbihexium Ubh	n/a	n/a	n/a
127	Unbiseptium, Ubs	n/a	n/a	n/a

* The longest lived isotope of fermium is fermium-257 with a half-life of 101 days.

** These should be treated as provisional.

*** There have been claims for half-lives of roentgenium-261 and 265 of 10^8 years, copernicium-272 of 47 days, and unbibium-292 of 10^8 years. These claims have yet to be proven.

As the Table above shows, transfermium elements have very unstable nuclei, and as the atomic number increases they becomes progressively harder to make and detect. The half-lives of the longest-lived isotope become shorter until we reach element 111, when they begin to increase. The elements 114 to 126 are in fact referred to as the 'island of stability' although the stability is relative and only to be compared to those elements immediately preceding it. This island of stability is so called because these elements were predicted by physicists to have longer half-lives because of closed, or near-closed, nuclear shells having 'magic' or 'near magic' numbers of protons and neutrons. No transfermium element is stable and all that can be said about elements in this zone is that they are less unstable than those elements surrounding it. The term 'island of stability' is merely a memorable and convenient way of indicating this.

On the island of stability there should be a peak corresponding to element 114, and especially its isotope 298. This would have 114 protons and 184 neutrons, and nuclear theory has it that this would be particularly long-lived because it corresponds to energy levels within the nucleus being complete and hence less unstable. As yet, it remains undiscovered. Element 114 would come below lead in group 14 of the periodic table.

Elements adjacent to 114 might also have isotopes with reduced instability of the nucleus, so that element 112 (below mercury), element 113 (below thallium), and element 115 (below bismuth) might have isotopes that existed long enough for them to be collected, and from which chemical compounds might even be made. Element 126, with both proton and neutron shells complete is also expected to have a longer half-life.

There has been speculation about the highest possible atomic number and some say it might be at around 200. More realistically, Albert Khazan applied the mathematical principle of Hyperbolic Law to the periodic table and calculated that the highest atomic number possible is 155.

The transfermium elements are useful in the study of nuclear structures and radio-active decay, and their isotopes are routinely synthesised for this kind of research. Many of their isotopes decay by exotic decay modes discovered in the 1980s such as cluster decay and cold (neutronless) binary and ternary fission.

Transfermium elements up to element 126 might once have occurred naturally on planet Earth. Theoretical research starting in the 1960s suggested that traces of heavy and superheavy elements could explain extreme radiation damage in certain minerals caused by alpha-decay energies that supposedly could not be produced by any known element but could be emitted by elements up to 126. This stimulated widespread efforts to discover and synthesise superheavy elements. From the late 1960s to the early 1980s, attempts were made in all the major nuclear laboratories to produce transfermium elements, while researchers worldwide analysed mineral samples for traces of element 126.

The first extensive search for superheavy elements occurred from 1969 to 1972 at the Lawrence Berkeley Laboratory in California, where physicists studied more than 40 types of minerals for evidence of emissions from trace amounts of elements 110 to 114. They did most of this work in railway tunnels where they took their detector equipment and samples so as to reduce interference from cosmic rays. None of the hoped-for elements came to light. In 1969, the Russian Joint Institute for Nuclear Research (JINR) in Dubna carried out an extensive search for element 114 by analys-ing many minerals. They also were unsuccessful.

In the 1960s, it had been discovered that isomers of radioactive isotopes could have longer half-lives that their ground state as found with tantalum. The isotope tantalum-180 has a ground state half-life of 8.1 days but the deformed isomer tanta-lum-180m1 has a half-life of 1.2×10^{15} (1.2 million billion) years and occurs on Earth as a primordial isomer. Thus it was argued that there might be long-lived isomers of transfermium elements and it was calculated that these could have half-lives greater than 10^8 (100 million) years. These elements and their isomers would have been produced by a supernova and so were part of the material that formed the planets. Indeed there have been claims of the detection of several naturally occurring trans-fermium elements (hassium, roentgenium, unbibium), but more research is needed to confirm these.

The naming controversy

It takes a long time for a synthesised element to be given its official name. IUPAC is extremely demanding in the rules that it established for assigning priority of dis-covery. It does not want a repeat of what happened with element 102 when the hasty assignment of priority of discovery was proved incorrect by later research. There were also other controversies between the USA and Soviet Russia laboratories embroiled in Cold War rivalries.

In 1974, IUPAC and the International Union of Pure and Applied Physics (IUPAP) set up a committee to study the claims being made by physicists in the USA and Soviet Russia. It consisted of three members from the USA, three from Soviet Russia, and three from non-aligned countries. Due to Cold War animosities the group never met although several of its members did engage in correspondence.

The committees that eventually evaluated claims were the Transfermium Working Group (TWG), another joint committee set up in 1985 to study claims for all transfermium elements, and its successor, the Joint Working Party (JWP) set up in 1998 to study claims for elements 110–112 (later extended to 118). This used somewhat overly cautious criteria and would only award discovery priority to the laboratory that provided complete and convincing data, even though that element might have been discovered earlier by another laboratory whose data was not as complete. Some elements had been discovered and known for years, with many laboratories regularly producing and studying their isotopes, yet no discoverer had been recognised and those elements had no official names. In the 1970s two unofficial names were in use for elements 104 and 105.

In 1979, IUPAC introduced a system of temporary names based on Greco-Latin terms. Elements would be named solely on the basis of their atomic number: 1 was 'un', 2 was 'bi', 3 was 'tri', 4 was 'quad', 5 was 'pent', 6 was 'hex', 7 was 'sept', 8 was 'oct' and 9 was 'enn'. Thus these elements around which the disputes had arisen now became unnilquadium for element 104, unnilpentium for element 105, and unnilhexium for element 106.

In 1992, the TWG issued its report of assessments and recommendations. Its conclusions were somewhat vague, indicating that for elements 104–107 both contesting groups had some justification for their claims and that the discoveries should be shared. They did, however, assign priority to specific laboratories as having what they said was the 'more credible' claim. See dubnium (105), seaborgium (106), and bohrium (107) for more details.

In 1994, the LBNL caused a stir by proposing the name of seaborgium for element 106. IUPAC pointed out that the rules did not permit the approval of an element named for a living scientist. Later that year, IUPAC's Commission on Nomenclature of Inorganic Chemistry issued a suggested list of names for elements 104–109: element 104 was to be called dubnium, named after Dubna, the town where the JINR was located; element 105 was joliotium (symbol Jl) named after the French physicist Jean-Frédéric Joliot. (This name had originally been proposed by the JINR for element 102, but in honour of Irene Joliot-Curie and given the symbol Jo.) Element 106 was named rutherfordium, which was the LBL's proposed name for element 104; 107 was bohrium; and 108 was hahnium with the symbol changed from Ha to Hn. This was the name that LBNL had proposed name for element 105. The proposed name meitnerium for 109 was retained.

This suggested list of names angered almost everyone; discoverers' names were being disregarded, and name shuffling would result in confusion because several element names had become widely known, such as rutherfordium for 104 and hahnium for 105. Not accepting the name seaborgium angered both the Lawrence Berkeley National Laboratory (LBNL) and the American Chemical Society (ACS). The ACS defied IUPAC by approving the name seaborgium for element 106 at its annual meeting in 1995 and ordered it be used in all its publications. The confrontation led to two more years of negotiations. The LBNL too refused to relent and demanded that all of their proposed names for elements 104–106 be accepted. As a result, the process was being held hostage to their demands. The animosity between the LBNL and the JINR was particularly acrimonious with the LBNL insisting on its name proposals to the total exclusion of the JINR.

In an effort to achieve a compromise, an ad hoc committee of IUPAC in 1995 suggested a revised list of names which included seaborgium for element 106 to appease the LBNL and flevorium for element 102 in honour of the Russian nuclear physicist Georgy Flerov (1913–90), who discovered elements 102, 104, and 105. In addition, it had dubnium for 104, joliotium for 105, bohrium for 107, and hahnium for 108 with the originally-proposed symbol Ha re-instated. This now angered the German physicists at the Geselleschaft für Schwerionenforschung (Institute for Heavy-ion Research, GSI) because it disregarded their proposed name hassium. This new list was quickly rejected by everyone.

The final list that was eventually approved had rutherfordium for 104, dubnium for 105 to honour the JINR's many discoveries, seaborgium for 106, bohrium for 107, hassium for 108, and meitnerium for 109. As part of IUPAC's agreeing to accept seaborgium for 106, the LBNL had to agree to a compromise that named 105 in honour of Dubna, making it possible for names for elements 104 to 109 to be finally approved in 1997 by IUPAC.

Fuller accounts of the research that went into synthesising the transfermium elements and the background details behind the controversies in naming them are given with the various elements concerned.

As expected, none of the transfermium elements has a human role, a biological role, an economic role, or an environmental role. Even the chemical data are limited to atomic number, isotopes and their half-lives, and sometimes a little information about one or two compounds they form. Nevertheless they have existed, however briefly, and so cannot be ignored and they have been useful in the study of nuclear structures and radioactive decay; their isotopes are routinely synthesized for this research.

Tungsten or Wolfram

Pronounced tung-sten, the name is derived from the Swedish *tung sten*, meaning heavy stone, and the chemical symbol W is derived from wolfram, the old name of the tungsten mineral wolframite. The International Union of Pure and Applied Chemistry (IUPAC) prefers wolfram as the name for the element, but accepts tungsten as an alternative. The element got the name wolfram ('wolf dirt') from German tin miners who found that the presence of certain stones greatly interfered with the smelting of tin and produced a lot more slag than usual; they explained this in terms of the tin being 'devoured,' like a wolf devours sheep. Tungstate refers to the ion WO_4^{2-}.

French *tungstène*; German *Wolfram*; Italian *wolframio* (*tungsteno*); Spanish *wolframio*; Portuguese *tungsténio*.

Essential element

Tungsten alloys are some of the strongest metals known. Tungsten is replacing lead in applications which require a heavy metal, such as in bullets.

Element of life

Tungsten in the human body	
BLOOD	1 p.p.b.
BONE	0.2 p.p.b.
TISSUE	no data
TOTAL AMOUNT IN BODY	20 micrograms

There has been a suggestion that some plants may have a biological need for tungsten, in that it might substitute for molybdenum in nitrogen-fixing enzymes, but the evidence is slim. On the other hand, it is known that some bacteria need tungsten, making this metal the heaviest one to have a role in living cells. The bacteria contain an enzyme which incorporates tungsten and this can reduce a carboxylic acid molecule (RCO_2H) to an aldehyde (RCHO).[199]

Tungsten is only mildly toxic, although tungsten dust is a skin and eye irritant. The estimated daily intake of tungsten is around 12 µg per day but only a small amount of this is absorbed, as shown by tests using a radioactive tungsten tracer. That which is absorbed tends to deposit in the bones and spleen.

Plants will pick up tungsten from the soil in which they grow and some trees in the Rockies have been found to have up to 100 p.p.m. Food plants generally have very

---

[199] The R refers to the rest of the molecule.

little—and some none at all—and when it is detectable it rarely exceeds 0.1 p.p.m. (dry weight). It was observed that watering with sodium tungstate solution increases the growth and yields of grapes and alfalfa. Tests using radioactive tungsten-185 showed that barley removed a large part of that which was applied to a sample of soil and it appeared to absorb it in the form of the tungstate ion.

Element of history

In China, porcelain manufacturers of the 1600s produced a unique peach colour with the aid of a tungsten pigment, although without realising the nature of what it was.[200]

Minerals containing tungsten came to the attention of investigators in the mid-1700s. In 1761, the German chemist Johann Gottlob Lehmann (1719–1767) analysed wolframite without recognising that it contained two unknown metals, tungsten and manganese. In 1779, the Irish chemist Peter Woulfe (1727–1803), examined the mineral called *tung sten*, which was to be found in Sweden and in Germany, and concluded that it probably contained a new metal, but got no further with his research. Two years later the eminent Swedish chemist, Wilhelm Scheele (1742–86), investigated it and succeeded in isolating an acidic white oxide which he concluded was not the molybdenum that he had thought the mineral might contain. He rightly deduced that it was the oxide of a new metal. However, Scheele ended his investigations before he could show conclusively that it was an element.

The credit for discovering tungsten went to two Spanish brothers, Juan José Elhuyar (1754–96) and Fausto Elhuyar (1755–1833) who came from northern Spain. They were interested in mineralogy and studied at some of the main European centres in Sweden and Germany. It was at the Seminary at Vergara, in Spain, that they carried out their researches into *tung sten* and wolframite in 1783, showing that both yielded the same acidic metal oxide, some of which they were even able to reduce to the metal itself by heating it with carbon. Fausto wanted to call the new element volfram, which is still its name in Sweden, while Juan preferred tungsten, the name by which the ore was better known, and this became this preferred name for the metal in England and France. In Germany, Spain, and Italy it was called wolfram.[201]

Element of war

Just as adding tin to copper produced bronze, which made stronger swords and axes, so adding tungsten to iron produced a superior type of steel, and this too was much used in weaponry in the wars of the last century. This discovery occurred in 1864, when the Englishman, Robert Forester Mushet (1811–91) added about 5 percent of tungsten to steel and produced a metal that was not only harder and stronger, but

---

[200] Tungsten is still used in pigments as bright yellow tungsten oxide or bright white barium and zinc tungstates.

[201] Juan Elhuyar became Director of Mines in the Spanish colony of Colombia where he died in 1796, while Fausto Elhuyar was appointed to the same position in Mexico. He set up the School of Mines in Mexico City, and oversaw the development of mining in that country, before returning to Spain in the early 1820s, where he became Director General of Mines for the whole of Spain.

could withstand red heat without deforming. Mushet's self-hardening steel, as it was called, first found use in machine tools enabling metal cutters to work much faster and to last longer.

Soon, tungsten steels were being used to make guns and cannons, and in World War I the German armament makers perfected some that could fire up to 15,000 rounds without becoming unserviceable, compared to Russian and French weapons that became unserviceable after firing about half this number of shells or bullets. As the benefits of tungsten steel were appreciated, the demand rose rapidly, so that whereas only a few hundred tonnes of the metal were produced in the late 1800s, by 1918 demand had soared to 35,000 tonnes creating a chronic shortage of the metal in war-torn Europe.

The far-sighted German gun makers had stockpiled the metal early in the 1890s against such an eventuality, but this was soon used up. The shortage became so acute that even the slag heaps of old tin mines and smelters were reworked to extract the tungsten that had so bedevilled the Medieval miners of Germany. In World War II, the mines in neutral Portugal, at Panasqueira and Borralha, became major producers.

Today, tungsten heavy metal, also known as heavimet or densalloy, is produced for armaments and this is made from tungsten sintered with either iron-copper or iron-nickel and it has a density of around 18g/cm^3. This is used for high-velocity penetrating shells designed as anti-tank weapons. Tungsten is now replacing lead in bullets especially in the US Army as part of the Green Ammunition Program.

Economic element

There are several tungsten minerals, such as ferberite, $FeWO_4$, scheelite, $CaWO_4$, and wolframite, $(Fe,Mn)WO_4$, of which the last two are most important. World production of tungsten is around 74,000 tonnes per annum of which China accounts 65,000 tonnes, Russia for 4,000 tonnes, Canada for 3,000 tonnes, Rwanda for 2,000 tonnes and Austria for 1,000 tonnes. Other countries have tungsten mines but these produce only small amounts. (Portugal still mines around 800 tonnes per year.) Reserves are estimated to be around 5 million tonnes, most of which is in China (3.5 million tonnes), Canada (600,000 tonnes), and the USA (400,000 tonnes), with the Canadian deposits as yet virtually untouched. Tungsten is also recycled and this meets 30 percent of demand.

Tungsten is generally obtained as a dull grey powder, which is difficult to melt. As the pure metal, it is easily worked, can even be cut with a hacksaw, and is very ductile (a gram of the metal can be drawn into a wire 400 m long) but tungsten generally contains a small amount of carbon and oxygen and this makes it much harder.

Pure metallic tungsten was used as the filaments in old-style incandescent light bulbs which were a common form of lighting during most of the last century. In 1903, W. D. Coolidge succeeded in preparing tungsten wire. By reducing tungsten oxide to the metal and then pressing the tungsten powder he obtained thin rods which he was able to draw out into fine wire, and this became the ideal material for filaments. Not only was it high melting, it has the lowest coefficient of expansion of all metals and the same as that of glass, so it could be sealed into the glass. Moreover, its vapour pressure is the lowest of all metals, which meant that the filament did not evaporate and redeposit on the cooler parts of the light bulb. However, incandescent light bulbs waste energy by producing much more heat than light, although some are still used for specialty lighting such as for airport runway markers and for stage lighting.

Tungsten is needed for electric contacts, arc-welding electrodes, and for heating elements in high temperature furnaces. Tungsten does not easily volatilise because it has the highest boiling point of any material at 5,700°C.[202] Tungsten is used in alloys, such as steel, to which it imparts great strength. So-called high-speed steels contain around 7 percent tungsten and other metals, and these are used for various tools such as milling cutters, gear cutters, saw blades, punches, and dies. However, these grades of steel are declining as cemented carbide tools take their place.

Cemented carbide, also known as hardmetal or cermet, is the most important use for tungsten and its main component is tungsten carbide (WC). It is made by mixing tungsten powder with pure carbon powder and heating to 2,200°C in a furnace. Various grades of cemented carbides are made, containing between 75 and 96 percent of tungsten carbide. Cemented carbide accounts for 40 percent of the world production of tungsten. It was invented in 1923 by Kart Schroter, who combined cobalt and WC. Cemented carbide has the strength to cut cast iron and it makes excellent cutting tools for the machining of steel, and indeed it revolutionised productivity in many industries when it was introduced in the 1930s. For example, drills with cemented carbide tips increased the lifetime of rock drills by a factor of 10 and these were particularly useful in drilling for oil. At the other end of the scale are the ultra-high speed tungsten-tipped dental drills, and the drills used to bore very fine holes in printed circuit boards, which are tipped with a particularly fine grade of cemented carbide.

X-ray tubes for medical use have a tungsten anode from which the rays are emitted when this is excited by an electrical discharge.

A few tungsten compounds have found limited use. Tungsten oxide catalysts are used for hydrocracking oil, and the removal of sulfur and nitrogen compounds from oil-based products. It helps to increase the yields of gasoline and other light hydrocarbons in crude oil processing thereby making the products more environmentally acceptable. Tungsten disulfide is a high temperature lubricant, able to act at temperatures higher than the more usual molybdenum disulfide.

Because it is such a heavy metal, tungsten has been used as ballast for yachts keels, aircraft tails, and F1 racing cars.

Environmental element

Tungsten in the environment	
EARTH'S CRUST	1 p.p.m.
	Tungsten is the 58th most abundant element.
SOILS	1–2.5 p.p.m.
SEAWATER	90 p.p.t.
ATMOSPHERE	virtually nil.

[202] It would only just be molten on the surface of the sun.

In soil, tungsten metal oxidises to tungstate, WO_4^{2-}. Very little tungsten has been detected in the few soils that have been analysed, although around an ore-processing plant in Russia levels as high as 2000 p.p.m. (0.2 percent) were found. Even so, this element poses no real threat to the environment. Indeed, in one respect it has been entirely beneficial: fishing weights made from plastic-covered tungsten have now replaced the lead sinkers previously used. These caused the death of bottom feeding birds, such as swans, which scooped them up. The sinkers became lodged in their stomachs and, as the lead dissolved, the birds slowly died.

Chemical element

Data file	
CHEMICAL SYMBOL	W
ATOMIC NUMBER	74
ATOMIC WEIGHT	183.84
MELTING POINT	3,407°C (highest among metals)
BOILING POINT	5,700°C
DENSITY	19.3 g/cm³
OXIDES	WO_2, W_2O_5 and WO_3

Tungsten is a lustrous and silvery-white metal that is a member of group 6 (row 5d) of the periodic table, a group it shares with the lighter metals chromium and molybdenum. Tungsten is the metal with the highest melting point, and the highest tensile strength at temperatures above 1,300°C when it is red hot. The bulk metal resists attack by oxygen, acids and alkalis. The most common oxidation state of tungsten is +6, as in the oxide WO_3, the chloride WCl_6 and in the tungstate ion, WO_4^{2-}. In solution tungstate forms more complex structures such as the ion $[W_{12}O_{41}]^{10-}$. Other oxidation states are known, namely +2, +3, +4 and +5, as in the chlorides: WCl_2, WCl_3, WCl_4, and WCl_5.

There are 33 known isotopes of tungsten with masses 158 to 190, and 11 known isomers. There are five naturally occurring isotopes of tungsten: tungsten-184 accounts for 30.6 percent, tungsten-186 (28.4 percent), tungsten-182 (26.5 percent), tungsten-183 (14.3 percent), and tungsten-180 (0.1 percent). All these have been found to be radioactive but all with incredibly long half-lives. Tungsten-180 has a half-life of 1.8×10^{18} years (1.8 *billion* billion years) while the others have half-lives as yet unmeasured but calculated to be up to 10^{32} years. A more typically radioactive isotope is tungsten-181 with a half-life of 121 days.

Tungsten has the distinction of being the element used as the target material during the first so-called 'cold fusion' method of making elements beyond the end of the periodic table. This took place in 1971 when Amnon Marinov bombarded a tungsten target with strontium projectiles and claimed to have synthesised element 112—see copernicium.

575

Element of surprises

1. There are several ways in which tungsten is used at the cutting edge of objects designed to impact on things, such as the tips of ball-point pens, darts, armour-penetrating shells, and the nozzles of Polaris submarine nuclear rockets.

2. In 2010, chemists discovered a tungsten molecule with the unique capability of breaking the carbon-carbon bonds of benzene-type rings.

Un-named elements[203]

These are the elements which have yet to be given their names and they have atomic numbers from 113 upwards. Here they are listed in numerical order, not alphabetical order. Their names are based on a Greco-Latin code for their atomic number so, for example, element 116 is un-un-hex-ium (one-one-six-ium). Un-names are only temporary and will be replaced by names chosen by their discoverers when the element has been confirmed. The name will then need to be approved by the International Union of Pure and Applied Chemistry (IUPAC).

As the atomic number, that is the number of protons in the nucleus, increases, so does the instability of the nucleus. There are a few exceptions to this generalisation such as element 43 (technetium) which is unstable, but inherent instability does not really manifest itself until we go beyond bismuth (element 83), and yet some elements beyond this are relatively stable such as thorium (element 90) and uranium (element 92).

Elements above atomic number 100, the so-called transfermium elements, are all short-lived but there is hope that some of these superheavy elements might have half-lives long enough for them to be of interest beyond the world of nuclear physics. Indeed, it was suggested by Glenn Seaborg (1912–99) and others that there would be nuclei which were less unstable even though they contained many protons. The longer-lived of these 'magic number' nuclei were predicted to be 114, 120, and 126. The elements adjacent to these would also show increased stability and together these were an 'island of stability'.

Superheavy elements have been of interest since the 1960s when it was suggested that traces of them might occur naturally and their high-energy decay could be the cause of extreme radiation damage (the giant radio-haloes) in certain minerals such as the biotite inclusions in the rare earth mineral monazite. Biotite is a complex silicate mineral of composition $K(Mg,Fe)_3AlSi_3O_{10}(F,OH)_2$. The alpha-decay energies calculated to have caused the damage were not produced by any known radioactive element. Because known superheavy elements all had short half-lives, it was suggested that some of them must have long-lived isomers.

A paper published in 1976, postulated that several elements, and particularly those of atomic number 116, 124, 126, and 127, were the possible cause of the unexplained radiation damage. This stimulated worldwide research activity, both physical and theoretical. Some scientists analysed mineral samples in an effort to detect element 126 (and possibly other superheavy elements) while others tried to synthesise superheavy elements in their accelerators. From 1976 to 1983 many papers were published by those looking for traces of these superheavy elements. Of those studying mineral samples, most reported negative results but a few claimed the detection of high-energy emissions that were inconclusive and needed further research.

---

[203] For elements of atomic numbers 100–112 see transfermium elements.

Un-named elements

To occur naturally, the half-life of a superheavy isotope would have to be longer than 100 million years but this is not unrealistic because deformed isomers of isotopes can have longer half-lives than their ground state, and this was known by the 1960s. An outstanding example is the isotope tantalum-180. This has a ground state half-life of only 8.1 days whereas the deformed isomer tantalum-180m1 has a half-life of 1.2×10^{15} years. (1.2 *million* billion years). Consequently tantalum-180m1 occurs naturally as a primordial isomer.

Several efforts to make superheavy elements were made in the late 1970s using the UNILAC accelerator at Darmstadt. In some experiments, a bismuth target was bombarded with xenon in an effort to make element 137; in another experiment a uranium target was bombarded with xenon to make elements 146; and they even bombarded a uranium target with uranium in an effort to make element 184. All failed.

Although these superheavy elements could not be made, others have been produced at various nuclear research facilities around the world, and these institutions are listed in the box.

Nuclear Research Facilities

The sites where the un-named elements have been made are known by their acronyms, which are as follows (in alphabetical order):

CERN stands for Conseil Européen pour la Recherce Nucléaire (European Council for Nuclear Research). This is now called the European Organization for Nuclear Research but is still officially referred to as CERN. It is located on the border between France and Switzerland.

GANIL stands for Grand Accélérateur National d'Ions Lourds (Large National Accelerator of Heavy Ions) and is the French centre for advanced nuclear research and is based at Caen.

GSI stands for Gesellschaft für Schwerionenforschung (Institute for Heavy-ion Research) which is the German centre for nuclear research and is near Darmstadt.

JINR stands for the Joint Institute for Nuclear Research at Dubna, Russia, where there are eight laboratories of which the Flerov Laboratory of Nuclear Resarch (FLNR) is the best known.

LANL stands for Los Alamos National Laboratory, and it is part of the US Department of Energy and is located at Los Alamos, New Mexico.

LBNL stands for the Lawrence Berkeley National Laboratory, which is also part of the US Department of Energy and is located at Berkeley, California.

LLNL stands for Lawrence Livermore National Laboratory. It was founded by the University of California and part funded by the US Department of Energy. It is located at Livermore, California.

PSI stands for the Paul Scherrer Institute which is located in Villigen, Switzerland.

RIKEN stands for the Institute of Physical and Chemical Research, and it is located in Wako, near Tokyo, Japan.

Ununtrium

Pronounced un-un-tree-um,[204] chemical symbol Uut.

There are six known isotopes of ununtrium with mass numbers 278, and 282 to 286, with no known isomers. The longest lived isotope is ununtrium-286 with a half-life of 19.6 seconds, although it is theoretically calculated that ununtrium-287 would have a half-life of 20 minutes.

Element 113 was first sought from 1969 to 1972 when the Lawrence Berkeley Laboratory analysed mineral samples for traces of possible naturally occurring superheavy elements—see ununquadium for details. In 1998, the GSI used the 'cold fusion' method and bombarded bismuth-209 with zinc-70 atoms for six weeks in an unsuccessful attempt to make this element. In fact it was discovered in 2003 at the JINR by a team led by Yuri Oganessian and Vladimir Utyonkov. It was a product of the radioactive decay of the newly-discovered element 115 emitting an alpha particle. This was the first time that an element had been discovered as the decay product of another, newly discovered, element. Isotopes ununtirum-284 and ununtrium-283 were detected.

The first direct production of ununtrium took place in 2004 at RIKEN when it was indeed synthesised by bombarding bismuth-209 with zinc-70. This yielded one ununtrium-278 atom in an experiment lasting eight months. The physicists at RIKEN submitted a claim to IUPAC and have indicated that if they were awarded the discovery they would propose that the element be called japonium (Jp).

Ununtrium is the highest atomic number to have been synthesised by the 'cold fusion' method. Nuclear theory and calculations indicate that this method is not useful for higher atomic numbers.

From its position in the periodic table, in group 13 below thallium, this element should have the physical properties of a main group metal and, were it long-enough lived, it should be possible for it to have two kinds of chemistry corresponding to oxidation states +1 and +3, the lower one being the more stable and having compounds such as UutO and UutF. The corresponding higher oxidation forms would be Uut_2O_3 and $UutF_3$.

---

[204] IUPAC would like such names to be pronounced oon-oon- etc. because the 'un' is derived from the Latin word 'uno'.

Ununquadium

Pronounced un-un-kwad-ee-um; chemical symbol Uuq.

There are four known isotopes of ununquadium with mass numbers 286 to 289, with no known isomers. The half-life of the longest-lived known isotope, ununquadium-289, is 2.6 seconds. (No isotope has yet been theoretically calculated to have a longer half-life.)

It has long been predicted that an atom with 114 protons and 184 neutrons would be particularly stable. That particular isotope of ununquadium has yet to be made but several other isotopes of this element have been produced.

The first attempts to discover this element occurred in 1969 when it was simply referred to as element 114. The JINR tried to detect it in mineral samples and the LBL, as the LBNL was then called, tried to produce it in an accelerator. Both these little-known efforts produced nothing. From 1969 to 1972, the LBL carried out an extensive search for elements 110 to 114 in minerals but again they found nothing. In 1977, physicists at the JINR bombarded ploutonium-244 with calcium-48 in an unsuccessful attempt. These early efforts were stimulated by research into the possibility of there being many naturally occurring superheavy elements. However, the intense interest in element 114 was a result of a 1966 published paper which theorised that 114 might be a nuclear 'magic number' with a possible half-life in excess of a hundred million (10^8) years.

In 1998, a team led by Yuri Oganessian and Vladimir Utyonkov at JINR, with the assistance of Kenton Moody of the LLNL which supplied the plutonium target material, tried again, bombarded plutonium-244 with the rare, naturally occurring isotope, calcium-48. This attempt to produce isotope-289 was successful and a single atom was synthesised. They calculated that 5 billion billion (5×10^{18}) atoms of calcium had been fired at the target during the course of the 40-day experiment.

The JINR did a re-run in 1999 with the same target and projectile materials, synthesising an additional two atoms and confirming their earlier discovery.

From its position in the periodic table, in group 14 below lead, element 114 should have the physical properties of a heavy metal and were it long-enough lived it, should be possible for it to have two kinds of chemistry corresponding to oxidation states +2 and +4, with the former more stable. First indications are that it is reluctant to interact with other atoms suggesting that even the lower oxidation state might not be stable and that ununquadium prefers to remain chemically aloof, rather like a noble gas.

Ununpentium

Pronounced un-un-pent-ee-um; chemical symbol Uup.

There are four known isotopes of ununpentium with mass numbers 287 to 290, with no known isomers. The longest-lived isotope is ununpentium-289 with a half-life of 220 milliseconds. Ununpentium-291 is theoretically calculated to have a half-life of one minute.

An attempt was made in 1977 at the JINR to produce element 115 by bombarding plutonium-244 with calcium-40. Not a single atom of element 115 was produced. Ununpentium was first synthesised there in 2003 by a team led by Yuri Oganessian and Vladimir Utyonkov using the so-called hot fusion method in which americium-243 was bombarded with calcium-48. This produced four atoms of ununpentium, three of isotope 288 and one of isotope 287 from beams of 248 MeV and 253 MeV respectively. (Element 113, ununtrium, was discovered as a product of the decay of these atoms).

From its position at the bottom of the periodic table, in group 15 below bismuth, this element should have the physical properties of a heavy metal, and were it long-enough lived to be examined chemically, it should in theory have two kinds of chemistry corresponding to oxidation states +3 and +5 with the former being much more stable. Ununpentium might even prefer a lower oxidation state of +1 judging by trends down this group of the periodic table.

Ununhexium

Pronounced un-un-hex-ee-um; chemical symbol Uuh.

There are four known isotopes of ununhexium with mass numbers ranging from 290 to 293, and no known isomers. The half-life of the longest-lived known isotope is ununhexium-293 with a half-life of 61 milliseconds.

The first attempt to synthesise element 116 took place in 1976–77 at the LBNL where a collaboration of researchers from the LLNL, LBNL, and GSI bombarded curium-248 with calcium-48 in an unsuccessful run. In 1977, the JINR bombarded curium-246/curium-248 with calcium-48 in another unsuccessful attempt. In 1984, there were two further unsuccessful attempts using curium-248 with calcium-48 in a joint effort at both the LBNL and GSI. In 1997, the GSI bombarded lead-208 with selenium-82 in another failed attempt. The early attempts from 1976 to 1984 were stimulated by a 1976 paper which suggested that naturally occurring superheavy elements caused unexplained radiation damage in minerals, specifically mentioning element 116.

Finally, a team of physicists at JINR led by Yur Oganessian, Vladimir Utyonkov, and Kenton Moody synthesised element 116. In 2000 they bombarded curium-248 target with calcium-48 nuclei and got isotope-292. This was done in collaboration with the LLNL who supplied the target material. At the time, the researchers simply thought that they were synthesising a new isotope of an already discovered element.

In 1999, the LBNL had announced the discovery of elements 118 and 116, the latter being produce by the alpha decay of the former. Victor Ninov reported it in *Physical Review Letters* saying that he had discovered ununhexium-289 as part of the decay chain of element 118, and that it had a half-life of 0.64 milliseconds. However, the LBNL withdrew its claim in 2001 after five other experiments at LBNL and various other laboratories around the world had attempted to produce element 118 but failed to confirm its existence. Re-analysis of the original LBNL data exposed the earlier claim as a fraud perpetrated by Ninov.

Finally research at JINR in 2001 really did produce ununhexium-293. Since then, around 25 atoms of ununhexium have been made.

From its position in the periodic table, in group 16 below polonium, this element should have the physical properties of metal and given a long-lived isotope it should be possible to make compounds of oxidation states +4 and +6, such as the oxides $UuhO_2$ and $UuhO_3$, and the fluorides, $UuhF_4$ and $UuhF_6$. In view of the trend in properties down group 16, it is possible that these higher oxidation states would be unstable and that the +2 lower oxidation state might be preferred.

Ununseptium

Pronounced un-un-sept-ee-um; chemical symbol Uus.

There are two known isotopes of element 117: ununseptium-293 and ununseptium-294, and no known isomers. The latter is the longer-lived isotope with a half-life of 78 milliseconds.

In 2009, a team led by Yuri Oganessian at the JINR discovered element 117 by bombarding berkelium-249 with calcium-48 in a 250-day run, and got the two isotopes. The experiment was performed at two energies, synthesising five atoms of 293 and one of 294 from beam energies of 252 MeV and 247 MeV respectively. The 293 isotope underwent alpha-decay via a chain of seven elements to lawrencium-265; the 294 isotope alpha-decayed by a chain of three elements of roentgenium-280.

The experiment was run in cooperation with the Oak Ridge National Laboratory (ORNL) who supplied the berkelium target material while physicists at LLNL assisted with analysing the data.

From its position in group 17 of the periodic table, below iodine and astatine, this element should have the physical properties of a halogen or a metalloid and, were it long-enough lived, the element should exist as a volatile M_2 molecule, form salts with many of the other elements, as does iodine, including acids, oxy acids, and fluorides such as $UusF_5$ and $UusF_7$.

When it is given a name, this should have the ending 'ine' to fit with the other members of group 17 of the periodic table, the halogen elements, although that will be decided by IUPAC. However, its nomenclature rules now say that all new elements should end in 'ium'.

Ununoctium

Pronounced un-un-okt-ee-um; chemical symbol Uuo.

There is one known isotope of ununoctium with mass number 294, and no known isomers. It has a half-life of 0.89 milliseconds. (No isotope has yet been theoretically calculated to have a longer half-life.)

In 1999, the LBNL bombarded lead-208 with krypton-86, claiming to have discovered element 118 as isotope-293; they also claimed to have discovered element 116 as the alpha-decay product. Five re-runs to confirm this failed to produce any element 118; two by the LBNL in 2000 and 2001. Efforts by the GSI, RIKEN, and GANIL also failed to produce any. In 2001, LBNL retracted its claim and started an investigation which uncovered a fraud that had been perpetrated by head researcher Victor Ninov. He had faked data to show that three atoms of 118 has been produced—see element 116 for details.

Element 118 was discovered in 2002 by a team led by Yuri Oganessian and Vladimir Utyonkov at the JINR. Using the 'hot fusion' method, they bombarded californium-249 with calcium-48, synthesising two atoms of isotope-294 in a run lasting 2,300 hours. In 2006, they published a paper detailing a successful re-run that they had done the previous year and filed a discovery claim.

Ununoctium completes row 7p of the periodic table and its position is at the bottom of group 18 below radon. Consequently this element should have the physical properties of a noble gas and, were it long-enough lived it should be possible to make a few compounds of it, such as the fluorides $UuoF_2$ and $UuoF_4$. It might also be expected to take a name ending in '-on' in accord with the noble gas elements above it in the periodic table. (Its boiling point has been calculated as likely to be around 50°C so on that basis it would not qualify as a gas.)

Ununennium

119

Uue

Pronounced un-un-en-ee-um; chemical symbol Uue.

In 1985, physicists at the LBNL bombarded einsteinium-254 with calcium-48 in an unsuccessful attempt to make ununennium and as yet it remains undiscovered. Even when it is made its isotopes are likely to have extremely short half-lives.

From its position in row 8s of the periodic table, in group 1 below francium, this element should have the physical properties of an alkali metal and, were it long-enough lived, the element should display the singly-charged ion M^+ as its most favoured chemical state.

Unbinilium

Pronounced un-by-nil-ee-um; chemical symbol Ubn.

There are no known isotopes of this element, but compound nuclei 296, 298–300, and 302 have been produced. Compound nuclei are purposely produced to study nuclear bindabiltiy.[205]

In 2003, the JINR bombarded plutonium-244 with iron-58, and uranium-238 with nickel-64 to get element 120 isotope 302 as compound nuclei. These nuclei lasted about 10^{-18} second and decayed by spontaneous fission because of excess excitement energy, before the nucleons could assume their nuclear shell positions to become a fully bound system, from which the hot excited nucleus could then evaporate excess neutrons and cool to a ground-state isotope. The compound nuclei were identified by the high-atomic-number fission fragments.

In 2004, GANIL bombarded a natural nickel target (mixed isotopes) with uranium-238 to get various compound nuclei of element 120 (196, 298–300, 302). These also lasted about 10^{-18} second and decayed by spontaneous fission.

The first effort to produce an isotope took place in 2007 at the JINR where they bombarded plutonium-244 with iron-58 but were unsuccessful. The same year, the GSI made two unsuccessful attempts, bombarding uranium-238 with nickel-64. In 2008, the GSI made two more unsuccessful attempts, again bombarding uranium-238 with nickel-64.

Its position in the periodic table places it below radium in group 2 (row 5s) which makes it one of the so-called alkaline earth metals. Consequently if a long-lived isotope were to be made it would display the doubly-charged ion M^{2+} as its most favoured chemical state. This being so, then there would be an oxide UbnO, hydroxide $Ubn(OH)_2$, and salts like the chloride, $UbnCl_2$.

---

[205] A compound nucleus is a jumble of particles made from a target nucleus plus a projectile nucleus jammed together by the force of a collision in a particle accelerator. It has no internal structure and it quickly flies apart into fragments of various sizes and this happens before the protons and neutrons can occupy their nuclear shells to become an isotope.

Unbiunium

Pronounced un-bi-oon-ee-um; chemical symbol Ubn.

This would be the first element in a new row (5g) of the periodic table so in theory it is entering unknown territory chemically, but it is most unlikely to be very different from elements in the 4f block. To date, no attempt has been made to make element 121.

Unbibium

Pronounced un-by-bee-um; chemical symbol Ubb.

There is one claimed isotope, unbibium-292, as a naturally occurring primordial isotope with a half-life in excess of 100 million years. The compound nucleus unbibium-306 has also been produced.

In 1972, Georgy Flerov (1913–90) of the JINR bombarded uranium-238 with zinc-66 in an unsuccessful attempt to make this element. In 1978, researchers at the GSI bombarded erbium (natural mixed isotopes) with xenon-136 in another unsuccessful attempt. These early attempts at a high atomic number were stimulated by research suggesting the possibility of naturally occurring superheavy elements. In 2000, the GSI bombarded uranium-238 with zinc-70 in an unsuccessful attempt.

In 2003, the JINR bombarded curium-248 with iron-58 and plutonium-242 with nickel-64 to get compound nuclei of unbibium-306.[206] These lasted about 10^{-18} seconds and decayed by spontaneous fission because of excess excitement energy. They were identified by the high atomic number fission fragments.

In 2007, Amnon Marinov claimed detection of one atom of naturally occurring unbibium-292 in a purified thorium sample as a super-deformed/hyper-deformed isomer with a half-life of 100 million years or greater. Other researchers the following year claimed that they were unable to detect any unbibium in a similar, but not exact, experiment. There is a possibility that not all samples would contain traces of primordial unbibium which might exist only in scattered trace amounts.

Unbibium's chemistry should be like that of thorium, exhibiting several oxidation states.

---

[206] See element 120 (footnote) for a description of a compound nucleus.

Unbitrium

Pronounced un-by-tree-um or un-bit-ree-um; chemical symbol Ubt.

As with unbiunium (121), no attempt has yet been made to make this element, but were its atoms to be made then isotope unbitrium-326 is predicted to be the least unstable.

Fantasy element

Element 123 featured in an episode of the US TV series *Star Trek* when it was called jamesium and its atomic weight was actually given as 326. Clearly whoever wrote the script appears to have had it checked by a nuclear physicist.

Unbiquadium

Pronounced un-by-kwad-ee-um; chemical symbol Ubq.

In 2004, at GANIL, physicists bombarded a natural germanium target (mixed isotopes) with uranium-238 and got compound nuclei of unibiquadium-308, 310, 311, 312, and 314.[207] These lasted about 10^{-18} second and decayed by spontaneous fission. They were identified by the high-atomic number fission fragments.

[207] See element 120 (footnote) for a description of a compound nucleus.

Unbipentium

Pronounced un-by-pent-ee-um; chemical symbol Ubp.

As with unbiunium (121) and unbitrium (123) no attempt has yet been made to make this element, but were its atoms to be made then isotope unbipentium-332 should be the least unstable.

Unbihexium

Pronounced un-by-hex-ee-um; chemical symbol Ubh.

Element 126 is a nuclear 'magic number' with the protons forming a closed nuclear shell, making for less instability and a longer half-life. It is possible that an isomer of isotope-126 might have an extremely long half-life, and consequently several researchers have attempted to make it. It has been calculated that isotopes unbihexium-310 and unbihexium-322 would be of particular interest.

In 1971, at CERN, Réne Bimbot and John M. Alexander tried unsuccessfully to synthesise element 126 by bombarding thorium-232 with krypton-84. Also working at CERN in 1971 was Amnon Marinov who claimed discovery of element 112 as deformed nuclei 272 with a half-life of around 47 days (see copernicium). Marinov had attempted to demonstrate that neutron-deficient deformed isomers of heavy/ superheavy elements could have long half-lives.

Many research papers were published in the period from 1976 to 1983 by scientists who analysed mineral samples in an effort to detect element 126. This was stimulated by a 1976 published paper which suggested element 126 as one of several elements that could be the cause of unexplained damage in minerals.

Several researchers analysing radiation-damaged mineral samples claimed to have detected alpha particle radiation with the appropriate energy to have caused the damage, suggesting the presence of element 126; other researchers analysing other samples claimed not to have detected any. It is possible that not all samples would contain traces of primordial element 126. Element 126 might exist only in scattered remnant trace amounts. Of course, element 126 might not be on planet Earth today in any case, and not detecting it would not rule out the possibility that it was once present and had long since decayed after causing radiation damage in minerals in ages long ago.

The chemistry of unbihexium is likely to be that of a metal displaying various oxidation states, even as high as +8.

Unbiseptium

Pronounced un-by-sept-ee-um; chemical symbol Ubs.

In 1978, researchers using the Darmstadt UNILAC accelerator bombarded tantalum (natural mix of isotopes) with xenon-136 in an unsuccessful attempt to make element 127.

Whether there is an upper limit to atomic number and hence the periodic table has been discussed by physicists for more than half a century. In the early days of this debate it was stated that no element higher than 110 could exist but this upper limit has been revised upwards. Richard Feynman is reputed to have said that element 137 (untriseptium) should be the limit because this number has cosmic significance and is related to the so-called fine-structure constant (α) that defines the strength of the electromagnetic interaction and is linked to the charge on the electron. The inverse of this basic constant, $1/\alpha$, is 137. Others have suggested higher limits, such as Albert Khazan who thought it would be 155 (unpentpentium). Then, in 2010, Pekka Pyykkö at the University of Helsinki used a highly accurate computational model to predict electronic structures and therefore the positions in the periodic table for elements up to atomic number 172 (unseptbium). Whether any of these limits will ever be tested practically remains to be seen. Meanwhile it appears that the practical limit is a chemically reassuring element 118 which completes the 7p row of the conventional periodic table and fills the last vacant slot in the bottom right hand corner.

There is now growing evidence that the element with the highest atomic number will be in the 120s with 128 being the most likely limit. Elements above 118 all resist being formed and attempts to make them have so far been unsuccessful. Even elements below 118 have taken months of accelerator time to make only one or two of atoms, and these have incredibly short half-lives. It will be even harder to make elements as we approach element 126 (unbihexium) which might be slightly less unstable because it is part of the so-called 'island of stability' and had a 'magic number' of protons. Isotope-310 of this element would be of particular interest because it is doubly 'magic' having a magic number of neutrons as well, i.e. 184 of them. The nuclear shell of element 126 might also confer some relative stability on neighbouring elements so that possibly 127 and 128 might exist, but this latter one (unbioctium) would probably mark the end of the periodic table.

Uranium

Pronounced yoo-ray-nee-um, it is named after the planet Uranus, which was itself named after the Uranus of Greek mythology, who was the God of the sky.
French *uranium*; German *Uran*; Italian *uranio*; Spanish *uranio*; Portuguese *urânio*.

Essential element

Uranium is much more abundant and widespread in the environment than generally realised. It is the fuel of nuclear reactors, the explosive of atomic bombs, and the heavy metal of anti-tank ammunition.

Cosmic element

In 2001, a team of astronomers at the European Southern Observatory in Chile detected uranium-238 in the spectrum of a distant star, named CS31082-001, at the edge of our galaxy. This was the first time uranium had been observed in such a distant object and by comparing the amount of uranium to that of thorium it was possible to date the star has having formed 12.5 ± 3 billion years ago.

Uranium is produced in supernova and in the ratio of one part uranium-235 to 1.65 parts uranium-238, and since the former decays more rapidly than the latter then on Earth, which was formed 4.57 billion years ago, their relative amounts are now very different. Of 1,000 atoms of uranium-235 when the Earth was formed there would now be fewer than five today, whereas of the heavier isotope there would still be around 500.

A significant part of the internal heat of the Earth comes from the radioactive decay of uranium atoms.

Element of life

Uranium in the human body	
BLOOD	0.5 p.p.b.
BONE	varies 0.2–70 p.p.b.
TISSUE	varies 1–3 p.p.b.
TOTAL AMOUNT IN BODY	0.1 mg (range 0.01–0.4 mg).

There is no biological role for uranium, nor is there ever likely to have been because of its radioactivity. However, this does not prevent certain microorganisms from absorbing it to the extent of concentrating it to levels 300 times higher than the surrounding environment, although why this happens is not understood.

(The leaves of some plants have higher levels of uranium than expected.) The bacterium *Citrobacter* readily absorbs uranyl ions (UO_2^{2+}) provided it is supplied with a source of organic phosphate such as glycerol phosphate at the same time. The bacteria form crystals of uranyl phosphate until they become encrusted with them, and within a day 1 gram of *Citrobacter* will grow 9 grams of these crystals around itself. There have been suggestions that these microbes might be used to decontaminate water.

Plants absorb a little uranium from the soil in which they grow. The ash from burnt wood can have up to 4 p.p.m of this element. The range for most plants is 5–60 p.p.b. (dry weight) but food plants tend to have less, for example, corn and potatoes have 0.8 p.p.b. (dry weight). Some uranium invariably enters the food chain and explains why we have trillions of uranium atoms in our body.[208]

The amount of uranium taken in with the diet is around 1–2 micrograms daily. The amount of uranium that is absorbed from food is around 5 percent if it is in the soluble uranyl ion but about 0.5 percent if it is present as insoluble compounds, such as the oxide. Most of the uranium which enters the blood stream eventually becomes lodged in the skeleton because of the affinity of uranium for phosphates, and once in the bone it will reside there for many years. Tests on dogs showed that sodium citrate solution boosted the excretion of uranium via the urine.

Deadly element

Normally it is the radiation emitted by uranium which is generally seen as the overriding need to avoid contact with this element, but it is poisonous in any case. The immediate symptoms of a large dose of a uranium salt are vomiting and diarrhoea. Uranium is a heavy metal poison which can be absorbed by the body and which will then affect the working of the kidneys, liver, heart, ovaries, and testicles. Uranium compounds may cause irreversible kidney damage because uranium salts lodge in its tubules, eventually causing renal failure with the production of proteins and glucose in the urine. Soluble compounds, which can even be absorbed through the skin, pass quickly through the body causing little damage whereas insoluble ones, and particular those which lodge as dust in the lungs, pose a more serious threat.

Particles of uranium metal that are lodged in the body appear to be relatively harmless. These might come from fragments of exploding ammunition if this is made of depleted uranium. More than 9,500 rounds of tank shells were fired in the Gulf War of 1991 scattering around 47 tonnes of depleted uranium to the environment. A group of Gulf War veterans were examined 10 years later and there were men who had small fragments of uranium still retained in their body, and while their urine showed they were discharging this element slowly, there was no evidence that they were being affected by the radiation which some had feared would cause an increase in various cancers. In fact, none of the men had suffered from leukaemia, bone cancer, or lung cancer.

---

[208] The average human contains 0.1 mg of uranium which is equivalent to 10^{17} atoms, or put another way, is 100,000 trillion atoms.

Element of history

In the late Middle Ages, the mineral pitchblende (uranium oxide, UO_2) often turned up in the silver mines of Joachimsthal, which is now part of the Czech Republic, but was then part of the Kingdom of Bohemia. While the ore was thought to contain zinc or iron, it was nevertheless of little use. The name was derived from *pech*, meaning pitch (or ill luck) and *blende* meaning mineral. In 1789, Martin Heinrich Klaproth (1743–1817), a pharmacist who had his own experimental laboratory in Berlin, Germany, investigated pitchblende and found it dissolved in nitric acid and precipitated a yellow compound when the solution was neutralised with sodium hydroxide.[209] He was convinced it was the oxide of a new element and when he heated the precipitate with charcoal and obtained a black powder, he assumed he had produced the metal itself. He based its name on that of the recently discovered planet Uranus and gave it the name uran, which later became uranium.

Klaproth's metal was really one of the oxides of uranium, and it fell to Eugène Peligot (1811–90), who was Professor of Analytical Chemistry at the Central School of Arts and Manufactures in Paris, France, to isolate the first sample of uranium metal which he did in 1841, by heating uranium tetrachloride with potassium.

For most of the 1800s, uranium was not regarded as particularly dangerous and commercial uses were found for it. For example, in 1855 a factory was set up in Austria to manufacture uranium pigments for colouring pottery and glass. Glass to which uranium oxide has been added has a fluorescent yellow-green colour. Antique glass containing uranium is still available and prized by collectors.

The discovery that uranium was radioactive came only in 1896 when Henri Becquerel (1852–1908) in Paris found that a sample of uranium left in a drawer on top of an unexposed photographic plate caused this to become 'fogged' as if it had been partly exposed to light. From this he deduced that uranium was radiating invisible rays. Thus began a new chapter in science—and human history.

However, during the first decades of the last century, when radium treatments were seen as beneficial, this was extracted from uranium ores and the uranium was seen as a by-product, but there was only 0.4 grams of radium per tonne of uranium. A lot of uranium was processed and more of it began to find an outlet in pottery glazes for tiles because of the range of colours it could provide.

Element of war

The realisation that an atomic weapon was possible came to Leó Szilárd (1898–1964) as he was waiting for the traffic lights to change on Southampton Row in London on 12 September 1933. He had been musing on a report he had read in *The Times* newspaper which quoted the great Ernest Rutherford's remark that while releasing energy from atoms was possible in theory, it could never be a practical proposition.

Szilárd realised that it was possible and worked out how it could be done, and he filed a patent for a nuclear reactor the following year. He deduced that if an atom was

---

[209] The precipitate was probably sodium diuranate ($Na_2U_2O_7$).

struck by a neutron and then underwent fission, thereby releasing energy, it would at the same time release two neutrons which could then go on to split more atoms, releasing more energy and more neutrons. He could even see that a run-away release of energy in the form of a nuclear bomb was possible. When it appeared that World War II was inevitable, Szilárd realised that there was every likelihood that the Nazis would be able to produce such as weapon and that he must inform the US President, Franklin D. Roosevelt. He decided that the person to do that was the world's most famous scientist, Albert Einstein who did indeed put his name to a letter which was handed to the President. The rest, as they say, is history.

The first atomic bomb used in warfare was a uranium bomb, code named Little Boy, which was dropped on the Japanese city of Hiroshima at 08:16 on the morning of 6 August 1945. This bomb contained enough of the uranium-235 isotope to start a runaway chain reaction, which in a fraction of a second caused a large number of the uranium atoms to undergo fission, thereby releasing a fireball of energy.[210] It had the destructive power of 16,000 tonnes of TNT and it destroyed almost 50,000 buildings and killed about 75,000 people, who either died immediately or later that month as a result of their burns.

Naturally occurring uranium is 99.28 percent uranium-238 and 0.71 percent uranium-235, but only the latter can sustain a chain reaction and be used to make a bomb. Uranium-235 was separated from the more abundant isotope, by converting the metal to uranium hexafluoride (UF_6), a volatile liquid which boils at 57°C. This compound is made from uranium oxide and hydrogen fluoride, which forms uranium tetrafluoride (UF_4) and this is then reacted with fluorine gas to form UF_6. Because the two isotopes differ slightly in molecular weight they diffuse at different rates through a membrane. The filtration barrier through which UF_6 has to pass consisted of micro-tubules of nickel about 10 nanometres in diameter, and the lighter isotopes passes through them quicker than the heavier isotope. A cascade of such separators will eventually produce a vapour of the enriched isotope.

After repeated diffusions, a sufficient concentration of uranium-235 was obtained to provide the so-called 'critical mass' which ensures that the majority of released neutrons are captured by other atoms, rather than escaping to the environment. The critical mass for a runaway chain reaction is only a few kilograms, and as little as 7 kg of uranium-235 is needed to make a bomb. In the Little Boy bomb, explosive charges forced together two sub-critical pieces of uranium from opposite ends of a tube to create such a critical mass.[211] This was called the gun-type or barrel type arrangement.

Separating uranium isotopes by means of a gas centrifuge is much cheaper and this was first proved possible by an Austrian engineer, Gernot Zippe (1917–2008) who found himself in the Soviet sector of East Germany after World War II. He developed the idea when he worked in Georgia. In a high speed gas centrifuge, the lighter UF_6 concentrates near the centre of the centrifuge. This is a much cheaper method and most uranium is now enriched this way. A cascade of such centrifuges will produce the desired enrichment of uranium. Zippe was eventually repatriated to Austria and developed his idea further in the West. The use of a centrifuge as a

---

[210] Most of the uranium atoms are there to provide the critical mass, and they do not experience fission.

[211] A uranium bomb can also be used as the trigger for a hydrogen bomb—see page 231.

means to separating isotopes had been demonstrated on carbon tetrachloride vapour at the University of Virginia in 1934 when it was used to separate the chlorine–35 and chlorine–37 isotopes.

Another method of separating UF_6 isotopes exposes this vapour to laser radiation of precise energy to sever the uranium-fluorine bond of the lighter isotope but not of the heavier one, the result being that uranium-235 metal deposits out from the mixture.

As mentioned above, depleted uranium is also used in ammunition. This is the uranium from which almost all of the uranium-235 has been extracted. It is one of the densest of metals and for this reason it delivers a much heavier blow as ammunition and it can be used in armour-piercing shells against tanks. Not only that, but when it strikes its target it also ignites. A 120mm tank shell contains about 5 kg of depleted uranium.

Economic element

Uranium occurs in a variety of mineral forms, the chief one of which is the oxide ore, uranite (aka pitchblende, U_3O_8). Other minerals are brannerite (uranium silicate), coffinite (uranium titanate), autunite (calcium uranyl phosphate), and carnotite (potassium uranyl vanadate). Some varieties of samarskite (a complex oxide of many metals of which tantalum is the most important) may contain up to 23 percent uranium. Torbernite (calcium uranyl phosphate) is a minor ore. Whatever its source, the extracted uranium is converted to the purified oxide form known as yellow cake which is U_3O_8.

Large uranium ore deposits occur in Australia, the USA, Canada, Gabon, Congo, South Africa, Russia, and China. Uranium is extracted in several ways: by open pit mining, underground mining, by leaching low grade ores, or as a by-product of processing other ores. For example, phosphate ores contain 100–200 p.p.m. of uranium oxide and this can be selectively extracted with organic esters dissolved in kerosene (paraffin). The esters form compounds with uranium that are soluble in this hydrocarbon solvent. In China there are plans to extract uranium from the fly ash of coal burning power stations. This ash is calculated to contain around 200 grams of U_3O_8 per tonne.

The main mining countries are Canada, which produces a quarter of all uranium, followed by Australia, Kazakhstan, Niger, Russia, and Namibia. Production in Kazakhstan is rising rapidly and this might soon overtake Canada as the main source of uranium. World production is 41,000 tonnes per annum and reserves are thought to exceed 3 million tonnes, with some estimates putting it at 30 million tonnes, and this does not include the uranium tied up in phosphate ores. Monazite sands also contain enough uranium to make them a potential source of the element. Although the concentration of uranium in seawater is low, the total amount there is around 4.5 *billion* tonnes, and it could be extracted using ion exchangers. Such a method was shown to be feasible in Japan in the 1980s. The world market for uranium is expected to reach 60,000 tonnes or more by 2015.

Uranium metal itself is obtained by converting the oxide into uranium chloride and then reacting that with sodium metal, or by reacting the oxide directly with calcium or aluminium.

Most uranium (95 percent) is used to generate electricity in nuclear power stations. There are currently around 450 nuclear power plants in the world which together contain about 1,600 nuclear reactors. The number of plants is rising and is expected to reach 750 in the near future. These require the uranium to be enriched with uranium-235, and the chain reaction to be controlled so that the energy is released in a more manageable way. In theory a kilogram of uranium-235 can release around 20 trillion joules of energy (20×10^{12}) equivalent to more than 1,500 tonnes of coal. The controlled release of this energy can be done by increasing the proportion of uranium-235 to around 3 percent, or by using a moderator to slow down the neutrons that are released so more of them are captured by other uranium atoms. Heavy water is an ideal moderator.

Nuclear powered naval vessels have been built by six countries: the USA, Russia, the UK, France, China, and India. Altogether 535 nuclear-powered vessels of all kinds have been constructed including aircraft carriers and cruisers. More than 485 nuclear submarines have been built, mostly by the USA and Soviet Russia. The USA built the first nuclear-powered vessel, the submarine *Nautilus*, which was launched in 1954. The USA has 12 nuclear-powered aircraft carriers and France has one. Non-navel nuclear-powered vessels include icebreakers built by Soviet Russia for scientific research in the Arctic. The USA built the first nuclear-powered cargo ship the *NS Savannah* which was launched in 1959, but it was not a commercial success, and the same was unfortunately true of similar ships built in Germany, Japan, and Soviet Russia.

At least eight nuclear submarines are sitting on the bottom of the world's oceans as a result of accidents, along with their nuclear reactors. Dozens of other reactors, scrapped from nuclear-powered vessels have also been dumped in the sea.

Nuclear reactors use rods of uranium or uranium oxide that are enriched with uranium-235 and these are clad in zirconium metal. As the fuel is consumed, the dissociation products accumulate and these may absorb the neutrons that sustain the reactor, so after about five years a rod has to be replaced. The spent fuel is sent for reprocessing, but only after being left for a year to allow the intense radiation from some of the products to decay away.

Depleted uranium is used in ammunition, also as ships' ballast, as counterweights for aircraft, and (as the uranyl ion) to add colour to ceramic glaze on tiles. Depleted uranium is preferred over other dense metals because it is easily cast and machined.

Environmental element

Uranium in the environment	
EARTH'S CRUST	2 p.p.m.
	Uranium is the 48th most abundant element.
SOILS	range 0.7–11 p.p.m.*
SEAWATER	3 p.p.b.
ATMOSPHERE	virtually nil.

* The addition of phosphate fertilizers has increased the level of uranium on most farmed land to around 15 p.p.m.

Although uranium is radioactive, it is not particularly rare and is more abundant than other metals such as tin, and 10 times more abundant than silver and mercury together. It is widely spread throughout the environment and so it is impossible to avoid uranium. Soils near fossil fuel power stations and phosphate fertilizer plants can be enriched in uranium due to emissions, since both types of raw material tend to have relatively high levels of uranium as impurities.

While uranium itself is not particularly dangerous, some of its decay products do pose a threat, and especially that from the radioactive gas radon, which can build up in confined spaces such as basements. Even the building material gypsum, used to make plaster, has been found to be contaminated with uranium if it had been made from the waste products of phosphate mining.

When a uranium-235 atom splits in two, it produces many other elements, but most tend to be in the weight ranges of 90–105 (krypton to ruthenium) and 130–145 (technetium to europium). Among the former is strontium-90, which was particularly worrying when the majority of above-ground nuclear weapons testing was carried out in the 1950s and early 1960s by the USA, Soviet Russia, the UK, and France.[212] This element is easily absorbed by humans because it mimics calcium, and yet is dangerously radioactive. Radioactive iodine-131 is equally worrying. This was widely scattered over the North of England in 1957 by an accident at the Windscale (now called Sellafield) nuclear power plant, and over Europe by the Chernobyl nuclear plant explosion in the Ukraine in 1986. Again, because iodine is an element essential to human health, this isotope is a problem when absorbed by the body.

Although uranium oxides are very insoluble, when they are oxidised to uranyl ions UO_2^{2+} they become very soluble, and very mobile in the environment. In one locality, Oklo in Gabon, in Africa, this allowed uranium to concentrate to such a level that it formed 16 natural nuclear reactors. This was thought to have occurred when oxygen levels first built up in the Earth's atmosphere, thereby allowing uranium to be converted to uranyl ions. The Oklo event occurred at a time when the natural level of uranium-235 was probably around 3 percent and high enough to sustain a chain reaction especially with water present. The reason we know this happened is that the amount of uranium-235 in the uranium ore mined in Gabon is much less than that found with other uranium ores, showing that at some time in the past it has undergone fission, and there are still some of these fission-produced isotopes around.

These natural nuclear reactors started up about 1.8 billion years ago and functioned continuously for several hundred thousand years until the concentration of uranium was too low to sustain them any longer. They produced transuranium elements with atomic numbers 93 to 100 in a series of neutron-capture and beta-decay reactions. Organic-rich water that percolated through the rocks probably helped to moderate the reactors, and this eventually became tar-like as it decomposed. Traces of this bitumen can still be found. The radioactive waste products from the Oklo reactors were not entirely safely locked away during the millennia that followed; they appear to have been disturbed by a volcanic eruption around 720 million years ago.

---

[212] Three countries, France, Israel, and China, continued above-ground tests throughout the 1970s with the last atmospheric tests taking place in 1981. From 1945 to this date, around 525 atmospheric explosions took place.

Chemical element

Data file	
CHEMICAL SYMBOL	U
ATOMIC NUMBER	92
ATOMIC WEIGHT	238.0289
MELTING POINT	1,132°C
BOILING POINT	3,754°C
DENSITY	19.0 g/cm³
OXIDES	U_2O_3, UO_2, U_3O_8, UO_3

Uranium is a silvery, ductile, malleable, radioactive metal and a member of the actinide (row 5f) series of the periodic table. It tarnishes in air and is attacked by steam and acids, but not by alkalis. It is one of the three fissile elements, the others being thorium and plutonium. It can exhibit several oxidation states such as +3 (the compounds of which are generally red, such as UF_3), +4 (green, UF_4), +5 (rare), and +6 (yellow, UF_6). The last of these oxidation states in solution exists as the uranyl ion UO_2^{2+}.

There are 26 known isotopes of uranium with mass numbers ranging from 217 to 242, and seven known isomers. There are nine uranium isotopes that occur naturally on planet Earth, namely uranium-232 to 240. However, most of the uranium which exists on Earth consists mainly of three radioactive isotopes: uranium-238 which accounts for 99.3 percent and has a half-life of 4.6 billion years, uranium-235 (0.7 percent, 700 million years), and uranium-234 (0.005 percent, 245,000 years). Two of these are primordial, namely uranium-235 and uranium-238, while uranium-234 is produced from uranium-238.[213]

The amount of uranium and lead in various rocks has been used to date them, and indeed to date the Earth itself to 4.57 billion years old.

Element of surprises

1. In 1912, R. T. Gunther of Oxford University excavated a first century AD Roman villa on Cape Posilipo on the Bay of Naples, and came across a mosaic in which there were curiously coloured glass pieces. When these were analysed it was

---

[213] Uranium-234 and the other non-primordial uranium isotopes are produced from the elements plutonium, neptunium, protactinium, thorium and even uranium itself by neutron-captures followed by beta-decays and alpha-decays. Uranium-234 is produced in the decay of uranium-238 which first loses an alpha particle to form thorium-234, which then emits a beta particle and goes to protactinium-234, and it is this which loses another beta particle to form uranium-234.

Uranium-238 captures a neutron to produce uranium-239. Uranium-232 is produced in minute traces by the rare double decay of thorium-232. Naturally occurring primordial plutonium-244 alpha-decays to produce uranium-240 which then undergoes beta-decay to neptunium-240, which beta-decays to plutonium-240, which alpha-decays to produce uranium-236. Uranium-237 is produced in trace amounts by the alpha decay of plutonium-241; this plutonium isotope undergoes bi-modal decay and also produced uranium-233 when it beta decays to americium-241, which alpha decays to neptunium-237, which alpha-decays to protactinium-233, which beta-decays to uranium-233.

discovered that their tint was due to the presence of 1 percent uranium oxide, and that this mineral must have been deliberately added to colour the glass.

2. In 1998, researchers from the Natural History Museum, London, discovered that the lichen *Trapelia involuta* was happily growing on spoil heaps from an abandoned uranium mine in Cornwall, UK. X-ray investigations showed that the lichen was absorbing, and storing, uranium in the walls of its outer fruit cells, although for what purpose is not known. The lichen appear to be unaffected by the uranium.

3. There is an urban myth that an experiment once took place which might have triggered a nuclear explosion in the middle of Chicago. The precise time at which it occurred was 3.25 on the afternoon of 2 December 1942. Physicist Enrico Fermi (1901–54) had gathered several tonnes of uranium and wanted to see if a sustained nuclear reaction was possible in which the neutrons from fissioning uranium atoms could then trigger a chain reaction by hitting other uranium nuclei causing them to undergo fission releasing yet more neutrons that would activate yet more uranium and so on—exactly what happens in a uranium atomic bomb.

   In fact, his atomic pile of uranium had been constructed so that it was interspersed with rods of cadmium, a metal which absorbs neutrons. These were then withdrawn and as predicted the pile of uranium began to get warm as matter was converted to energy. However, there was never any danger of the pile exploding because Fermi and his assistants were careful never to allow the neutron level to become anywhere near critical and in any case they had a fail-safe set of cadmium control rods that would fall into the uranium pile were things to appear to be getting out of control.

4. Depleted uranium can be used in place of lead as protection against radiation from more intense sources.

5. Uranium mono-nitride, UN, has a very high melting point and thermal conductivity and is being researched as an ideal type of nuclear fuel. What is curious about this compound is that the two elements are held together by a triple bond. This chemical curiosity is being researched and the Los Alamos National Laboratory in New Mexico has found that it can activate carbon-hydrogen bonds in a way that is similar to the enzyme *cyctochrome P450* whose role is to oxidise substances, such as converting hydrocarbons to alcohols.

Vanadium

Pronounced van-ay-dee-um, the element is named after Vanadis, the Scandinavian goddess of love and beauty. Other names were suggested by its original discoverer—see below. Oxygen-containing negative ions, such as VO_4^{3-}, are known as vanadates, while the positive ion, VO^{2+} is known as vanadyl.

French *vanadium*; German *Vanadium*; Italian *vanadio*; Spanish *vanadio*; Portuguese *vanádio*.

Essential element

Vanadium is an essential component of certain human enzymes. The metal itself is most likely to be encountered as alloys which are used for springs, tools, and jet engine parts.

Element of life

Vanadium in the human body	
BLOOD	0.02–0.9 p.p.b.
BONE	3.5 p.p.b.
TISSUE	av. 20 p.p.m.
TOTAL AMOUNT IN BODY	*ca.* 20 mg.

Vanadium is essential to some species, including humans. Feeding tests on chickens and rats showed that vanadium has a growth-promoting effect, and the same may be true for humans, although we are unlikely to be short of this metal—see below. Some micro-organisms require vanadium for an enzyme which converts atmospheric nitrogen to ammonia in order for them to make amino acids.

Vanadium compounds are not regarded as a serious health hazard, and indeed this element appears to stimulate metabolism. However, workers exposed to vanadium pentoxide (V_2O_5) dust were found to suffer severe eye, nose, and throat irritation and such workers must now wear protective gear.

Food element

The average daily intake of vanadium is 1–2 micrograms, which is more than enough to meet the body's requirements; indeed there are no cases of anyone ever suffering vanadium deficiency. The absolute requirement has been estimated to be as little as 2 micrograms per day.

Only about 0.5 percent of dietary vanadium is actually absorbed by the body, and most of that which is, is rapidly excreted. The element does not accumulate in any organ, but nevertheless there is a lot more vanadium in the human body than it appears to need. Its role is thought to be as a regulator of one of the enzymes that govern the way sodium operates in the body, but it may have other roles as well.

Vanadium first came to the attention of nutritionists in 1977 when some commercial preparations of ATP (adenosine triphosphate) were found to disturb the sodium-potassium balance that operates the nervous system. The reason turned out to be traces of vanadium which was acting as an enzyme inhibitor of ATPases, and this it appears to do by substituting vanadate for phosphate. This discovery sparked off an interest in this element, and yet it is still not completely clear why vanadium is essential for humans.

Typical vanadium levels in foods are between 0.01 and 0.1 p.p.m., for example, wheat has 0.08 p.p.m. and potatoes 0.02 p.p.m. (fresh weight), although rice has relatively high levels up to 0.8 p.p.m., and lettuce can even have 1 p.p.m. if it is grown on contaminated soil. Plants that produce linoleic acid, such as sunflowers, absorb a lot of vanadium, up to 50 p.p.m. (dry weight). Seafood and liver are the sources of dietary protein that contain most vanadium.

Element of history

Vanadium was discovered twice. The first time was in 1801 by Andrés Manuel del Rio (1764–1849) who was Professor of Mineralogy in Mexico City. He found it in a specimen of *plomo pardo de Zimapan* (brown lead ore of Zimapan), a mineral now known as vanadite, $Pb_5(VO_4)_3Cl$, which came from a mine in Hidalgo. He was surprised at the varied colours of its salts and so he called it panchromium (meaning 'all colours'), then finding that all its salts turned red on heating or treating with acid, he changed the name to erythronium (red).

Del Rio gave some of his material to a visitor, the well-known traveller Alexander von Humbolt (1769–1859), who sent it to the French Institute in Paris, with a brief note saying how like chromium the new element was, on account of its multicoloured salts. (A longer letter describing its chemistry in more detail never arrived, the ship carrying it being wrecked in a storm.) The result was that a French chemist was asked to examine the new specimen, and he came to the conclusion that it was really a chromium mineral and published his report. Del Rio was persuaded that all he had done was to rediscover chromium.

The second time vanadium was discovered was 30 years later, in 1831, by the Swedish chemist Nil Gabriel Selfström (1787–1845) who was Professor of Chemistry at the Karolinska Institute at Stockholm. He separated it from a sample of cast iron made from ore that had been mined at Småland. The iron makers were puzzled by the fact that sometimes their cast iron was brittle, while at other times it was strong. Part of the sample dissolved more readily in acid than the bulk of the metal and Selfström concentrated on this material. Although he had only a tiny amount, he was able to show that it was a new element, and in so doing he beat a rival chemist, Friedrich Wöhler (1800–82), to the discovery.

Wöhler had been working on a sample of the mineral from Zimapan, but had been badly affected by hydrogen fluoride gas and had been ill for several months, during which time Selfström announced his results. The young Wöhler, however, was also aware of del Rio's work of 30 years previously, and his work confirmed the original discovery. Nevertheless, Selfström's name for the element, vanadium, was the one which became accepted.

Early attempts to extract vanadium from its compounds produced an indifferent material. The first reasonably pure sample was produced by the English chemist Henry Roscoe (1833–1915) in Manchester, in 1869, and he showed that previous samples of the metal must have been vanadium nitride (VN). Even his sample was not very pure, and had around 4 percent of impurities. A purity of 99.99 percent was only achieved in the 1920s by J. W. Marden and M. N. Rich of the Westinghouse Lamp Company in the USA.

Element of war

Vanadium steel was stronger and lighter that ordinary steel and it made it possible for the first war plane to be equipped with a canon, rather than just a machine gun. This was produced for the French air force in World War I, and caused consternation when it first appeared. The warfare of the last century required extra protection for combatants and vehicles. Nickel steel was no use, being easily penetrated by bullets and shrapnel. Vanadium steel, on the other hand, was found to offer better protection and it became the standard material from which soldiers' helmets and other forms of protective armour were made.

Economic element

There are more than 60 known vanadium ores, of which the more common ones are vanadite (lead chloride vanadate, $Pb_5(VO_4)_3Cl$), patronite (vanadium sulfide, VS_4), and carnotite (potassium uranyl vanadate, $K_2(UO_2)_2(VO_4)_2$). The largest reserves of vanadium are to be found in South Africa and Russia. World production of vanadium ore is around 70,000 tonnes a year and it is mined in Russia (25,000 tonnes), South Africa (24,000 tonnes), China (19,000 tonnes), and Kazakhstan (1,000 tonnes). Vanadium is also abundant in Venezuelan oil, to the extent that the ash from it is a commercial source of vanadium pentoxide (V_2O_5).[214]

Production of the metal itself comes to about 7,000 tonnes per year. This is extracted from the red-brown oxide by heating with calcium in a pressure vessel. It cannot be extracted by heating with carbon, which reacts with the metal. Very pure vanadium is obtained by reacting vanadium trichloride (VCl_3) with magnesium, and this gives a metal that is much more ductile. Since most vanadium is added to steel, it is not necessary to use such high grade metal, and it is usually added as ferrovanadium which is produced from vanadium pentoxide by heating with ferrosilicon.

Vanadium is used mainly as alloys, especially for vanadium-steels, which contain 0.1–3 percent, the amounts depending on the intended use. Such steels are rustproof, shock-resistant and vibration-resistant, and are used for springs, tools, jet engines,

---

[214] The ash from Venezuelan oil is 45% vanadium pentoxide. It has been suggested that plants living at the time when these oil deposits were laid down absorbed much more vanadium than today's plants.

as well as armour plating. Most vanadium goes into tools. Vanadium alloys are also used in nuclear reactors because vanadium has low neutron-absorbing abilities and it does not deform by 'creeping' under high temperatures.

The oxide, V_2O_5, is used in ceramics to produce a golden colour, and is added to glass to produce a green or bluish tint. Industry uses the oxide as a catalyst for the oxidation of sulfur dioxide (SO_2) to sulfur trioxide (SO_3) which is the precursor to sulfuric acid. Vanadium compounds are also used as catalysts in polymer production.

Environmental element

Vanadium in the environment	
EARTH'S CRUST	160 p.p.m.
	Vanadium is the 19th most abundant element.
SOILS	av. 100 p.p.m., ranging from 10–500 p.p.m.
SEAWATER	1.5 p.p.b.
ATMOSPHERE	0.02 nanogram per cubic metre over the oceans, but around 1mg per cubit metre over populated regions, due to the burning of heating oil.

Weathering is an important way in which vanadium is redistributed around the environment because vanadates are generally very soluble, and the annual movement of vanadium this way is thought to exceed 2 million tonnes. This metal is not regarded as an environmetal threat.

Vanadium is abundant in most soils, in variable amounts, and is taken up by plants at levels that reflect its availability. Some species such as toadstools, take up a lot of vanadium. The fly agaric mushroom *Amanita muscaria* has 180 p.p.m. (dry weight).

Marine worms absorb vanadium, and one species, *Ascidia nigra,* has blood cells with 1.5 percent vanadium. It has the ability to concentrate vanadium to levels 10,000 times higher than the surrounding seawater. The sea cucumber, *Holothuroidea,* has vanadium making up 10 percent of its blood cell pigment. There was once a scheme in Japan to harvest such ascidians for their vanadium, from the vast tracts which live on the coastal shelf of that country.

Chemical element

Data file	
CHEMICAL SYMBOL	V
ATOMIC NUMBER	23
ATOMIC WEIGHT	50.9415
MELTING POINT	1,887°C
BOILING POINT	3,377°C
DENSITY	6.1 g/cm³
OXIDES	V_2O_5 (most stable), VO, V_2O_3 and VO_2.

Vanadium is a shiny, silvery metal, which comes at the head of group 5 (row 3d) of the periodic table. It resists corrosion due to a protective film of oxide on the surface. The metal is attacked by concentrated acids, but not by alkalis, not even when these are molten. Its preferred oxidation state is +5, which in the older terminology led to the easily memorable coincidence of V(V). Vanadium exhibits oxidation states +2, +3 and +4 as well as +5, as shown by the compounds VCl_2, VCl_3, VCl_4, and VF_5. In solution the most stable form at high pH is the vanadate ion, VO_4^{3-} but at lower pHs there are various complex vanadates in addition the simple VO_3^- which can easily be precipitated as its ammonium salt, NH_4VO_3. The colours of the solutions of vanadium in its different oxidation state are: lilac (+2), green (+3), blue (+4), and yellow (+5).

There are 26 known isotopes of vanadium with masses 40 to 66, and five known isomers. There are two naturally occurring isotopes of vanadium: non-radioactive vanadium-51 which accounts for 99.8 percent, and weakly radioactive vanadium-50 (0.2 percent). This has a half-life of around 150 quadrillion years (150×10^{15}), which is 30 million times the age of the Earth.[215] The next longest-lived radioactive isotope is vanadium-49 with a half-life of 330 days.

Element of surprises

1. The famous Damascus steel was due to the presence of vanadium. This steel was fashioned into the much prized weapons of daggers, scimitars, and swords that were unsurpassed in their day. Research by John Vanhoeven, a materials scientist and engineer, showed that minute traces of vanadium were responsible for giving Damascus steel its great strength and durability, with the especial advantage of taking and holding a lasting sharp edge. Amazingly, as little as 0.02 percent of vanadium was responsible for the steel's high quality, yet its presence was not appreciated by the Damascus steelmakers. The vanadium just happened to be an impurity in the source of iron ore they were using, and when the supply of this ran out, with it went their reputation of making the world's finest steel. The steelmakers could not understand why their tried-and-tested methods no longer worked, and by 1800s Damascus steel was no more.

2. 'But for vanadium there would be no automobiles!' So said Henry Ford who based his popular T-model motor car on lightweight vanadium steel. He came across this alloy in 1905 when he attended a racing car event at which there was a crash that completely wrecked a French vehicle. When Ford picked up a fragment, which was part of a valve spindle, he was surprised how hard, yet light, it was. He sent it for analysis and was told that it was steel which contained a few percent of vanadium.

   Although such steel was not available in the USA, he decided nevertheless to make it an essential part of his new car and in 1908 the first new model-T was produced. Mass production began in 1913 and, by 1927, when the model was discontinued; around 15 million had been made, all with components such as gears, axles, shafts, springs and suspension made from vanadium steel.

---

[215] In a 1 gram sample of vanadium, only two atoms a week undergo radioactive decay.

3. The best alloy for the kinds of wheelchairs used by those who wish to play sports is one made of vanadium, titanium, and aluminium which has a high strength to weight ratio. That is needed because some sports require fast manoeuvring, like basketball, and this puts great strains on the chair.

4. The price of vanadium is affected most by earthquakes in China. When these occur then the price of vanadium automatically shoots up because often these quakes occur in areas where it is mined, causing mines to be shut down, and the quakes also damage the railway lines which transport the ore to the outside world.

Xenon

Pronounced zee-non, the name is derived from the Greek *xenos*, meaning stranger. French *xénon*; German *Xenon*; Italian, *xeno*; Spanish *xenón*; Portuguese *xénon*.

Essential element

Xenon provides the powerful flash for cameras; it is part of thin screen technology; and it is the safest anaesthetic.

Cosmic element

Xenon is relatively rare in the Sun's atmosphere but in the atmosphere of Jupiter it is much more abundant, albeit still a minor component. The atmosphere of Mars has about the same ratio of xenon as that of the Earth, although with more of the xenon-129 isotope. This particular isotope has been detected in gas wells in New Mexico and is thought to derive from when the planet was first formed. Higher than expected levels of it are to be found in some stony meteorites, and there it came from the radioactive decay of iodine-129 which has a half-life of 17 million years.

Element of life

The amount of xenon in the human body is purely that which has dissolved in the blood and that is very small. It can have no biological role, and the gas is harmless, although it could asphyxiate if it were to exclude oxygen from the lungs.

Element of history

Xenon was discovered on 12 July 1898 by William Ramsay (1852–1916) and Morris Travers (1872–1961) at University College London. They had already isolated neon, argon, and krypton from liquid air, and wondered if there were others still to be found. Their researches had aroused a lot of interest not least that of the industrialist Ludwig Mond (1839–1909), who gave them a new liquid-air machine. They used it to extract more of the rare gas krypton from liquid air and, by repeated fractional distillation of this, they isolated an even heavier gas: xenon. When they examined it in a vacuum tube it gave a beautiful blue glow and they realised at once that it was yet another member of the 'inert' group of gaseous elements as they were then called. Because it had proved so elusive they named it xenon.

In 1920, the French chemists Charles Moureu (1863–1929) and Adolphe Lepape found that all the noble gases were present in the natural gas of Alsace-Lorraine and in many spring waters found in France, including the most famous: Vichy water.

Medical element

Xenon was first successfully used as an anaesthetic for surgery in 1951. However, it is expensive but because it produces fewer side effects, it is increasingly being used despite its cost; but cost still rules out its widespread use. Recycling the gas should be possible in future, thereby making xenon anaesthetic more widely available.

Radioactive xenon-133, with a half-life of 5.25 days, is produced in nuclear reactors and this can be used in the medical investigative technique known as photon emission tomography to image the heart, lungs, and brain, and to measure blood flow.

Economic element

The only commercial source of xenon is as a by-product from liquid-air plants. Xenon production worldwide probably amounts to more than 10,000 cubic metres (m^3), equivalent to six tonnes per year, but even so it is 10 times more expensive than any other noble gas and more than a hundred times more expensive than neon. Reserves of xenon gas in the atmosphere amount to 2 billion tonnes.

Xenon has several practical uses: in lamps, in semiconductor manufacture, in research laboratories, and in surgery—about 15 percent of xenon is now used an anaesthetic.

When an electric discharge is passed through the gas it produces a gentle blue light, which extends into the 'safe' ultraviolet region as well, and so can be used for sun beds and the biocidal lamps seen in food preparation areas. Xenon 'blue' headlights and fog lights are used on some vehicles, and they are said to be less tiring on the eyes. They also illuminate road signs and markings better than conventional lights.

Other types of lamps use xenon's ability when pulsed with a high voltage to deliver an intense burst of light, which is very like daylight. It is part of the built-in flash of modern cameras, as well as being essential for the strobe lights used in high speed photography, for example, to record such events as a bullet emerging from the barrel of a gun or striking a target. It was in the 1930s that Harold Egerton (1903–90) invented the xenon flash lamp. In this a burst of electric current passes through a tube filled with xenon gas and this produces an intense flash lasting only a microsecond. This made high speed photography possible. Xenon flash lamps are also used to activate ruby lasers.

Xenon difluoride (XeF_2) is available commercially and used to etch silicon microprocessors and in the manufacture of the anticancer drug 5-fluorouracil.

Xenon is employed as a supercritical fluid in research laboratories.[216] Its critical temperature is 16°C at 58 atmospheres pressure (290K at 5.84 MPa) above which it has remarkable solvent properties. Indeed its lack of chemical reactivity makes it an ideal medium for studying molecules and its lack of chemical bonds make it transparent to infrared, visible, and ultraviolet light, and suitable as a solvent for magnetic resonance analysis.

---

[216] A solvent enters the supercritical state when its temperature exceeds the point at which even the highest of pressures is unable able to keep it in a liquid form. A supercritical fluid ensues, which has the properties both of a liquid and a gas, and it can then dissolve all kinds of materials.

Zipping around in space with xenon

Xenon makes the best fuel for the ion engines of space craft. In these a beam of ions is accelerated by an electric or electromagnetic field and then expelled from the vehicle at 30 km per second (about 100,000 k.p.h.) thereby giving it a powerful thrust in the opposite direction. An ion engine is 10 times more efficient than a conventional propulsion unit. To be suitable for an ion drive, atoms must be easily ionised (that is, lose an electron, thereby acquiring a positive charge) and have as high a mass as possible. Mercury, caesium, and xenon are the most likely contenders for ion engines, of which xenon is the safest to handle. Being a gas, it poses problems of storage on Earth, but not in space, where conditions are cold enough to freeze it solid. Xenon is preferable to caesium, which is corrosive, and to mercury, which poses a threat to those exposed to its vapour.

Xenon ion propulsion systems are known as XIPS and pronounced 'zips'. They are now in operation on several orbiting satellites and are used to keep them in the correct orbit, with each one having enough xenon fuel on board to last more than 12 years. The space probe Deep Space I is powered by a XIPS. The lifespan of TV satellites can be extended for up to 25 years with a XIPS because it needs less fuel to adjust its position.

XIPS engines are used in Europe's SMART-1 spacecraft which was launched in 2003 to map unexplored parts of the moon and analyse the elements of its surface rocks. NASA's Dawn spacecraft which was launched in September 2007 is also powered by XIPS and its mission is to study the asteroid Vesta in 2011 and the dwarf planet Ceres in 2015.

Environmental element

Xenon in the environment	
EARTH'S CRUST	2 p.p.t.
	Xenon is the 85th most abundant element, i.e. it is very rare.
SOILS	nil.
SEAWATER	100 p.p.t.
ATMOSPHERE	90 p.p.b.

Xenon is one of the rarest elements on Earth. Geochemists have a problem with this because there is much less of it in the atmosphere than of the other noble gases—which explains why it is so much more expensive than neon. Some geochemists suspect the missing xenon is in the Earth's crust, either hiding in water clathrates and even locked-up in silicate rocks because it may bond chemically to quartz when exposed to high temperatures and pressures. Analysis of the xenon isotopes in rocks from deep in the Earth's mantle appears to show that it originated from the xenon dissolved in seawater, and that it has being carried into the mantle by the subduction of one tectonic plate sliding beneath an adjacent one.

The xenon of the air, and that which is dissolved in the sea, poses no threat to the environment. However, the radioactive xenon-133, with a half-life of five days, is a potential danger because it continually leaks from nuclear power and reprocessing plants, albeit in tiny amounts. It was released in larger quantities during nuclear accidents, such as that at Chernobyl, Ukraine, in 1986. Xenon-133 and xenon-135 are by-products of nuclear explosions and they are monitored to ensure compliance with test ban treaties and to detect whether rogue states are testing such weapons.

Chemical element

Data file

CHEMICAL SYMBOL	Xe
ATOMIC NUMBER	54
ATOMIC WEIGHT	131.29
MELTING POINT	−112°C
BOILING POINT	−107°C
DENSITY	5.9 g per litre
OXIDES	XeO_3 (explosive) and XeO_4 (unstable)

Xenon is a colourless, odourless gas, and one of the noble gas elements of group 18 (row 5p) of the periodic table. It was regarded as completely inert until in 1962, when at the University of British Columbia, a young chemist named Neil Bartlett produced a compound of it, namely $XePtF_6$. This was a discovery which took chemists by surprise because it overturned the theory that all the gases of this group of the periodic table were totally inert, and indeed up to then they had been labelled the inert gases. Bartlett had made PtF_6 gas which he knew was a molecule that was avid to grab electrons and he deduced that it would do this from a xenon atom. At 7.00 on the evening of Friday 23 March 1962 he allowed the two gases to come together and they reacted instantly to form a yellow solid, $XePtF_6$. A whole new element chemistry was about to begin.

Other xenon compounds soon followed namely XeF_2, XeF_4, and XeF_6, and today there are more than 100 xenon compounds including oxides, acids, and salts, some of which are commercially available. Even compounds with xenon-hydrogen, xenon-sulfur, and xenon-gold bonds have been made although these are stable only at low temperatures. HXeOXeH has been reported. The most exotic xenon compound to date has four xenon atoms attached to a central gold atom and this ion, $AuXe_4^{2+}$, will form stable crystals below −40°C,[217] This is a relatively high temperature for something so intrinsically unstable. Another curious compound of xenon is $Ca_2(XeF_2)_9(AsF_6)_4$ in which the two calcium ions are bonded only to the xenon atoms of the XeF_2 molecules.

There are 39 known isotopes of xenon with mass numbers 110 to 148, and 12 known isomers. There are nine main naturally occurring isotopes of xenon of which six are

---

[217] The counterbalancing negative ions are two antimony fluoride ions, $[Sb_2F_{11}]^-$.

stable. The most abundant isotope is xenon-132 (relative abundance 27 percent), followed by xenon-129 (26.5 percent), xenon-131 (21 percent), xenon-134 (10.5 percent) which is radioactive with a half-life of 11×10^{15} years, xenon-136 (9 percent) which is also radioactive with a half-life of 9.3×10^{19} years, xenon-130 (4 percent), xenon-128 (2 percent), and xenon-126 (0.1 percent). There are also traces of xenon-124 with a half-life of 48×10^{15} years.

Xenon has the most naturally occurring isotopes of any element, at least 17. They consist of those mentioned above plus radioactive isotopes produced as fission products in uranium-bearing deposits and they come from uranium, plutonium and other transuranium elements found there. The others are produced by the decay of radioactive isotopes of adjacent atomic number elements, themselves produced as fission products. These fission-generated and radioactive-generated isotopes, together with the primordial stable and long-lived radioactive isotopes, include xenon isotopes 124, 126, and 128 to 142.

Element of surprises

1. When solid xenon is subjected to a pressures of 140 GPa (equivalent to 1.3 million times atmospheric pressure) then it takes on the properties of a metal and has a bright blue colour.

2. Imax projectors use 15 kilowatt xenon short-arc lamps which are filled with the gas at such high pressures that they have to be handled in special enclosures by people wearing protective clothing because of the risk of explosion.

3. When a mixture of xenon and hydrogen gas was put under extremely high pressure in the laboratory of the Carnegie Institution of Washington in the USA, then they combined to form crystals in which there are as many as seven hydrogen pairs clustering around each xenon, forming $Xe(H_2)_7$. Not only that, but the crystals appear to be stable.

Ytterbium

Pronounced it-terb-ee-um, it is named after Ytterby, Sweden. For more details see 'The village of the four elements' under yttrium.

French *ytterbium*; German *Ytterbium*; Italian *itterbio*; Spanish *yterbio*; Portuguese *itérbio*.

Ytterbium is one of 15 chemically similar elements referred to as the lanthanoid or rare earth elements, which extend from element atomic number 57 (lanthanum) to element atomic number 71 (lutetium). The term rare earth is a misnomer because some are not rare. The minerals from which they are extracted, and the properties and uses they have in common, are discussed under Lanthanoids.

Essential element

Ytterbium is very much an element that has found it role this century in specialist uses like memory devices and tuneable lasers.

Element of life

The amount of ytterbium in the human body is not known but is very small. It has no known biological role, but it has been noted that ytterbium salts stimulate metabolism. Its soluble salts are mildly toxic by ingestion, but insoluble salts are non-toxic. Ytterbium is a skin and eye irritant.

It is difficult to separate out the various amounts of the different lanthanoids in the human body, but they are present, and the levels are highest in bone, with smaller amounts being present in the liver and kidneys. There is no estimate of the amount of ytterbium in an average adult, and no one has monitored diet for ytterbium content, so it is difficult to judge how much we take in every day but it is probably only a milligram or so per year. Ytterbium is not taken up by plant roots to any extent, so does not get into the human food chain, and the amount found in vegetables can be as low as 0.08 p.p.b. (dry weight). Some plants, however, such as lichens can absorb ytterbium and have 900 p.p.b.

Element of history

Ytterbium was isolated in 1878 by the Swiss chemist Jean Charles Galissard de Marignac (1817–94) at the University of Geneva. The story begins with yttrium which, unknown at the time it was first reported, was contaminated with traces of other rare earth elements. In 1843, erbium and terbium were extracted from it, and then in 1878, Galissard de Marignac separated ytterbium from erbium. He had heated erbium nitrate until it decomposed and then extracted the residue with water and obtained two oxides: a red one which was erbium oxide, and a white one which he knew must be a new element, and this he named ytterbium oxide.

Ytterbium oxide was also discovered by Karl Auer (1858–1929) about the same time and he named the new element aldebaranium. (The ytterbium oxide that he and Galissard de Marignac had isolated was itself to yield another rare earth oxide, in 1907, when George Urbain extracted lutetium oxide from it.)

Ytterbium metal was first made in 1937 by heating ytterbium chloride and potassium together but it was not pure enough for its properties to be reliably measured. This had to wait until 1953 when a pure sample of the element was finally obtained.

Economic element

Ytterbium is extracted commercially from monazite which contains 0.1 percent, whereas the other common rare-earth ore, bastnäsite, has only 0.0006 percent ytterbium. The main mining area is China, and there are large deposits also in the USA, Brazil, India, Australia, Greenland, and Tanzania. Total reserves of ytterbium are estimated to be around a million tonnes. Other ores are euxenite, a mixed oxide ore found in Greenland and Brazil, and xenotime, which is mainly yttrium phosphate. Ytterbium is also present in gargarinite, another yttrium mineral discovered in Kazakhstan in 1961, and which was named after the first astronaut Yuri Gagarin. Its composition is $NaCaY(F,Cl)_6$.

World production of ytterbium is around 50 tonnes per year, reflecting the fact that it finds little commercial application, although it has been shown to have some benefits, such as strengthening stainless steel, and some alloys have even been suggested as useful in dentistry. Like other rare earth elements, it can be used to dope light-emitting phosphors, or for ceramic capacitors, and garnet bubble memory devices. These consist of a thin film of magnetic material within which are areas known as 'bubbles' and which can store data. Ytterbium is used as a doping agent to create high power and tuneable solid state lasers.

Ytterbium can act as an industrial catalyst and its salts are being introduced into the chemicals industry as catalysts in place of ones that are regarded as toxic and polluting.

Ytterbium has a single absorption band at 985 nm in the infrared and this has been used to convert radiant energy into electrical energy in devices that couple it to silicon photocells.

Environmental element

Ytterbium in the environment	
EARTH'S CRUST	3 p.p.m.
	Ytterbium is the 43rd most abundant element.
SOILS	*approx.* 2 p.p.m. (dry weight), range 0.08–6 p.p.m.
SEAWATER	1.5 p.p.t.
ATMOSPHERE	virtually nil.

Ytterbium is one of the rarer rare earth elements although it is still twice as common in the Earth's crust as tin. It poses no environmental threat to plants or animals.

Chemical element

Data file	
CHEMICAL SYMBOL	Yb
ATOMIC NUMBER	70
ATOMIC WEIGHT	173.04
MELTING POINT	824°C
BOILING POINT	1,193°C*
DENSITY	6.97 g/cm³
OXIDES	Yb_2O_3 and YbO (less stable)

* Ytterbium has the shortest liquid range of any metal.

Ytterbium is a soft, silvery-white metal and one of the lanthanoid group (row 4f) of the periodic table. It is slowly oxidised by air, but this forms a protective layer on the surface. The metal reacts slowly with water, and dissolves readily in dilute acids. Its preferred oxidation state is +3 as in compounds like YbF_3 and $YbCl_3$, although there is a lower +2 oxidation state which manifests itself in compounds like $YbCl_2$. In solution it exists as the ion Yb^{3+} in association with nine water molecules. The solution is colourless, as are ytterbium salts.

There are 34 known isotopes of ytterbium with mass numbers 148 to 181, and 12 known isomers. There are seven naturally occurring isotopes of ytterbium: ytterbium-174 which is the most abundant and accounts for 32 percent, ytterbium-172 (22 percent); ytterbium-173 (16 percent), ytterbium-171 (14.5 percent), ytterbium −176 (12.5 percent), ytterbium-170 (3 percent), and ytterbium-168 (0.1 percent). Two of these are now known to be radioactive but with incredibly long half-lives: ytterbium-173, which has a half-life of 1.6×10^{17} years, and ytterbium-168, which has a half-life of 1.3×10^{14} years. The next longest-lived radioactive isotope is ytterbium-169 with a half-life of 32 days, and this finds use as a portable source of gamma rays.

Element of surprise

The electrical conductivity of ytterbium varies rather oddly as pressure on the metal increases. At 16,000 atmospheres it becomes a semiconductor and thereafter its electrical resistance increases as the pressure mounts until it reaches 40,000 atmospheres when the trend suddenly reverses and it starts to become more conducting again. This feature of ytterbium was put to use in stress gauges designed to monitor the intense shock waves of nuclear explosions.

Yttrium

Pronounced it-ree-um, it is named after Ytterby, Sweden—see 'The village of the four elements' below.

French *yttrium*; German *Yttrium*; Italian *ittrio*; Spanish *ytrio*; Portuguese *itrio*.

Yttrium is often classified as one of the rare earth elements because it is found in all rare earth minerals and it is very like them chemically. Even so, it is not one of this group, and it is certainly not rare, being twice as abundant as lead.

Essential element

Yttrium-containing crystals are part of various electronic devices such as lasers.

Cosmic element

Most yttrium has been formed from lighter elements capturing neutrons and this occurs inside red giant stars. Some is formed during supernova explosions. Samples of material from the Moon, amounting in all to 300 kg of dust, rock, and drilled cores, were found to have a relatively high yttrium content. These were brought back from the six Apollo moon landings, which took place between July 1969 and December 1972.

Element of life

The amount of yttrium in the human body is about 0.5 mg. No biological role has been identified for yttrium although it occurs in humans, where it concentrates in the liver and bone. Indeed there is evidence that it is to be found in every living organism, sometimes in relatively high amounts, and it is associated with nucleic acids. Human breast milk contains 4 p.p.m.

Edible plants can have quite a range of yttrium, from 20–100 p.p.m. (fresh weight) with the higher values being recorded for cabbage. The seeds of woody plants have the highest amounts of all: 700 p.p.m. Although coal contains yttrium (7–14 p.p.m.) this is not thought to indicate its selective absorption by those plants and trees which were growing at the time it was formed.

Yttrium salts are regarded as mildly toxic if they are soluble, such as the nitrate, and non-toxic if they are insoluble, such as the sulfate or oxide.

Medical element

The radioactive isotope, yttrium-90, has been used in several types of medical treatment, generally for cancers of the ovaries, rectum, and pancreas, as well as leukaemia. It attaches itself to monoclonal antibodies and these attach themselves to cancer cells.

In this way the cells get a deadly dose of the intense beta-radiation that this isotope emits as it decays. Needles of yttrium-90 have been also used by surgeons as a precision alternative to the scalpel when severing pain-transmitting nerves in the spinal cord.

Yttrium-90 is produced from radioactive strontium-90. This is a fission product of nuclear reactors and has a half-life of 29 years, decaying to yttrium-90 which has a half-life of 64 hours. For this to be used in medical treatment it has to be separated from the strontium because this would be retained by the body. In the 1990s, chemical separation techniques were developed that could achieve complete separation of the yttrium-90 from the strontium-90.

Element of history

Yttria (yttrium oxide, Y_2O_3) was first isolated and recognised as that of a new element in 1794 by the chemist Johan Gadolin (1760–1852) at the University of Åbo, Finland. For the story of how this came to be discovered and the other elements it contained see 'The village of the four elements' below.

The village of the four elements

In 1787, a 30-year-old army lieutenant, Karl Axel Arrhenius (1757–1824) came across an odd piece of black rock which looked like a lump of coal, in an old quarry near the village of Ytterby, located near Vauxholm. (The name Ytterby simply means 'outer village' and signifies merely its location, suggesting it has no noteworthy feature otherwise. The island on which it is located is now part of Stockholm.)

Arrhenius knew it was too heavy to be coal and he thought he had found a new ore of tungsten, a metal which had been discovered in Spain only four years previously. He gave the specimen over to the chemist, Johan Gadolin of the University of Åbo, (now called Turku) in Finland, who investigated the new mineral and in 1794 announced that it contained a new 'earth' which made up 38 percent of its weight. He called the new earth yttria, from the name of the village, and though he spelt it wrongly, the name stuck.

Yttria was in fact yttrium oxide, and was called an earth because it could not be separated from the oxygen it contained by the conventional method of heating it strongly with charcoal. The metal itself was first isolated in 1828 by Friedrich Wöhler (1800–82) by heating yttrium chloride with potassium metal.

The mineral from which yttrium oxide came was named gadolinite in honour of Gadolin. Gadolinite is essentially a silicate mineral of beryllium, iron and yttrium, and although it also contained beryllium this element was missed by Gadolin and was discovered by Nicolas-Louis Vauquelin (1763–1829) in Paris four years later. In any case we now realise that Gadolin's new earth must have been contaminated with many other elements, but how many were present in his sample of yttrium oxide we shall never know because his collection was lost when the University of Åbo was destroyed by fire in 1827.

In 1843, Carl Gustav Mosander (1797–1858) investigated yttrium oxide more thoroughly and found that it was made up of three oxides: yttrium oxide, which was white; terbium oxide, which was yellow; and erbium oxide, which was rose-coloured. The names of these two new elements, erbium and terbium, were again derived from the name Ytterby. The fourth element to be based on the name of the village was ytterbium. This was discovered by Jean Charles de Marignac (1818–94) in 1878.

Economic element

The yellow-brown ore xenotime can contain as much as 50 percent yttrium phosphate (YPO_4), and is mined in China and Malaysia. Yttrium is found in the rare earth mineral monazite of which it makes up 2.5 percent. The other abundant rare earth mineral, bastnäsite, contain much less. Yttrium is also part of fergusonite, a glassy black ore found in Madagascar, and of samarskite—see samarium, page 464. The output of yttrium is about 9,000 tonnes per year, measured as yttrium oxide, and world reserves are estimated to be around nine million tonnes. Yttrium metal is produced by heating yttrium fluoride with calcium metal, but only a few tonnes are made each year.

Yttrium is used in alloys and it improves magnesium castings, and gives a finer grain to chromium, molybdenum and zirconium metals. When added to cast iron it makes the metal more workable. Although metals are generally very good at conducting heat, there is an alloy of yttrium with chromium and aluminium which is heat resistant.

Yttrium oxysulfide, doped with europium, was used as the standard red phosphor of old-style colour televisions. Yttrium oxide in glass makes it heat and shock resistant, and this is used for camera lenses. Yttrium oxide is suitable for making superconductors, which are metal oxides that conduct electricity without any loss of energy, and one of the best known of these is the mixed yttrium, barium, copper oxide, $YBa_2Cu_3O_{7.9}$, commonly referred to as '1-2-3'. This was one of the first superconducting materials to work at liquid nitrogen temperatures (−196°C).

Yttrium-iron garnet (YIG, formula $Y_3Fe_5O_{12}$) can be made into large crystals for optics, or deposited as a thin layer and as such is used in magnetic recording. It is also a transducer of acoustic energy, in other words turning sound into electrical energy, and is a component of microwave filters for radar.

Yttrium-aluminium garnet, known as YAG, is used in lasers and these can operate at high power output and can even cut and drill through metals. YAG is very hard and makes a sparkling diamond-like gemstone. When YAG is doped with cerium then it is used in white LED lights.

In 2006 a new alloy of aluminium-yttrium-nickel was produced by the US Department of Energy's Ames Laboratory which can significantly reduce the weight of an aircraft. The new alloy was made by blasting a stream of its molten components through a nozzle to produce a fine powder. When this is hot pressed and extruded it creates a metal that is much stronger than current aircraft-grade aluminium. It is destined for use in the F-35, the main fighter aircraft of the US Air Force and Navy.

Environmental element

Yttrium in the environment	
EARTH'S CRUST	30 p.p.m.
	Yttrium is the 28th most abundant element.
SOILS	av. 23 p.p.m. (dry weight)*, range 10–150 p.p.m.
SEAWATER	9 p.p.t.
ATMOSPHERE	virtually nil

* Cultivation lowers this to around 15 p.p.m.

Yttrium is one of the rarer rare earth elements although it is still 400 times as common as silver in the Earth's crust. It poses no environmental threat to plants or animals.

Chemical element

Data file	
CHEMICAL SYMBOL	Y (formerly Yt until the 1920s)
ATOMIC NUMBER	39
ATOMIC WEIGHT	88.90585
MELTING POINT	1,522°C
BOILING POINT	3,338°C
DENSITY	4.47 g/cm³
OXIDE	Y_2O_3

Yttrium is a soft, silvery-white metal that is a member of group 3 (row 4d) of the periodic table. It is stable in air because it is protected by the formation of a resistant oxide film on its surface. It burns if ignited, and is attacked by water which it decomposes to release hydrogen gas. Its only oxidation state is +3 as in such salts as YCl_3 and $Y(NO_3)_3$, and Y^{3+} ions in solution.

There are 34 known isotopes of yttrium with masses 76 to 109, and 28 known isomers. There is only one main naturally occurring isotope and that is yttrium-89, which is not radioactive.[218] The longest-lived radioactive isotope is yttrium-88 with a half-life of 107 days.

Element of surprise

When a mirror coated with yttrium metal is exposed to hydrogen gas it becomes transparent. What happens is that the hydrogen is taken up by the yttrium and this prevents visible light from interacting and being reflected, so it passes through the mirror unaffected.

[218] There are also traces of radioactive yttrium isotopes as fission products in uranium minerals.

Zinc

Pronounced zink, the name is derived from the German *zinke*. The word may have originated in the Persian word *sing* meaning stone.

French *zinc*; German *Zink*; Italian *zinco*; Spanish *cinc*; Portuguese *zinco*.

Essential element

Zinc is essential for all forms of living things as well as providing remarkable protection for steel and rubber products.

Cosmic element

Zinc nuclei are too large to form in normal stars; they are generated during a supernova. Even so, there is much less zinc in the universe than theory predicts, and this seemed to be confirmed because nearby galaxies have less than 20 percent of the zinc that they should have. Then, in 2006, Varsha Kulkarni at the University of South Carolina, using the Very Large Telescope Interferometer on Mount Paranal in Chile, discovered a Lyman-alpha type of galaxy 6.3 billion light years away in which the missing zinc appears to be concentrated.

Element of life

Zinc in the human body	
BLOOD	7 p.p.m.
BONE	75–170 p.p.m.
TISSUE	50 p.p.m. (muscle)
TOTAL AMOUNT IN BODY	2–3 g.

Zinc is generally thought to be non-toxic, and is an essential element for animals, humans, plants, and microbes. In plants, it is believed to protect them against drought conditions and diseases. In humans, zinc is at its highest concentrations in the eye, prostate, muscle, kidney, and liver. Semen is particularly rich in zinc, and there is evidence that a lack of zinc can lower sperm counts in men. The brain requires zinc to modulate its activity as well as to help with learning, and to activate those parts of the brain that govern taste and smell.

Zinc is important for two groups of proteins: enzymes and transcription factors.[219] There are more than 200 kinds of zinc-containing enzymes, and transcription factors appear to be equally numerous, some having protrusions referred to as zinc fingers.

---

[219] Transcription is the process whereby RNA is synthesised from a DNA template.

Zinc enzymes regulate growth, development, longevity, and fertility. They are also important in digestion, nucleic acid synthesis, and the immune system. One of the most important ones is *carbonic anhydrase,* without which the body could not rid itself of carbonate which it converts to carbon dioxide. Zinc is especially needed where there is rapid turnover, as with the immune system, bone marrow, and the lining of the gut.

Surplus zinc can be stored in the bones and spleen, but it is not easily released from these even though there is a lack of this element in the diet. Zinc is lost from the body at about 1 percent of the total per day, excreted via the intestines (90 percent), urine (5 percent) and sweat (5 percent). Zinc differs from metals like iron and copper in that it exists only as Zn^{2+} ions. As such it is non-toxic and it can attach itself to many other kinds of molecules.

Dangerous element

Zinc metal is a skin irritant and breathing the fumes of hot zinc leads to fume fever with sore throat, cough, and sweating, progressing to more severe symptoms if the exposure is heavy.

Too much zinc can also be damaging when it is consumed. It causes vomiting, stomach cramps, diarrhoea and fever, and mass poisonings by zinc have occurred following the storage of fruit juices in galvanised containers.

Food element

Zinc in plants varies according to the level in the soil. In soil with adequate zinc levels, the food plants with the highest levels are wheat with 20–60 p.p.m., sweet corn (a type of maize) with around 20 p.p.m., and lettuce with 12 p.p.m. Fruits have low levels, with apples and oranges having only 1 p.p.m. or less (all fresh weights).

The World Health Organisation is aware that 2 billion people have zinc-deficient diets and it advocates that zinc supplements be made available. It is estimated that almost half the world's population is at risk of zinc deficiency, especially those whose diet is mainly based on cereals and beans, because these contain a lot of phytic acid and this forms insoluble zinc phytate which cannot be absorbed. Also at risk are those whose diet consists mainly of edible tubers, such as cassava or potatoes, because these have low levels of zinc. In some countries where the population relies on such diets, foods are fortified with zinc, and this occurs in Mexico, Indonesia, and Peru.

Even in a modern Western diet, zinc is more likely to be deficient than iron. Red meats are the best source of zinc. The average adult takes in 5–40 mg of zinc per day, depending on their liking for meat. The minimum intake of zinc needed to avoid deficiency symptoms is 2–3 mg per day and, assuming a 30 percent absorption rate, the dietary intake for men should be around 7.5 mg per day and for women 5.5 mg. The average man takes in around 10 mg and the average woman about 8 mg. Research shows that during pregnancy this amount is also sufficient, because a higher zinc absorption rate comes into operation, and this continues during lactation. A breast-feeding mother produces around 850 ccs of milk a day containing the 2 mg of zinc that her baby needs. (Cow's milk that is used to make formula feed for babies has to be fortified with zinc.)

Oysters have more than 7 mg of zinc per 100g, liver 6 mg, beef 4 mg, wheat 4 mg, cheese 3 mg, shrimp 2 mg, eggs 1 mg, and milk 0.4 mg. Fruit generally has relatively little zinc, the average being around 0.15 mg per 100g (fresh weight). Strict vegetarians can get their zinc from sunflower or pumpkinseeds, brewer's yeast, maple syrup, and bran. All nuts and seeds contain zinc.

Zinc salts can be purchased over-the-counter in many health food shops and pharmacies, and are recommended by dietary advisers for conditions which do not respond to conventional treatments, such as anorexia nervosa, pre-menstrual tension, post-natal depression, acne, and the common cold. However, there is no robust evidence in support of zinc remedies being successful in curing these conditions.

Medical element

Although few zinc medications are now prescribed by doctors, this was not always so, and down the centuries zinc compounds played a part in the treatment of illness: zinc sulfate was used as an emetic, zinc chloride as an antiseptic, zinc oxide as astringent (to stop bleeding), and zinc stearate as an ointment. So-called calamine lotion, which is a mixture of zinc and iron oxides, has been sold by pharmacies since the 1600s and used to treat skin conditions. Zinc ointments were made by mixing zinc oxide with plant oils, wool oil (lanolin), or even mineral oil (aka liquid paraffin), and were applied to eczema, nappy rash, scalds, and sunburn.

Self-medication with zinc-containing remedies is still popular and these are said to be beneficial in treating a range illnesses such as the common cold, sore throat, and upset stomach. However, a nasal spray containing zinc has been banned because it causes damage to that sense organ making it impossible to detect smells correctly. Zinc is a key ingredient in multivitamin pills.

In the last century zinc oxide was used as a sunblocker against damaging ultraviolet rays from the sun which can cause skin cancer. It was applied to the face, and especially the nose and lips of sportsmen like cricketers, but it was not popular with the general public because it was too white a pigment. (Colourless nanoparticle titanium dioxide creams are now used instead.)

The importance of zinc for certain enzymes was realised early in the last century, but it was not until 1968 that the first cases of human zinc deficiency came to light, in Iran. This was identified by Ananda Prasad who had carried out zinc deficiency audits on animals, and was then confronted with the same symptoms, of stunted growth and retarded sexual development, in humans. Prasad was to become the leading world authority on human zinc deficiency, finding many cases in Egypt where the soil is lacking in this nutrient, and he proved that supplementing the diet with zinc sulfate was an effective treatment.

Lack of zinc was also shown to be the cause of the genetic disorder acrodermatitis enteropathica which was previously considered to be fatal for all babies born with it, but which can now be cured with doses of zinc.

Zinc also plays a role in the way the body metabolises alcohol. This is broken down in the liver by an enzyme called *alcohol dehydrogenase* which has a zinc atom at its centre. Excessive consumption of alcohol damages the liver, and it has long been known that those with cirrhosis have lower than normal levels of zinc in that vital organ. Again, if the damage is not severe, zinc supplements can restore the liver's function.

Element of history

According to Western tradition, zinc was discovered in Europe by the German chemist Andreas Marggraf (1709–82) in 1746. It would be more correct to say that he was the first to identify it as an element, because zinc had been known and used long before that date.

Strabo (66 BC–24 AD) of Asia Minor mentioned an ore from Cyprus that, when refined, would distil 'mock silver' and that when this metal was alloyed with copper it produced brass. Alternatively, brass was made by heating powdered calamine, which is zinc silicate, with charcoal and copper, a process that continued for almost two thousand years. The Roman writer Pliny the Elder (23–79 AD), who wrote an encyclopaedia, *Historia Naturalis,* mentioned an ointment that was used for healing wounds and sore eyes and which was probably zinc oxide. Even a statuette made of zinc has been found in the Roman province of Dacia (modern Romania) and may possibly pre-date the Roman occupation.

Marco Polo (*c.* 1254–1328) described the manufacture of zinc oxide in Persia in the 1200s, but it was almost certainly in India that zinc was first recognised as a metal in its own right. Care had to be taken not to lose the zinc, which is quite volatile, during the refining process and a method was developed in which organic materials, such as wool, were heated with zinc ores in a closed crucible. Waste from a zinc smelter at Zawar, in Rajasthan, testifies to the large scale on which the metal was refined, with one estimate being that more than a million tonnes of metallic zinc and zinc oxide were produced there from the 1100s to the 1500s.

The knowledge of zinc refining spread eventually to China where it was smelted on a large scale by the 1500s, and imported from there into Europe. An East India Company ship which sank off the coast of Sweden in 1745 was carrying just such a cargo of zinc and analysis of reclaimed ingots showed them to be almost the pure metal.

In 1668, a Flemish metallurgist, P. Moras de Respour (1614–99), reported the extraction of metallic zinc from zinc oxide. Chemists too began to take an interest in zinc around 1700 when Geoffroy the Elder (1672–1731) described it, and spoke of it as if it were something new. He described how it would condense as curious yellow crystals—which were clearly zinc oxide—on bars of iron placed above the ore that was being smelted. Zinc could also turn up a by-product of the lead industry, sometimes being collected as the metal from crevices in the chimneys of furnaces, and referred to as *zinc* or *counterfeit.*[220]

Marggraf investigated calamine from Poland, England, Germany, and Hungary in 1746 and obtained zinc by heating the ores with carbon in closed retorts. He realised that he was dealing with a new metal element. He was unaware that three years

---

[220] This had even been remarked on in the 1500s by Georgius Agricola and Paracelsus.

earlier the Swedish chemist Anton von Swab (1703–68) had also distilled zinc from calamine and even extracted some from zinc blende (zinc sulfide). Meanwhile in Bristol, in England, a zinc smelter was already in operation by the end of 1743, producing the metal at the rate of about 200 tonnes per year.

Economic element

The dominant ore is zinc blende which is also known as sphalerite and is zinc sulfide (ZnS). Other important zinc ores are wurtzite (which is another form of zinc sulfide), smithsonite (zinc carbonate, $ZnCO_3$), and hemimorphite, which is also known as calamine and is a zinc silicate ($ZnSiO_4$). Hydrozincite, which is basic zinc carbonate, is also mined for the metal when it is found in economic amounts.

World production of zinc exceeds 11 million tonnes a year and commercially exploitable reserves exceed 180 million tonnes, of which Australia holds 42 million, China 33 million, and Peru, 18 million. There are another 250 million tonnes that could be mined were it to become economic to do so. The main zinc mining areas are China (3 million tonnes), Australia (1.5 million tonnes), Peru (1.4 million tonnes), the USA (800,000 tonnes) and Canada (600,000 tonnes) although many other countries mine it, such as Iran, India, Mexico, and Ireland, all with production levels in excess of 400,000 tonnes per year.

Most zinc metal is smelted in China (3.7 million tonnes), Canada (800,000 tonnes), South Korea (700,000 tonnes), Japan (600,000 tonnes) and Australia (500,000 tonnes). This is done by roasting the sulfide ore to form the oxide and then reducing this by heating it in a furnace with coke, under an inert atmosphere to prevent the zinc from oxidising as it distils off. An alternative method of winning the metal is to convert the oxide to zinc sulfate and then extract the zinc electrolytically. More than 30 percent of the world's need for zinc is met by recycling.

Zinc is used both as the metal and the oxide. More than 50 percent of metallic zinc goes into galvanising steel, while other large uses are in die-casting of components, such as carburettors, and for making brass. Galvanising is done by dipping a steel object into molten zinc, or depositing a layer of zinc electrolytically. The zinc coating prevents oxygen and moisture from reaching the steel and even when the coating is breached it is the zinc which corrodes, and not the steel. For a similar reason, a disc of zinc bolted to an iron rudder will protect it—at least until all the zinc has corroded away and zinc is used as a sacrificial anode for underground pipelines made of steel. Galvanised steel is used for car bodies, fencing, motorway guard rails, street lighting posts, suspension bridges, heat exchangers, and roofing tacks.[221]

The metal is also used in alloys and batteries. The first dry batteries had a zinc anode, a carbon cathode, an alkaline electrolyte of ammonium chloride paste, and delivered 1.5 volts. Brass and bronze account for 17 percent of zinc production, and other alloys for 17 percent. The chemical industry takes 6 percent. The main zinc alloy is brass, which consists of 33 percent zinc and 67 percent copper, but there are also the nickel silvers (typically with 20 percent zinc, 60 percent copper, and 20 percent nickel) having names like China silver, German silver, nickel brass, white

---

[221] Small items like tacks are galvanized by tumbling them in zinc dust at 400°C.

copper, etc., which are often used for tableware. The alloy Prestal (78 percent zinc, 22 percent aluminium) is almost as strong as steel but as easy to mould as plastic.

As the oxide, zinc's biggest uses are in the rubber industry and in pigments. As a pigment it is used in plastics, cosmetics, photocopier paper, wallpaper, and printing inks, while in rubber production its role is to act as a catalyst during manufacture, and a heat disperser in the final product. In rubber and plastics it also prevents damage to the polymer from ultraviolet rays. Zinc oxide will conduct electricity when light shines on it, and use has been made of this in some photocopiers.

Zinc sulfide, doped with other metals, is used as a phosphor in X-ray screens and fluorescent lighting. When the phosphor is bombarded with electrons its atoms are energised and as they return to their original state they emit visible light.

Zinc compounds have been used as fire retardants, anti-fouling paints, and wood preservatives.

Environmental element

Zinc in the environment	
EARTH'S CRUST	75 p.p.m.
	Zinc is the 24th most abundant element.
SOILS	av. 64 p.p.m., range 5–770 p.p.m.
SEAWATER	30 p.p.b.
ATMOSPHERE	ranges from 0.1–4 mg per cubic metre, and much of it comes from the wearing away of motor tyres.

Some crops deplete the zinc in soil to such an extent that a subsequent crop may suffer, and this has been observed with maize when it follows sugar beet. Soils in parts of India and China can lack zinc, and fertilizers with this element added can produce improved crop yields and more diseases-resistant plants. For most soils the amount of zinc which arrives via the atmosphere generally exceeds that which is lost by leaching. Zinc is soluble as the zinc ion (Zn^{2+}) but this is easily absorbed by minerals, clays and organic matter, and strongly held. In the past, rivers passing through industrial and mining districts could have zinc in excess of 20 p.p.m., but in most areas sewage treatment now removes it and levels are much reduced, for example, the Rhine has only 50 p.p.b.

Some soils are heavily contaminated with zinc, and these are to be found in areas where zinc has been mined or refined, or sewage sludge from industrial areas, has been used as fertilizer. Such soil can have several grams of zinc per kg of dry matter. At levels in excess of 500 p.p.m., zinc in soil interferes with the uptake of other essential metals such as iron and manganese, and also of boron. Yet there are soils with levels of zinc in excess of 2,000 p.p.m. and a few with levels of 80,000 p.p.m. (8 percent) and even 180,000 p.p.m. (18 percent) have been recorded, the former in the US, the latter being in Belgium. However, for these it is the cadmium, which invariably accompanies the zinc, which poses more of a threat.

Fruit and nut-bearing trees in the Western states of the USA were afflicted by disease in the 1800s, and the reason was thought to be that they were growing on soil deficient in copper. Curiously, some trees responded to the application of a copper solution while others did not. It eventually transpired that those which did benefit were actually responding to the addition of zinc, which was coming from some of the galvanised buckets in which the copper solution was prepared.

Chemical element

Data file	
CHEMICAL SYMBOL	Zn
ATOMIC NUMBER	30
ATOMIC WEIGHT	65.39
MELTING POINT	420°C
BOILING POINT	907°C
DENSITY	7.1 g/cm³
OXIDE	ZnO

Zinc is a bluish-white metal that heads group 10 (row 3d) of the periodic table. Zinc tarnishes in air, which eventually forms an impervious layer of zinc hydroxide carbonate. The metal is attacked by acids and alkalis. Its preferred oxidation state is +2 as in almost all of its compounds and in solution as Zn^{2+} ions.

There are 32 known isotopes of zinc with masses 54 to 85, and 10 known isomers. There are five naturally occurring isotopes: zinc-64 accounts for 49 percent, zinc-66 (28 percent), zinc-68 (19 percent), and zinc-67 (4 percent), and zinc-70 (0.6 percent). In fact the most common and least common of these are radioactive: zinc-64 has a half-life of 2.3×10^{18} years, and zinc-70 has a half-life of 1.3×10^{16} years.[222] The next longest-lived radioactive isotope is zinc-65 with a half life of 244 days.

Element of surprise

Rolled zinc sheeting has been used in roofing, especially in Paris where it was made mandatory in the 1860s when the city underwent a transformation, which resulted in the attractive city roofscape that became famous and inspired many paintings. Even in Roman Pompeii, destroyed when Mount Vesuvius erupted in 79 AD, a zinc-capped fountain has been discovered showing that zinc has been a weatherproof material of long standing. Indeed, zinc roofing has a long and serviceable life, although it is attacked by acid rain and sulfur dioxide.

---

[222] In addition there are trace amounts of other radioactive isotopes produced as fission products in uranium-bearing deposits.

Zirconium

Pronounced zer-koh-nee um, it is named from the Arabic word *zargun* meaning gold coloured.

French *zirconium*; German *Zirconium*; Italian *zirconio*; Spanish *circonio*; Portuguese *zircónio*.

Essential element

Zirconium makes useful alloys, while its oxide make ultra-strong ceramics, and this also creates beautiful gem stones known as cubic zirconia or CZ.

Cosmic element

Zirconium appears to be relatively abundant in certain types of stars, known as S-type stars in which heavier elements are formed by neutron capture followed by beta-decay, and it is also present in the spectrum of the sun. Meteorites contain zirconium and samples of rock brought back from the moon by the Apollo space missions were found to have a surprisingly high zirconium content.

Element of life

Zirconium in the human body	
BLOOD	10 p.p.b.
BONE	less than 0.1 p.p.m.
TISSUE	av. 0.1 p.p.m.
TOTAL AMOUNT IN BODY	1 mg.

Zirconium has no known biological role, and its salts are generally regarded as being of low toxicity. The estimated daily dietary intake is around 50 micrograms (although for one individual it was measured at more than 100 micrograms). Most passes through the gut without being absorbed, and that which is absorbed tends to accumulate slightly more in the skeleton than in tissue. Zirconium cannot be detected in urine.

While aquatic plants have a rapid uptake of soluble zirconium, land plants have little tendency to absorb it, and indeed 70 percent of plants that have been tested showed no zirconium to be present at all. However, some plants do absorb a little from the zirconium that is present in all soils, and the leaves of deciduous trees can have as much as 500 p.p.m. (ash weight), but food plants have less than 2 p.p.m. (dry weight) and often as little as 5 p.p.b.

Medical element

In the USA, lotions containing zirconium carbonate used to be sold as treatments for poison ivy. The zirconium combined with the irritant urushiol from the plant and rendered it inactive, but they were discontinued in the 1960s after a few people appeared to react badly to them. However, a group of 22 workers who had been exposed for up to five years to the fumes of molten zirconium were monitored and even years later were found not to have suffered any adverse effects.

Element of history

Gems that contain zirconium were known in biblical times and called hyacinth, jacinth, jargon and zircon. Colourless varieties were thought to be an inferior kind of diamond, but that was shown to be false when a German chemist, Martin Heinrich Klaproth (1743–1817),[223] analysed a zircon in 1789 and discovered zirconium. He ran a pharmacy in Berlin, and investigated a specimen of jargon, a semi-precious stone from Ceylon,[224] and he began by heating it with alkali. From the product of that reaction he extracted a new oxide, which he called zirconia, and which he realised must be that of a new element, zirconium. In the year he discovered zirconium, he also discovered uranium, and this coincidental link between the two was to be strangely echoed 150 years later when the fates of these two elements became truly linked in the nuclear power industry—see below.

Klaproth was unable to isolate the pure metal itself, and sadly he did not live to see this achievement. Humphry Davy, in London in 1808, tried to separate zirconium from its oxide by electrolysis, which was the way that he had successfully isolated potassium and sodium metals, but he failed. It was not until 1824 that the element was isolated, when the Swedish chemist Jöns Jacob Berzelius (1779–1848) heated potassium hexafluorozirconate (K_2ZrF_6) with potassium metal in a sealed iron tube, and he obtained it as a black powder. Even so, little use was found for zirconium; it had no commercial application as a metal, and its chemical compounds were without any noteworthy features.

As the bulk metal, impure zirconium is hard and brittle whereas the pure metal is easily worked and can be drawn into wire. Metal of this quality was first produced only in 1925 by the Dutch chemists Anton Eduard van Arkel (1893–1976) and Jan Hendrik de Boer (1899–1971), by the decomposition of zirconium tetraiodide (ZrI_4). In the 1940s, William Justin Kroll (1889–1973) of Luxembourg developed a commercial process for making pure metal and this involved heating zirconium tetrachloride ($ZrCl_4$) with magnesium.

Element of war

The cluster bomb unit known as CBU-87/B contains zirconium which is part of the incendiary component of this weapon. The bomb weighs 430 kg overall and contains armour penetrating missiles, anti-personnel shrapnel, and incendiary bomblets.

---

[223] Klaproth eventually became the first Professor of Chemistry at the University of Berlin when he was 60.
[224] This is a form of zircon (zirconium silicate).

There are 202 of these bomblets each weighing 1.5 kg and holding 287 g of a mixture of cyclotol (70 percent) and zirconium sponge (30 percent) which breaks up into burning particles that can saturate the target area and is intended to be used against ammunition dumps, convoys, or parked aircraft. Cyclotol is a mixture of the explosives RDX (1,3,4-trinitro-1,3,5-triazacyclohexane) and TNT (2,4,6-trinitrotoluene).

Economic element

The chief ore is zircon ($ZrSiO_4$) but some baddeleyite (zirconium oxide, ZrO_2) is also mined. World production is in excess of 1.5 million tonnes per year of zircon, which comes mainly from Australia (600,000 tonnes), South Africa (400,000 tonnes), China (200,000 tonnes), Indonesia (100,000 tonnes), and the USA (100,000 tonnes). Brazil mines mainly baddeleyite (26,000 tonnes). The estimated reserves exceed a billion tonnes. Australia (East and West Coasts), South Africa (Richards Bay), India (Kerala State), Sri Lanka and the USA (Trail Ridge, Florida) have vast deposits of zircon and zirconia sands. Often the zircon is a by-product of the mining in these regions, which is primarily directed at recovering titanium from mineral-rich sands. Around 1,200 tonnes of zirconium metal itself are produced.

Zirconium in its various forms finds use in ceramics, refractories, foundries, glass, chemicals, and alloys. Zircon sand is traditionally a refractory material, making heat resistant linings for furnaces, giant ladles for molten metal, and foundry moulds. It is used because it resists molten metal penetration and it gives a uniform finish. While these long established markets have declined, others have developed to take their place and ceramic glazes account for almost half of zircon consumption. It was previously used to make the glass for televisions. Zircon mixed with vanadium or praseodymium makes blue and yellow pigments for glazing pottery, and zircon glazes are particularly employed for sanitary ware. Zirconium oxide is used for grinding wheels and special kinds of sandpaper for grinding down metals.

Zirconia, the common name for zirconium oxide, only melts above 2,500°C, and is used to make heat resistant crucibles,[225] ceramics and abrasives. World production of pure zirconia is almost 25,000 tonnes per year, and it also goes into various chemicals that end up as cosmetics, antiperspirants, food packaging, and even fake gems—see also below. Zirconium hydroxychloride is an anti-perspirant component of roll-on deodorants.[226] The paper and packaging industry is finding zirconium compounds make good surface coatings because they have excellent water resistance and strength. Equally important is their low toxicity, and zirconium carbonate has been approved for treating wrapping that comes in contact with food.

The most dramatic use of zirconium, as zirconia, is in ultra-strong ceramics. The research which led to this was the need to develop a tank engine that was not made of metal, and so would need neither lubricating oil nor a cooling system. From this research came a new generation of tough, heat resistant ceramics that are stronger and sharper than toughened steel and which make excellent high speed cutting tools

---

[225] A red-hot zirconia crucible can be plunged into cold water without cracking.
[226] At one time it looked as if zirconium compounds would replace aluminium compounds in many applications because aluminium was believed, wrongly, to be the cause of Alzheimer's disease.

for industry. They are also sold as knives, scissors and golf irons. Bio-ceramics for joint replacement parts are also made of zirconia, and these are formed in precision presses and then fired at 1,700°C.

Zirconium is the basis of two kinds of gem stone, one natural (zircon, zirconium silicate) and one synthetic (cubic zirconia, zirconium oxide). Larger crystals of zircon are sold as a semi-precious gems, and they come in a variety of colours of which the stones with a golden hue are most admired. When zircon is cut and polished it shines with an unusual brilliance because it has a high refractive index and colourless stones look very like diamonds. Cubic zirconia (aka CZ) has the same crystal structure as diamond and even outsparkles it. Gem quality stones were first made in Soviet Russia in the early 1970s and commercial production began in 1976 using specially designed furnaces. CZ may not be so hard as diamond but it is as hard as other gem stones and is at the upper end of the scale of hardness: 8.5 Mohs compared to diamond's 10. Although it is hard, it is also somewhat brittle. What distinguishes cubic zirconia from diamond is its density of 5.68g/cm^3 compared to diamond's 3.52. CZ contains around 10 percent of calcium oxide which is needed to induce it to grow crystals of the right shape and for these to be stable. CZ can be made even more like diamond by coating it with a surface layer of diamond using chemical vapour deposition techniques. CZ can also be coloured by incorporating small amounts of other metal oxides in the melt from which it is grown. For example, chromium produces green gems, cerium will give red ones, and those using neodymium are purple.

A coating of zirconium and yttrium oxide ceramic is added to the blades of jet engines and turbines which are run at high temperatures and it provides a thermal barrier coating.

Zirconium metal had some hidden assets which suddenly brought it to prominence in the 1940s; it was found to be the ideal metal for inside nuclear reactors and nuclear submarines. It does not corrode at high temperatures, nor absorb neutrons to form radioactive isotopes. Even today the nuclear industry buys almost all of the metal that is produced and some nuclear reactors have more than 100,000 metres of zirconium tubing. Zirconium metal, as Zircaloy alloy (consisting of 98 percent zirconium with 1.5 percent tin plus 0.1 percent of iron, chromium and nickel), is used to make the cladding for uranium oxide fuel elements. As mined, zirconium contains 1–3 percent of hafnium, which is chemically very similar, and although it is difficult to separate the two elements this has to be done for the zirconium that is used by the nuclear industry because hafnium absorbs neutrons very strongly.

Zirconium metal is produced by heating zircon with carbon and chlorine to convert it to zirconium chloride, then separating this and heating it with magnesium. The metal is protected by an oxide layer which forms instantly on its surface by reaction with the oxygen of the air or with water vapour. This oxide film is impervious to both strongly acidic and alkaline media and mechanical attack, even up to 400°C, so that it finds use for handling corrosive chemicals in industry.[227] It is also used for

---

[227] Some reagents will attack it, such as hydrofluoric acid, chlorine gas, and solutions of iron(III) chloride and copper(II) chloride.

surgical implants and prosthetic devices because it is compatible with body fluids. The metal resists high temperatures, so it was used for space-vehicles like the Shuttle which heat up on re-entry to the Earth's atmosphere.

Zirconium metal is used in alloys—it makes steel stronger and improves its machinability. In 1998, the world's largest petrochemical processing unit, 3 metres in diameter, 50 metres tall, and costing $6 million, was installed at a plant near Houston, Texas. It was made from zirconium alloy and is claimed will last for more than 30 years, far longer than any stainless steel tower.

Early bulbs for flash photography used magnesium or aluminium foil. These were ignited electrically, but they were prone to exploding. In the USA, the Speed Midget bulb was developed which used zirconium metal, and was much more reliable. As photographic film became more sensitive, the size of the bulbs became smaller, and from 1960 to the 1980s the zirconium flash bulbs were the only ones in use, requiring tonnage quantities of the metal. With the advent of electronic discharge bulbs, built into cameras, such flash bulbs became obsolete.

Zirconium metal is used as a 'getter' in vacuum tubes to remove traces of gases because it combines readily with oxygen, hydrogen, and even nitrogen.

Environmental element

Zirconium in the environment	
EARTH'S CRUST	190 p.p.m.
	Zirconium is the 18th most abundant element.
SOILS	range 35–550 p.p.m.; 200–270 p.p.m. in garden soils.
SEAWATER	9 p.p.t.
ATMOSPHERE	traces.

Zirconium is not a particularly rare element but because its most common mineral, zircon, is highly resistant to weathering it is only slightly mobile in the environment. Zirconium is more than twice as abundant as copper and zinc, and over 10 times more abundant than lead. Zirconium is regarded as completely non-toxic and environmentally benign, hence its use will continue to grow. For example, it is being added to paints to replace the small amounts of lead compounds that were included to improve drying.

Zircon is one of the most widely dispersed of all minerals and is present as tiny crystals in many rocks. It is an important geological material because it permits the dating of rocks via its traces of uranium and thorium, the reason being that zircon crystals are highly stable and survive all kinds of geological processes. Zircon has shown that life might have started much earlier than once thought. Some rocks found in Australia in 2000 were 4.4 billion years old and their oxygen isotope ratio of O^{16}/O^{18} showed they could only have been formed when there was liquid water on the surface of the Earth, and this was nearly 500 million years earlier than previously assumed.

Chemical element

Data file	
CHEMICAL SYMBOL	Zr
ATOMIC NUMBER	40
ATOMIC WEIGHT	91.224
MELTING POINT	1,852°C
BOILING POINT	4,377°C
DENSITY	6.5 g/cm³
OXIDE	ZrO_2

Zirconium is a hard, lustrous, silvery metal, known as a transition metal and it is a member of group 4 (row 4d) of the periodic table. It is very resistant towards corrosion due to an oxide layer on the surface. Zirconium does not dissolve in acids (except hydrofluoric acid) or alkalis. Powdered zirconium is black and it will burn in air if ignited, and zirconium dust is regarded as a dangerous fire hazard.

There are 35 known zirconium isotopes with masses 78 to 112, and six known isomers. Five main naturally occurring isotopes of zirconium are zirconium-90, which accounts for 52 percent, zirconium-92 (17 percent), zirconium-94 (17 percent), zirconium-91 (11 percent), and zirconium-96 (3%). Two of these isotopes are radioactive: zirconium-96 has a half-life of 2.4×10^{19} years, which is more than five billion times the age of the Earth, and zirconium-94 which has a slightly shorter half-life of 1.1×10^{17} but still long in terms of the age of this planet. Otherwise the longest-lived zirconium isotope is zirconium-93 which has a half-life of 1.5 million years.

There are also a number of radioactive isotopes which occur naturally, in minute amounts in uranium minerals, produced by the spontaneous fission of atoms of uranium.

Element of surprises

1. There are two zirconium materials with extreme properties, one which it displays when very cold, the other when it is heated to high temperatures. The first is a zirconium-niobium alloy which becomes superconducting below 35 Kelvin (−238°C) in other words it will conduct electricity with no loss of energy. The second is zirconium tungstate (ZrW_2O_8) which actually shrinks when heated, at least until it reaches 700°C when it decomposes into ZrO_2 and WO_3.

2. A zirconium zinc alloy, of composition $ZrZn_2$, has the unusual property of becoming both superconducting and magnetic at low temperatures (below −238°C).

The Periodic Table

The Russian chemist Dimitri Mendeleyev (1834–1907) drew up the first successful periodic table of the chemical elements one winter's morning in 1869. Today he is rightly seen as its *discoverer* because he based it on scientific principles, rather than its being merely a table of the elements. Chemists had been groping towards a classification of the elements before this date, as they found more and more of them, and some came close to discovering the periodic table itself, as we shall see.

Since the time of Mendeleyev, other chemists have redesigned his periodic table, sometimes in response to new elements being discovered or made artificially, sometimes in response to advances in our knowledge about the nature of atoms. Of course it was inevitable that a periodic table would be devised sooner or later as the science of chemistry reached maturity, and once the nature of the atom had been revealed then it would have been an imperative. The surprise was that Mendeleyev's work preceded these advances by more than 30 years.

Three sub-atomic particles determine the chemistry of an atom: protons, electrons and neutrons, and atoms determine the chemistry of the elements and consequently the layout of the periodic table.

The building blocks of atoms

Protons: the number of positive protons in the nucleus of an atom is the so-called atomic number, and it determines the element itself: hydrogen has one proton, helium 2, lithium 3, beryllium 4, boron 5, and so on up element 118 and maybe beyond—see un-named elements.

Electrons: there are the same number of negative electrons in the orbits that surround the nucleus as there are protons in the nucleus, but it is the electrons in the outermost orbits that are responsible for the chemical behaviour of an element. When the outer electron arrangement is the same, then those elements resemble one another chemically. This explains the periodic repetition of chemical properties and behaviour that is observed in the groups of the periodic table. Electrons weigh very little and contribute less than 0.05 percent to the mass of an atom.

Neutrons: the neutrally charged neutrons in the nucleus are also important—they have the same mass as protons, so contribute to the overall atomic weight. They are the key to explaining fractional and anomalous atomic weights and isotopes, and were discovered in 1930 by James Chadwick (1891–1974) who earned a Nobel Prize in 1935 for his work.

The remarkable achievement of Mendeleyev was that he produced his periodic table 27 years before Joseph Thomson (1956–40) discovered the first sub-atomic particle, the electron, in 1896, and 42 years before Ernest Rutherford (1871–1937) discovered the positively charged atomic nucleus in 1911. When Mendeleyev heard of the discovery of electrons in 1897, he rejected the idea that they came from atoms. He believed that atoms were indivisible. Indeed he predicted that the electron would disappear from science in the way that phlogiston had. This may strike us as mildly eccentric now, but in those days some eminent scientists were still not convinced of the existence of atoms, never mind electrons. One such sceptic was the respected electrochemist Wilhelm Ostwald (1853–1932). Today we can photograph atoms.

Atoms and elements

The idea of atoms can be traced back to the ancient Greek philosophers who deduced that the world must be composed of them, and they did this by applying arguments of logic, one of which went as follows. Consider what happens when you slice horizontally through a cone. If matter is continuous then both faces of the slice must be exactly the same size, but since they are part of a cone then they must be different. The lower face must be larger, but that can only be so if we have sliced between two planes of atoms. The chief atomists were Leucippus (around 440 BC) and his pupil, Democritus, and their ideas were developed by Epicurus (341–270 BC). Leucippus said atoms must be incredibly small because we cannot see them, and they are indivisible, which is why he gave them the name 'atom' which means irreducible. We know of their theories because of the book *De rerum natura* (*Natural things*) that was written around 55 BC by the Roman poet Titus Lucretius. The book is still in print, thanks to a single copy that survived the Dark Ages.

Another philosophical debate concerned the concept of elements. Again we owe the idea to the Greek philosophers who speculated that although the world appeared extremely complex, all that we were seeing was just different forms of some primordial element. What might it be? For example, Heraclitus (died 460 BC) thought it was water, while others suggested air, fire, or earth. Empedocles (who also lived in the fifth century BC) combined them and said that there were four elements, and this view was developed by the greatest Greek philosopher of them all, Aristotle (384–322 BC).[228]

It seemed a reasonable theory, and one that was supported by common sense observations such as what happens when a stick burns in a fire. It can be seen to break down into the four elements: fire, steam, gases and ash. For 2,000 years Aristotle's ideas were accepted in Europe, almost without question, until the dawn of modern science in the seventeenth century.

Chemistry emerged from alchemy around 1700 AD and as it developed so too came the discoveries of new metals and gases, and the slow realisation that the Greek concept of four elements was fundamentally wrong. There were elements, but these had to be defined in terms of their chemistry. Indeed, had the ancient philosophers but known it, they already were familiar with 10 of them:

---

[228] In the East, Chinese philosophers came to the same conclusion, but with the addition of a fifth element: wood.

Carbon, which as charcoal came with fire.

Sulfur, deposits of which were discovered near volcanoes.

Copper, the first metal to be worked, around 5000 BC.

Gold and silver objects, which were first produced about 3000 BC.

Iron smelting led to the Iron Age, which began around 2500 BC.

Tin as such was smelted around 2100 BC, but it has been used earlier for bronze.

Antimony objects date from 1600 BC, although this metal was rarely used.

Mercury was first obtained from its sulfide ore, cinnabar, about 1500 BC.

Lead appeared around 1000 BC and was the most important metal domestically.

And there things rested for two millennia. In the West, the Roman Empire concentrated more on technological developments rather than scientific ones, although alchemy flourished for a while until the Emperor Diocletian suppressed it around 300 AD. He feared that if the alchemists were to find a way of turning lead into gold, it would undermine the Empire's currency. However, the Dark Ages were about to descend in Europe and progress came to an end for 500 years. This traditionally began with the Sack of Rome in 410 AD, and mass migration of peoples from the north, moving south as the Earth entered one of its cooler periods.[229] Science and technology in the West declined, and its study became the province of the Arabs and only returned to Europe as they advanced westwards, bringing with them their store of ancient wisdom and their new knowledge of the world.

Alchemy still had it adherents and some elements were discovered by them during this period. Arsenic appears to have been isolated by the German alchemist and philosopher, Albertus Magnus (c.1200–80), in the mid-1200s; bismuth appeared towards the end of the 1400s; phosphorus was made by Hennig Brandt of Hamburg in 1669. This last element had amazing properties: it burst into flames spontaneously when exposed to air and it glowed in the dark. The discovery of phosphorus is seen as a turning point between alchemy and chemistry, or rather its investigation turned the English alchemist Robert Boyle (1627–91) into the first modern chemist. Meanwhile the development of mining had uncovered zinc and the exploration of the New World brought platinum to people's attention.

The 1700s saw the emergence of chemistry in Europe and the discovery of several new metals: cobalt (1735), nickel (1751), magnesium (1755), manganese (1774), chromium (1780), molybdenum (1781), tellurium (1783), tungsten (1783), zirconium (1789), uranium (1789) and a few gases: hydrogen (1766), nitrogen (1772), oxygen (1772), and chlorine (1774).

It was left to the great French chemist, Antoine Laurent de Lavoisier (1743–94), to bring order to the fledgling science, which he did with his remarkable book *Traité Elémentaire de Chemie* (*Elements of Chemistry*) in 1789. In this he defined a chemical element as something that could not be further broken down (decomposed) and he listed 33 substances that came within the terms of his definition. He classified them into four categories: gases, non-metals, metals and earths:

---

[229] The Dark Ages lasted until around 1000 AD when the Earth warmed up again and the Middle Ages began.

Gases	Non-metals	Metals	Earths
[Light]	Sulfur	Antimony	Lime
[Heat]	Phosphorus	Arsenic	Magnesia
Oxygen	Carbon	Bismuth	Barytes
Nitrogen	Chloride	Cobalt	Alumina
Hydrogen	Fluoride	Copper	Silica
	Borate	Gold	
		Iron	
		Lead	
		Manganese	
		Mercury	
		Molybdenum	
		Nickel	
		Platinum	
		Silver	
		Tin	
		Tungsten	
		Zinc	

Lavoisier had given the idea of chemical elements scientific respectability. While his list contains some components that were never destined to be considered elements, such as 'light' and 'heat', it is still remarkable because he included things he called 'radicals' which he knew were non-metal elements but which could not be separated from other components, that is, chloride, fluoride and borate. He also had a group called 'earths' which likewise were oxygen compounds but from which it had proved impossible to remove the oxygen using the methods then available. We now know that lime is calcium oxide, magnesia is magnesium oxide, barytes is barium oxide, alumina is aluminium oxide and silica is silicon dioxide.[230]

Another of Lavoisier's far-reaching contributions to chemistry was to advocate that chemicals should be named after the elements of which they were composed. This de-mystified the language of chemistry of its alchemical names, and the result was that chemicals like 'butter of antimony' became antimony chloride, and 'lunar caustic' became silver nitrate.

A milestone in the development of chemistry came with the discovery of oxygen, the gas given off when calx of mercury (mercuric oxide) was heated. The discovery is attributed to Joseph Priestley (1733–1804) in 1774, but oxygen had already been isolated by the Swedish chemist Carl Wilhelm Scheele (1742–86) two years earlier, although his work was not reported until 1777. Oxygen delivered the greatest blow to Aristotle's four elements when, in 1781, Priestly demonstrated that water was formed from hydrogen and oxygen.

[230] This convention of naming oxides simply by replacing the 'um' or 'ium' ending of a metal with an 'a' continues to this day in names like alumina, titania, and zirconia.

Another intellectual revolution was initiated by John Dalton (1766–1844) who was a science teacher living in Manchester, England. He pondered on the atomic nature of matter and realised that the Law of Fixed Proportions, which recognised that elements combined with one another in fixed ratios of weights, could only be explained if they were composed of atoms. In 1803 Dalton presented a paper to the Manchester Literary and Philosophical Society entitled 'On the absorption of gases by water and other liquids'. In this he attempted to explain the different solubilities of gases by comparing the relative weights of them. This led him to propose a scale of atomic weights and when he published his talk in the newly launched *Proceedings* of the Society in 1805, he explored the idea further and appended a list of 20 elements with their atomic weights, and suggested atomic symbols—see Figure 1 below. At a stroke Dalton's theory not only proposed the existence of atoms, but suggested that atoms of each element had individual weights and that these could be calculated relative to one another.

By 1810, chemistry was an established science, producing remarkable discoveries almost every year, not least of which was a supply of new elements. In the years between the publication of Lavoisier's *Traité* and Dalton's talk, another 11 elements had been discovered: titanium (1791), yttrium (1794), beryllium (1797), vanadium (1801), niobium (1801), tantalum (1802), rhodium (1803), palladium (1803), osmium

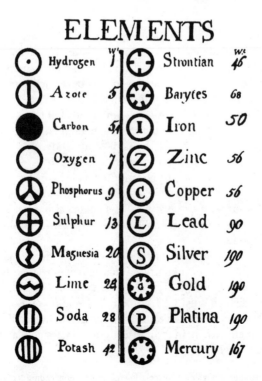

Fig. 1 Dalton's table of the elements.

(1803), iridium (1803), and cerium (1803). This pace was to continue, especially when Humphry Davy showed that some of Lavoisier's 'earths' could be decomposed by electrolysis and he isolated the metals potassium (1807), sodium (1807), calcium (1808) and barium (1808). In the same year, 1808, Adair Crawford (1748–95) discovered strontium.

Boron was also produced in 1808, and then came iodine in 1811. The year 1817 was particularly fruitful with thorium, lithium, selenium, and cadmium being announced.

Order out of chaos

By now the chemical similarity of groups of elements began to be commented on. In 1829 Johann Wolfgang Döbereiner (1780–1849) announced his Law of Triads, in which he noted that many elements came in groups of three ('triads'), and of which the weight of the middle element was the average of the lighter and heavier members.

Lithium-sodium-potassium were one such triad and others were chlorine-bromine-iodine and sulfur-selenium-tellurium. By the time Leopold Gmelin published the first edition of his compilation of essential chemical data, *Handbuch der Chemie*, in 1843 he noted there were 10 triads, three tetrads and even a pentad consisting of nitrogen-phosphorus-arsenic-antimony-bismuth. Today we recognise this pentad as group 15 of the modern periodic table.

Occasionally there were curious insights into the underlying nature of matter—and one in particular was well in advance of its time. This was the hypothesis put forward by William Prout (1785–1850) in 1815. He proposed that all the elements were multiples of the atomic weight of hydrogen, which is the lightest. If this is taken as 1, then it explained why all the other elements had weights that were whole numbers, or nearly so.

Prout's hypothesis was tantalisingly almost correct in assuming that hydrogen was the element of elements. In a way he was right. Most hydrogen atoms consist of a single proton, and it is the number of protons that determines the nature of all the elements. The nucleus of an atom is where 99.98 percent of its weight resides. Because this is composed of protons and neutrons, which have exactly the same weight as protons, and because most elements have one dominant isotope, then it is not surprising that their weights will tend to whole numbers. Most elements have atomic weights that fall within the limits ± 0.1 of a whole number, with very few having fractional numbers. Nevertheless it was these few exceptions that 'proved' Prout's rule, that is, tested it, and led to its rejection. It appeared to be merely a coincidence that many elements had almost whole number atomic weights, because there were several quite important exceptions, notably magnesium (24.3), chlorine (35.5), nickel (58.7) copper (63.5), zinc (65.4) and tin (118.7).

The problem of atomic weights was to dog chemistry for several decades. Clearly atoms were too small to be weighed, or even counted, and so only comparative weights could be meaningful. But against what standard were they to be compared? The best one appeared to be oxygen, which forms compounds with almost all of the other elements. Oxygen was taken to be 8, which followed from a comparison with hydrogen, and the 1:8 ratio of these elements in water. At the time the composition of water was assumed to be HO, not H_2O.

In fact the true chemical formula of water could have been deduced as early as 1800 when William Nicholson (1753–1815) and Anthony Carlisle (1768–1842) electrolysed water and showed it decomposed into two volumes of hydrogen and one of oxygen. The significance of this was not realised, even though in 1811 Amadeo Avagadro (1776–1856) suggested that equal volumes of gases contained equal numbers of particles (molecules). Eventually it was realised that the formula for water was H_2O, and the weight of oxygen was correspondingly adjusted to 16.

From 1830 to 1860 only three new elements appeared: lanthanum (1839), erbium (1842) and terbium (1843). These metals were later to yield other metals because they were part of the group known as the rare earths (aka lanthanoids), all of which have very similar properties, making them difficult to separate from one another.

The next era of element discovery came with the atomic spectroscope, which revealed that each element had a characteristic fingerprint pattern of lines in its visible spectrum. The technique was to spray a tiny sample of a solution of the element into the colourless flame of a Bunsen burner and observe the colours that ensued by viewing them through a prism so that they became separated across the spectrum, when it could be seen that they consisted of many bands and lines of colour. Because each element always gave the same pattern of lines, no matter its source, and that each gave a different pattern of lines compared to any other element, it was realised that here was a technique for uncovering new elements. Merely submitting a new mineral to atomic spectroscopy, as it was called, immediately showed whether a new element was present. As a result rubidium, caesium, thallium and indium were announced in the years 1860–63.

By then the total of known elements exceeded 60 and chemists were beginning to ask whether there was a limit to the number. To answer this question required a theory that could explain the known, and predict the unknown. The 'triads', 'tetrads' and 'pentad', were governed by regular increases in relative weight. Clearly chemists might use this universal property to rank all the elements, but the needed relative *atomic* weights rather than the more generally used comparative weights (known as equivalent weights). Equivalent weights depended upon other factors and, for example, if a metal had two kinds of oxide it had two equivalent weights.

By 1850, researchers were groping towards a logical arrangement of the elements. The German chemist, Max von Pettenkofer (1818–1901) noted that there was a connection between atomic weight and chemical behaviour but he was hamstrung by the lack of reliable atomic weights so could not verify his theory. In 1853, the British chemist John Hall Gladstone (1827–1902) arranged all the known elements in order of their relative atomic weights and noted relationships between them. In the 1850s, the American chemist Josiah Parsons Cooke (1827–94), the German chemist Ernst Lenssen, and the French chemist Jean-Baptiste Dumas (1800–84) also mused on the problem and indeed the last of these observed what we would now regard as horizontal relationships between metals of the periodic table, and he published a paper on this in 1851. However, all were thwarted by the lack of reliable atomic weight data.

Atomic weights and the first periodic tables

In 1858, the Italian chemist Stanislao Cannizzaro (1826–1910) published his 'Outline of a course of chemical philosophy' in which he showed how it was possible to assign an *atomic* weight to an element if Avogadro's Law was accepted. This said that equal volumes of different gases had the same number of molecules. Cannizzaro argued that this law led directly to the atomic weights of gaseous elements, and thence to other elements. A given volume of oxygen gas is 16 times heavier than the same volume of hydrogen. If the atomic weight of hydrogen is taken as 1 then it followed that the atomic weight of oxygen was 16. He presented a paper on his ideas at the First International Chemical Congress, which was held at Karlsruhe in 1860, and they were quickly accepted. Conference members eagerly sought copies of his table of atomic weights. One fell into the hands of a young Russian student, Dimitri Mendeleyev, who was then engaged in postgraduate research in Germany, and he took it back to St. Petersburg.

For the first time the elements could be arranged in an ascending scale from the lightest to the heaviest. The first attempt to arrange them in a regular pattern was made in 1862 by a French geologist Alexandre Emile Becuyer de Chancourtois (1820–86). He wrote the list on a piece of tape and then wound this spiral-like around a cylinder. The cylinder surface was divided into 16 parts, based on the atomic weight of oxygen. Chancourtois noted that certain triads came together down the cylinder, such as the alkali metals, lithium, sodium and potassium whose atomic weights are 7, 23 (7+16), and 39 (23+16). This coincidence was also true of the tetrad oxygen-sulfur-selenium-tellurium. He called his model the *Vis Tellurique* (Telluric Screw) and published it in 1862 but parts of it may no sense at all. For example, one of his groups consisted of boron, aluminium, nickel, arsenic, lanthanum and palladium.

What Chancourtois had discovered was the *periodic* nature of the elements. In other words, elements with the same properties occurred at regular periodic intervals, in this case when the atomic weights differed by 16, as with the alkali metals, or if they were multiples of 16, or nearly so, such as we find with that group of elements which includes oxygen at 16, sulfur at 32 (= 2×16), selenium at 79 ($5 \times 16 = 80$) and tellurium = 128 (= 8×16). Clearly there appeared to be an underlying numerical rhythm to the elements, even though we now know that it was purely by chance that the elements of the oxygen-tellurium tetrad are multiples of 16.

The Danish chemist Gustav Detle Hinrichs (1836–1923) started in 1855 to arrange the elements in some sort of order. He then emigrated to the USA in 1861 and took up a post at the University of Iowa. He too arranged the elements in a spiral form and this he published in 1867.

Another attempt to classify the elements was made by an Englishman, 27-year-old John Alexander Reina Newlands (1837–98). In 1864 he read a paper entitled 'The Law of Octaves' to a meeting of the London Chemical Society, and in it he pointed out that there was a periodic similarity of elements at intervals of 8. The title of his paper was chosen by analogy with octaves in music. This was an ill-judged choice, and it is said that one member of his audience sarcastically asked Newlands whether he had ever thought of arranging the elements alphabetically instead. The society's journal refused to publish his talk as a paper.[231] But he wrote an account for the *Chemical News*, in 1865, so we know what he was proposing. He had arranged 56 elements into groups and noted that there seemed to be a periodic repetition of properties with every eighth element.

Although Newlands was moving in the right direction, and some of the groups in his table we can now see in a modern periodic table, he was clearly not aware of any underlying imperative for what he had done, nor did he build on it. For example, he noticed that silicon and tin should be part of a triad, but he left no vacant place in the table for the missing element between them. Nor had he any qualms about putting two elements together in some of the boxes in his table.

William Odling (1829–1921), successor to Michael Faraday (1791–1867) at the Royal Institution in London, was another chemist to speculate about relationships among the elements, and he published a paper entitled: 'On the proportional numbers of the elements' in the first volume of the *Quarterly Journal of Science* in 1864. His arrangement of the elements came surprisingly close to that of Mendeleyev, and he left gaps where there were missing elements. In fact he even left gaps that were later to be filled by the noble gases helium and neon, although he thought these would be lighter elements of a pentad that expanded the triad of zinc-cadmium-mercury. Despite his realisation that the table could have predictive power, it was little more than a convenient way of classifying the elements.

Another scientist to come near to discovering the periodic table was the German, Julius Lothar Meyer (1830–95). He published a table of 49 elements with their 'valencies' in 1864.[232] Meyer drew a graph of atomic volumes of the elements versus their atomic weights in 1868 which shows a periodic rise and fall, and from which he deduced a periodic table. He passed this paper to a colleague, Professor Adolf Remelé (1839–1915), for his comments. Unfortunately these were slow in coming, and before he could submit it for publication Mendeleyev's definitive paper was published.

Dmitri Mendeleyev (1834–1907)

It still strikes one as odd that, in 1869, a relatively unknown Russian professor of chemistry, Dimitri Ivanovich Mendeleyev, produced the first successful periodic table of the elements. Mendeleyev was born in Tobolsk in western Siberia on 8 February 1834, the fourteenth child of a local schoolmaster. However, it was his mother, Maria, who raised the children after her husband became blind. She came from a family with interests in glassworks and paper mills and these she ran while struggling to educate her favourite son, Dimitri, whom she recognised as highly intelligent. She

---

[231] Nevertheless, the Royal Society of London awarded him its prestigious Davy Medal in 1887 in belated recognition of his achievements.

[232] Valency is a term usually reserved for non-metals, so that if such an element forms a compound of formula MCl_x we would say it was x-valent, e.g. in PCl_3 phosphorus has a valency of 3 and in PCl_5 it has a valency of 5. When it comes to talking about metals the preferred term is oxidation state rather than valency and this is shown by adding a Roman number suffix to the name of the element, so we would call $PtCl_2$ platinum(II) chloride and $PtCl_4$ platinum(IV) chloride. Nevertheless, valency was an essential part of chemical theory until relatively recently, and it is still a term that is used.

was determined that he should have a university education and when he was 18 she took him to Moscow, but failed to secure him a place at the university there.

Undaunted, they travelled on to St Petersburg, where he was allowed to enrol at the Central Pedagogic Institute to study physics and mathematics. Mendeleyev graduated with a gold medal for excellence in scholarship, and in 1859 went to study for his doctorate, first to Paris, working under Henri-Victor Regnault (1810–78), and then to Heidelberg, where he spent time with the great German chemist, Robert Bunsen (1811–99).

In 1861 he returned to St Petersburg, that great city on the Baltic, which was at the leading edge of reform in Russia and where science and the arts were encouraged in an atmosphere that was well ahead of its time; for example, women were encouraged to become educated and take up professions. In this atmosphere of enlightenment, chemistry flourished.

Mendeleyev became a professor of chemistry in 1867, and began to write his textbook *The Principles of Chemistry*, which ran to many editions and was translated into French, German, and English. He wondered how best to deal with the many elements with their diverse properties, and in the end he decided to group them according to their valencies. What followed was to transform a large part of chemistry from a disorganised jumble of facts into a disciplined science.

Mendeleyev's movements on the day of his great discovery are well documented. For him it was 17 February but for the rest of the Western world it was 1 March. Reform in Russia had not yet changed the yearly dating system from the Julian calendar of the Roman Empire to the Gregorian that we use today. Mendeleyev had planned to visit a cheese factory that morning, but the weather was bad so he stayed at home and worked on the manuscript for his book.

He had written the name of each element on separate piece of card, on which he also wrote its atomic weight, a few physical properties and the formulae of any hydrides and oxides it formed, which revealed its preferred valency. These cards he began to arrange like those in a game of patience (solitaire) with rows of elements having the same valency in descending order of their atomic weights. Suddenly things began to fall into place and he produced one arrangement that particularly impressed him and wrote it down on to the back of an envelope. This can be considered his first periodic table, and that envelope still exists.

After his mid-day meal Mendeleyev took a nap, and on waking he decided that a vertical arrangement of groups was a better way of depicting the periodic table. This event was to give rise to the romantic notion that the periodic table came to him in a dream. In any case he re-drew his table and this has remained the standard format to this day. What makes Mendeleyev's table so important, was that he realised that he had stumbled on an underlying order to the elements and this gave him the confidence to make a few predictions, most of which were soon to be proved right.

Mendeleyev realised that if his table of the elements meant anything, then he had to leave gaps where elements, yet unknown, should come. Below boron, aluminium and silicon were such vacant spaces. These elements he named eka-boron, eka-aluminium and eka-silicon. He predicted their atomic weights, melting points, densities, and said what the chemical composition of their oxides and other compounds would be.

```
                                        Ti—50        Zr—90       ?—180.
                                        V—51         Nb—94       Ta—182.
                                        Cr—52        Mo—96       W—186.
                                        Mn—55        Rh—104,4    Pt—197,4
                                        Fe—56        Ru—104,4    Ir—198.
                                  Ni—Co=59           Pl—106,6    Os—199.
        H—1                             Cu—63,4      Ag—108      Hg=200.
              Be—9,4     Mg—24          Zn—65,2      Cd—112
              B—11       Al—27,4        ?—68         Ur—116      Au—197?
              C—12       Si—28          ?—70         Sn—118
              N—14       P—31           As—75        Sb—122      Bi—210
              O—16       S—32           Se—79,4      Te—128?
              F—19       Cl—35,5        Br—80        I—127
        Li—7  Na—23      K—39           Rb—85,4      Cs—133      Tl—204
                         Ca—40          Sr—87,6      Ba—137      Pb—207.
                         ?—45           Ce—92
                         ?Er—56         La—94
                         ?Yt—60         Di—95
                         ?In—75,6       Th—118?
```

Fig. 2 Mendeleyev's first periodic table.

The most convincing demonstration of the correctness of Mendeleyev's periodic table was the discovery of these three missing elements. Eka-aluminium was discovered in 1875 at Paris by Paul-Émile Lecoq de Boisbaudran (1838–1912) and he called it gallium after the Latin name for his native France. He measured its properties, including the density, which he said was 4.7 g/cm³. He was informed that the new element was Mendeleyev's eka-aluminium and that most of its properties had already been foretold. However, Boisbaudran was aware of the discrepancy between his density and that predicted by Mendeleyev, which was 5.9 g/cm³, so he checked his measurements, discovered he had made an error, and reported a corrected value of 5.956 g/cm³.

In 1879 Lars Fredrik Nilson (1840–1899) discovered eka-boron at Uppsala, Sweden, and this too was named after the region of its discovery, scandium, from the Latin for Scandinavia. It too had the properties Mendeleyev predicted. For example its atomic weight was 44 (predicted 44) and density 3.86 g/cm³ (predicted 3.5 g/cm³).

Finally, in 1886, Clemens Alexander Winkler (1838–1902) discovered eka-silicon at Freiberg in Germany, and he also named it after the Latin name for his native land. Germanium was almost exactly as Mendeleyev had predicted, right down to the density of its oxide which he said would be around 4.7 g/cm³ and turned out to be 4.703 g/cm³, and the boiling point of its chloride which he said would be a few degrees below 100°C, and which turned out to be 86°C.

Mendeleyev had also noted that his group IV, the titanium group, was lacking a member and urged that this be sought among titanium ores. It only came to light in 1923 when the D. Coster and George von Hevsey found it in a zirconium ore and named it hafnium. Mendeleyev had predicted its atomic weight would be 180—it was measured as 178.5.

Sometimes Mendeleyev got it wrong. For example, he placed mercury in a group with copper and silver, even though this meant placing it in the wrong order of atomic weights.[233] His difficulty with tellurium and iodine, whose atomic weights were in the wrong order, is noted on page 532, yet despite this, he knew he must put them in the correct group of his table so sure was he that the periodic table was correct. And it was correct. What surprises us today is that he was able to deduce it long before chemists were even aware of the fundamental entities that shape it: the protons in an atom's nucleus and the electrons surrounding it. Mendeleyev based his table on the two properties, atomic weight and valency, which are reasonable approximations: atomic weight closely mirrors the number of protons, and valency is linked to the accessibility of electrons.

What distinguishes Mendeleyev was his ability to grasp that the known elements fitted into a scheme that was predetermined. The elements were not being arranged to *make* a periodic table, but to fit *the* periodic table. The difference may seem trivial to us today, but for its time the mental jump was truly one of genius. The certainty that he was right gave Mendeleyev the confidence not only to forecast the discovery of new elements, but he even predicted that a whole row of his periodic table awaited discovery because of the gap in atomic weights between cerium (140) and tantalum (182). These we now recognise as the lanthanoids, only two of which, erbium and terbium, were correctly known in his day.

The development of the periodic table

Considering his isolation from the main centres of chemistry, Mendeleyev's discovery might easily have gone unnoticed. However, he was determined that it should be publicised and he immediately wrote a paper, entitled 'The relationship of properties and the atomic weights of the elements', which was published in the very first edition of a new journal, *The Russian Journal of General Chemistry*. When copies of this reached Germany later that year, Mendeleyev's article was summarised in a German journal *Zeitschrift für Chemie*, thereby ensuring it a wide circulation among the chemical community. The importance of the discovery was immediately recognised, and a German translation of Mendeleyev's paper was published that same year. The periodic table that this contained is shown in Figure 3 below and it remained the standard version for almost a century.

Mendeleyev's table had eight columns labelled with the Roman numerals I to VIII, corresponding to the chemical valencies (oxidation states) of the elements. This property is revealed by the chemical formula of the highest oxide. Yet it brought together elements that may be quite dissimilar in other ways, such as metals and non-metals. For example in group V we find vanadium and phosphorus, which have almost no chemistry in common. Mendeleyev consequently split the columns of his periodic table into two sub-groups labelled A and B. Vanadium was in VA, phosphorus in VB.

---

[233] These elements share a common valency of 1, but this is deceptive as a way of classifying them because mercury's dominant valency is 2.

Reihen	Gruppe I. — R²O	Gruppe II. — RO	Gruppe III. —. R²O³	Gruppe IV. RH⁴ RO²	Gruppe V. RH³ R²O⁴	Gruppe VI. RH² RO³	Gruppe VII. RH R²O⁷	Gruppe VIII. — RO⁴
1	H = 1							
2	Li = 7	Be = 9,4	B = 11	C = 12	N = 14	O = 16	F = 19	
3	Na = 23	Mg = 24	Al = 27,3	Si = 28	P = 31	S = 32	Cl = 35,5	
4	K = 39	Ca = 40	— = 44	Ti = 48	V = 51	Cr = 52	Mn = 55	Fe = 56, Co = 59, Ni = 59, Cu = 63.
5	(Cu = 63)	Zn = 65	— = 68	— = 72	As = 75	Se = 78	Br = 80	
6	Rb = 85	Sr = 87	?Yt = 88	Zr = 90	Nb = 94	Mo = 96	— = 100	Ru = 104, Rh = 104, Pd = 106, Ag = 108.
7	(Ag = 108)	Cd = 112	In = 113	Sn = 118	Sb = 122	Te = 125	J = 127	
8	Cs = 133	Ba = 137	?Di = 138	?Ce = 140	—	—	—	— — — —
9	(—)	—	—	—	—	—	—	
10	—	—	?Er = 178	?La = 180	Ta = 182	W = 184	—	Os = 195, Ir = 197, Pt = 198, Au = 199.
11	(Au = 199)	Hg = 200	Tl = 204	Pb = 207	Bi = 208	—	—	
12	—	—	—	Th = 231	—	U = 240	—	— — — —

Fig. 3 Mendeleyev's periodic table as published in Germany in 1870.

The same pattern was repeated in the other columns with the exception of group VIII which contained metals which were very similar and which occurred in three sets of three. These were iron-cobalt-nickel, ruthenium-rhodium-palladium and osmium-iridium-platinum.

Within two years of Mendeleyev's table appearing, other chemists produced alternative versions and these have continued to appear year-by-year ever since until today there are well in excess of 600 variants, although many are very similar to one another. Most tables are two-dimensional representations although there have been several three dimensional versions in the shapes of cylinders, pyramids, spirals and even trees, which make excellent displays for museums and exhibitions.

Basically there have been two approaches to devising a periodic table: the first lists all the elements in a continuous line, rather like the numbers on a tape measure, and this is then looped in such a way that similar elements come together. The second version chops the tape into segments and stacks these in rows, again bringing similar elements together. The former approach is what Chancourtois used in 1862, and what many others have done since. The others are direct descendants of Mendeleyev's table.

Circular versions of the continuous type of table have also been proposed, but despite their elegance and the tantalising analogy with electrons in shells around a nucleus (see Figure 4 below) they all suffer the drawback of being difficult to read and to abstract the information they contain, because they tend to crowd together the more important elements at their centre while giving the less important elements more room at the periphery. One particularly popular spiral table is that devised by Philip J. Stewart of Oxford and published in 2007, and this is both elegant and easy to read.

The preferred way of arranging the elements is still the matrix format that Mendeleyev used, but now there is more purpose to it. The modern table denotes each row as the filling of an electron orbit, which finally results in a noble gas element when

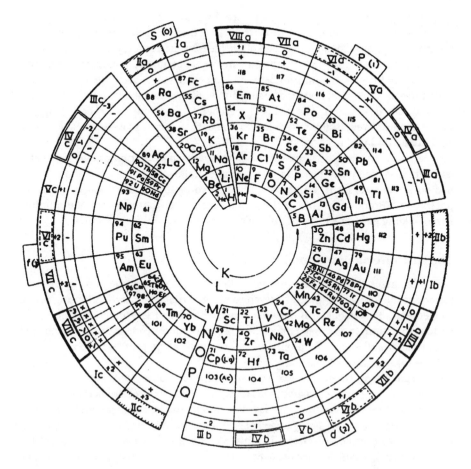

Fig. 4 A circular periodic table.

the orbit is full. These noble gases have a special place in the development of chemical theory and the layout of the periodic table.

Noble gases

Although Mendeleyev did not realise it, there was a group missing from his periodic table. These were the noble gases, and when they were discovered 30 years later, they were to exert an influence on the way the table was perceived.

The lightest noble gas, helium, had in fact been reported the year before Mendeleyev produced his table, by two astronomers, Pierre Janssen and Norman Lockyer, who detected it in the spectrum of light from the Sun in 1868. The new element was called helium (from the Greek word *helios*) but it was generally assumed that it would not be found on Earth.

Then, in 1894, Lord Rayleigh and William Ramsay separated a new gas, argon, from air. Helium too was discovered a few months later. Having measured the atomic weight of helium as 4 and that of argon as 40, Ramsay was able to see that there was a missing group in the periodic table, in which case there must be another gaseous element of atomic number about 20, and others heavier than argon with atomic weights around 82 and 130. These would fill gaps in the list of atomic weights between bromine (80) and rubidium (85), and between iodine (127) and caesium (133).

By 1899 they had extracted the missing three gases: neon (atomic weight 20), krypton (84) and xenon (131). The heaviest member of the group is radioactive radon, whose longest-lived isotope is radon-222 and this was discovered inside sealed ampoules of radium, in 1899.

We now see the noble gases as *completing* the rows of the periodic table, but chemists were first inclined to place them as group 0 at the start of the table because they had no tendency to form compounds, in other words their valency was 0. Today we put them at the end of the rows of the periodic table since they represent the culmination of adding electrons to a particular electron shell.

The structure of atoms and the periodic table

The noble gases completed the format of the periodic table as we know it (see end paper). In the long form there are seven rows, or periods, with 2, 8, 8, 18, 18, 32 and 32 elements. This pattern can be understood in terms of the underlying electronic structure of atoms, and indeed the pattern is really that of electron orbits, which hold 2, 6, 10 and 14 electrons. Combinations of these numbers give rise to 8 (= 2+6), 18 (= 2+6+10) and 32 (= 2+6+10+14).

This underlying structure could only be understood once the nature of the atom was revealed. This puzzle began to unravel with the discovery of the electron, in 1897, by J.J. Thompson (1856–1940). A few year after this, in 1903, Japanese physicist Hantaro Nagaoka (1865–1950) hypothesised that the atom had a positively-charged centre that was orbited by electrons. This was the so-called planetary model of the atom. In 1909, Ernest Rutherford (1871–1937) bombarded thin gold foil with alpha particles, and discovered that atoms of gold must consist of a tiny, positively-charged nucleus in which almost all the mass was concentrated.

In 1912, Dutch physicist Antonius van den Broeck (1870–1926) hypothesised that the periodic table should be ordered according to the positive charge on the nucleus, and published this in early 1913. Stimulated by this, and later that year, Henry G.J. Moseley (1887–1915) used X-ray spectroscopy to formulate the concept of atomic number. He showed that the sequence of elements in the periodic table was really the order of their atomic number, which we now know is the number of protons of positive charge in the nucleus. Also in that same productive year, Niels Bohr (1885–1962) published an article which linked the form of the periodic table to the atomic structure of atoms and applied quantum theory to explain the arrangement of electrons in energy levels surrounding the nucleus. He published a periodic table based on electron energy levels in 1922.

If the elements are arranged in rows of increasing atomic number, and in columns having the same electron outer shell, then we arrive at the long form of the periodic table. Across a row of the periodic table we are adding electrons to a particular shell until that shell is full when we arrive at one of the noble gases. Consequently these represent a natural break in the table.

Despite these advances in atomic theory, Mendeleyev's eight-column periodic table remained the common type for almost a hundred years. Then the extended or long form slowly displaced it as the man-made elements were announced and interest centred on the final row. The long form had earlier been championed by the great Swiss inorganic chemist Alfred Werner (1866–1919) in 1905.

To understand the periodic table, it is necessary to realise that atoms have arrays of electrons in shells and orbitals around the nucleus. We can imagine these being filled one-by-one with electrons, starting at the levels of lowest energy nearest the nucleus. Studies of the spectra of atoms showed there to be lines corresponding to the energies of electromagnetic radiation (ultraviolet, visible, and infrared) that represented electrons jumping between these various levels.

The shells are numbered 1, 2, 3, etc. from the nucleus outwards. The orbitals are given the letters s, p, d, f, g, h, etc. (the first four of these were derived from words used to describe the lines in the spectrum which corresponded to electrons moving in and out of the orbitals, these are sharp (s), principal (p), diffuse (d), and fundamental (f); beyond f they are listed alphabetically. The orbital nearest the nucleus is just a single orbital and labelled 1s, the next is a pair of orbitals labelled 2s and 2p, the next is a trio of orbitals, 3s, 3p, and 3d, and so on. However, as these fan out from the nucleus they begin to overlap with orbitals even further out so that the next one, 4s, is actually occupied before the 3d orbital, and so it goes on. The quantum-allowed number of electrons in each of the electron shells is 2, 8, 18, 32, 50, 72, etc. according to the formula $2n^2$ where n is the shell number (1, 2, 3, etc.)

The orbitals can hold increasing numbers of electrons. The s ones can hold 2, the p ones 6, the d ones 10, the f ones 14, and the g ones 18. These are the basis of the various blocks of the periodic table. The s-block consists of two groups (numbered 1, the alkali metals, and 2, the alkaline earths), the p-block elements consist of six groups (numbered 12 to 18), and the d-block elements have 10 groups (numbered 3 to 11). The f-block elements consist of two rows 4f and 5f, and the g block elements consist of one row of 5g; these are not given group numbers.

The changeover from the eight-column periodic table to the modern form was not without its difficulties. When Mendeleyev's periodic table of eight groups was turned into the long form of 18 groups, the Europeans numbered the groups on the left half IA to VIII, and on the right hand half they were numbered IB to VIIB with the noble gases being group 0. The Americans, on the other hand, kept the IA/IB system of Mendeleyev so their long form of the table was numbered IA, IIA, etc for the s and p block elements, and IB, IIB, etc. for the d block. Both systems numbered the alkali metals groups IA, and the alkaline earth metals IIA but after that they diverged.

Eventually the International Union of Pure and Applied Chemists (IUPAC) suggested that the groups simply be numbered 1 through to 18 across the periodic table and this has been widely accepted. The f block and g block do not fit into the numbering system but this poses no problem since the 4f, 5f, and 5g periods of elements are best dealt this as rows of the periodic table rather than groups. This is illustrated below as a minimalist periodic table:

The Periodic Table

Periodic table of electron groupings

1s				
2s				2p
3s				3p
4s			3d	4p
5s			4d	5p
6s		4f	5d	6p
7s		5f	6d	7p
8s	5g	6f	7d	8p

Or we could have a periodic table which just indicated electron totals even beyond the elements already known:

Periodic table of electron numbers

	s	g	f	d	p	
	2	18	14	10	6	Total
1	1 and 2					2
2	3 and 4				5–10	8
3	11 and 12				13–18	8
4	19 and 20			21–30	31–36	18
5	37 and 38			39–48	49–54	18
6	55 and 56		57–70	71–80	81–86	32
7	87 and 88		89–102	103–112	113–118	32
8	119 and 120	121–138	139–152	153–162	163–168	50

Problem elements

There are four elements that pose a problem for any periodic table. These are hydrogen (atomic number 1), helium (2), lanthanum (57), and actinium (89).

Hydrogen has an electron configuration 1s, which should place it in the s-block of the table above lithium (2s) and this is where it is to be found in some tables. But hydrogen is not a metal like lithium. Indeed by the same logic the noble gas helium should therefore be placed above beryllium, but no form of the table places it so because it is clearly a noble gas and must go above neon even though this has the electron configuration of a filled p shell.

Some tables place hydrogen by itself, or with helium, in the very centre of the table, floating free above the other elements. Others place hydrogen above fluorine, although it shares little in common with the halogen gases. Some tables give it double billing and place it above both lithium and fluorine. The periodic table at the end of this book has it next to helium and above fluorine and neon, but put both of them in a separate block, labelled 1s, as a mark of their uniqueness.

The other problem concerns the f-block of elements which comes between the s-block and d-block at the bottom of the table. For reasons of economy of space this block is generally written below the d-block. Which elements belong to the f-block?

Traditionally, and mainly for historical reasons, lanthanum and actinium are placed in group 3, below scandium and yttrium. After lanthanum come the other so-called rare earths or lanthanides, which are in fact the upper row of the f-block and this begins with cerium. The weight of evidence is that lanthanum and actinium are the first members of their respective rows of the f-block. These rows must have 14 elements and so the final ones in each are ytterbium and nobelium respectively.[234] If lanthanum is the first lanthanoid element then lutetium is the element to occupy the group 3 slot of row 5d in the d-block of the periodic table. Similarly actinium should be the first of the actinides of row 5f and lawrencium be in group 3.

As long as chemistry is studied there will be a periodic table. Even if someone we communicate with another part of the Universe, we can be sure that one thing both cultures will have in common is an ordered system of the elements that will be instantly recognisable by both intelligent life forms. Meanwhile it has inspired artists to construct periodic tables in which the elements are symbolised by flowers and foods. Perhaps the most dramatic periodic table is that by the Glasgow artist, Murray Robertson, which is entitled 'Visual Elements' and for which he had created a stunning computer graphic for each element. This can be accessed, via the Internet, on the Royal Society of Chemistry's website. And if you want to access a more dramatic version online then there is the Periodic Table of Videos produced by Nottingham University (<http://www.periodic videos.com/>) with presenters Martyn Poliakoff and Pete Licence. Also accessible as an app is the periodic table of Theodore Gray which is also available in book format. Finally, if you want a real hands-on periodic table, see the one available from RGB Research and produced by Max Whitby and Fiona Barclay (<http://www.periodictable.com>). This can be purchased and comes complete with a sample of each element that is legally available. The larger versions of this table of so-called 'Element Collections' can be seen as several institutions, such as the Chemical Heritage Foundation in Philadelphia, Pennsylvania, and at company sites such as the Dow Chemical headquarters in Michigan.

---

[234] In 1982 William B. Jensen argued cogently for this change to be made to periodic tables in an article in the *Journal of Chemical Education* (volume 59, page 634).

Appendix

The discovery of the elements in chronological order

What constitutes the discovery of an element? For some it is the first time it is identified as something previously unknown; for others it is the first time it is obtained in a pure form so its properties can be recorded; and in more modern times it is the first time it has been synthesised. And how do we decide on who is the discoverer? The person who first realises it exists? The person who first isolates it? Or the person who first reports it in the scientific literature? Thus there can be more than one contender to the claim to be a discoverer and in the list which follows there may be two or more people named, if they independently discovered the element around the same time. For some elements I have them listed twice, the earlier date [in italics] being the first report, which was generally unconfirmed at the time, and the later date being the one when the element was officially certified by IUPAC. More details can be found in the *Element of History* sections, or in the case of the more recent elements under the transfermium elements and un-named elements sections.

Elements known to the ancient world

Date	Element	Discoverer	Place
pre-history	Carbon	not known	not known
pre-history	Sulfur	not known	not known
pre-history	Copper	not known	not known
c.3000 BC	Silver	not known	not known
c.3000 BC	Gold	not known	not known
c.2500 BC	Iron	not known	not known
c.2100 BC	Tin	not known	not known
c.1600 BC	Antimony	not known	not known
c.1500 BC	Mercury	not known	not known
c.1000 BC	Lead	not known	not known

Elements discovered in the Middle Ages

Date	Element	Discoverer	Place
c.1200	Zinc	not known	India
c.1250	Arsenic	Magnus	Germany
c.1500	Bismuth	not known	Central Europe

1669	Phosphorus	Brandt	Hamburg
Pre-1700	Platinum	not known	Central America

Elements discovered in the eighteenth century

Year	Element	Discoverer	Place
1735	Cobalt	Brandt	Stockholm
1751	Nickel	Cronstedt	Stockholm
1755	Magnesium	Black	Edinburgh
1766	Hydrogen	Cavendish	London
1771/4	Oxygen	Scheele/Priestley	Edinburgh
1772	Nitrogen	Daniel Rutherford	Uppsala/Wiltshire
1774	Chlorine	Scheele	Uppsala
1774	Manganese	Grahn	Stockholm
1780	Chromium	Vauquelin	Paris
1781	Molybdenum	Hjelm	Uppsala
1783	Tellurium	von Reichenstein	Sibiu, Romania
1783	Tungsten	Elhuijar & Elhuijar	Vergara, Spain
1789	Zirconium	Klaproth	Berlin
1789	Uranium	Klaproth	Berlin
1791	Titanium	Gregor/ Klaproth	Cornwall/Berlin
1794	Yttrium	Gadolin	Åbo, Finland
1797	Beryllium	Vauquelin	Paris

Elements discovered in the nineteenth century

Year	Element	Discoverer	Place
1801	Vanadium	del Rio	Mexico
1801	Niobium	Hatchett	London
1802	Tantalum	Ekeberg	Uppsala
1803	Rhodium	Wollaston	London
1803	Palladium	Wollaston	London
1803	Osmium	Tennant	London
1803	Iridium	Tennant	London
1803	Cerium	Berzelius & Hisinger	Vestmanland, Sweden
1807	Potassium	Davy	London
1807	Sodium	Davy	London
1808	Boron	Lussac & Thenard/Davy	Paris/London
1808	Calcium	Davy	London
1808	Strontium	Crawford	Edinburgh

1808	Ruthenium	Sniadecki	Vilno, Poland
1808	Barium	Davy	London
1811	Iodine	Courtois	Paris
1815	Thorium	Berzelius	Stockholm
1817	Lithium	Arfvedson	Stockholm
1817	Selenium	Berzelius	Stockholm
1817	Cadmium	Davy	London
1824	Silicon	Berzelius	Stockholm
1825	Aluminium	Oersted	Copenhagen
1825/6	Bromine	Löwig/Balard	Heidelberg/ Montpellier
1839	Lanthanum	Mosander	Stockholm
1842	Erbium	Mosander	Stockholm
1843	Terbium	Mosander	Stockholm
1860	Caesium	Bunsen & Kirchhoff	Heidelberg
1861	Rubidium	Bunsen & Kirchhoff	Heidelberg
1861	Thallium	Crookes	London
1863	Indium	Reich & Richter	Freiberg
1868	*Helium*	*Janssen/Lockyer*	*Guntur, India/Vijaydurg, India*
1875	Gallium	de Boisbaudran	Paris
1878	Holmium	Cleve/ Delafontaine & Soret	Uppsala/Geneva
1878	Ytterbium	de Marignac	Geneva
1879	Scandium	Nilson	Uppsala
1879	Samarium	de Boisbaudran	Paris
1879	Thulium	Cleve	Uppsala
1880	Gadolinium	de Marignac	Geneva
1882	Helium	Palmieri	Naples
1885	Praseodymium	von Welsbach	Vienna
1885	Neodymium	von Welsbach	Vienna
1886	Germanium	Winkler	Freiberg
1886	Fluorine	Moissan	Paris
1886	Dysprosium	de Boisbaudran	Paris
1894	Argon	Rayleigh & Ramsay	London & Bristol
1895	Helium	Ramsay	London
1898	Krypton	Ramsay & Travers	London
1898	Neon	Ramsay & Travers	London
1898	Xenon	Ramsay & Travers	London
1898	Polonium	Curie (Marie)	Paris

1898	Radium	P. Curie & M. Curie	Paris
1899	Actinium	Debierne	Paris
1899	Radon	Rutherford & Owen/ P. Curie & M. Curie	Montreal/ Paris

Elements discovered and synthesised in the 1900s

Year	Element	Discoverer	Place
1901	Europium	Demarçay	Paris
1905	Rhenium	Ogawa	London
1907	Lutetium	Urbain/von Welsbach/ James	Paris/Germany/New Hampshire USA
1913	Protactinium	Fajans & Göhring	Karlsruhe
1923	Hafnium	Coster & Hevesey	Copenhagen
1925	Rhenium	Noddack, Tacke & Berg	Berlin
1925	Technetium	Noddack & Tacke	Berlin

Year	Element	Synthesised by	Place
1936	Francium	Hulubei & Cauchois	Modavia, Romania
1937	Technetium	Segré & Perrier	Palermo
1939	Francium	Perey	Paris
1940	Neptunium	McMillan & Abelson	Berkeley, California
1940	Astatine	Corson, Mackenzie & Segré	Berkeley, California
1940	Plutonium	Seaborg & colleagues	Berkeley, California
1944	Curium	Seaborg & colleagues	Berkeley, California
1944	Americium	Seaborg & colleagues.	Chicago
1945	Promethium	Marinsky, Glendenin & Coryell	Oak Ridge, USA
1949	Berkelium	Ghiorso & Seaborg	Berkeley, California
1950	Californium	Ghiorso, Seaborg & colleagues	Berkeley, California
1952	Einsteinium	Ghiorso & colleagues	Berkeley, California
1953	Fermium	Ghiorso & colleagues	Berkeley, California
1955	Mendelevium	Ghiorso & colleagues	Berkeley, California
1956	Nobelium	Flerov & colleagues	Dubna, Russia[235]
1958	Lawrencium	Ghiorso & colleagues	Berkeley, California
1963	Nobelium	Flerov & colleagues	Dubna, Russia
1964	Rutherfordium	Flerov & colleagues	Dubna, Russia

[235] At this time of the discovery, and until 1991, Russia was officially known as the USSR.

Appendix

1965	Lawrencium	Flerov & colleagues	Dubna, Russia
1968	Dubnium	Flerov & colleagues	Dubna, Russia
1970	*Seaborgium*	*Ghiorso & colleagues*	*Berkeley, California*
1971	*Copernicium*	*Marinov*	*Switzerland*
1974	Seaborgium	Ghiorso & colleagues	Berkeley, California
1975	*Bohrium*	*Oganessian & colleagues*	*Dubna, Russia*
1978	*Hassium*	*Oganessian & colleagues*	*Dubna, Russia*
1981	Bohrium	Armbruster & Münzenberg	Darmstadt, Germany
1982	Meitnerium	Armbruster & Münzenberg	Darmstadt, Germany
1984	Hassium	Armbruster, Münzenberg	Darmstadt, Germany
1991	*Darmstadtium*	*Ghiorso & colleagues*	*Berkeley, California*
1994	Darmstadtium	Armbruster & Münzenberg	Darmstadt, Germany
1994	Roentgenium	Armbruster & Münzenberg	Darmstadt, Germany
1996	Copernicium	Hofmann & colleagues	Darmstadt, Germany
1998	Ununquadium	Oganessian & Utyonkov	Dubna, Russia

Elements first synthesised this century

Year	Element	Discoverer	Place
2000	Ununhexium	Oganessian & Utyonkov	Dubna, Russia
2002	Ununoctium	Oganessian & Utyonkov	Dubna, Russia
2003	Ununpentium	Oganessian & Utyonkov	Dubna, Russia
2003	Ununtrium	Oganessian & Utyonkov	Dubna, Russia
2009	Ununseptium	Oganessian & Utyonkov	Dubna, Russia

Bibliography

In compiling this reference work I have drawn on more than 20 years' accumulation of hundreds of magazine and newspaper articles culled from a wide range of sources. Many of these were taken from *Chemical and Engineering News*, published by the American Chemical Society, *Chemistry in Britain*, published by the Royal Society of Chemistry, *Chemistry and Industry*, published by the Society for Chemical Industry, London, and *New Scientist*. More recently I have drawn upon newsletters and websites published by industry associations, which share a common interest in a particular element or metal. These are not only highly informative but carry details of current usages and production figures. And, of course, Wikipedia, whose element pages are well organised and up-to-date and often with information that is not easily available elsewhere.

Despite all this wealth of information, I have found that books are still the most comprehensive sources of information with older data that is not otherwise available, and the following ones have proved invaluable (listed in alphabetical order of first named author):

N.K. Aras and O.Y. Ataman, *Trace Element Analysis of Food and Diet*, RSC Publishing, Cambridge, 2006.

H.-D. Belitz and W. Grosch, *Food Chemistry*, Springer-Verlag, Heidelberg, Germany, 1987.

H.J.M. Bowen, *Environmental Chemistry of the Elements*, Academic Press, London, 1979.

S.S. Brown and Y. Kodama (eds.), *Toxicology of Metals*, Ellis Horwood, Chichester, England, 1987.

W. Büchner, R Schliebs, G. Winter and K.H. Büchel, *Industrial Inorganic Chemistry*, VCH, Weinheim, Germany, 1989.

S. Budavari (ed.), *The Merck Index*, 11th edn., Merck & Co. Inc., Rahway NJ, USA, 1989.

F. Cardarelli, *Materials Handbook: a Concise Desktop Reference*, Springer, London, 2000.

P.A. Cox, *The Elements: Their Origin, Abundance and Distribution*, Oxford University Press, Oxford, 1989.

P.A. Cox, *The Elements on Earth*, Oxford University Press, Oxford, 1997.

M. Chown, *The Magic Furnace*, Jonathan Cape, London, 1999.

E. Crabb and E.A. Moore, *Metals and Life*, RSC Publishing, Cambridge, 2009.

D. Crystal (ed.), *The Cambridge Encyclopedia*, Cambridge University Press, Cambridge, 1992.

J. Daintith and D. Gjertsen (eds.), *Oxford Dictionary of Scientists*, Oxford University Press, Oxford, 1999.

J.A. Dean (ed.), *Llange's Handbook of Chemistry*, 14th edn., McGraw-Hill Inc., New York, 1992.

K. Diem and C. Lentner (eds.), *Scientific Tables*, 7th edn., Documenta Giegy, Basle, Switzerland, 1970.

Dietary Reference Values for Food Energy and Nutrients for the UK, Dept of Health, HMSO, London, 1991.

J. Emsley, *The Elements*, 3rd edn, Clarendon Press, Oxford, 1998.

D.H. Evans (ed.), *Episodes from the Discovery of the Rare-Earth Elements*, Kluwer Academic Publishers, Berlin, 1996.

Bibliography

J.E. Fergusson, *The Heavy Elements: Chemistry, Environmental Impact and Health Effects*, Pergamon Press, Oxford, 1990.

H.D. Foth, *Fundamentals of Soil Science*, 8th edn., John Wiley & Sons, New York, 1990.

H.B. Gray, J.D. Simon and W.C. Trogler, *Braving the Elements*, University Science Books, Sausalito, California, 1995.

T. Gray, *The Elements: A visual exploration of every known atom in the universe*, Black Dog & Leventhal Publishing, New York, 2009.

N.N. Greenwood and A. Earnshaw, *The Chemistry of the Elements*, 2nd edn, Pergamon Press, Oxford, 1997.

W. Greiner and R.J. Gurpta (eds.) *Heavy Elements and Related New Phenomena*, (2 vols.) World Scientific Publishing, Singapore, 1999.

G. Hawley (ed.), *The Condensed Chemical Dictionary*, 10th edn., Van Nostrand Reinhold, New York, 1981.

D.C Hoffman, A. Ghiorso, and G.T. Seaborg, *The Transuranium People: the inside story*, Imperial College Press, London, 2000.

S. Hoffman, *On Beyond Uranium; journey to the end of the periodic table*, CRC Press, Boca Raton, Florida, 2002.

E. Hubbard, M. Stephenson and D. Waddington (eds.) *The Essential Chemical Industry*, 4th edn., Chemical Industry Education Centre, University of York, UK, 1999.

D. Hunter, *The Diseases of Occupations*, 5th edn, Hodder and Stoughton, London, 1976.

A.J. Ihde, *The Development of Modern Chemistry*, Harper & Row, New York, 1966.

A. Isaacs, J. Daintith, E. Martin (eds), *Oxford Dictionary of Science*, 4th edn., Oxford University Press, Oxford and New York, 1999.

A.M. James and M.P. Lord, *Macmillan's Chemical and Physical Data*, Macmillan, London, 1992.

L.K. James (ed.), *Noble Laureates in Chemistry, 1901–1992*, American Chemical Society and the Chemical Heritage Foundation, USA, 1993.

H.-D. Jakubke and H. Jeschkeit, *Concise Encyclopedia Chemistry*, trans. by M. Eagleston, Walter de Gruyter, Berlin, 1993.

A. Kabata-Pendias and H. Pendias, *CRC Trace Elements in Soils and Plants*, 2nd edn., CRC Press, Boca Raton, USA, 1991.

U. Kaldor and S. Wilson (eds.), *Theoretical Chemistry and Physics of Heavy and Superheavy Elements*, (6 vols.), Springer, New York, 2006.

G.W.C. Kaye and T.H. Laby, *Tables of Physical and Chemical Constants*, 15th edn., Longman Scientific & Technical, London, 1993.

A. Khazan, 'Upper limit in the periodic table of elements', *Progress in Physics*, vol.1, pp 38–41, 2007.

A. Khazan, 'Effect from hyperbolic law in periodic table of elements', *Progress in Physics*, vol.2, pp 83–86, 2007.

A. Khazan, 'Upper limit of the periodic table and synthesis of superheavy elements', *Progress in Physics*, vol. 2, pp 104–109, 2007.

J.A. Kent (ed.), *Riegel's Handbook of Industrial Chemistry*, 9th edn., Van Nostrand Reinhold, New York, 1992.

B.T. Kilbourn, *A lanthanide lanthology*, vols. A-L and M-Z, Molycorp Inc., Fairfield NJ, 1993.

P.K. Kuroda, *The Origin of the Chemical Elements and the Oklo Phenomenon*, Springer-Verlag, Berlin, 1982.

R.J. Kutsky, *Handbook of Vitamins, Minerals and Hormones*, 2nd edn., Van Nostrand Reinhold, New York, 1981.

G.J. Leigh, *Principles of Chemical Nomenclature*, RSC Publishing, Cambridge, 2010.

D.R. Lide (ed.), *CRC Handbook of Chemistry and Physics*, 75th edn., CRC Press, Boca Raton, USA, 1995.

K.Lodders and B. Fengley Jr., *Chemistry of the Solar System*, RSC Publishing, Cambridge, 2010.

T.D. Luckey and B. Venugopal, *Metal Toxicity in Mammals*, Vol. 1, Plenum Press, New York, 1977.

O.K. Manuel (ed.) *Origin of Elements in the Solar System: implications of post-1957 observations*, Kluwer Academic Publishers, Berlin, 2001.

P. Marshall, *The Philosopher's Stone*, Macmillan, London, 2001.

E. Merian, (ed.), *Metals and Their Compounds in the Environment*, VCH, Weinheim, 1991.

L.R. Morss and J. Fuger (eds.), *Transuranium elements: a half century*, American Chemical Society, Washington DC, 1992.

L.R. Morss, N.M. Edelstein, and J. Fuger (eds.), *Chemistry of the Actinide and Transactinide Elements*, (6 vols.), Springer, New York, 2006.

A.J. Moses, *The Practicing Scientist's Handbook*, Van Nostrand Reinhold, New York, 1978.

H. Muir (ed.), *Larousse Dictionary of Scientists*, Larousse, Edinburgh, 1994.

I. Nechaev and G.W. Jenkins, *The Chemical Elements*, Tarquin Publications, Diss, England, 1998.

E.H. Nickel and M. C. Nichols, *Mineral Reference Manual*, Van Nostrand Reinhold, New York, 1991.

J. Pearce (ed.), *Gardner's Chemical Synonyms and Trade Names*, 9th edn., Gower Technical Press, Aldershot, England, 1987.

D. N. Poenaru, *Nuclear Decay Modes*, Institute of Physics Publishing, Bristol, 1996.

D. N. Poenaru, and W. Greiner, *Handbook of Nuclear Properties*, Oxford Studies in Nuclear Physics, 17, Oxford University Press, Oxford, 1996.

D.N. Poenaru, M. Ivascu, and W. Greiner (eds.), *Particle Emission from Nuclei*, (3 vols.), CRC Press, Boca Raton, Florida, 1989.

A.S. Prasad (ed.), *Trace Elements in Human Health and Disease*, Volume II, Essential and Toxic Elements, Academic Press, New York, 1976.

D. Purves, *Trace Element Contamination of the Environment*, revised edn., Elsevier, Amsterdam, 1985.

H.-J. Quadbeck-Seeger (ed.) *World Records in Chemistry*, Wiley-VCH, Weinheim, Germany, 1999.

W.L. Roberts, T.J. Campbell, and G.R. Rapp Jr., *Encyclopedia of Minerals*, Van Nostrand Reinhold, New York, 1990.

H. Rossotti, *Diverse Atoms: Profiles of the Chemical Elements*, Oxford University Press, Oxford, 1998.

N.I. Sax and R.J. Lewis Sr., *Dangerous Properties of Industrial Mateirals*, 7th edn., Van Nostrand Reinhold, New York, 1989.

K. Schwochau, 'Extraction of metals from sea water', *Topics in Current Chemistry*, no. 124, p. 91, Springer Verlag, Berlin, 1984.

G.T. Seaborg, and W.D. Loveland, *The Elements Beyond Uranium*, Wiley Interscience, New York, 1990.

B. Selinger, *Chemistry in the Marketplace*, 5th edn., Harcourt Brace, Sydney, 1998.

E.E. Scerri, *The Periodic Table: its story and its significance*, Oxford University Press, Oxford, 2007.

M. Schädel (ed.), *The Chemistry of the Superheavy Elements*, Kluwer Academic Publishers, 2003.

Bibliography

C.H. Snyder, *The Extraordinary Chemistry of Ordinary Things*, John Wiley & Sons, Inc., New York, 1992.

A. Stwertka, *A Guide to the Elements*, Oxford University Press, New York, 1996.

S. I. Venetsky, *On rare and scattered metals: tales about metals*, trans. by N.G. Kittel, Mir Publishers, Moscow, 1981.

A. Wade (ed.), *Martindale, the Extra Pharmacopoeia*, The Pharmaceutical Press, London, 1977.

R. Wayne and A.M. Holloway, *Atmospheric Chemistry*, RSC Publishing, Cambridge, 2010.

M.E. Weeks and H.M. Leicester, *Discovery of the Elements*, 7th edn., Journal of the American Chemical Society, Easton, PA, 1968.

J.P. Williams and J.J.R. Fraústo da Silva, *The Natural Selection of the Chemical Elements*, Clarendon Press, Oxford, 1996.

J.P. Williams and J.J.R. Fraústo da Silva, *The Biological Chemistry of the Elements*, Clarendon Press, Oxford, 1991.

For some of the elements there are individual monographs devoted entirely to the chemistry of that element, and sometimes even to one aspect of its chemistry. These were of secondary important in compiling this book, but those that proved helpful were (listed in alphabetical order of the element concerned):

A. Sprague, *The ecological significance of boron*, US Borax, California, 1972.

N.J. Travis and E.J. Cocks, *The Tincal Trail: a History of Borax*, Harrap, London, 1984.

B.T. Kilbourn, *Cerium: a guide to its role in chemical technology*, Molycorp Inc., Fairfield NJ, 1992.

R.P. Foster (ed.), *Gold Metallogeny and Exploration*, Blackie, Glasgow, 1991.

I. Grinberg, *Aluminium: light at heart*, (trans. By A. Keens), Pechiney, Paris, 2003.

B.S. Hetzel, *The story of iodine deficiency*, Oxford University Press, Oxford, 1989.

C.H. Evans, *Biochemistry of the Lanthanides*, Plenum, NY, 1990.

T.B. Griffin and J.H. Knelson, *Lead*, Georg Thieme Publishers, Stuttgart, Germany, 1975.

P.A. D'Itri and F. M. D'Itri, *Mercury contamination: a human tragedy*, Wiley Interscience, New York, 1977.

S. Mitra, *Mercury in the Ecosystem*, Trans Tech Publications, Switzerland, 1986.

E.R. Braithwaite and J. Haber (eds.), *Molybdenum: an outline of its chemistry and uses*, Elsevier Science, Amsterdam, 1995.

P.H. Abelson, 'Discovery of neptunium' in L.R. Morss and J.Fuger (eds.), *Transuranium elements*, American Chemical Society, Washington DC, 1992.

G.F. Combs Jr and S.B. Combs, *The Role of Selenium in Nutrition*, Academic Press, Orlando, 1986.

A. Lewis, *Selenium, The Essential Element You Might Not be Getting Enough of*, Thorsons, Wellingborough UK and Rochester USA, 1982.

R.J. Shamberger, *Biochemistry of Selenium*, Plenum, New York, 1983.

R.A. Zingaro and W.C. Cooper (eds.), *Selenium*, Van Nostrand Reinhold, New York, 1974.

I.C. Smith and B.L. Carson, *Trace Metals in the Environment: Silver*, Ann Arbor Science, Ann Arbor Michigan, 1977.

I.C. Smith and B.L. Carson, *Trace Metals in the Environment: Thallium*, Ann Arbor Science, Ann Arbor Michigan, 1977.

P.G. Harrion (ed.), *Chemistry of Tin,* Blackie Publishing Group, Glasgow, Scotland, 1989.

C. T. Horovitz, 'Two hundred years of research and development of yttrium' *Trace Elements and Electrolytes,* vol.12, p.153, 1995.

For further reading on the periodic table consult:

P. Atkins, *The Periodic Kingdom,* Weidenfeld & Nicolson, London, 1995.

B. Bensaude-Vincent, 'La genèse du tableau de Mendeleyev' in *La Recherche,* 15, 1206, 1984.

W.H. Brock, *The Fontana History of Chemistry,* Fontana Press, London, 1992.

H. Cassebaum and G.B. Kauffman, *Isis,* vol. 62, p. 314, 1971.

J. Emsley, 'Mendeleyev's dream table' in *New Scientist,* 7 March 1984.

J. Emsley, 'The development of the periodic table of the chemical elements' in *Interdisciplinary Science Reviews,* vol. 12, p. 23, 1987.

E.S. Gilreath, *Fundamental concepts of inorganic chemistry,* McGraw-Hill, New York, 1958.

N.E. Holden, 'Mendeleyev and the periodic classification of the elements' *Chemistry International,* no. 6, 18, 1984.

E.G. Mazurs, *Graphic representations of the periodic system during one hundred years,* University of Alabama Press, Tuscaloosa, 1974.

D.H. Rouvray, 'Turning the tables on Mendeleyev' in *Chemistry in Britain,* May 1994, p. 373.

R.T. Sanderson, *Chemical Periodcity,* Reinhold Publishing Corp., New York, 1967.

G.T. Seaborg, *The Elements Beyond Uranium,* John Wiley & Sons Inc., New York, 1990.

E.R. Scerri, 'Plus ça change...' in *Chemistry in Britain,* May 1994, p. 379.

J.W. Van Spronsen, *The periodic system of chemical elements: a history of the first hundred years,* Elsevier, Amsterdam, 1969.

P. Strathern, *Mendeleyev's Dream: the Quest for the Elements, Hamish* Hamilton, London, 2000.

P.J. Stewart, 'A century on from Dmitri Mendeleev: tables and spirals, noble gases and Nobel prizes', *Foundations of Chemistry,* vol. 9, p. 235, 2007.

F.P. Venables, *The development of the periodic law,* Chemical Publishing Co., Easton, PA, 1896.

M.E. Weeks and H.M. Leicester, *Discovery of the elements* (7th edn.), Chapter 14, published by Journal of Chemical Education, Easton, PA, 1968.

Chemical elements in alphabetical order of chemical symbol, with atomic numbers

Ac	Actinium	89	He	Helium	2	Rf	Rutherfordium	104
Ag	Silver	47	Hf	Hafnium	72	Rg	Roentgenium	111
Al	Aluminium	13	Hg	Mercury	80	Rh	Rhodium	45
Am	Americium	95	Ho	Holmium	67	Rn	Radon	86
Ar	Argon	18	Hs	Hassium	108	Ru	Ruthenium	44
As	Arsenic	33	I	Iodine	53	S	Sulfur	16
At	Astatine	85	In	Indium	49	Sb	Antimony	51
Au	Gold	79	Ir	Iridium	77	Sc	Scandium	21
B	Boron	5	K	Potassium	19	Se	Selenium	34
Ba	Barium	56	Kr	Krypton	36	Sg	Seaborgium	106
Be	Beryllium	4	La	Lanthanum	57	Si	Silicon	14
Bh	Bohrium	107	Li	Lithium	3	Sm	Samarium	62
Bi	Bismuth	83	Lr	Lawrencium	103	Sn	Tin	50
Bk	Berkelium	97	Lu	Lutetium	71	Sr	Strontium	38
Br	Bromine	35	Md	Mendelevium	101	Ta	Tantalum	73
C	Carbon	6	Mg	Magnesium	12	Tb	Terbium	65
Ca	Calcium	20	Mn	Manganese	25	Tc	Technetium	43
Cd	Cadmium	48	Mo	Molybdenum	42	Te	Tellurium	52
Ce	Cerium	58	Mt	Meitnerium	109	Th	Thorium	90
Cf	Californium	98	N	Nitrogen	7	Ti	Titanium	22
Cl	Chlorine	17	Na	Sodium	11	Tl	Thallium	81
Cm	Curium	96	Nb	Niobium	41	Tm	Thulium	69
Co	Cobalt	27	Nd	Neodymium	60	U	Uranium	92
Cn	Copernicium	112	Ne	Neon	10	Ubb	Unbibium	122
Cr	Chromium	24	Ni	Nickel	28	Ubh	Unbihexium	126
Cs	Caesium	55	No	Nobelium	102	Ubn	Unbinilium	120
Cu	Copper	29	Np	Neptunium	93	Ubo	Unbiquadium	124
Db	Dubnium	105	O	Oxygen	8	Uuh	Ununhexium	116
Ds	Darmstadtium	110	Os	Osmium	76	Uuo	Ununoctium	118
Dy	Dysprosium	66	P	Phosphorus	15	Uup	Ununpentium	115
Er	Erbium	68	Pa	Protactinium	91	Uuq	Ununquadium	114
Es	Einsteinium	99	Pb	Lead	82	Uus	Ununseptium	117
Eu	Europium	63	Pd	Palladium	46	Uut	Ununtrium	113
F	Fluorine	9	Pm	Promethium	61	V	Vanadium	23
Fe	Iron	26	Po	Polonium	84	W	Tungsten	74
Fm	Fermium	100	Pr	Praseodymium	59	Xe	Xenon	54
Fr	Francium	87	Pt	Platinum	78	Y	Yttrium	39
Ga	Gallium	31	Pu	Plutonium	94	Yb	Ytterbium	70
Gd	Gadolinium	64	Ra	Radium	88	Zn	Zinc	30
Ge	Germanium	32	Rb	Rubidium	37	Zr	Zirconium	40
H	Hydrogen	1	Re	Rhenium	75			

The chemical elements in order of their atomic numbers, with chemical symbol

1	Hydrogen	H		42	Molybdenum	Mo		83	Bismuth	Bi
2	Helium	He		43	Technetium	Tc		84	Polonium	Po
3	Lithium	Li		44	Ruthenium	Ru		85	Astatine	At
4	Beryllium	Be		45	Rhodium	Rh		86	Radon	Rn
5	Boron	B		46	Palladium	Pd		87	Francium	Fr
6	Carbon	C		47	Silver	Ag		88	Radium	Ra
7	Nitrogen	N		48	Cadmium	Cd		89	Actinium	Ac
8	Oxygen	O		49	Indium	In		90	Thorium	Th
9	Fluorine	F		50	Tin	Sn		91	Protactinium	Pa
10	Neon	Ne		51	Antimony	Sb		92	Uranium	U
11	Sodium	Na		52	Tellurium	Te		93	Neptunium	Np
12	Magnesium	Mg		53	Iodine	I		94	Plutonium	Pu
13	Aluminium	Al		54	Xenon	Xe		95	Americium	Am
14	Silicon	Si		55	Caesium	Cs		96	Curium	Cm
15	Phosphorus	P		56	Barium	Ba		97	Berkelium	Bk
16	Sulfur	S		57	Lanthanum	La		98	Californium	Cf
17	Chlorine	Cl		58	Cerium	Ce		99	Einsteinium	Es
18	Argon	Ar		59	Praseodymium	Pr		100	Fermium	Fm
19	Potassium	K		60	Neodymium	Nd		101	Mendelevium	Md
20	Calcium	Ca		61	Promethium	Pm		102	Nobelium	No
21	Scandium	Sc		62	Samarium	Sm		103	Lawrencium	Lr
22	Titanium	Ti		63	Europium	Eu		104	Rutherfordium	Rf
23	Vanadium	V		64	Gadolinium	Gd		105	Dubnium	Db
24	Chromium	Cr		65	Terbium	Td		106	Seaborgium	Sg
25	Manganese	Mn		66	Dysprosium	Dy		107	Bohrium	Bh
26	Iron	Fe		67	Holmium	Ho		108	Hassium	Hs
27	Cobalt	Co		68	Erbium	Er		109	Meitnerium	Mt
28	Nickel	Ni		69	Thulium	Tm		110	Darmstadtium	Ds
29	Copper	Cu		70	Ytterbium	Yb		111	Roentgenium	Rg
30	Zinc	Zn		71	Lutetium	Lu		112	Copernicium	Cn
31	Gallium	Ga		72	Hafnium	Hf		113	Ununtrium	Uut
32	Germanium	Ge		73	Tantalum	Ta		114	Ununquadium	Uuq
33	Arsenic	As		74	Tungsten	W		115	Ununpentium	Uup
34	Selenium	Se		75	Rhenium	Re		116	Ununhexium	Uuh
35	Bromine	Br		76	Osmium	Os		117	Ununseptium	Uus
36	Krypton	Kr		77	Iridium	Ir		118	Ununoctium	Uuo
37	Rubidium	Rb		78	Platinum	Pt		120	Unbinilium	Ubn
38	Strontium	Sr		79	Gold	Au		122	Unbibium	Ubb
39	Yttrium	Y		80	Mercury	Hg		124	Unbiquadium	Ubq
40	Zirconium	Zr		81	Thallium	Tl		126	Unbihexium	Ubh
41	Niobium	Nb		82	Lead	Pb				

Periodic Table

1s		
1 H 1.008		**2** He 4.003

s-block

1 / I	2 / II
3 Li 6.941	**4** Be 9.012
11 Na 22.990	**12** Mg 24.305
19 K 39.098	**20** Ca 40.078
37 Rb 85.468	**38** Sr 87.62
55 Cs 132.905	**56** Ba 137.327
87 Fr (223)	**88** Ra 226.025
119 ?	**120** ?

d-block

3 / III	4 / IV	5 / V	6 / VI	7 / VII	8 / VIII	9 / VIII	10 / VIII	11 / I	12 / II
21 Sc 44.956	**22** Ti 47.867	**23** V 50.942	**24** Cr 51.996	**25** Mn 54.938	**26** Fe 55.845	**27** Co 58.933	**28** Ni 58.693	**29** Cu 63.546	**30** Zn 65.39
39 Y 88.906	**40** Zr 91.224	**41** Nb 92.906	**42** Mo 95.94	**43** Tc 98.906	**44** Ru 101.07	**45** Rh 102.906	**46** Pd 106.42	**47** Ag 107.868	**48** Cd 112.411
71 Lu 174.967	**72** Hf 178.49	**73** Ta 180.948	**74** W 183.84	**75** Re 186.207	**76** Os 190.23	**77** Ir 192.217	**78** Pt 195.08	**79** Au 196.967	**80** Hg 200.59
103 Lr (262)	**104** Rf (261)	**105** Db (262)	**106** Sg (266)	**107** Bh (267)	**108** Hs (273)	**109** Mt (268)	**110** Uun (281)	**111** Rt (281)	**112** Cp (286)

p-block

13 / III	14 / IV	15 / V	16 / VI	17 / VII	18 / VIII
5 B 10.811	**6** C 12.011	**7** N 14.007	**8** O 15.999	**9** F 18.998	**10** Ne 20.180
13 Al 26.982	**14** Si 28.086	**15** P 30.974	**16** S 32.066	**17** Cl 35.453	**18** Ar 39.948
31 Ga 69.723	**32** Ge 72.61	**33** As 74.922	**34** Se 78.96	**35** Br 79.904	**36** Kr 83.80
49 In 114.818	**50** Sn 118.710	**51** Sb 121.760	**52** Te 127.60	**53** I 126.904	**54** Xe 131.29
81 Tl 204.383	**82** Pb 207.2	**83** Bi 208.980	**84** Po (209)	**85** At (210)	**86** Rn (222)
113 Uut (286)	**114** Uuq (286)	**115** Uup (289)	**116** Uuh (293)	**117** Uus (294)	**118** Uuo (294)

f-block

57 La 138.906	**58** Ce 140.115	**59** Pr 140.908	**60** Nd 144.24	**61** Pm 144.913	**62** Sm 150.36	**63** Eu 151.965	**64** Gd 157.25	**65** Tb 158.925	**66** Dy 162.50	**67** Ho 164.50	**68** Er 167.26	**69** Tm 168.934	**70** Yb 173.04
89 Ac (227)	**90** Th 232.038	**91** Pa 231.036	**92** U 236.029	**93** Np 237.048	**94** Pu 244.064	**95** Am 243.061	**96** Cm 247.070	**97** Bk 247.070	**98** Cf 251.080	**99** Es (254)	**100** Fm (257)	**101** Md (258)	**102** No (259)

III

The next element beyond the above table would occupy position 119 and come below francium (element 87) in group 1, to be followed by element 120 which would come below radium (element 88) in group 2. Subsequent elements would then begin a completely new row of the periodic table, the 5g row. This would have 18 vacant slots and lie below the f-block.

Index

Index